13th AUN/SEED-Net Regional Conference on Geological and Geo-Resource Engineering (RCGeoE 2021)

IOP Conference Series: Earth and Environmental Science
Volume 1071

Online
20-21 December 2021

Editors:

Tsutomu Sato
Subagyo Pramumijoyo
Inthanongsone Inthavongsa
Kamar Shah Ariffin
Chandoeun Eng
Kotaro Tonezu

ISBN: 978-1-7138-6027-3
ISSN: 1755-1307

Printed from e-media with permission by:

Curran Associates, Inc.
57 Morehouse Lane
Red Hook, NY 12571

Some format issues inherent in the e-media version may also appear in this print version.

Institute of Physics
Dirac House, Temple Back
Bristol BS1 6BE UK

Phone: 44 1 17 929 7481
Fax: 44 1 17 920 0979

techtracking@iop.org

Additional copies of this publication are available from:

Curran Associates, Inc.
57 Morehouse Lane
Red Hook, NY 12571 USA
Phone: 845-758-0400
Fax: 845-758-2633
Email: curran@proceedings.com
Web: www.proceedings.com

TABLE OF CONTENTS

DISASTER MANAGEMENT AND DISASTER RISK REDUCTION

GEOLOGY AND GEOPHYSICS

GEO-RESOURCES: EXPLORATION AND OPTIMIZATION

ENVIRONMENTAL SCIENCE AND GEOTECHNICAL ENGINEERING

IOP Conf. Series: Earth and Environmental Science 1071 (2022) 011001 doi:10.1088/1755-1315/1071/1/011001

TECHNICAL PROGRAMME BOOK

The 13th AUN/SEED-Net Regional Conference on Geological and Geo-Resource Engineering
RCGeoE 2021

Geological and Geo-Resource Engineering for Sustainable Earth

20-21 December 2021
Virtual Conference

Geological Engineering Department
Universitas Gadjah Mada
Jl. Grafika No 2, Mlati, Sleman
Yogyakarta, 55281
INDONESIA

IOP Conf. Series: Earth and Environmental Science 1071 (2022) 011001 doi:10.1088/1755-1315/1071/1/011001

ABOUT RCGeoE 2021

The Regional Conference on Geological and Geo-Resource Engineering (RCGeoE) is part of the Regional Conference (RC) series, an academic event involving ten thematic engineering fields, supported by the JICA Project for AUN/SEED-Net, a network to strengthen collaborative efforts between several universities in ASEAN and Japan region. Each RC event is organized annually by Member Institutions (MIs) which is appointed through a selection process, with technical assistance from Japanese Supporting Universities (JSUs) and financial support from JICA Project for AUN/SEED-Net. The first AUN/SEED-Net Regional Conference was done in 2003, hosted by Geological Engineering Department of Universitas Gadjah Mada and sponsored by JICA Projects for AUN/SEED-Net.

For almost two decades, the program has successfully served as a platform for networking and exchanging current progress in engineering and science among AUN/SEED-Net members. The conference promotes and also serves as a milestone for academic collaboration among ASEAN and Japanese universities, initiating future collaborations in the engineering field with the government and private sectors.

This conference, the Regional Conference on Geological and Geo-Resource Engineering 2021 (RCGeoE), is one of those ten thematic fields. This year's RCGeoE 2021 is the 13th conference event held in the field of Geological and Geo-resource Engineering. Geological Engineering Department of Universitas Gadjah Mada (one of the AUN/SEED-Net Member Institutions) is selected to host the 13th RCGeoE. The theme for this year's RCGeoE is: "Geological and Geo-resource Engineering for Sustainable Earth". Conference was held fully virtual through Zoom platform, run and managed by organizing committee from Geological Engineering Department, Universitas Gadjah Mada, Yogyakarta, Indonesia, with full support from the MI member representatives.

The event was successfully held virtually in 20-21 December, 2021, presenting two plenary sessions, a special lecture series, a scientific workshop, and 47 presentations in 9 parallel sessions.

IOP Conf. Series: Earth and Environmental Science 1071 (2022) 011001 doi:10.1088/1755-1315/1071/1/011001

FOREWORD

Conference Chair

Dr. Wahyu Wilopo
Assoc. Prof.
Geological Engineering Department
Universitas Gadjah Mada
Yogyakarta, Indonesia

It is an honour for the Geological Engineering Department of Universitas Gadjah Mada to host the 2021 Regional Conference on Geological and Geo-resources Engineering, an annual conference program sponsored by JICA Project for ASEAN University Network/ Southeast Asia Engineering Education Development Network (AUN/SEED-Net).

UGM has been part of AUN/SEED-Net for almost 2 decades, working on various events and programmes that help students and members of ASEAN Universities in setting forth opportunities for exchanging knowledge, experiences, and insight - in advancing geology and geo-resources engineering in Indonesia and ASEAN.

The 13th AUN/SEED-Net RCGeoE presents 'Geological and Geo-Resource Engineering for Sustainable Earth' as this year's theme. With the everchanging Earth due to climate crisis, geological and geo-resource engineering could be very much redundant if not done properly. Other than that, advancements in this field should be executed for better sustainable development. We deem that it's important for us, as geologist and geo-engineers to harness the Earth while also protecting it. During the pandemic, the fully online RCGeoE provides the opportunity for all participants to initiate conversations while expanding network, as a way for the community to achieve future collaborative and coordinated approach in a regional and international level.

I hope the success of this event will strengthen future collaboration between Japan and ASEAN universities. Once again, we thank you for your contribution to and participation in RCGeoE 2021 and I hope you had a great time and wonderful experience.

EDITORS

Prof. Tsutomu Sato
Hokkaido University, Japan

Prof. Subagyo Pramumijoyo
Universitas Gadjah Mada, Indonesia

Prof. Kotaro Yonezu
Kyushu University, Japan

Assoc. Prof. Dr. Kamar Shah Ariffin
Universiti Sains Malaysia, Malaysia

Dr. Chandoeun Eng
Institute Technology of Cambodia, Cambodia

Dr. Inthanongsone Inthavongsa
National University of Laos, Laos

REVIEWERS

Assoc. Prof. Dr. Makoto Kagabu
Nagasaki University, Japan

Dr. Rita Susilawati
Center for Mineral, Coal, and Geothermal Resources, Geological Agency, Indonesia

Dr. Chandoeun Eng
Institute Technology of Cambodia, Cambodia

Dr. Inthanongsone Inthavongsa
National University of Laos, Laos

Dr. Gayatri Indah Marliyani
Universitas Gadjah Mada, Indonesia

Dr. Sarju Winardi
Universitas Gadjah Mada, Indonesia

Dr. Rahmadi Hidayat
Universitas Gadjah Mada, Indonesia

Dini Nurfiani, PhD
Post-doctoral, National Research and Innovation Agency, Indonesia

Dr. Anastasia Dewi Titisati
Universitas Gadjah Mada, Indonesia

IOP Conf. Series: Earth and Environmental Science 1071 (2022) 011001 doi:10.1088/1755-1315/1071/1/011001

ORGANIZING COMMITTEE

Chairman : Dr. Wahyu Wilopo
Co-Chairman : Dr. Esti Handini
Secretary & Treasurer : Wita Dyaswati
Sponsorship : Manida Pakpian

Program Committee

Dr. Gayatri Indah Marliyani
Najma Aurora B I
Narendra Mulya
Radista Saga

Publication Team

Dr. Sarju Winardi
Vidya Syarifah
Aqnaa Furqon A F
Geika Pramana Surya

Technical Committee, Website, IT, & media

Rocky Tirajean S
Antonius Chandra W
Widyatmoko
Eko Hendrawan Arianto
Riyantono

Prepared by

Publication Team RCGeoE 2021

Geological Engineering Department
Universitas Gadjah Mada
Jl. Grafika No 2, Kampus Teknik UGM
Senolowo, Sinduadi, Mlati, Sleman
Yogyakarta, Indonesia, 55281
Website: https://geologi.ugm.ac.id/

IOP Conf. Series: Earth and Environmental Science 1071 (2022) 011002 doi:10.1088/1755-1315/1071/1/011002

Peer Review Statement

All papers published in this volume have been reviewed through processes administered by the Editors. Reviews were conducted by expert referees to the professional and scientific standards expected of a proceedings journal published by IOP Publishing Publishing.

- **Type of peer review:** Double Anonymous
- **Conference submission management system:** Submission and registration to our conference were done through a third-party platform and a website, managed by a technical committee. Submission and review processes were all done at the following link https://easychair.org/conferences/?conf=rcgeoe2021, unless a difficulty arose in any step of the processes, the communication was done through conference website at https://rcgeoe2021.ugm.ac.id/ and/or by email at rcgeoe2021@ugm.ac.id.
- **Number of submissions received:** 51
- **Number of submissions sent for review:** 45
- **Number of submissions accepted:** 37
- **Acceptance Rate (Submissions Accepted / Submissions Received × 100):** 72.5
- **Average number of reviews per paper:** 1.3
- **Total number of reviewers involved:** 9
- **Contact person for queries:**
  Name: Assoc. Prof. Dr. Wahyu Wilopo
  Email: wilopo_w@ugm.ac.id
  Affiliation: Universitas Gadjah Mada

Monitoring lava dome extrusion of Merapi Volcano during 2018-2019, using low-cost UAV application

Raditya Putra[1,2], Agung Harijoko[1*], and Agus Budi Santoso[2]

[1] Geological Department, Universitas Gadjah Mada,
Jl. Grafika Bulaksumur no 2 Yogyakarta Indonesia
[2] Geological Agency, Center of Volcanology and Geological Hazard Mitigation,
Jl. Cendana no 15 Yogyakarta Indonesia
*Corresponding author's e-mail: aharijoko@ugm.ac.id

Abstract. Merapi volcano has a well-known eruption type, namely Merapi type, in which an extruded lava dome collapses and is accompanied by pyroclastic density current (PDC). This type of eruption makes morphological monitoring of the lava dome crucial in the hazard mitigation process. After the VEI 4 eruption in 2010, a new lava dome of Merapi appeared on top of the 2010 lava dome in August 2018 and continuously grew. In November 2019, the lava dome started to collapse outward the crater area. We reported the lava dome morphological monitoring using a UAV (Unmanned Aerial Vehicle) photogrammetry conducted from August 2018 to February 2019. This UAV monitoring provides processed aerial photo data in Digital Terrain Model (DTM) and orthophoto with low operating costs and short data acquisition time. The lava domes erupted from the same eruptive canter within this period and grew evenly in all directions. The 2018-2019 Merapi lava dome has basal ratio of 0.183 to 0.290 with height of 11 to 41 m, respectively. Volume changed from 33,623 m^3 in August 2018 to 658,075 m^3 in February 2019, suggesting growth rate at ~3,500 m^3/day. The lava base filled the crater base area (0.21 km^2) and started to collapse outward in November 2019.

1. Introduction

Mt. Merapi is a stratovolcano-type volcano with andesite to Basaltic-andesite magma composition with an altitude of 2968 meters above sea level. This Quaternary volcano is one of the largest volcanoes in the Sunda Arc. Merapi is a volcano with a world-famous eruption type, called the "Merapi type," when a lava dome collapse mechanism generating by Pyroclastic Density Current [1]. Consequently, monitoring the lava dome morphology is necessary to provide early warning related to the eruption hazard assessment.

The lava dome monitoring of Merapi using a UAV (Unmanned Aerial Vehicle) has been carried out since 2018. This monitoring was done using a low-cost UAV Quadcopter model following [2]. This UAV technology allows us to obtain detailed morphological data of the summit with a Digital Terrain Model (DTM) resolution of 0.5 m [3].

This article report the monitoring of Merapi lava dome growth using a UAV conducted from August 2018 to February 2019 (6 months). During this period, the lava dome appeared for the first time after the 2010 Merapi eruption until it grew and collapsed. We presented morphological change, volume estimation, the growth rate of the lava dome from the early formation up to before it collapses out of the crater.

2. Method

The aerial photography was done using DJI Phantom 4 PRO, a Quadcopter with a diagonal size of 350 mm and a payload of less than 1.3 kg. Despite its small size, the Phantom 4 PRO comes with a very stable flight controller, a sensitive gimbal, and a high-quality camera (20 megapixels camera, 4 K video capability). The flight station is around Pasar Bubar with an altitude of ± 2500 m.a.s.l., with a flight path target around the summit with an area of 0.5 km^2. Each aerial photo already has a coordinate system (referenced) and has 4864 x 3648 pixels, with a horizontal resolution of 72 dpi. A flight path is made with the following parameters: flying height of 350 meters, front overlap 80%, side overlap 80%, flight direction - 180° mapping flight speed 15 m/s, aerial photos produced are 40 - 60 pieces, area 50 ha.

This high-resolution digital aerial photo was created by merging 3D Point Clouds. This photo-processing produces DTM, which has a reasonably high accuracy, less than 0.5 meters. According to the guidance issued by the Geospatial Information Agency [4], cloud coverage as noise has to be less than 10% in aerial photo data collection.

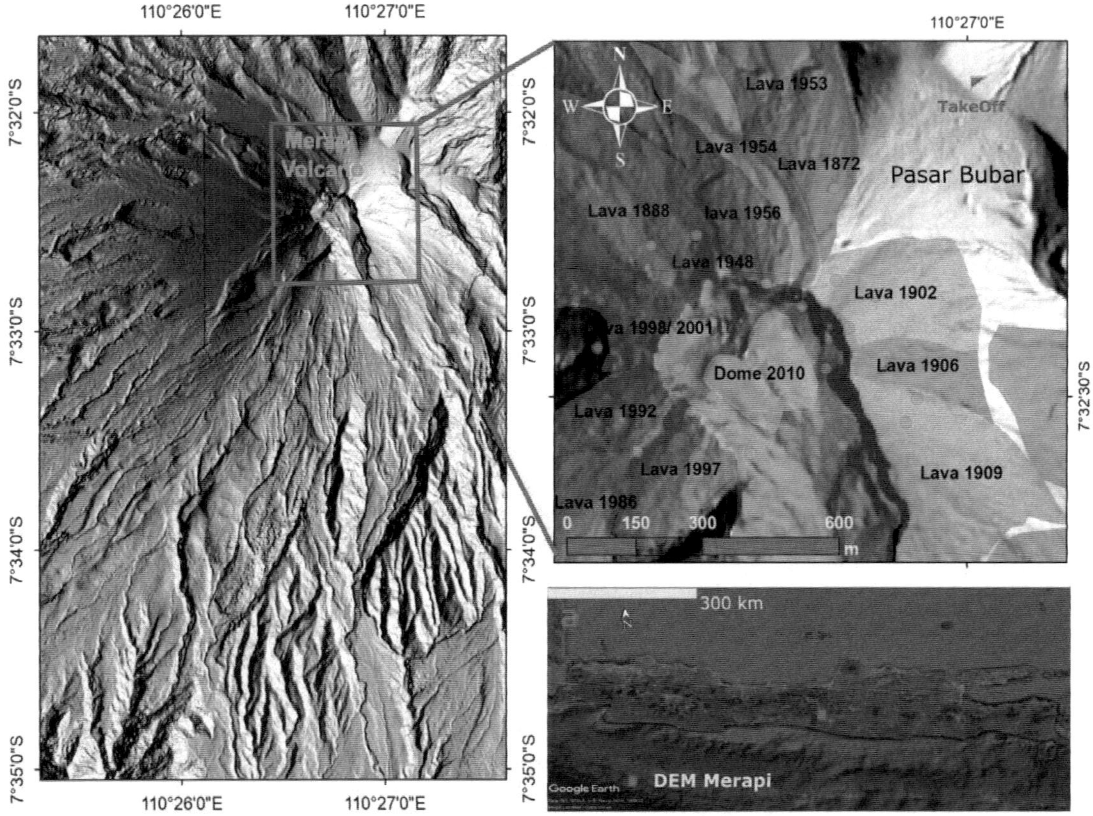

Figure 1. (a). Index map Java Island, map downloaded as a picture from Google Earth, added information area of (Digital Elevation Model) DEM Merapi (b). DEM of Merapi volcano using GIS processing with hillshade operation (data from Geological Agency of Indonesia) (c). Inset map showing summit morphology containing historical lava modified from map [5]. The UAV (Unmanned Aerial Vehicle) take-off point was done at Pasar Bubar at 2500 m.a.s.l. Green dot showGCP (Ground Control Location) around Merapi summit.

The data acquisition was processed using Agisoft Metashape Professional 64 Bit. Ground Control Point (GCP) was carried out in two methods. The first method is measuring GCP coordinates using Real-Time GPS installed at Merapi, Pasar Bubar, and Sadel Kawah Mati (Lava 1956, Figure 1c). The second method used Digital Image Matching [6]. The GCP was used for taking aerial photos of the stairs on 28 December 2018, and the points used were 22 points (Figure 1c). Further processing of 3D morphology analysis used commercial software Global Mapper - Blue Marble. This software provides easy to use DTM analysis and volume dimension calculation.

Volume was calculated using two input surfaces, DTM before Initial Condition (IC) it's mean before lava dome appear and after lava dome growth displaying the areas and volumes of surface materials that have been modified by the removal or addition of surface material [7]. The volume (V) was obtained from the following equation:

$$V = \sum_i \Delta x^2 \Delta zi \quad (1)$$

From equation 1 where Δx is the grid cell side, and Δzi is the height difference between surface DTM before (IC) and after lava dome growth for each pixel within the lava dome.

3. Result

The Merapi lava dome observed from 12 August 2018 to 9 September 2019 shows significant changes. The growth and development of this lava dome are at least divided into 2 phases, namely Initial Conditions (12 August 2018) and Dome Growth (25 August 2018 - 4 February 2019).

3.1. Initial Condition (IC).

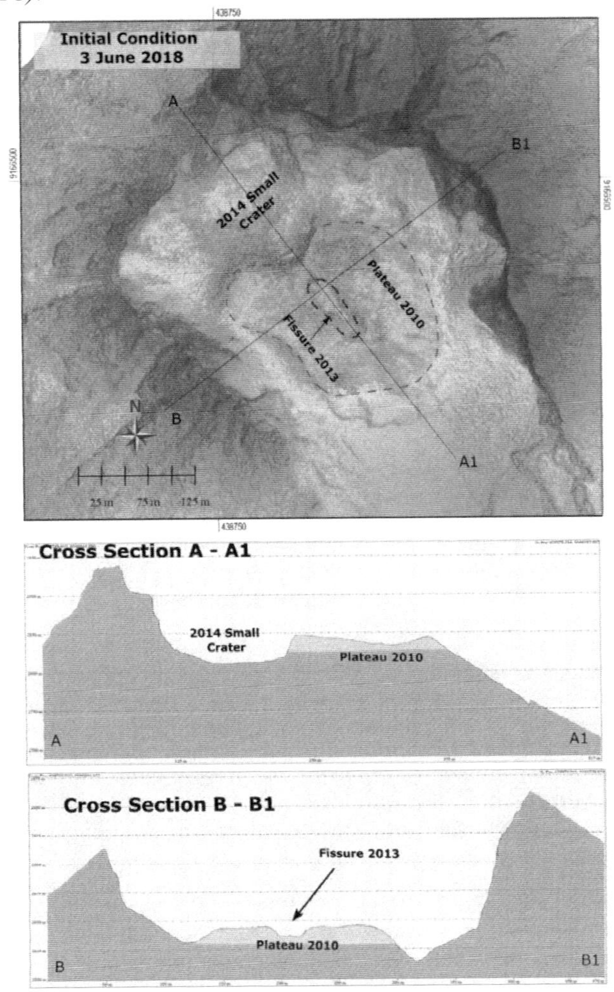

Figure 2. Digital Terrain Model was taken on 3 June 2018, showing initial morphological features inside the crater of Merapi before the appearance of the 2018 lava dome.

Aerial photo obtained from UAV monitoring conducted on 3 June 2018 (Figure 2) shows the size of Merapi summit crater is with the length of 400 m, width of 350 m, and depth of 120 m. The basal area of this crater is approximately101,500 m^2. This crater is attributed to a 2010 lava dome and small circular depression. The 2010 lava dome is 115 m in diameter and 115 m high, forming a plateau morphology with a relatively flat surface. This lava dome was formed during the Warning State of Merapi 2010 eruptive period when voluminous lava (1.5 x 10^6 m^3) extruded in a relatively short period of 1 - 4 November 2010 [8]. Warning State is Major volcanic eruption is imminent, underway, or suspected but it poses limited hazards to aviation because of no or minor volcanic-ash emissions (e.g., an eruption with only substantial lava flows and no risk of ash production) [1]. A small fissure crack (length of 80 m, width of 25 m, and height of 120 m) was observed in the middle of the 2010 lava dome. This feature is related to the explosive phreatic eruption event on 18 November 2013. In this study, we named it to fissure 2013. A small circular depression on the north of the 2010 lava dome has a diameter of 95 m and 20 m depth. This circular depression is located beneath the lava of 1848 and was formed during several explosive phreatic eruptions in 2012 - 2014 [3]. In this study, we named this circular depression as a 2014 small crater.

3.2. Growth Phase (25 August 2018 - 5 March 2019)

After a series of phreatic eruptions in Merapi that occurred in May - June 2018 [9], a UAV mission discovered that on 12 August 2018, a new lava dome appeared inside the crater. The photo shows the presence of a lava dome that extruded in the 2013 fissure (Figure 3). This lava extrusionutilized the same fissure crack of the preceding phreatic eruptions.

Figure 3. UAV data shows the presence of new material that appears in the middle of the post-2010 lava dome fissure 2013.

UAV aerial photographs of the early stage of the lava dome (12 August 2021) were adjusted following the method of [10] to obtain an optimum picture of the dome structure. Figure 4 shows that the thallus is radial and darker than the carapace, while the viscous fluid core part of the lava dome cannot be observed using aerial photographs. We recommend combining this UAV aerial photograph with a thermal camera to identify the lava dome structure better.

Figure 4. (a) Image of the early stage lava dome taken from UAV monitoring on 18 August 2021. (b) Adjusted image of the lava dome following the method of [10] shows the features of the talus and carapace. (c) Sketch of lava dome internal structure consists of core, carapace, and talus modified after [11].

During the 176 days of monitoring period between 12 August 2018 to 4 February 2019the dimension of the lava dome grew from length of 60 m in the initial stage to 156 m, width of 47 m to 152 m, height of 11 m to 41 m. he volume of the lava dome changed from 33,623 m^3 in the initial stage to 658,075 m^3. Figure 5 top part shows the DTM which reflects the Merapi lava dome's growth. The bottom part image shows orthophoto images of the crater taken in its original condition shown. Detailed dimension data and volume of each monitoring period are presented in Table 1.

Table 1. Summary of estimation of the lava dome volume changed from each UAV monitoring mission.

		Lava Dome Growth							Lava dome collapsed to K. Gendol**	
Date	Cumulative Days	Length (m)	Width (m)	Height (m)	Ratio H/L	Ratio W/L	Volume (m^3)	Extrusive Rate (m^3/day)	Volume (m^3)	Distance (m)
12/8/2018	0	0	0	0			0			
25/8/2018	13	60	47	11	0.18	0.78	33,623	2,586		
19/9/2018	38	82	65	20	0.24	0.79	133,159	3,504		
18/10/2018	67	108	98	28	0.26	0.91	244,486	3,649		
28/12/2018	138	131	127	38	0.29	0.97	578,998	4,196		
29/01/2019									20,000	1,400
29/01/2019									17,000	1,300
29/01/2019									5,000	500
4/2/2019	176	156	152	41	0.29	0.97	658,075	3,739		

* Calculation of the volume of the lava dome is obtained from the processing and correction of DTM data.
** daily report BPPTKG

The increase in lava dome size is shown in Figure 6. The lava dome was still on top of the 2010 plateau up to 25 August 2018. Monitoring on 19 September 2018 shows that the northwestern part of the lava dome overflown the 2010 plateau and filled the 2014 small crater. Data on 28 December 2018 shows that the north-eastern part of the lava dome has overflown the 2010 plateau. This data also indicates that the lava dome has reached the outer rim of the 2010 crater towards the Gendol River (Figure 6).

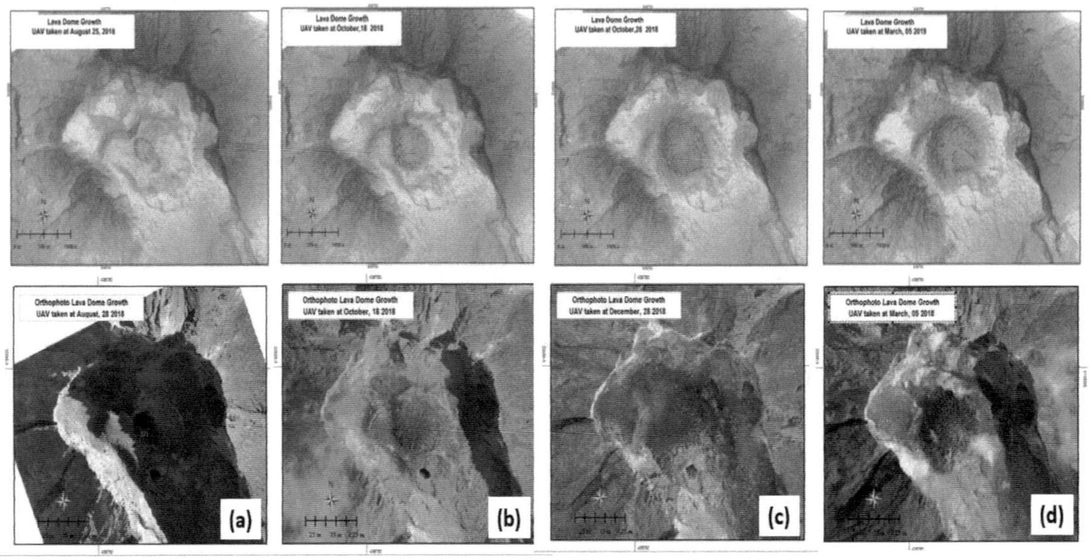

Figure 5. The top shows the DTM which reflects the growth of the Merapi lava dome, the bottom image shows orthophoto images of the crater taken in its original condition. (a) Dome growth from DTM and Orthophoto on 25 August 2018 volume is 33,623 m3 (b) 19 Sept 2018 volume is 133,159 (c) 28 December 2018 volume is 244,486 (d) 4 February 2019 volume is 658,075.

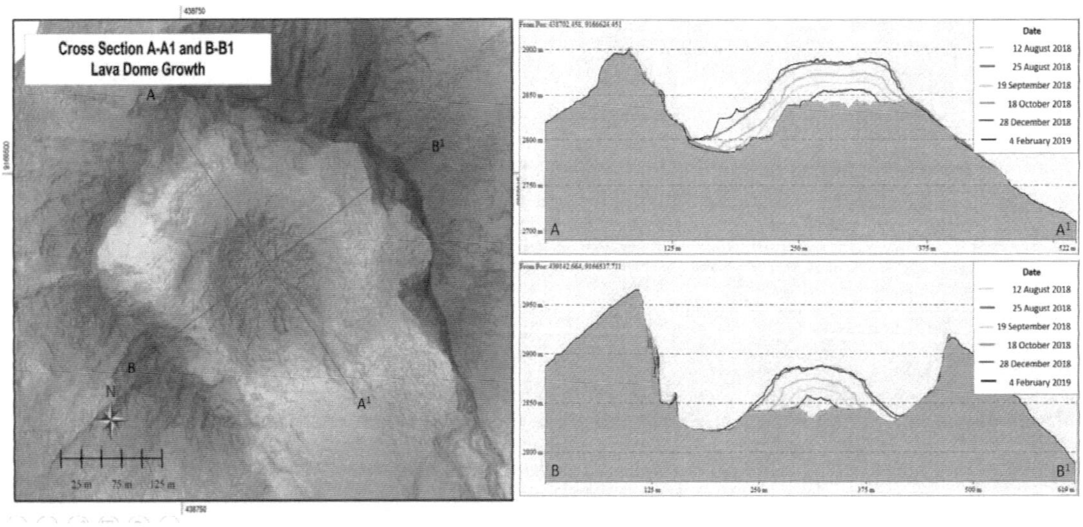

Figure 6. The volume of the lava dome in this growth phase shows an increase in height and diameter (long vs. wide) in each period.

4. Discussion

4.1. Morphometric Changes of the Lava Dome

The dimensions of the Merapi lava dome from 12 August 2018 to 4 February 2019 (176 days) were continuously monitored by UAV (Table 2). The height of the lava dome in that period was between 11 - 41 m, the length between 60 - 156 m, and the width of 47 - 152 m. The basal ratio (width/length) of the lava dome changes from 0.78 - 0.97. This basal ratio suggests the dome transformation from an ellipsoid into a half-sphere. The 2013 fissure might contribute to the initial shape of the lava dome.

Carapace and talus fill the fissure with a northwest-southeast direction, following the shape of the 2013 fissure, creating an ellipsoid-shaped initial stage of the lava dome. A shape change was observed on 18 October 2018 - 4 February 2019, where the length was almost proportional to the width (W/L ratios 0.91 - 0.97). This shape suggest that the lava dome grew in all directions. Lava dome started to pour towards K. Gendol on 29 January 2019. However, the basal ratio of the lava dome remains constant between 28 December 2018 and 4 February 2019. It is worth noting that the basal area of the lava dome on 4 February 2019 is about 21.780 m^2, much less than that of the 2010 crater (101,500 m^2). In this state, the talus of the lava dome has reached the outer rim of the 2010 crater towards K. Gendol. Considering the growth pattern of the lava dome, it is most likely that the southeastern flank of the lava dome will partially collapse towards K. Gendol in response to the growth of the lava dome.

Table 2 The dimensions of the Merapi lava dome were continuously monitored by UAV.

		Lava Dome				Cross Section B-B1 Fig 6				Cross Section A-A1 Fig 6			
Date	Cumulative Days	Length (m)	Width (m)	Height (m)	H/L	Summit width (m)	Basal width (m)	SW/ BW	Slope (^{0})	Summit width (m)	Basal Width (m)	SW/ BW	Slope (^{0})
12 August 2018	0	0	0	0	0								
25 August 2018	13	60	47	11	0.18	27	49	0.55	58	29	60	0.48	36
19 September 2018	38	82	65	20	0.24	46	88	0.52	56	65	99	0.66	41
18 October 2018	67	108	98	28	0.26	53	112	0.47	53	50	116	0.43	49
28 December 2018	138	131	127	38	0.29	66	153	0.43	51	71	142	0.50	58
4 February 2019	176	156	152	41	0.26	78	164	0.48	49	75	150	0.50	62

4.2. Lava Dome Extrusion Rate

The volume of the Merapi lava dome from 12 August 2018 to 4 February 2019 (176 days) changed from 33,623 m^3 to 658,075 m^3. The extrusion rate of the lava dome ranged from 2,586 m^3/day to 4,196 m^3/day. Record of Merapi lava dome volume, extrusion rate, and related paramaeter is presented in Table 3. The extrusion rate of 2018-2019 lava dome is similar to lava dome Merapi 1980 (3000 m^3/day) but much less than the extrusion rate of lava dome 1930 (200,000 m^3/day), 1992 (200,000 m^3/day), 2006 (285,000 m^3/day), and 2010 (2,000,000 m^3/day). These high extrusion rates are mostly associated with PDC run-out distance from 7 to 15 km, potentially causing high casualties. Exceptions come from the 1992 lava dome, which remains stable despite the high extrusion rate (. However, lava dome 1992 with total volume 9,1 x 10^6 m^3 be additional material PDC for the next event, the 1994 lava dome, with an extrusion rate only of 17,000 m^3/day, produced a 6 km distance PDC.

The 2018-2019 lava dome might not produce far-flowing distance PDC considering the growth rate, crater volume, and the past behavior of Merapi. However, the continuous collapse of the lava dome to seoutheastern valleys (e.g., Gendol River) may reduce its accommodation space, producing longer distance PDC. Therefore, we highly recommend that lava dome monitoring using UAV photogrammetry be conducted frequently.

4.3 UAV Application Limitation

Although the UAV method provides fast and relatively low-cost aerial photo data, this method has several limitations. The UAV survey cannot be conducted during the specified period due to the wind and clouds that often occur at the summit of Merapi. Morphological analysis for flight paths needs to be taken seriously to avoid accidents.

The DJI Phantom 4 Pro has a limited flying range, and personnel must approach hazardous areas. During the crisis, it is recommended that we use a UAV able a long-range mission, for example, a fixed-wing model. Before a flight mission, the team should coordinate with the local aviation authority. It must be carried out to obtain a NOTAM (notice to airmen).

Table 3. Recent lava dome extrusion with impact phenomena over the last century [9] with the addition of lava dome data in 2018.

Year	Extrusion Rate (m^3/day)	Volume (10^6m^3)	PDC Run Out (km)	Phenomena
1890	3,000	-	-	Rockfalls accompanied Dome growth. Explosions also took place, some of which produced ash fallout and lava bombs.
1930	200,000	6.0	15	No effusive eruption, paroxysmal eruption
1992	200,000	9.1	-	Remained stable.
1994	17,000	2.6	6	Gravitational PDC without explosion (killed 60 people) the accumulation of the dome thickness was 50 m.
2006	285,000	5.3	7	Gravitational PDC without explosion (killed2 people) the accumulation of the dome thickness was 100 m.
2010	2,000,000	-	15	Sub Plinian - Plinian eruption
2018	3,500	0.5	1.4	Remained stable small PDC

5. Conclusion

The Low-Cost UAV method is very effective to be applied on Merapi Volcano, especially volcanoes with lava domes. Inaccessible morphological monitoring, the UAV provides DTM and Orthophoto data with a resolution of 0.5 – 1 m. With this data inform maximum extrusion volume during the 2018-2019 period is 658,075 m^3 with an average extrusion rate of 3,500 m3/day. The lava dome grew in all directions from the basal ratio W/L with a value of 0.91 - 0.97. For some of those facts mitigation steps need to be taken, considering that there is already PDC to the K. Gendol channel.

References

[1] Voight B, Constantine E, Siswowidjoyo S and Torley R 2001 Historical eruptions of Merapi volcano, Central Java, Indonesia, 1768–1998 *J. of Volcanology and Geothermal Research*, 100(1-4), pp.69-138.

[2] Granados B, Quesada A and Alvarado G 2021 Low-cost UAV applications in dynamic tropical volcanic landforms *J. of Volcanology and Geothermal Research*, 410, p.107143.

[3] Darmawan H, Walter T, Sri B, Kirbani, Subandriyo, Made A 2018 *M*orphological and structural changes at the Merapi lava dome monitored in 2012–15 using unmanned aerial vehicles (UAVs) *J. of Volcanology and Geothermal Research*, 349, 256-267.

[4] Regulation of the Geospatial Information Agency 2020 Standar *Pengumpulan Data Geospasial Dasar Untuk Pembuatan Peta Dasar Skala Besar* KEPALA BIG RI.

[5] Solikhin A, Thouret J, Liew S, Gupta A, Sayudi D, Oehler J and Kassouk, Z 2015 High-spatial-resolution imagery helps map deposits of the large (VEI 4) 2010 Merapi Volcano eruption and their impact *Bulletin of volcanology*, 77(3), pp.1-23.

[6] Carrilho A, dos Santos C, Pessoa, G and Galo M 2020 Comparison between airborne LiDAR data and digital image matching for Surface Reconstruction *In 2020 IEEE Latin American GRSS & ISPRS Remote Sensing Conference (LAGIRS)* (pp. 161-166).

[7] De Beni E, Cantarero M, and Messina A 2019 *U*AVs for volcano monitoring: A new approach applied on an active lava flow on Mt. Etna (Italy), during the 27 February –02 March 2017 eruption *J. of Volcanology and Geothermal Research*, 369, pp.250-262.

[8] Pallister J, David J, Griswold J, Keeler P, Ronald, Burton, William C, Christopher N, Ratdomopurbo, A 2013 Merapi 2010 eruption—Chronology and extrusion rates monitored with satellite radar and used in eruption forecasting J. *of Volcanology and Geothermal Research*, 261, 144-152.

[9] Kelfoun K, Budi S, Thierry L, Martial B, Ilham N, François B 2021 Growth and collapse of the 2018–2019 lava dome of Merapi volcano Bulletin *of volcanology,* 83, 8

[10] Andaru R, Syahbana D, Prayoga A and Purnamasari H 2021 The use of UAV remote sensing for observing lava dome emplacement and areas of potential lahar hazards: An example from the 2017–2019 eruption crisis at Mount Agung in Bali *J.Volcanology and Geothermal Research*, 415, p.107255.

[11] Harnett C and Heap J 2021 Mechanical and topographic factors influencing lava dome growth and collapse J. *of Volcanology and Geothermal Research*, 420, p.107398.

Acknowledgements

We wish to acknowledge Geological Department Gadjah Mada University to arrangement event RC-GeoE2021. We would also like to thank the reviewers for their insightful reviews. Thank for the efforts of the BPPTKG (Center for Research and Technological Development, Yogyakarta) staff for support and help during the field. We are grateful to “Drone Team” Alzwar Nurmanaji, Ahmad Sopari, Yulianto and Dwi Julianto hard work in the field. Thanks to “Lanud Adisutcipto” as local aviation authority give permit to flight around Merapi. Thanks to “Seribu Bintang” and “Micro UAV” give suggestion and discussion during flight path planning.

Landslide susceptibility zone in steep slope landscape at Gunung Kidul Regency, southeast part of Yogyakarta, Indonesia

Phoulid Saenkang[1*] **and Hendy Setiawan**[1]

[1]Department of Geological Engineering, Universitas Gadjah Mada, Yogyakarta, Indonesia
*Corresponding author's e-mail: phoulids@gmail.com

Abstract. Gunung Kidul is located in Central Java Province, Indonesia. which is vulnerable to geological hazard such as floods and earthquake. However, it is also susceptible to landslide in some areas of Gunung Kidul which contributes to damage and loss. The study area is covers sub districts such as: Patuk, Gedangsari, Nglipar, Ngawen, Playen. It covers an area of 1,485.36 km2 and had a population of 747,161 at the 2020 Census. This research aims to produce a landslide susceptibility map, construct the detailed geological map, consider geomorphology aspect, lithology aspect, soil-rock masses, hydrogeology/groundwater aspect, structural geology aspect, land use. The parameters that caused the occurrence of landslide were determined and the landslide susceptibility map was produced using Weightage Overlay Method (WOM) in ArcGIS software. Results showed that the susceptibility map was classified into three zones which is low, moderate and high zone. The factor that triggered the landslide were identified which is heavy rainfall intensity and earthquake. As a conclusion, the ability to detect landslide susceptibility led to a better understanding of landslide mechanisms for the research area, thus leading to an enhanced identification of the most likely failure sites within a landslide-prone area.

1. Introduction

Indonesia is a complicated country since it's where many tectonic plates collide and is located on the ring of fire called pacific plate and the Indo-Australian Below the Eurasian continental plate is a continent plate that has been pushed down [1]. Since Indonesian region is tectonically active, a complex geological context emerges.

Landslides are natural disasters that occur often in mountainous areas of the globe, resulting in the death and destruction of property. It has the potential to wreak a great deal of harm and loss. The human and environmental factors are the two main factors that can cause a landslide to occur. Despite the fact that One of the most vulnerable countries is Indonesia in terms as a result of natural calamities, such as landslides, planning and mitigation efforts in many areas are failing. However, these operations are required to lessen the geohazard effect is a term that refers to a phenomenon that occurs when.

Yogyakarta is the majority of disasters strike at the same time, beginning with an earthquake, followed by landslides and flooding, especially in 2017, when the city was proclaimed a state of emergency. High-intensity rainfalls and earthquakes, particularly in the mountains of the south, could cause a landslide. Landslides are one of the most common natural dangers that occur after an earthquake, according to the majority of incidents recorded globally. As a result, Java has been known to be prone to landslides, particularly in the south portion districts.

To be able to create a landslide susceptibility map, GIS methods can be employed in a variety of ways. The spatial data from the previous investigation was organized in GIS SPRING program [2]. It's feasible to determine the failure of the slope by developing a slope inclination map for the research area using both methods.

Rodeano and colleagues [3]. A landslide hazard zoning map was created using the Method of Weighted Overlay (WOM) and the map of the Landslide Susceptibility Level (LSL). The most harmful factors, such as lithology, drainage, land use, were weighted in the study. Gunung Kidul is a place with a high mountainous terrain compared to Yogyakarta City and neighbouring areas.

2. Methodology

The research area is in the Gunung kidul regency, southeast part of Yogyakarta, the Almost the entire research area is covered 40 km2, which coincided with latitude 8'51'' 7'51''S to 8'51'' 7'51''S and longitude 110° 20' 110.40''E to 110° 20' 110.40''E. To produce the landslide susceptibility map, The field of study's geomorphology takes the form of corrugated mountains and claystone units, andesite units, and alluvial units make up the plain morphological units of stratigraphy, as well as structures of the Gunung kidul regency. The use of secondary data such as topographical data, Shuttle Radar Topography Mission (SRTM) year 2021 DEM and Digital Elevation acquired as of the United States survey of geology is critical. To be able to create the map of landslide susceptibility, the criteria that produced landslides were considered, such as slope, lithology, drainage density, vegetation, land use, aspect are all factors to consider. and these parameters were used for Weight Overlay Method [4]. The parameters were chosen and then implemented [3]. After that, parameters were constructed using DEM data from the year 2021. To calculate the landslide susceptibility area, the average weightage score of each criterion was used. The expected output of the map of landslide susceptibility was then created with the use of a geographic information system (GIS) overlay and raster calculation. Those factors were obtained landslide susceptibility levels ranging from very low to very high, according to the topographic database, indicating the likelihood of a landslide occurring in the research area. The landslide was precipitated by heavy rains, which changed the intensity of the land and soil. Layering approach using raster input was weighted out of all the data. The study area's base map is shown in Figure 1. A field investigation was carried out in numerous locations where landslides had occurred. In order to back up the outcome, obtained the use of a weighted overlay approach, the truth occurrence and situation on the field producing Based on the identified parameter, landslides were detected, allowing the landslide is at risk region to be calculated. The landslide began with a failure in the higher a section of the cliff, with the a flow of residue downstream in relation to gravity. Weathered lava that isn't covered by vegetation makes up the lithology of the area. The porosity of soil gets filled up as a result of high precipitation and continuous rainfall, until it can no longer contain the water and subsequently slides downwards. It also has weathered lava lithology from the Gunung kidul Formation, with a relatively steep slope of 24°. The lithology is sandstone carbonate sedimentary rock. The stability of the rock becomes weakened as a result of the high concentration of rain. The following are the parameters of this study: Slope, Drainage density, Lithology, Aspect, and Land use/cover.

2.1 Lithology

Among the parameters utilized in estimating the lithology map shows the study area's susceptibility to landslides. Where the terrain is prone or low sensitivity to landslides, lithological differences had a vital effect. The properties of rock and soil are distinguished by their lithologic state and formation. Figure 1 depicts the lithology of the research area, showing from north (oldest) to south (youngest) rock distribution. The Formation, which consists of pyroclastic and epiclastic breccia units, dominated the northern area.

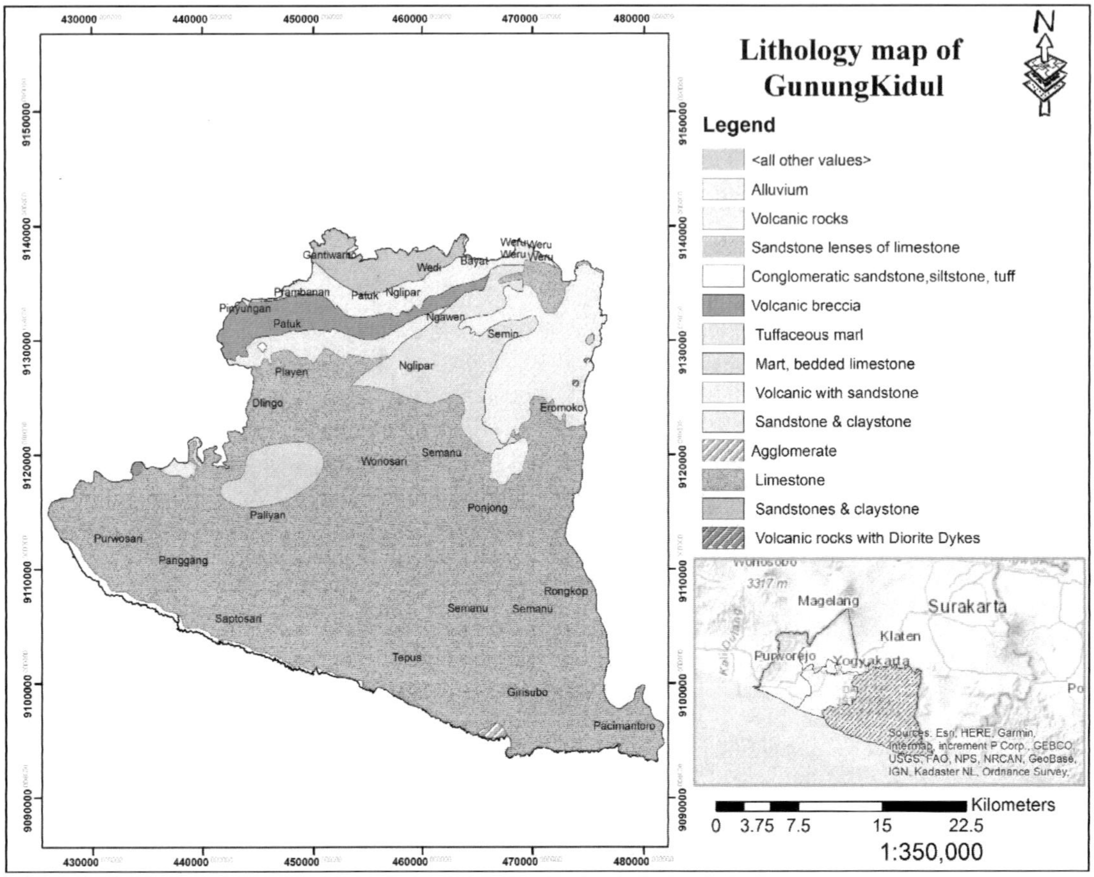

Figure 1. Study Area with Lithological Map.

2.2 *Slope*

The slope is a key factor in assessing the susceptibility of the study region to landslides. According to the works, the slopes were divided into Nine groups. The research area was separated into two sections: a mountainous region in the north and a flat region in the south. The northern section of the research area has high to extremely steep slopes of 29.97° – 35.96° and 35.96° and more, whereas the southern part has a moderate to mild grade. Figure 2 depicts the study area's slope map.

2.3 *Aspect*

DEM data was used to create an aspect map. The slope direction due to the north was classified into eight categories on the aspect map. The parameter was utilized to observe the slope direction, which resulted in gravitational force differences. It is possible to determine the slope's instability and direction. As indicated in Figure 3, it is measured in degrees counter clockwise from 0 to 360 degrees -referencing due North. North, East, Southeast, South, Flat, Southwest, West, Northwest, Northeast, and North are some of the slope direction classes that arise from the aspect direction. Because of wind exposure, the angle at which a slope faces the sun (aspect) has an effect on vegetation and architecture.

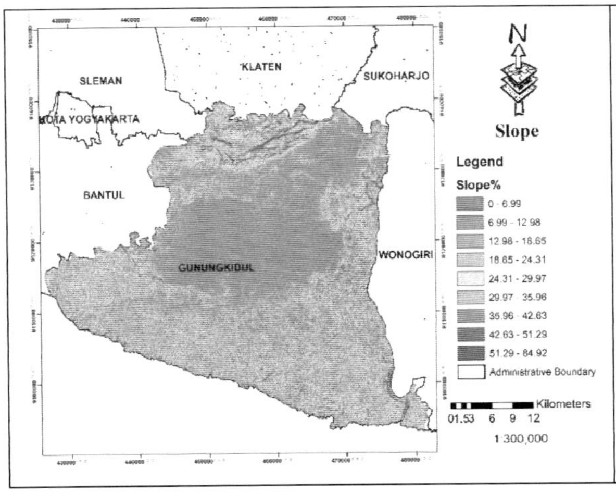

Figure 2. Slope map.

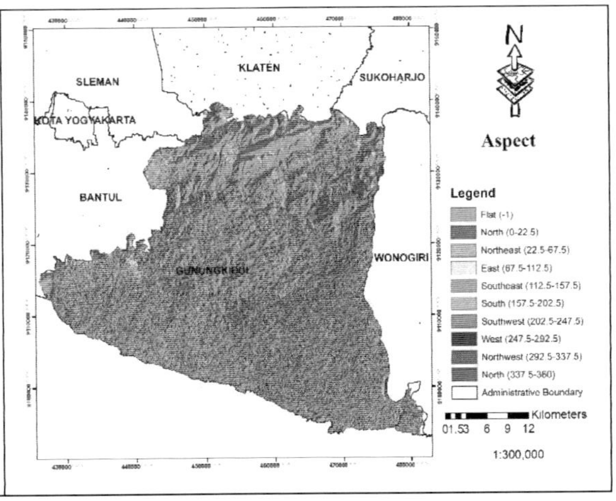

Figure 3. Aspect map.

2.4 Vegetation and Land use

Land use map was created using secondary data gathered from Esri Global land use/cover map [7]. Majority of the land was used for plantation purposes, such as paddy fields and eucalyptus groves grassland, flooded vegetable Some of them were engulfed by trees. Aside from that, settlements with a middling population density. The stability of the slopes is determined by the amount of land cover. Landslides can be accelerated and have a significant impact due to human activity. Human interference slope deforestation, elimination of slope support in road cuts, and other natural land cover changes and altering of drainage from the surface routes, has converted a feature trigger for the occurrence of landslides, causing environmental harm. Human activities can alter groundwater flow by destroying vegetation through logging, fires, or droughts, increasing the danger of landslides. Figure 4 depicts the research area's land use and vegetation maps.

2.5 Drainage Density

The entire length of streams and rivers in a drainage basin divided by the drainage basin's total area is known as drainage density. Rainwater penetration has caused a lot of debris flow and seepage. can be used to evaluate the landslide's vulnerable area. In ArcGIS, DEM data was used to create a drainage density map. Classifying the drainage density as shown in Figure 5 to obtain reference values.

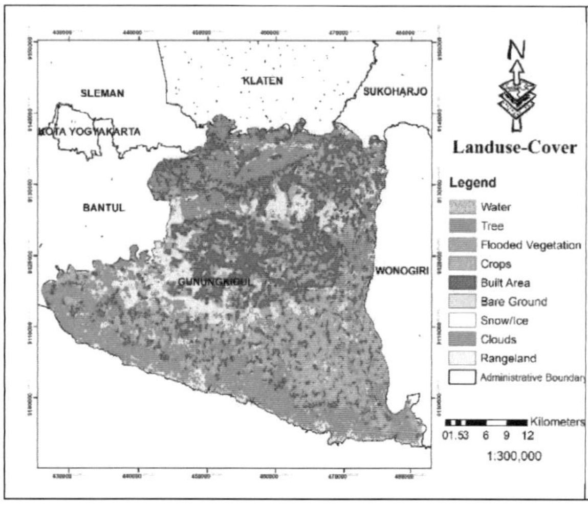

Figure 4. Vegetation and Land use map [5].

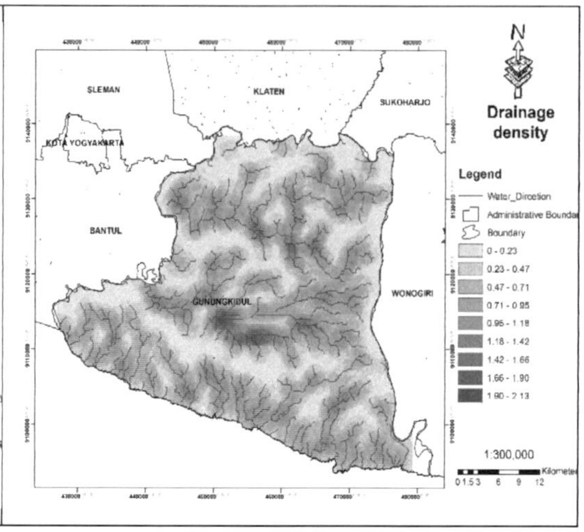

Figure 5. Drainage density map.

The presence of a high drainage density implies the likelihood of a landslide. Because of the large volume of surface runoff, this is the case.

2.6 Analysis of Landslide Susceptibility

All chosen parameters are transformed into raster data collection before processing and combining into landslide susceptibility. Reclassification is done upon weightage of raster data collection. These many methodologies use a variety of rating systems, weights, and regulating elements, all of which are very dependent on the subject area. As a result, these weights and ratings are crucial when creating susceptibility maps using any method. The landslide susceptibility map was created using ArcGIS software and the weightage overlay method (WOM). The lithological data, aspect, slope, vegetation & land use, and drainage density characteristics for landslide causative factors were chosen. The weights were applied to the parameters, and the sum of all factors was set to 100%. All of the resulting maps are classed using weighted values based on their landslide occurrence magnitude. Finally, using the Weighted Overlay Analysis Model, two landslide susceptibility maps were created.

Landslide Susceptibility by WOM = [Lithology*21 + Slope*29 + Aspect*13 + Vegetation & land use*7 + Drainage Density *30] .... (1)

The weightage overlay was used to create the landslide susceptibility map approach as described in Table 1, it is necessary to reclassify the data.

Table 1. Influence value and raster datasets

No	Raster Datasets	Influence
1	Lithology: Alluvium Volcanic Rocks Sandstone lenses of limestone Conglomeratic sandstone, Siltstone, tuff Volcanic breccia Tuffceous marl Mart, bedded limestone Volcanic with sandstone	21

	Sandstone & claystone	
	Agglomerate	
	Limestone	
	Sandstone & claystone	
	Volcanic rock with Diorite Dykes	
2	Slope:	29
	0 – 6.99°	
	6.99 – 12.98°	
	12.98 – 18.65 °	
	18.65 – 24.31°	
	24.31 – 29.97 °	
	29.97 – 35.96°	
	35.96 – 42.63°	
	42.63 – 51.29°	
	51.29 – 84.92°	
3	Aspect:	13
	Flat (-1)	
	North (0– 22.5)	
	North-East (22.5 – 67.5)	
	East (67.5 – 112.5)	
	South-East (112.5 – 157.5)	
	South (157.5 – 202.5)	
	South-west (202.5 – 247.5)	
	West (247.5 – 292.5)	
	North-West (292.5 – 337.5)	
	North (337.5- 360)	
4	Vegetation & Land use:	7
	Water	
	Tree	
	Flooded Vegetable	
	Crops	
	Built Area	
	Bare Ground	
	Snow/Ice	
	Clouds	
	Rangeland	
5	Drainage density:	30
	(0 – 0.23)	
	(0.23 – 0.47)	
	(0.47 – 0.71)	
	(0.71 – 0.95)	
	(0.95 – 1.18)	
	(1.18 – 1.42)	
	(1.42 – 1.66)	
	(1.66 – 1.90)	
	(1.90 – 2.13)	
	TOTAL	100

3. Results and discussion

The landslide susceptibility maps of Gunung Kidul and its environs are shown in Figure 6. According to the maps, the majority of the study region (39.84 %) is moderately susceptible to landslides. Low susceptibility to landslides makes up 28.26% of the research area, whereas high susceptibility to landslides makes up 31.54 %. The study's findings are derived from determining the map analysis results, which are divided into zones, including the three zones that are analysed, and then using the analysis values of the three zones as parameters in calculating the total area, we get the percentage of each zone as described in the analysis results.

The susceptibility class for the landslide hazard in the research area is shown in Table 2. In studies of landslide susceptibility, lithology, slope & drainage density parameters are relevant characteristics. Landslides are more likely to occur on slopes with a steeper angle. In addition, condition of rocks and soils that exist in the area are important to look at due to its porosity and void that saturates upon water contact, also, the lithology profile is correlated to landslide prospects. Most landslides might occur along streams or river system, as drainage density indicates that the hydrological factors can trigger landslides. The landslide phenomena are moderately influenced by vegetation, land use, and aspect.

The landslide is riskier near the stream and river, according to the landslide susceptibility map. This is most likely owing to the moderate to high density of the stream. It reveals that landslides are more likely to occur in northern areas with moderately steep to extremely steep slopes, volcanic rocks, and breccia lithologies. When weathered lava rocks became soil, landslides were common. Figure 6 Gunung kidul Area Landslide Susceptibility Map Table 2.

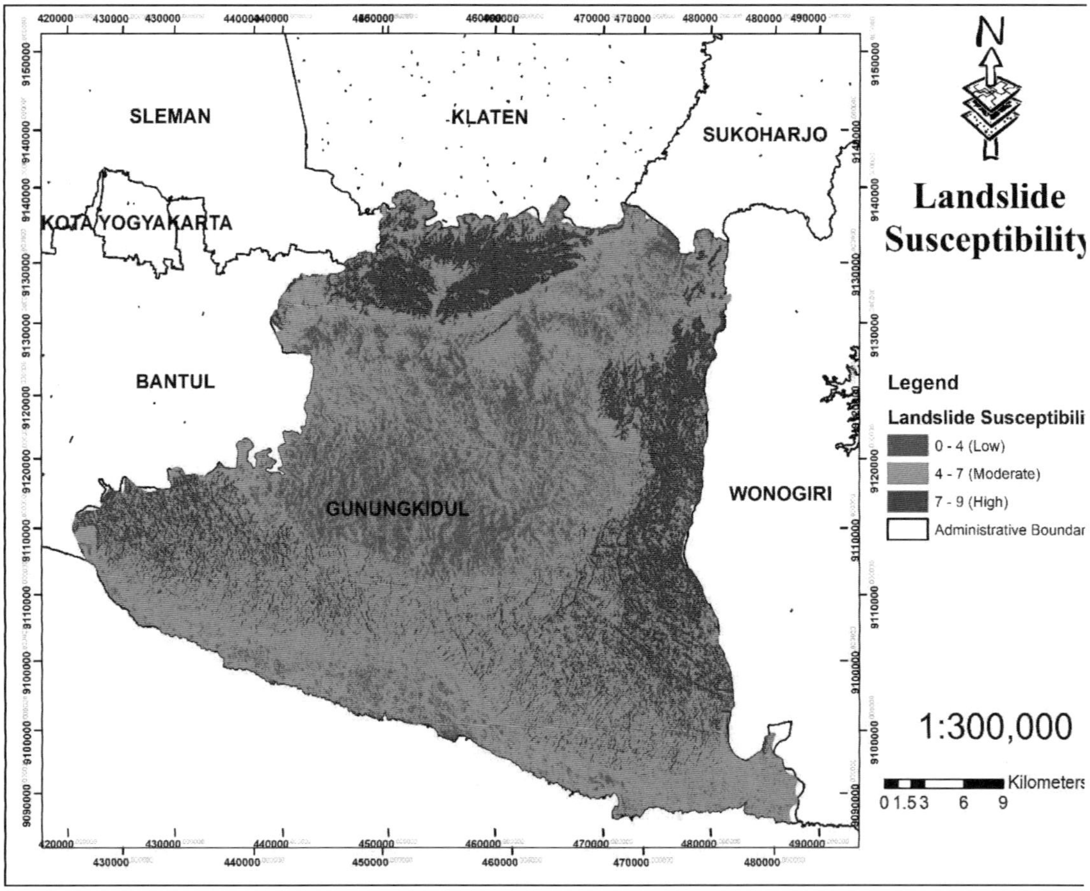

Figure 6. Landslide Susceptibility Map of Gunung Kidul Area.

IOP Conf. Series: Earth and Environmental Science 1071 (2022) 012002 doi:10.1088/1755-1315/1071/1/012002

Table 2. Values of Landslide susceptibility.

Susceptibility Class	Value	Area Percentage (%)
Low	<4	28.26
Moderate	4 - 7	39.84
High	>9	31.54

Susceptibility of landslide hazard in the research field dominantly moderate. The amount of time the surface is exposed of volcanic rocks contributes to being weathered in comparison to the rock that is beneath the earth's surface Water and wind act as weathering mechanisms, causing soil instability. The porosity between the soils was filled when significant rain fell in the study location. When the lowest layer of interflow is bedrock, an excessive amount of water flows and may cause debris slide. The breakdown of the land structure caused the earth's bulk to slide. The landslide was prompted by severe rainfall and an earthquake, primarily in the northern part of the country. and eastern part has a steep slope and a high elevation. Aside from that, the southern section of the research area reveals a high risk of landslide, particularly along the river. That region's lithology is characterized by granite by limestone and some carbonate bearing-sandstone, clay & silt. Although the locations were low in elevation, landslides in limestone and sandstone are the most common. River actively flows erode slope angles in river basements and lose Strength of the soil and rock as a result, the sites were recognized as landslide-prone.

4. Conclusion

In the research area, the intensity of rainfall and earthquakes have been as the most important factors causing landslides every year, cause rainfall alters the pressure within the slope, which leads to slope instability and by imposing inertial loads or creating a loss of strength in the slope materials, earthquakes can make a slope unstable. Rainfall that is too heavy might cause flooding cause landslides, especially in areas with dense soils and weather soil conditions. When heavy rains came down, filling the soil's porosity and voids, causing landslide because it leaves the large volume of water that flows through the bedrocks Clay with a lot of flexibility causes a soil with a high pore pressure a small shear angle which will cause landslides. Earthquakes can trigger the movement of land masses. The research area is close to major faults such as fault movements may be less than an inch per year but cause earthquakes. Landslide risk maps were developed for the research region utilizing the most relevant parameters such as lithology, slope, aspect, vegetation & land use, and drainage density. Susceptibility maps were created using the weighted overlay approach in ArcGIS and classed into three categories: low, medium, and high potential for landslide susceptibility. In general, the research areas were prone to landslides, with 28.26% of low susceptibility, 31.54 % of high susceptibility, and 39.84 % of medium susceptibility. Although Indonesia is a country that is constantly threatened by geohazards, mitigation efforts in many areas, such as Gunung Kidul, are failing. There is an Early Warning System in place for the high geohazard cases that have been documented, as well as potential locations (EWS). EWS aids in the monitoring and prediction of landslides. It's one of the mitigating techniques used in the research area. As a result of conducting this investigation, the lack of awareness among villagers might be reduced. due to the estimation of potential risk areas. Villagers' lack of awareness can be reduced due to the estimation of potential risk areas. With The precision of the generated map will aid decision-makers, contractors, and local residents in catastrophe mitigation and planning. It could be used by local governments to implement policies and programs aimed at reducing the risk of landslides. Any development decision should be based on spatial planning, especially in high-risk areas.

References

[1] Saputra, A., Gomez, C., Hadmoko, D. S., & Sartohadi, J. (2016). Coseismic Landslide Susceptibility Assessment Using Geographic Information System. *Geoenvironmental Disasters.* doi: 10.1186/s40677-016-0059-4

[2] DEM, G., Souza, R.C.M., Freitas, U.M., Garrido, J. and Mitsuo, F. (2021). Spring: Integrating Remote Sensing and GIS By Object-Oriented Data Modeling. Computers & Graphics, 20 (3) 395-403.

[3] Rodeano R., Alyvyn C. M., Norbert S., Mohd N. N. (2017). Landslide Susceptibility Analysis (LSA) Using Weighted Overlay Method (WOM) Along the Genting Sempah to Bentong Highway, Pahang. Malaysian *Journal of Geosciences*, 13-19.

[4] European Space Agency, 1920. Digital Elevation Model. Retrieved from: https://www. https://www.esa.int/

[5] Karra, Kontgis, et al. 2021. Global land use/land cover with Sentinel-2 and deep learning. IGARSS 2021-2021 IEEE International Geoscience and Remote Sensing Symposium. IEEE.

[6] W. S. Udin, N. A. S. Norazami, N. Sulaiman, N. A. Che Zaudin, S. Ma'ail and A. N. Mohamad Nor, "UAV Based Multi-spectral Imaging System for Mapping Landslide Risk Area Along Jeli-Gerik Highway, Jeli, Kelantan," *2019 IEEE 15th International Colloquium on Signal Processing & Its Applications (CSPA)*, Penang, Malaysia, 2019, pp. 162-167, doi: 10.1109/CSPA.2019.8695995.

Reducing groundwater use and mitigating floods through increasing surface water utilization in the Cimahi Region of West Java Indonesia

M Y Purnawan[1,2], H Hendrayana[1*] and L D Setijadji[1]

[1] Department of Geological Engineering, Universitas Gadjah Mada, 2 Grafika, Bulaksumur, Yogyakarta 55281, Indonesia
[2] Ministry of Public Works and Housing, Jakarta, Indonesia
*Corresponding author's e-mail: heruhendrayana20@gmail.com

Abstract. Population growth and economic development increase the demands for water and land in Indonesia from year to year. In addition, various spatial and temporal changes resulted from human activities in changes in of hydrology of areas around rivers that affect the water's quality and quantity. Groundwater conditions in the Cimahi area, West Java, show a significant groundwater decline in southern Cimahi, a problem that is now studied in this research. This research is conducted by using groundwater modelling from the Experimental Station for Groundwater in the Bandung-Soreang Groundwater Basin, then we further develop the previous model by reducing groundwater utilization and replacing it with surface water to assess the impact of this substitution on the groundwater level, and reduction of surface water runoff for flood mitigation during the modelling projection period 2021 - 2030. The study starts by identifying groundwater utilization in the research area, and then we analyze the feasibility of surface water quality and quantity as a substitute for groundwater utilization. This study concludes that groundwater utilization can be replaced by surface water utilization. Our study results show that surface water runoff can be reduced up to 2.48 m^3/s and positively increase groundwater level by 11.26 m.

1. Introduction

Water and water sources as well as the energy contained in water, these three are the definition of water resources. Water which located above or below the ground surface, including surface water, ground water, rain water, and sea water on land is a water resource. In the implementation of water resources management, there are various activities including planning, implementing, monitoring, and assessing the conservation of water resources, water resources utilization, and control of water damage[1]. The increasing of population along with the development of economic affect the water needs and area in Indonesia over time. Furthermore, quality and quantity of water is also affected by the changes in hydrology of the area arround water resources caused by human influence. [2].

The city of Cimahi, in combination with parts of Bandung, West Bandung, and Sumedang regencies are combined into the Greater Bandung Metropolitan Area of West Java, Indonesia. This region was merged due to community activities, built-up land, economic, and population [3].

Cimahi city has experienced changes in land use in residential and industrial areas, which cause a decrease in the ability of the land to absorb water, thereby increasing the volume of runoff. Conditions of drainage channels such as narrowing of capacity from upstream to downstream, drainage channels that are closed by buildings, and drainage channels that are not connected cause the function of drainage channels to accommodate runoff water to be disrupted. This situation makes the water difficult to manage and its overflows from the existing channels become a puddle/flood. The lack of availability of waste service facilities risks making piles of waste that are not accommodated carried away into drainage channels. This will cause the channel to become clogged so that the water overflows into a puddle/flood [4]. Accordingly, Cimahi city's critical problems of water resources are clean water and floods. Cimahi city flooded by cilember river which flows along the city that has maximum value of 41.2 m^3/s as Q50 flood discharge. The narrow river border along with local drainage that not function properly, resulting the fact that the river of cilember is now unable to accomodate the discharge [5].

Excessive groundwater utilization caused by population growth and rapid industrialization in Bandung-Soreang Groundwater Basin, including Cimahi city, resulting environmental issues such as enormous groundwater table decline and also land subsidence [6]. In 2020, the Experimental Station for Groundwater conducted research on condition of groundwater table in Bandung - Soreang groundwater basin by conducting modelling of groundwater for 2001 - 2020 and concluding that unconstrained groundwater consumption has developed in Citarum watershed, because of increasing water demand in certain area, mainly in industrial and urban areas. In addition, change of land-use in the upstream area worsens the problem because it cause decreasing the capacity of groundwater recharge. It is now known that the south Cimahi area has significant groundwater table decline/ cone of depression due to excessive groundwater utilization [7]. The increasing need for water for domestic and non-domestic activities in the community causes an increase in groundwater use and lowers the groundwater level. The current situation in the Cimahi area is studied in this research, with a focus on reducing groundwater utilization, mitigating flood disasters, and recovering the groundwater level in areas with high groundwater table decline. One solution for groundwater utilization reduction can be implemented by replacing clean water sources with surface water which is also beneficial for increasing groundwater levels [8].

Therefore, this study aims to assess the impact that can be given in increasing the surface water utilization to replace groundwater consumption in southern Cimahi and also to assess how much surface water run-off can be absorbed to implement flood mitigation in the Cimahi area. In this research, the development of the previous model was by reducing groundwater utilization and replacing it with surface water to assess the impact of this substitution on the groundwater level, and also its impact on the reduction of surface water runoff for flood mitigation during the modelling projection period 2021 - 2030

2. Scope and Methods of the study

The area of Cimahi city is included in a groundwater basin called the Bandung-Soreang groundwater basin which in a basin area covers the area of Cimahi City, Bandung City, West Bandung Regency, Sumedang Regency, and Bandung Regency. The research area is geographically located at 107° 30' 30''- 107° 34' 30'' E and 60° 50' 00 '' – 6° 56' 00'' S. Cimahi City has 4,025.73 ha and consists of 3 sub-districts namely North Cimahi district with 4 villages, Central Cimahi District with 6 villages, and South Cimahi District with 5 villages [9]. The research location for this study is limited to South Cimahi (Figure 1). This research has of four stages: study of previous groundwater modelling in research location, surface water quantity availability analysis, surface water quality analysis, and substitution of groundwater utilization by surface water utilization analysis.

2.1 Study of previous groundwater modelling

Groundwater modelling was previously conducted by the office of Groundwater's Experimental Station for of Indonesia's Ministry of Public Works at Bandung- Soreang Groundwater Basin by modelling the groundwater level conditions from 2001-to 2020. The Experimental Station of Groundwater conducted real-time research on groundwater modelling in the Bandung-Soreang groundwater basin to produce prototypes of groundwater modelling in industrial and urban areas in greater Bandung area. The groundwater modelling is conducted using a numerical groundwater flow model using IMOD (Interactive Modeling). The results of the groundwater modelling project have identified three regions with a cone of depression in the Bandung – Soreang groundwater basin, in which the affected area based on the modelling result is the south Cimahi sub-district (Figure 2). Cone of depression defined as states of groundwater is pumped and form a conical subsidence of groundwater level, for cone of depression in confined aquifers, conical subsidence showed as decreases of pressure head around the pump well.

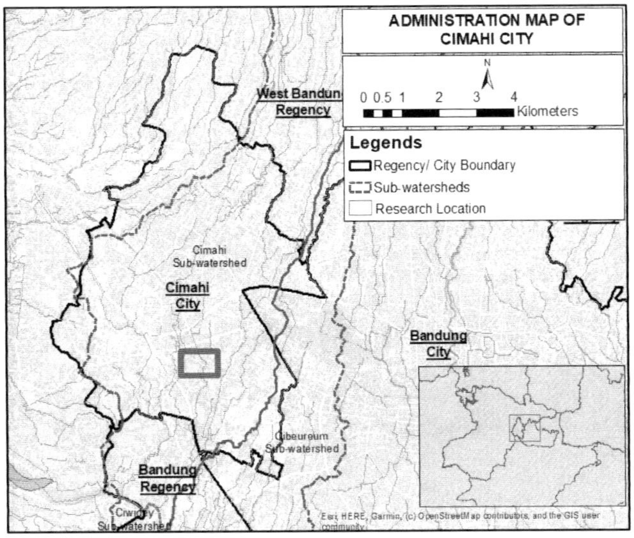

Figure 1. Research Location.

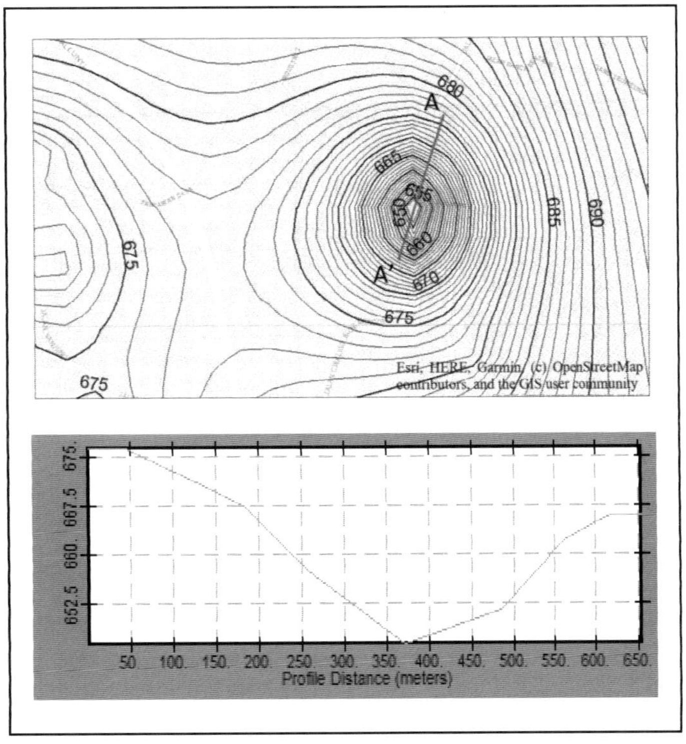

Figure 2. Groundwater table decline in southern Cimahi (Seizarwati, 2020).

IOP Conf. Series: Earth and Environmental Science 1071 (2022) 012003 doi:10.1088/1755-1315/1071/1/012003

2.2 Surface water quantity availability analysis
Dependable discharge is certain amount of water discharge calculated with certain probability or return period and reliable discharge of 90 % is used for raw water [10]. The dependable discharge of the Citarum sub-watershed acquired from the Hydrology and Water Environmental Agency of the Ministry of Public Works and Housing carried out the analysis of reliable discharge for the upper Citarum watershed in 2020. Furthermore, reliable discharge data was later used to analyze surface water quantity availability by comparing dependable discharge with existing utilization of surface water in the upper Citarum watershed. The analysis will explain the potential quantity of each sub-watershed around the Cimahi city to reduce groundwater utilization and mitigate flood disasters caused by surface water.

2.3 The quality analysis of surface water
This research use surface water as subsitute or to reduce groundwater utilization in research location, therefore, before the surface water can be use as substitute, the quality of the surface water must be analyzed. Data of surface water quality is aquired by taking water sample in particular sub-watersheds to determine the physical and chemical parameter of the sampel in laboratory. The sample is taken in the downstream of 7 sub-watershed and the laboratory test result is later used to determine the need of water treatment in particular sub-watersheds.

2.4 Substitution analysis of groundwater utilization to surface water
The analysis is done by using previous groundwater modelling in Bandung-Soreang groundwater basin, and updating its well input data also assuming that the other numerical parameters didn't change during ten years of simulation (2021-2030). Wells input data in groundwater modelling is acquired from the previous well input data after being reduced by the dependable discharge of surface water. the modelling is staged according to the construction time of the water treatment plant during the period of simulation. the groundwater table during the period of simulation is later observed yearly to identify the benefit of replacing groundwater utilization with surface water utilization in research location.

3. Result and Discussion

3.1 Surface water quantity availability analysis
Surface water in this study uses 7 river in 4 sub-watersheds namely Cibeureum, Cimahi, Cibolerang and Ciwidey sub-watersheds (Figure 3 and Table 1). The surface water availability is calculated from the dependable discharge combined with existing surface water utilization and later can be used to replace groundwater utilization in the research location. This substitution also helps to implement flood disaster risk reduction by reducing surface runoff and fulfil water demands in the Cimahi area.

3.2 The quality analysis of surface water
The quality of surface water sample is analyzed in each subwatershed. According to water quality standard in Minister of Health Regulation Number 32 of 2017 concerning Environmental Health Quality Standards and Water Health Requirements, the testing result shows that several parameters (manganese, turbidity, cadmium, colour, TDS, ammonia, iron) in sampling location didn't meet the clean water quality standards (Table 2).

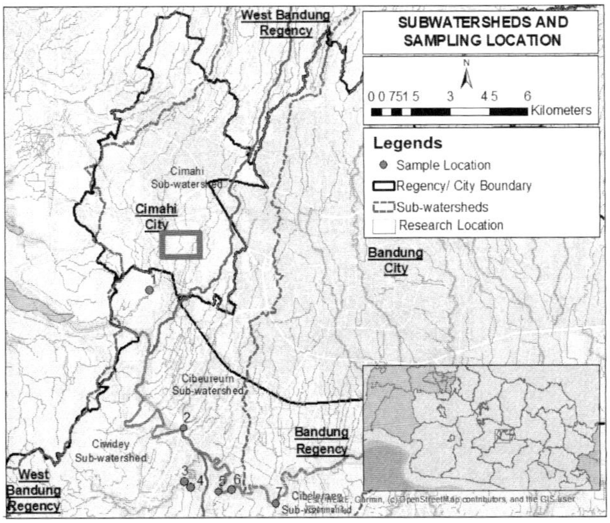

Figure 3. Sub-watersheds and sampling location.

Table 1. Surface water availability.

Sub Watershed	Dependable discharge (Q90) (m^3/s)	Existing intake (m^3/s)	Potential for utilization (m^3/s)	(l/s)
Cimahi	0.624968	0.571751	0.053218	53.218
Cibeureum	0.435572	0.114913	0.320659	320.659
Ciwidey	1.612476	0.00226	1.610216	1610.216
Cibolerang	0.533227	0.033738	0.499489	499.489
		Total	2.483581	2483.581

Table 2. The quality analysis of surface water

Parameters	Units	Water Quality Standards	Location						
			1	2	3	4	5	6	7
Odour	-	Odourless	Odourless	Odourless	Odourless	Odourless	Odourless	Odourless	Odourless
Colour	Pt. Co	15	69.5	97.6	52	36.1	34.7	40.2	81.7
Turbidity	NUT	5	25.6	58.2	26.2	30.4	15.8	14.6	40.5
TDS	mg/L	500	180	246	339	287	862	559	201
Taste	-	-	-	-	-	-	-	-	-
Ammonia	mg/L	1.5	1.1	2.52	2.25	1.33	3.32	0.89	0.08
Cadmium	mg/L	0.003	0.003	0.003	0.003	0.003	0.052	0.003	0.003
Total Chromium	mg/L	0.05	0.02	0.02	0.02	0.02	0.02	0.04	0.02
Chloride	mg/L	250	8.9	15.9	22.1	14.5	74.4	36.1	20.6
Cyanide	mg/L	0.07	0.009	0.009	0.009	0.008	0.003	0.009	0.008
Aluminum	mg/L	0.2	0.04	0.03	0.01	0.03	0.03	0.01	0.09
Copper	mg/L	2	0.17	0.2	0.14	0.12	0.01	0.1	0.29
Hardness ($CaCO_3$)	mg/L	500	66	110	79	92	30	33	68
Manganese	mg/L	0.4	0.33	0.92	0.34	0.16	0.61	0.16	0.5
Iron	mg/L	0.3	1.48	1.85	0.74	0.41	0.52	0.62	0.59
fluoride	mg/L	1.5	0.52	0.52	0.28	0.39	1.3	0.43	0.85
Nitrate	mg/L	50	1.6	1.5	1.6	4.5	3.4	1.3	3.2
Nitrite	mg/L	3	0.13	0.12	0.22	0.17	0.54	0.11	0.06
Zink	mg/L	3	0.22	0.16	0.12	0.12	0.07	0.14	0.08
Sulfate	mg/L	250	21	21	16	45	102	83	37
pH		6.5 – 8.5	6.98	7.05	7.62	7.2	7.3	7.36	7.36

Surface water samples are already tested in the laboratory which clarifies several quality parameters didnt meet the standards of clean water. Therefore, to utilize surface water as a substitute for groundwater consumption, water treatment is needed to improve surface water quality. The parameters of colour can be reduced by utilizing rapid sand filter, coagulation-sedimentation, and spesific chlorination. Turbidity can be decreased using plain settling, screening, rapid sand filter, and coagulation-sedimentation. TDS can be reduced with utilizing plain settling, pre chlorination, aeration, rapid sand filter, coagulation-sedimentation, slow sand filter, and spesific chemical treatment. Manganese and iron can be reduced by utilizing plain settling, pre chlorination, aeration, slow sand filter, coagulation-sedimentation, rapid sand filter, and spesific chemical treatment [11]. Zeolit can be utilized as adsorbent to decreased ammonia parameter [12]. PEFA (polyelectrolyte fly ash) can be utilize to decrease cadmium as an adsorbent [13].

Table 3. Water treatment needed to utilize surface water[11][12][13].

Parameter	Water Treatment needed to utilize surface water
Colour	Rapid sand filter, coagulation-sedimentation, and spesific chlorination [11]
Turbidity	Plain settling, screening, rapid sand filter, and coagulation-sedimentation [11]
TDS	Plain settling, pre chlorination, aeration, rapid sand filter, coagulation-sedimentation, slow sand filter, and spesific chemical treatment [11]
Manganese	Plain settling, pre chlorination, aeration, slow sand filter, coagulation-sedimentation, rapid sand filter, and spesific chemical treatment [11]
Iron	Plain settling, pre chlorination, aeration, slow sand filter, coagulation-sedimentation, rapid sand filter, and spesific chemical treatment [11]
Ammonia	Special Chemical Treatment – adsorbent zeolite [12]
Cadmium	Special Chemical Treatment – adsorbent PEFA (Polyelectrolit Fly Ash) [13]

3.3 Substitution analysis of groundwater utilization to surface water

The quantity and quality analysis is carried out in previous sub-chapter to make sure that the surface water is feasible to substitute groundwater utilization in research location.

Amount of available water in Table 1 is 2484 litre/second originating from four sub-watersheds surrounding the Southern Cimahi. Groundwater modelling will be carried out over ten years, considering the construction of a water treatment plant and then the groundwater levels will be observed during the modelling period. With a ten-year simulation (2021-2030), the groundwater level south of Cimahi is intented to increase by 11.2265 meter (Figure 4-14 and Table 4).

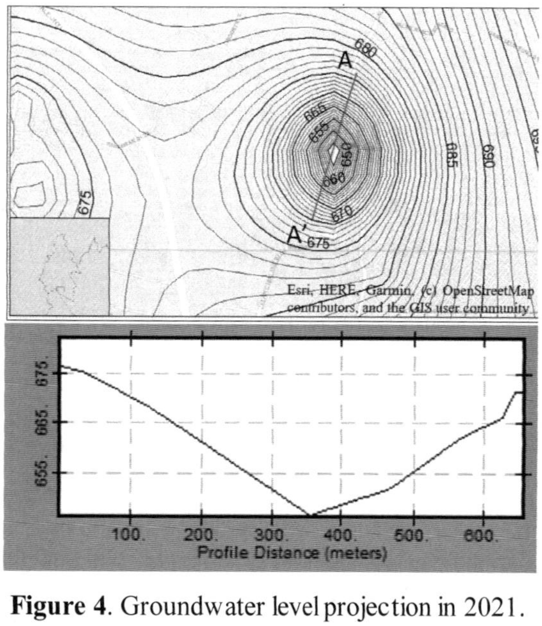

Figure 4. Groundwater level projection in 2021.

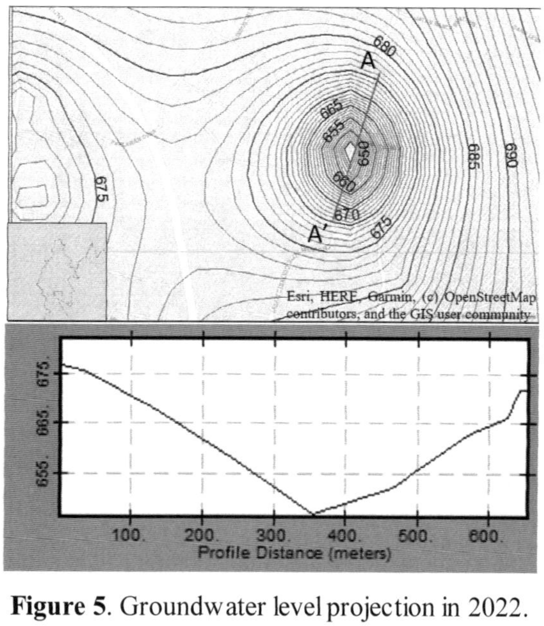

Figure 5. Groundwater level projection in 2022.

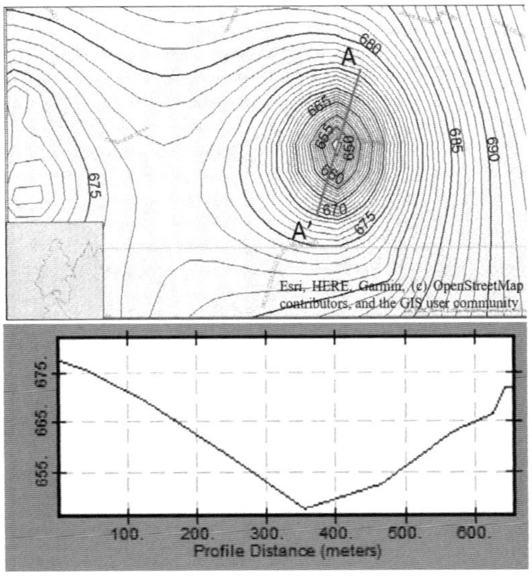

Figure 6. Groundwater level projection in 2023.

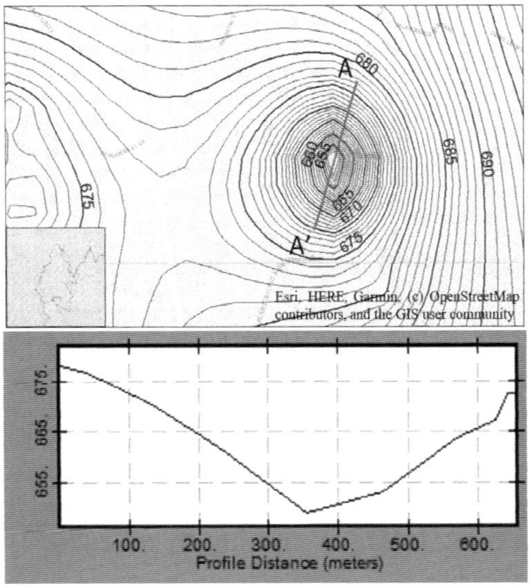

Figure 7. Groundwater level projection in 2024.

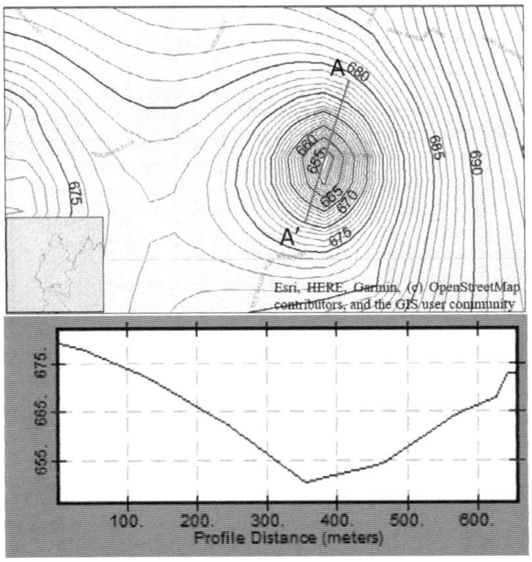

Figure 8. Groundwater level projection in 2025.

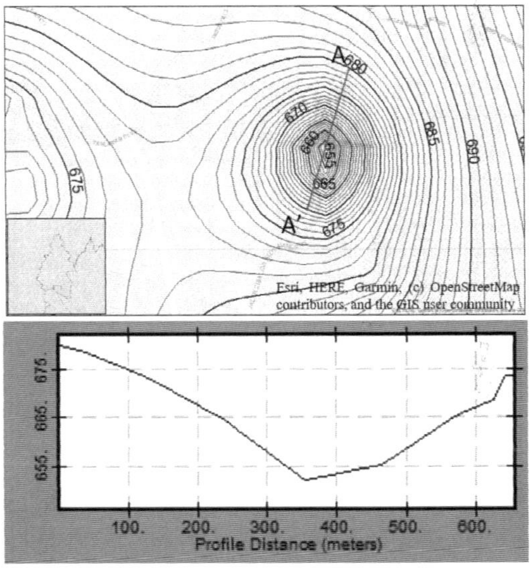

Figure 9. Groundwater level projection in 2026.

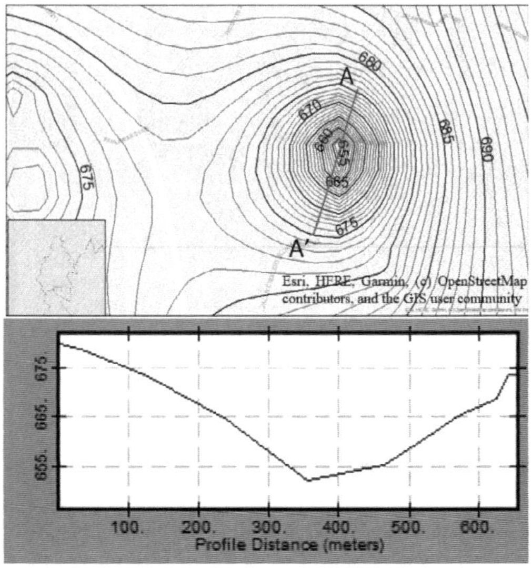

Figure 10. Groundwater level projection in 2027.

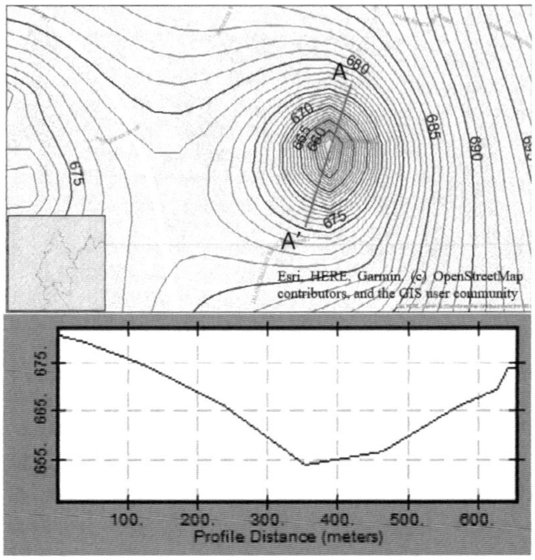

Figure 11. Groundwater level projection in 2028.

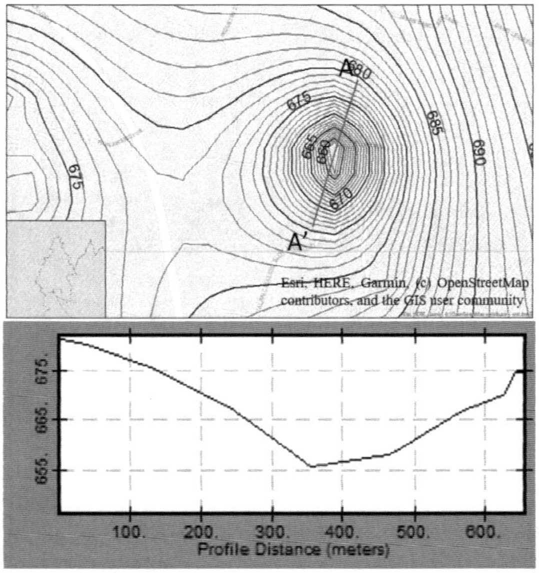

Figure 12. Groundwater level projection in 2029.

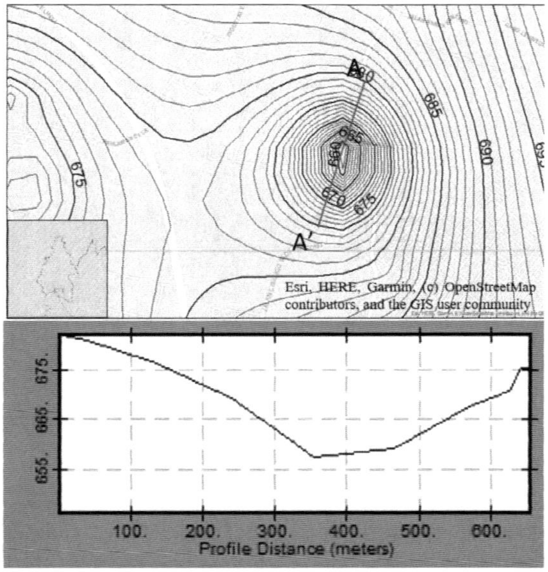

Figure 13. Groundwater level projection in 2030.

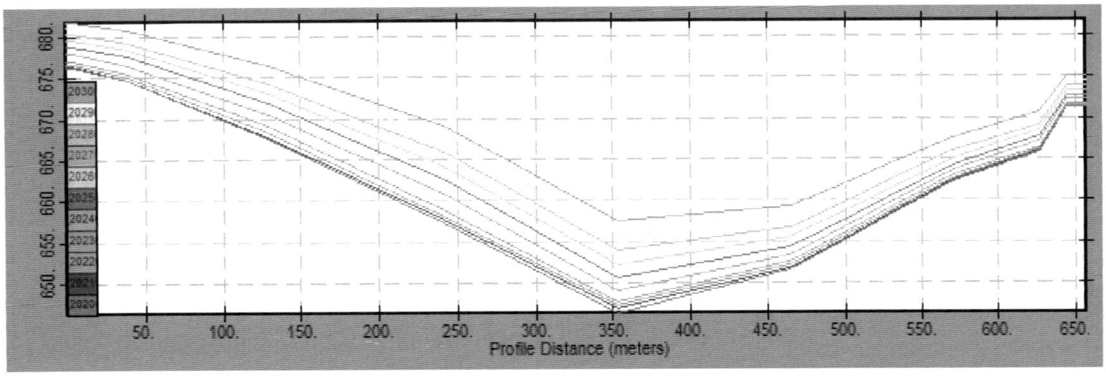

Figure 14. Groundwater level recovery projection in Southern Cimahi.

Table 4. Groundwater level projection in south Cimahi (2021-2030).

Year simulation	Surface Water Utilization		Predicted Groundwater level (in masl)
	(m^3/s)/ year	Accumulated (m^3/s)	
2020	0	0	646.2784
2021	0.110	0.110	646.7734
2022	0.120	0.230	647.1392
2023	0.150	0.380	647.7107
2024	0.350	0.730	649.0400
2025	0.350	1.080	650.4843
2026	0.350	1.430	652.1224
2027	0	1.430	652.1224
2028	0.350	1.780	653.8779
2029	0.350	2.130	655.7502
2030	0.350	2.480	657.5449

4. Conclusion

The analysis result shows that surface water in the sub-watersheds around the Cimahi area can be utilized in terms of its quantity. However, in terms of quality, the surface water still needs to be treated to improve some quality parameters that currently do not meet health standards. Therefore, simulations on the development of groundwater models include the use of surface water quantity as a replacement for groundwater utilization. Water treatment installation/ construction is considered to be carried out for the ten years (2021-2030) with total capacity of 2,450 litres/second of clean water and resulting a prediction that the substitution of simulated water utilization will positively increase the groundwater level up to 11.2665 meters in the year 2030. The volume of water in this research is limited by dependable discharge for raw water. Flood control is also carried out by building retention ponds in the Cimahi, Cibolerang, Ciwidey, and Cibeureum sub-watersheds. Retention ponds are also useful for increasing the volume of water that can be used to reduce groundwater consumption and expand access to clean water in Cimahi City.

References

[1] Republic of Indonesia Regulation number 17 2019 water resources, Jakarta.
[2] Fulazzaky M A 2014 Challenges of integrated water resources management in Indonesia Water (Switzerland) *Water* 2014-chapter 6 p 2000-2020
[3] West Java Provincial Regulation number 12 2014 Management of the development of metropolitan areas and growth centres in West Java Province, Bandung

IOP Conf. Series: Earth and Environmental Science 1071 (2022) 012003 doi:10.1088/1755-1315/1071/1/012003

[4] Indradhi F I, Eka W, and Lina A S 2019 *Identification The Causes of Flood in South Cimahi District. Seminar Nasional Cedekiawan,* chapter 1 p 1.2.1-1.2.6
[5] Safarina A B, Karnisah I, Permana A S, and Sufianto D 2019 *Green Cimahi watershed for balancing water supply and flood control* Cimahi IOP Conf. Series: Materials Science and Engineering 669 (2019)
[6] Taufiq A, Hosono T, Ide K, Kagabu M, Iskandar I, Effendi A J, and Shimada J 2018 Impact of excessive groundwater pumping on rejuvenation processes in the Bandung Basin (Indonesia) as determined by hydrogeochemistry and modelling. *Hydrogeology Journal*, 26(4), 1263–1279
[7] Seizarwati W, Rengganis H, Wardhana Y A W, Ahmad R D, Syahidah M, and Muthiya A, 2020 *Groundwater Model Prototype to Support SWMS in Citarum Watershed* Internal Report of the Center for Hydrology and Water Environment Bandung
[8] Purnawan M Y, Hendrayana H, and Setijadji L D 2021 *Analysis of surface water utilization for restoration of the groundwater table case study: Dayeuhkolot, Bandung-Soreang groundwater basin* Bandung IOP Conf. Series: Earth and Environmental Science 871 (2021)
[9] SCM 2021 Cimahi Municipality in Figures Statistics of Cimahi Municipality Cimahi
[10] NSAI 2015 SNI 6738:2015 Calculation of dependable discharge of river with discharge duration curve National Standardization Agency of Indonesia Jakarta
[11] Kawamura S 1991 *Integrated Design of Water Treatment Facilities* New York (US) : John Willey & Sons, Inc.
[12] Han B, Butterly C, Zhang W, He J Z, and Chen D 2021 Adsorbent materials for ammonium and ammonia removal: A review. *Journal of Cleaner Production* 283
[13] Olabemiwo F A, Tawabini BS, Patel F, Oyehan T A, Khaled M, and Laoui T 2017 Cadmium Removal from Contaminated Water Using Polyelectrolyte-Coated Industrial Waste Fly Ash. *Bioinorganic Chemistry and Applications* 2017

Acknowledgements
This research would not have been possible without the exceptional support from the Experimental Station of Groundwater and the Hydrology & Water Environmental Agency, which have been supported by providing data for the research process.

Land Subsidence Hazard Mapping in relation with Sinkholes in Saptosari District, Gunung Kidul, Indonesia

R R Fikrillah[1], S Say[1,2], D P E Putra[1]*

[1]Department of Geological Engineering, Faculty of Engineering, Universitas Gadjah Mada, Yogyakarta, Indonesia
[2]Faculty of Geo-resources and Geotechnical Engineering, Institute of Technology of Cambodia, Phnom Penh, Cambodia
*Corresponding author's e-mail : putra_dpe@ugm.ac.id

Abstract. Saptosari is one of the areas in the karst terrain of Gunung Kidul Regency, Yogyakarta, Indonesia. In this area, sinkholes are very common, especially in weathered and covered by soil overburden. It is known a lot of land subsidence due to sinkholes are found and reported by the local community. Therefore, the purpose of this study was to map the land subsidence hazard zone due to the presence of sinkholes. To achieve this goal, field observations and evaluation of secondary data were carried out to collect slope data, lithology, lineament of geological structures, and the location of sinkholes and subsidence that already existed. Analytical Hierarchy Process (AHP) was carried out to determine the weight of the geomorphological parameter, lithological parameter, and the distance to geological structure in order to predict the occurrence of sinkholes. The weighted results of the AHP method are then used to create a land subsidence hazard map by using the GIS-overlay method. The study reveals three subsidence hazard zones in the research area, namely (1) high susceptible area with subsidence probability of 92%, (2) moderate susceptible area with subsidence probability of 4.7%, and (3) low susceptible area zones with 0% of subsidence probability.

1. Introduction

A sinkhole by definition is an area on the ground that does not have natural surface drainage where when it rains, all water remains in this area and flows below the surface. Sinkholes can vary from a few meters to hundreds of hectares and from less than 1 meter deep to more than 30 meters deep. Some are shaped like shallow bowls or saucers while others have vertical walls; some retain water and form natural pools [1]. Furthermore, sinkholes are generally formed in the Karst terrain. Karst terrain is areas where bedrock can be dissolved by groundwater. The bedrock in the karst area is usually limestone, dolomite or gypsum. Karst areas are characterized by special features, namely springs, caves, underground rivers, and the presence of sinkholes [2]. Usually little more than a nuisance, sinkholes can cause substantial property damage and structural problems for infrastructures including buildings and roads due to subsidence [3]. Therefore, sinkholes are one of the triggers for land subsidence that occurs naturally[2]. Sinkhole subsidence events are often associated with the rainy season, although the dry season also has a role, where subsidence locations are commonly found with areas of large groundwater fluctuations [1]. In fact, the influence of human activities in the form of groundwater extraction in karst landscapes can trigger the accelerated subsidence of the land. Groundwater pumping will cause a decrease in the groundwater level and this will accelerate the process of erosion and dissolution of limestone downwards due to groundwater subsidence [3][4]. Either naturally or through the influence of human activities, the speed and rate of subsidence are controlled by lithological conditions, climate, vegetation and time and is triggered by several processes such as rock dissolution, subsurface mechanical erosion, compression or compaction, groundwater flow, vibration, hydrocompact and shrinkage [5][6][7].

Saptosari Subdistrict (See Figure 1) in Gunung Kidul Regency, Special Region of Yogyakarta, is growing rapidly by tourism activities [8]. This area is located in the karst morphology composed of limestone. In this sub-district, subsidence is common during the rainy season. The mapping of the hazards and risks of land subsidence in the Gunung Kidul karst area has been carried out in 2011 and 2014 particularly in the Rongkop and Semanu sub-districts [9][10]. In both studies, subsidence hazard mapping was carried out based on four parameters, namely elevation, slope, type of lithology and distance from the geological structure. The validation results show a good match with subsidence events in the two mentioned areas. In this regard, this paper aims to create a subsidence hazard map in Saptosari District, Gunung Kidul using a parameter approach to geomorphological, lithological and geological structure and applying parameter weighting based on the Analytic Hierarchy Process (AHP) method. In addition, with the subsidence hazard map, it is hoped that this research can provide better planning for the development of the Saptosari District, Gunung Kidul, Yogyakarta, Indonesia.

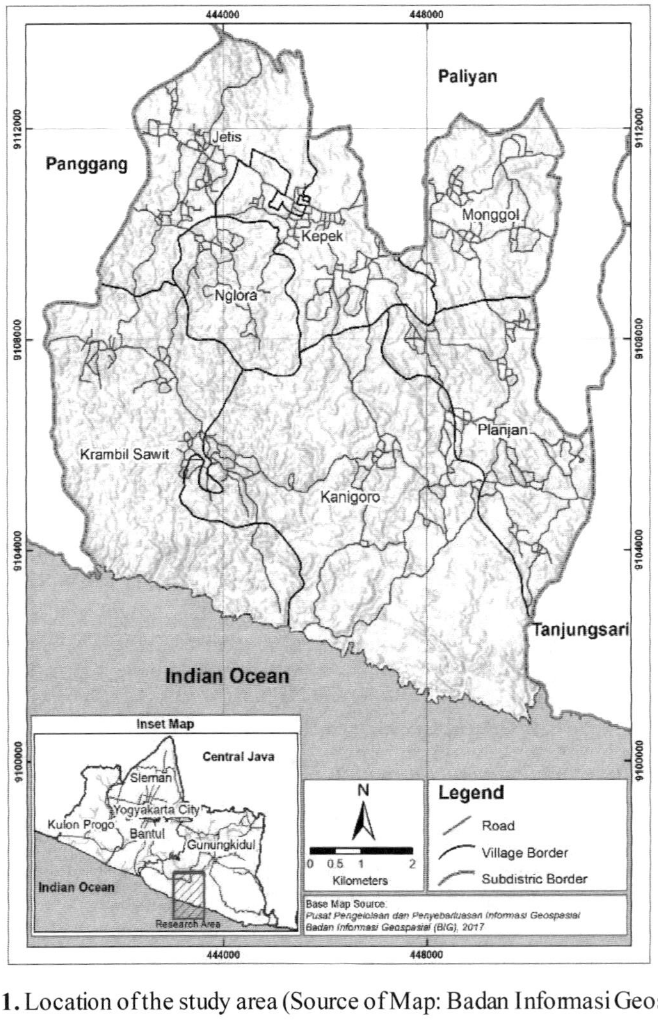

Figure 1. Location of the study area (Source of Map: Badan Infomasi Geospasial).

2. Research Method

In order to achieve the objectives of this research, some methods are used such as field data collection through lithological observation and slope observation. One more thing, secondary data collection was also done followed by multi criteria evaluation to develop the land subsidence hazard map. Observation of lineament was carried out by analysing topographic maps from Peta Rupa Bumi Indonesia and Digital Elevation Model (DEM) from DEMNAS provided by Geospatial Information Agency (Badan Informasi Geospasial, BIG) since there is no evidence of geological structures was found in the field. Observations and recordings of the locations of the distribution of subsidence and/or sinkholes were also carried out. Furthermore, the data is evaluated using the AHP method to determine the weight of the influence of each parameter, namely geomorphology which is represented by the degree of slope, type of lithology and lineament of geological structures in the relation to the occurrence of subsidence and/or sinkholes. Basically, the AHP is a multi-objective, multi-criteria approach that utilizes a comparison matrix. In making this comparison matrix, each factor is assigned a scale with a relative range of 1 – 9, this value depends on the value of importance [11].

Based on the AHP method, when a factor on the vertical axis is more important than the horizontal axis, the value of this factor is 1 – 9, otherwise, if the factor on the horizontal axis is more important than the factor on the vertical axis, the value will range from 1/9 – 1. By using matrix calculations, the largest eigenvalues will have eigenvector values that will represent factor weights. These eigenvector values are then added together, and factor weights are obtained by dividing the factor eigenvector value by the total eigenvector value [12]. In this study, the weight of the subsidence control parameter is determined in advance in terms of its relative importance by correlating the presence of subsidence/sinkhole with each class that exists in each parameter.

From a total of 33 subsidence/sinkhole location data obtained in the field, 20 randomly selected data were used to display on each of the parameter's map for the discussion and to see the percentage of subsidence/sinkholes location in each class of parameters. Then, all of 33 subsidence/sinkhole locations data were used to test the suitability of subsidence hazard zone with the existing subsidence/sinkhole to see whether the land subsidence susceptibility zonation correlates with the existence of all subsidence/sinkholes or not, and they were displayed on the Sinkhole's related land subsidence susceptibility map. To determine whether the weights are consistent and objective, a Consistency Ratio calculation is required. Consistency Ratio (CR) < 0.1 indicates that the weights are correct [13]. After the weight of each parameter is obtained with the smallest consistency index value, a subsidence hazard map is made using the overlay method on geographic information system software. To ensure the suitability of the subsidence hazard zone resulting from the overlay results, a comparison is made between the land subsidence susceptibility class and the number of subsidence and/or sinkholes obtained in the field.

3. Results and Discussion

a. Geology and The Occurrence of Sinkholes

Geomorphological aspects are one of the important aspects in geological mapping, especially thematic geology. The division of geomorphological units of the research area is based on the interpretation of DEM images, topographic maps, and field observations. In this research, the division of geomorphological units considers only aspects of morphometry. Thus, based on the morphometry unit of the slope, the geomorphological units in the study area are grouped into three classes based on the degree of slope, namely gentle slope (<20°), moderate slope (20° – 40°), and steep slope (>40°), as shown in Figure 2. After plotting 20 points of random subsidence/sinkhole locations on the geomorphological map, it can be concluded that there is a positive correlation between the slope class and the presence of subsidence/sinkhole, where the steeper the slope, the less subsidence locations/sinkholes are found where almost 95% of the sinkhole/sinkhole appears on the gentle slope unit, 5% is associated with the moderate slope unit and no subsidence/sinkhole is found on the steep slope unit (Figure 2). This can happen because, in sloping areas, water will easily flow on slope faces, while on gentle slope areas, water will tend to remain on the surface and flow downward thus increasing the possibility of weathering and dissolving underground lithology. The high intensity of weathering resulted in the thickening of the sediment overburden at that location. Then, intensive dissolution will result in the enlargement of the cavity in the limestone below the surface. Therefore, sinkholes and subsidence will be more often found in areas with gentle slopes.

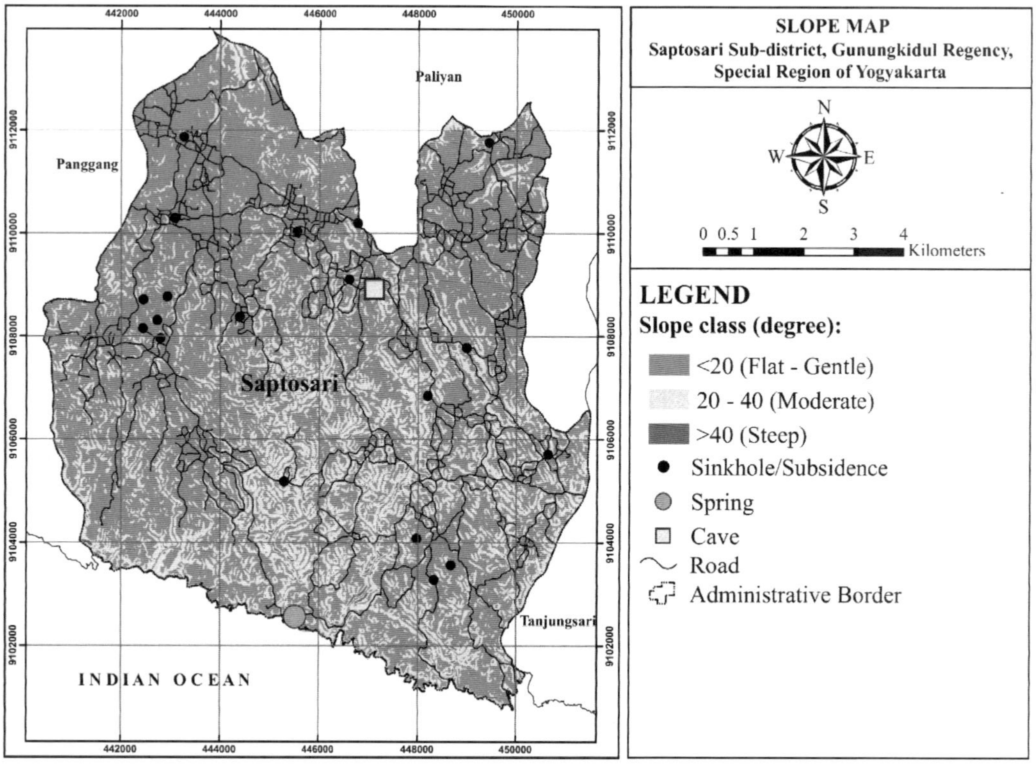

Figure 2. The slope distribution in the study area.

Based on the results of field observations, the surface lithology of the research area can be represented in thematic lithology maps which can be seen in Figure 3. The constituent lithology in the study area is limestone. On gentle slope units, the rock conditions have undergone weathered and form clay to sand overburden with limestone fragments ranging from 1 to 10 cm. Clay grain-sized overburden is scattered around watery areas such as ponds. Meanwhile, sand-grained overburden is found in dry valleys between karst hills. In relation to the process of subsidence/sinkhole events, it can be predicted that subsidence/sinkhole events in the fresh limestone zone will be related to the solution and collapse sinkhole, while in weathered limestone that forms the overburden, namely clay and sand materials, subsidence events/sinkholes will be related to buried sinkholes, dropout sinkholes, and suffocation sinkholes [2]. Galloway, et al., (1999) showed that subsidence occurs relatively in the form of sand overburden materials and very often occurs in clay overburden materials with thicknesses ranging from 10 to 60 m. From the plotting of 20 subsidence/sinkhole data on the thematic lithology map, it shows that almost all subsidence/sinkholes are in clay-sand overburden (95%) and the rest are in limestone units with no or very thin overburden (Figure 3).

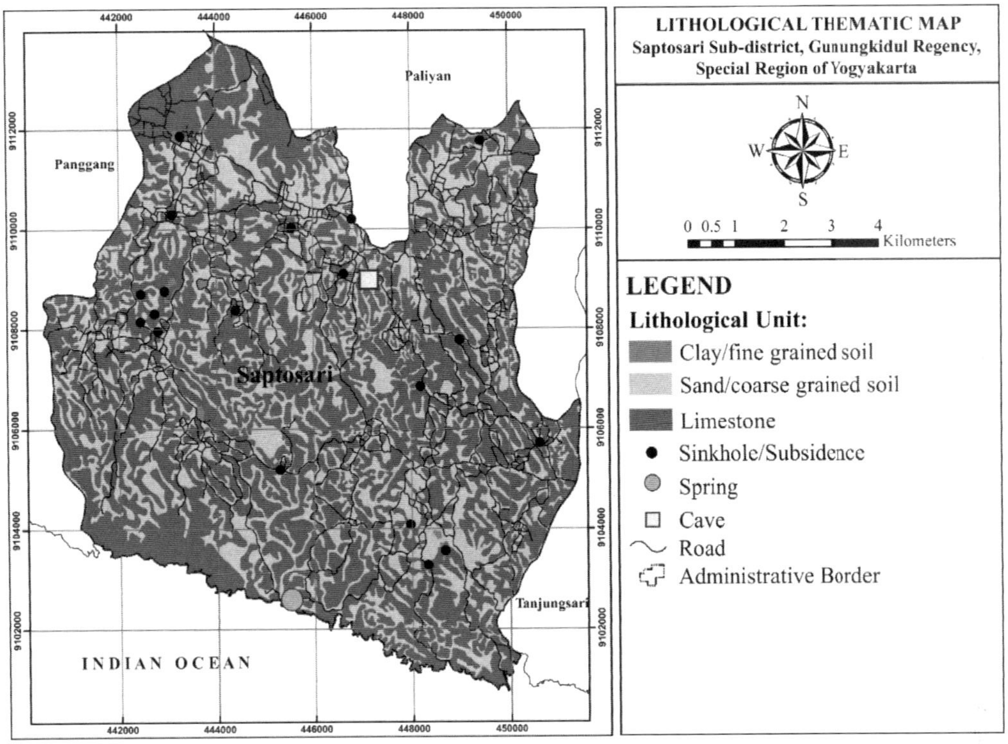

Figure 3. The distribution of lithological unit in the study area.

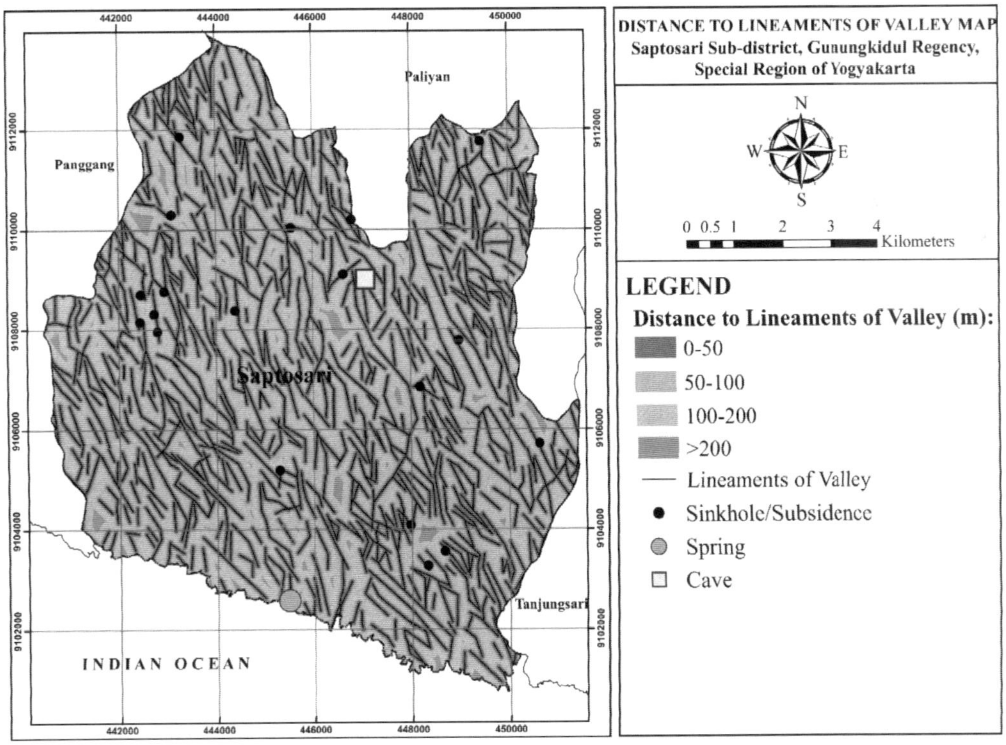

Figure 4. The distance to lineament (representing geological structure) in the study area.

To make a correlation between the existence of subsidence/sinkhole and the lineament of the geological structure, zoning of the distance from the straightness of the structure is made with the help of a geographic information system. The basis for dividing the distance from the lineament is taken from the results of Putra (2011) which divides the distance from the lineament related to subsidence in the karst area into four classes, namely <50 meters, 50-100 meters, 100-200 meters, and >200 meters. Plotting of 20 subsidence/sinkhole data with distance maps of geological structure lineages has a positive correlation, where about 75% of subsidence/sinkholes are associated with distance zones from lineaments < 50 m, 25% are associated with distance zones from lineaments of 50 - 100 m, and no sinkhole/subsidence was found in the area with a distance of >100 meters straight (Figure 4).

b. Determination of Weighting Factors

Before making a map of the susceptibility to subsidence, the parameter weights were determined using the AHP method. The research of Putra, et al., 2011, shows that the distance factor to the lineament/geological structure is more influential than the lithology and morphology types. In this research, a similar approach reveals a CR result of 0.1. The best value for CR is obtained with lithological parameters having a sufficient importance scale (3) for slope parameters and a strong importance scale (7) for geological structure lineament parameters. While the slope parameter gets a slightly more important importance scale (2) on the distance of the geological structure. Furthermore, the opposite importance scale (1/n) applies to the slope to lithology parameter, which is 1/3. While the distance of the geological structure to the lithology has importance of 1/7. Then, the distance of the geological structure to the slope has importance of 1/2. The description of the assessment can be seen in Table 1.

From Table 1, each sub-parameter column is added vertically which is then used as a divisor in the matrix normalization stage. Each row in the pairwise comparison matrix is divided by the total value of each column so that the susceptibility normalized pairwise comparison matrix is obtained. The next step is to calculate the maximum eigenvalue by adding up the multiplication results of each parameter weight with the sum of each column. After obtaining the maximum eigenvalue, the calculated CI value obtained a result of 0.002 and a CR value of 0.0035 (Table 2). Thus, based on the calculations in Table 2, the highest weight of the parameter controlling the presence of subsidence/sinkhole is the lithological parameter (68%), the slope parameter (22%), and the lowest weight is the distance parameter from the geological structure alignment (10%), as shown in Table 3. Compared with the research results of Putra et al. 2011, there is a difference in the order of importance of the controlling parameters that can indeed occur considering that there are many things that play a role in the sinkhole formation process.

Table 1. Pairwise comparison matrix for weight evaluation in the AHP method

Parameter	Lithology	Slope	Distance to geological structure (lineament)
Lithology	1	3	7
Slope	1/3	1	2
Distance to geological structure (lineament)	1/7	1/2	1
Total	1.48	4.50	10.00

Table 2. Normalized pairwise comparison matrix for weight evaluation.

Parameter	Normalization			Sum	Weight
Lithology	0.68	0.67	0.7	2.04	0.68
Slope	0.23	0.22	0.2	0.65	0.22
Distance to lineament	0.1	0.11	0.1	0.31	0.1
Total	1	1	1	3	1
Maximum *eigenvalue*	30.041				
CI	0.002				
CR	0.0035				

Table 3. Score and weight of each parameter for evaluation of weighted-overlay.

Parameter	Class	Score	Weight
Lithology	Clay overburden materials	3	68%
	Sand overburden materials	2	
	Limestone - no or very thin overburden materials	1	
Slope	< 20°	3	22%
	20° - 40°	2	
	> 40°	1	
Distance to lineament	< 50 m	4	10%
	50 - 100 m	3	
	100 - 200 m	2	
	> 200 m	1	

c. Sinkhole Subsidence Hazard Map

The determination of the subsidence hazard zone in the study area is then carried out by evaluating the weighted-overlay on the geographic information system by assigning a score and weight to each controlling parameter as shown in Table 3. The results of the weighted-overlay are then divided based on the normal distribution into three categories, namely low subsidence susceptibility, medium subsidence susceptibility and high subsidence susceptibility, as shown in Figure 5.

Table 4. Relationship between the existence of subsidence/sinkhole with subsidence susceptibility zone.

Land subsidence susceptibility	Area		Number of existing land subsidence	Probability *(%)*
	%	*(Km²)*		
Low	14	12.26	0	0
Moderate	48	42.15	2	4.7
High	38	33.42	31	92

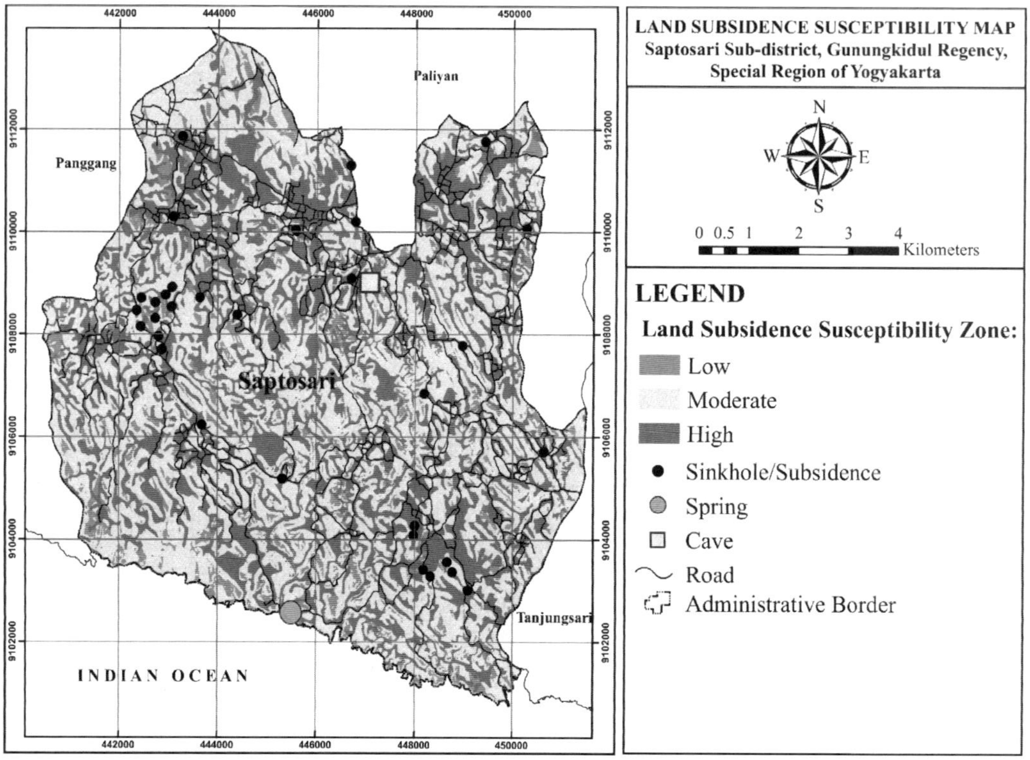

Figure 5. Sinkhole's related land subsidence susceptibility map in the study area.

Figure 5 provides the distribution of each zone of susceptibility or subsidence hazard. Based on the subsidence susceptibility map, the low subsidence susceptibility zone has the smallest distribution, which is about 14% of the total area of the study area. In general, this low susceptibility zone is associated with lithology in the form of limestone, with a slope of more than > 40°, and is far from the lineament (> 200 meters). About 48% of the total area of the study area is included in the zone of moderate subsidence susceptibility. This moderate susceptibility zone is associated with a condition of moderate slopes (20° - 40°), and a relative distance of 100 – 200 meters from the lineament. Meanwhile, the high subsidence susceptibility zone occupies approximately 38% of the study area and is associated with clay grain-sized and sand grain-sized overburden, gentle slopes (< 20°) and close to lineament (< 50 meters). To test the suitability of this subsidence hazard zone with the existing subsidence/sinkhole, all 33 points of subsidence/sinkhole data were used. The suitability test is carried out by calculating the probability of subsidence/sinkholes existence in a similar susceptibility area. Table 4 shows the tabulation of this probability and this table implies that the greater probability of subsidence will be obtained in the high subsidence susceptibility/hazard zone, so there is a very good match between the subsidence susceptibility/hazard zone and the possibility of subsidence events in the study area.

4. Conclusion

This study succeeded in producing a subsidence hazard map in relation to sinkholes in the Saptosari District, known as the karst terrain with the presence of sinkhole/ subsidence, cave and spring in Gunung Kidul, Indonesia. This area is dominated by limestone and also present of clay and sand, and slope range from 0 degree up to more than 40 degrees with the present of lineament valley. The subsidence hazard map is well-validated with the existence of subsidence that has occurred so far. Thus the results of the susceptibility/subsidence hazard zoning produced in the study are in accordance with the occurrence or presence of subsidence/sinkholes that can be observed in the field, where there is no potential for subsidence in the low hazard zone, then 4.7% of probability in the medium hazard zone, and in the high hazard zones reach 92% of subsidence/sinkhole's probability. Parameters or factors that control the presence of subsidence/sinkhole in the study area are influenced sequentially by lithology type (68%), slope degree (22%) and distance from geological structure (12%). However, these factors cannot be applied directly to other regions, for example, to predict the presence of subsidence/sinkholes in other karst terrains because apart from geological factors there are several triggering factors such as climate and seasons (including rainfall), groundwater fluctuation conditions, vibrations (including earthquakes) and other causes that have not been discussed in this study. Despite the weaknesses, it is hoped that with this subsidence hazard map, planning and construction of engineering construction to support development activities in the research area can pay attention to the existing subsidence hazard zones, so that the impact of losses due to sinkhole's related land subsidence can be minimized.

References

[1] Kaufmann, J.E., 2007, *Sinkholes*, US Geological Survey, Fachsheet 2007-3060.
[2] Waltham T., Bell F., and Culshaw M., 2005. *Sinkholes and Subsidence: Karst and Cavernous Rocks in Engineering and Construction*, Paris Publishing, Chichester, UK., 300p.
[3] Galloway, D., Jones, D.R., and Ingebritsen, S.E., (Ed.) 1999, *Land Subsidence in the United States*, US Geological Survey, Circular 1182.
[4] Orhan, O, Oliver-Cabrera, T., Wdowinski, S., Yalvac, S., and Yakar, M., 2021, Land Subsidence and Its Relations with Sinkhole Activity in Karapınar Region, Turkey: A Multi-Sensor InSAR Time Series Study. Sensors 2021, 21, 774.
[5] Allen A.S., 1984. *Types of Land Subsidence*, in: Poland, J.F., (ed.), Guidebook to Studies of Land Subsidence due to Groundwater Withdrawal, Studies and Report in Hydrology, UNESCO, p.133-141.
[6] Glopper, R.J., dan Ritzema, H.P., 1994, *Land Subsidence*, in: Ritzema, H.P., (ed.) Drainage Principles and Applications, International Institute for Land Reclamation and Improvement, The Netherlands, p.477-510.
[7] Foley, D., McKenzie, G.D., and Utgard, R.O., 2009. *Investigations in Environmental Geology*, 3rded, Pearson Prentice-Hall, New Jersey., 304p.
[8] BAPPEDA Kabupaten Gunungkidul. 2016. *Buku Putih Sanitasi Kabupaten Gunungkidul.* Gunungkidul: Pemerintah Kabupaten Gunungkidul, 121p.
[9] Putra, D P E, Setianto,A., Keokhampui K., and Fukuoka, H., 2011, *Land subsidence risk assessment in Karst region, case study: Rongkop, Gunung Kidul, Yogyakarta-Indonesia*, Mitteilungen zur Ingenieurgeologie und Hydrogeologie, Vol.104, Festschrift zum 60. Geburtstag von Univ.Prof Rafig Azzam, p.39-50.
[10] Widyaningtyas, P., and Putra, D P E., 2014, Pemetaan Bahaya Amblesan di Daerah *Karst* Kecamatan Semanu, Kabupaten Gunung Kidul, Propinsi Daerah Istimewa Yogyakarta, Prosiding Seminar Nasional Kebumian ke-7 dan Simposium Pendidikan Geologi Nasional, Desember 2014, Yogyakarta, p.138-148.
[11] Saaty, T.L., 1980. *The Analytic Hierarchy Process*, McGraw-Hill, New York. 287p
[12] Dai, F.C., Lee, C.F., and Zhang, X.H., 2001. *GIS-Based Geo-Environmental Evaluation for Urban Land -use Planning: A Case Study,* Engineering Geology, Elsevier, Vol.61, Issue 4, p.257-271.

Acknowledgments

The authors would like to thank the Department of Geological Engineering, Faculty of Engineering, Universitas Gadjah Mada and the CUBE ASEAN Project sponsored by JICA AUN/SEED-Net for the support on this research. Special thanks to Wahyu Wilopo, Wawan Budianta and Raja Susatio for discussion during the field observation.

IOP Conf. Series: Earth and Environmental Science 1071 (2022) 012005 doi:10.1088/1755-1315/1071/1/012005

Integrated mapping of the earthquake-induced liquefaction in Wates Groundwater Basin, Kulon Progo for land use planning consideration: A preliminary study

Muhamad Ilham[1], and Doni Prakasa Eka Putra[2*]

[1,2] Department of Geological Engineering, Faculty of Engineering, Universitas Gadjah Mada

*Corresponding author's e-mail : putra_dpe@ugm.ac.id

Abstract. Since the development of New Yogyakarta International Airport (NYIA), land use planning and development in Wates Groundwater Basin, Kulon Progo, Indonesia, becoming more crucial. A regional-scaled study led by Buana et al. (2019) pointed out the area as moderate-high susceptibility to liquefaction. This study provides updated land-use conditions and a detailed liquefaction susceptibility map using Geographic Information System and Multi-Criteria Analysis (GIS-MCA) from the previous research data by Maulana (2017). The result shows that the south and east part of the basin is classified as very high susceptibility to liquefaction while the other part is low to moderate susceptibility class. This preliminary study suggests the best recommendation for geophysics and Standard Penetration Test (SPT) data acquisition in further analysis.

1. Introduction

Liquefaction is a serious problem in the world, including in Indonesia. Historically, as of 2019, there are minimally 37 liquefaction disasters already happened in Indonesia [1]. One of the latest and biggest damages of liquefaction, in Indonesia, was the massive mudflows in Balaroa, Petobo, and Jono Oge, Central Sulawesi which happened in early October 2018 [2]. The 7.5 Mw tectonic earthquake that happened on 28 September 2018 was the major triggering factor responsible for the flow slides [3]. The liquefaction occurred after the big earthquake surrounding the Palu-Koro fault which was then followed by some shallow and small earthquakes. That multi-disaster event then made the researchers and the governments have fascinated much attention to the liquefaction potential research around Indonesia. The atlas that resulted from the vast research of liquefaction potential in Indonesia, which is displayed in Figure 1, shows that the Wates Groundwater Basin is classified as a highly susceptible area in the coastal area and majorly included in the moderately susceptible area for the others northern part [1].

Wates Groundwater Basin is located in the southern part of Kulon Progo District, which is shown in Figure 2. The development of New Yogyakarta International Airport (NYIA) in the coastal area of Kulon Progo district was the initial of further growth of the district. Several types of research have already been done in the coastal area of the Kulon Progo District. Dwiyanta (2016)[4] found that the NYIA area has majorly severe and moderate susceptibility to liquefaction hazards. Anggara et al (2020)[5] and Artati et al (2020)[6] stated that majorly, the coastal area of Kulon Progo district is highly susceptible to liquefaction if there is an earthquake with more than 6.3 Mw happened in the surrounding area.

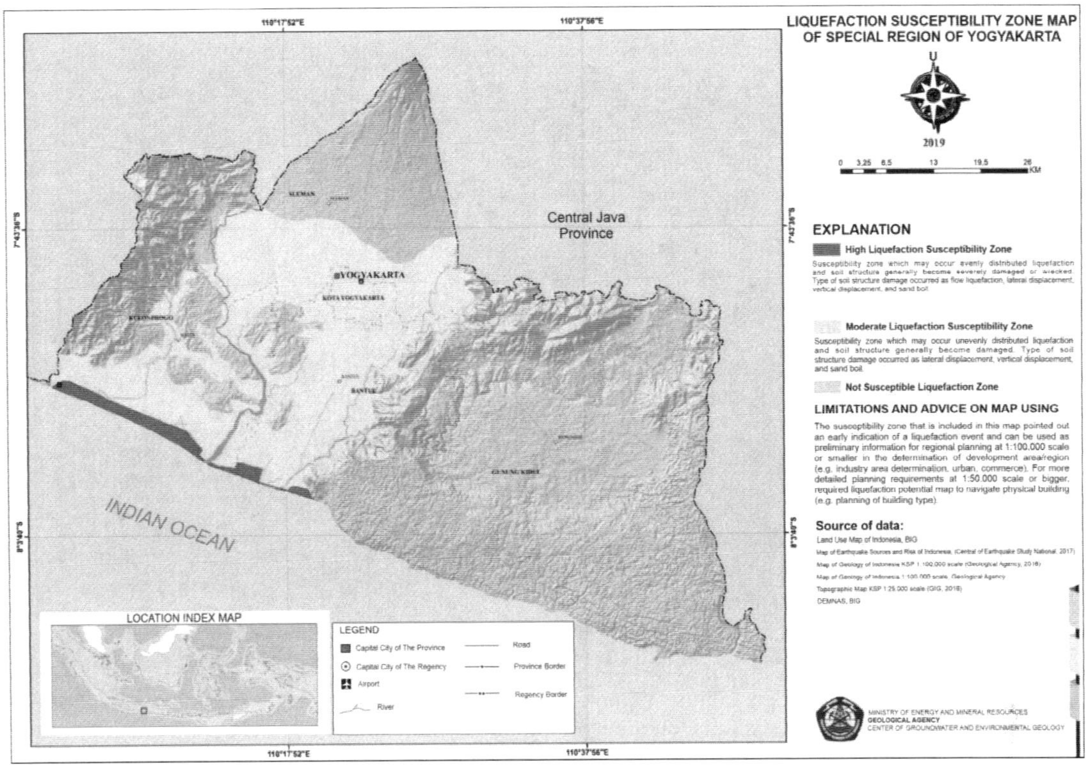

Figure 1. Liquefaction susceptibility map of the Yogyakarta area (modified from Buana et al, 2019).

The complete research is aimed at getting a detailed liquefaction potential map of the Wates Groundwater Basin area. Hence, the result could help the related stakeholders or government to decide on the growth of built-up areas. This preliminary research will solely discuss the estimated liquefaction hazard susceptibility of the area by using previous research data from Maulana (2017)[7], which consists of a detailed geological and hydrogeological map of the area. The recent land-use data shown here will only be a preview of the current development of the area. This research will be the reference for determining the sampling of field data in further research.

2. Geology of the study area

Wates groundwater basin with approximately 160 km square area is a productive groundwater basin that supplies most of the water needs of the Wates area. As shown in Figure 2, the basin ranged from the Bogowonto river in the west, the Serang river in the center, and the Progo river in the east. With the Indian ocean as the south of the border area, the groundwater direction generally flows from the north to the south with some influences from the river system [7].

2.1. Lithological and morphological condition

The regional geology map from Rahardjo et al (1995)[8] shows that the area is composed of quaternary alluvium deposits. A detailed lithological map of the area was already done by Maulana (2017)[7], which is shown in Figure 3, there are three types of alluvium deposits in the area. The fine sand deposits are distributed in the coastal area, while the sandy clay with sand lenses is majorly composed of the area and distributed in the northern part of the basin, and the gravelly sand deposits are located around the Progo which some materials are influenced by the deposition of river materials. The basement of the basin is consist of old andesite formation and limestone of Sentolo formation in some areas above the andesite. There is also a normal fault buried under the quarternary deposits near the northern border of the basin, which is included in the mountainous area of old andesite and Sentolo formation. The morphology of the area is generally flat with the sand dune in the south(coast) part and flood plain in the east part.

Figure 2. Wates groundwater basin border (red lines) which located in Kulon Progo Regency, Special Region of Yogyakarta, Indonesia (modified from Google Earth (2021)).

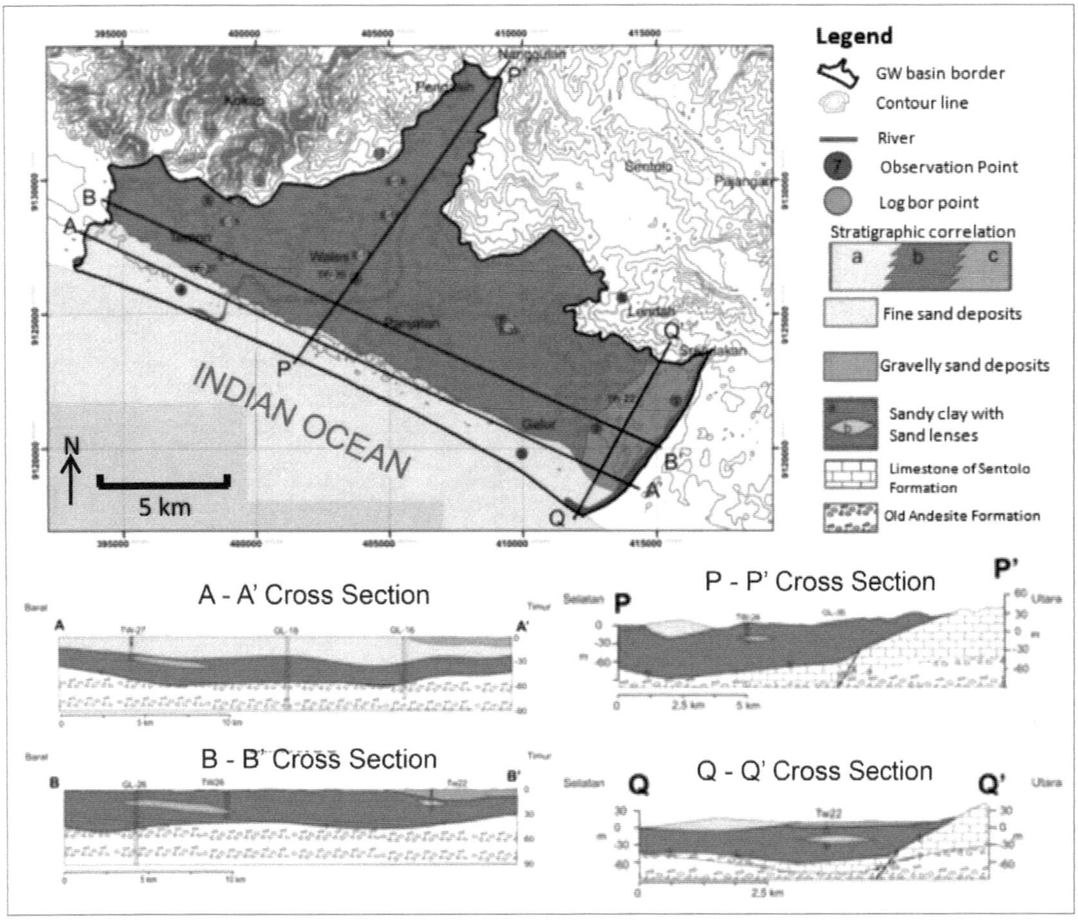

Figure 3. Lithological map and cross section of the area (modified from Maulana (2017)).

2.2. Seismicity of the study area

According to PusGeN (2017)[9], the Indonesia Center of Earthquake Study, the study area is generally located in very active seismicity with the source of the earthquake from the surrounding faults and the subduction zone in the south of Java island. Figure 4 (C, D,& E) is showing the history of the earthquake in and around the study area, from the shallow (red dots) to the deep (yellow dots) earthquakes in the time range 1900-2016. The history of earthquakes then becomes a database for developing the Peak Ground Acceleration (PGA) of an area. PGA data can be used to analyze and estimate the intensity and probability of the earthquake that might occur in the area. Figure 4 (A & B) shows that the PGA of the research area from the subduction source earthquake in 50 years of recurrence is ranged from 0.5 to 1.0, while the PGA from the source of shallow(faults) earthquake is ranged from 0.5 - 0.6. The latest shallow earthquake happened on 27 May 2006, in the Bantul area, which occurred due to the movement of the Opak Fault with 6.4 Mw. The higher PGA and the earthquake probability will lead to a higher susceptibility to liquefaction.

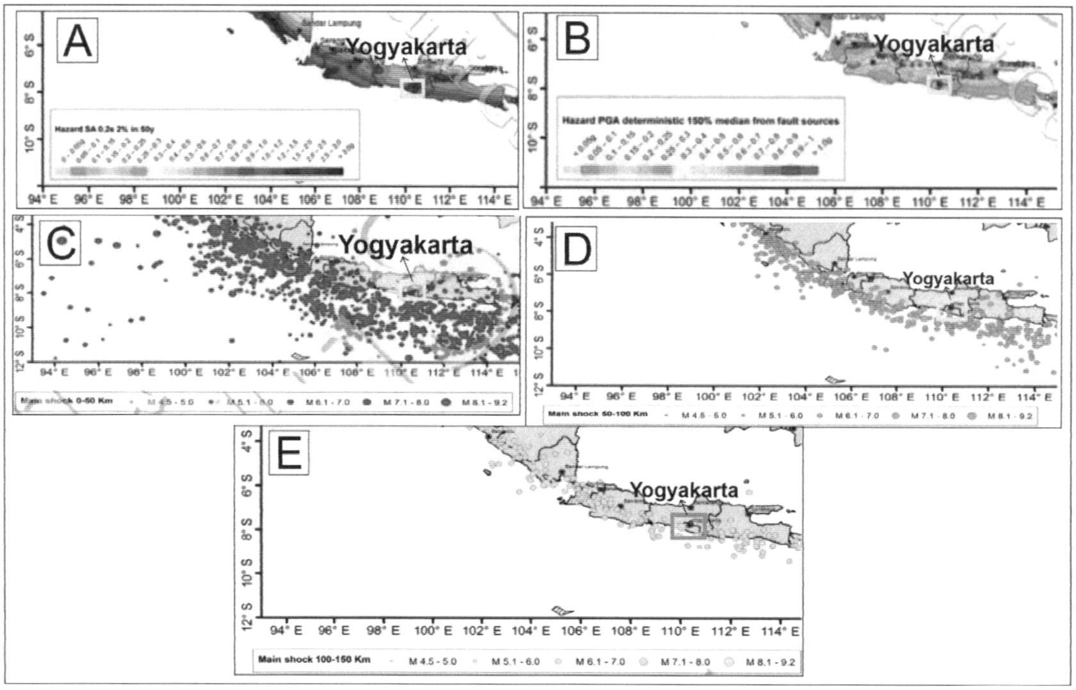

Figure 4. Peak Ground Acceleration (PGA) map of the reseach area due to subduction (A) & shallow earthquake (B) with reccurence in 50 years and Map of epicenter distribution with 0-50 km deep (C), 50-100 km deep (D), & 100-150 km (E) within 1900-2016 period (modified from PusGen (2017)).

3. Methodology

The methods used in this study are general delineation of the google earth image[10], to obtain the recent land use map, and an equal & different weight of the parameters to produce the estimated liquefaction susceptibility map of the area, which adopted from Elton et al (1995)[11]. The resulted maps will be plotted using the database and base map from the website of Badan Informasi Geospasial (BIG)[12].

3.1. Deliniation of google earth image

Delineation in this research means to classify the object shown in the google earth image of 2021[10]. The grouping will consist of (1) Settlement/Residential/Housing area, (2) Government/Primary Infrastructure area, and (3) Agriculture/other open space areas. The built-up zone area consists of the Residential and infrastructure area, while the undeveloped area is classified as agriculture or other open space areas. The use of this map is to know the most updated condition of the area development. This result will then be used for further research related to the liquefaction risk map of the area.

3.2. Liquefaction Susceptibility Index (LSI) Estimation

Some parameters are used in this research to estimate the liquefaction hazard potential. The parameters are selected from Elton et al (1995)[11] research which has done the liquefaction susceptibility evaluation using those parameters shown in table 1.

Table 1. Factors, weights, and criteria for rating potentially liquefiable sites [11].

Factor (Weight)	Rating Criteria	
	Condition	Grade
1. Type of deposits (A)	River channel	A
	Delta	B
	Collovium/talus	C
	dunes	D

2. Age of deposition (C)	Less than 500 years	A
	Late Holocene	B
	Holocene	C
	Pleistocene	D
	Pre-Pleistocene	E
3. Depth of water table (B)	Less than 1.5 m	A
	1.5 - 3.1 m	B
	2.1 - 6.1 m	C
	6.1 – 9.2 m	D
	More than 9.2 m	E
4. Grain-size distribution (B)	Uniform	A
	Fairly uniform	B
	Gap graded	C
	Well graded	D
5. Grain size (B)	Greater than 3 mm	A
	3 mm – 1 mm	B
	1 mm – 0.075 mm	C
	Smaller than 0.075 mm	D
6. N- value (A)	Less than 2	A
	2 – 5	B
	5 – 8	C
	8 - 11	D
7. Depth of burial (B)	Less than 1.5	A
	1.5 – 3.1 m	B
	3.1 – 6.1 m	C
	6.1-9.2 m	D
	More than 9.2 m	E
8. Ground slope (C)	More than 2 degree slope	A
	1 degree slope	B
	Level ground	C
9. Soil density (A)	Very loose	A
	Loose	B
	Medium density	C
	Dense	D
	Very Dense	E
10. Liquefiable layer thickness (C)	<0.15 m	E
	0.15 - 0.61 m	D
	0.61-1.22 m	C
	1.22 – 2.44 m	B
	Thicker than 2.44 m	A
11. Areal extend of liquefiable deposits (B)	Large (> 2 ha)	B
	Medium (0.4-2.0 ha)	C
	Small (<0.4 ha)	D
12. Capping low permeability layer (B)	Good cap	C
	Fair Capping effect	D
	No Capping effect	E

There are some differences in this research compared to the Elton et al (1995)[11] research. This research will only select some of the parameters from Elton et al (1995)[11] due to the limitation and unavailability of the data. The parameters number 1, 2, 3, 4, 5, 7, 8, and 9 will be used in this research by some assumptions for the classification. Elton et al (1995)[11] used the fuzzy set analysis to do the liquefaction susceptibility mapping, while this research will apply equal and different weights for each parameter and the ranged value will be included in its grade value. A will represent 5 scores while E will represent 1 score, and the highest total score will show the more susceptible the area. There will be no specific classification of the areas' susceptibility, but there will be only a comparison of one area's susceptibility to other areas' susceptibilities. The complete and integrated mapping will be done by further research while this research will only provide a general preview of the liquefaction potential of the area.

One of the Geographic Information System – Multi-Criteria Analysis (GIS-MCA) methods that are mostly used in geological research is Analytical Hierarchy Process (AHP) method [13]. After using the Equal Weight method for the parameters, the different weights were applied to test the sensitivity of the results. If the same susceptibility area is produced through these methods, the applied weight and parameters for the research are considered to be acceptable and consistent.

Another parameter that needs to be considered is the strength of an earthquake shaking, which is described in units of moment magnitude (Mw). According to Green and Bommer (2019)[14], soil liquefaction can occur starting on an earthquake magnitude scale of 4.5 Mw in swampy areas or river borders and 5.0 Mw at suitable locations for residential areas, with the condition that peak acceleration in bedrock > 1.5 g. While in the zone of moderate liquefaction susceptibility, it can experience an impact starting in an earthquake with a magnitude moment of 6 Mw. This study uses the magnitude moment assumption to be more than 6 Mw so that all of the areas will have the same impact by the earthquake.

4. Result and Discussion

The result of this study will reveal the recent condition of the land use in the Wates Groundwater Basin and its estimated liquefaction susceptibility map. Since this very rough delineation and estimation will be the reference for further detailed research and mapping, the result should be used very carefully.

4.1. Recent land use map

The land use map shown in figure 5 pointed out that the study area is still in an early stage of regional development. The agriculture or open space area (the green-colored) is still dominated the area, even the settlement area has also started to be developed. The NYIA development in the southern corner of the area is predicted to be denser in settlement and, consequently, people that will grow their business around the airport area.

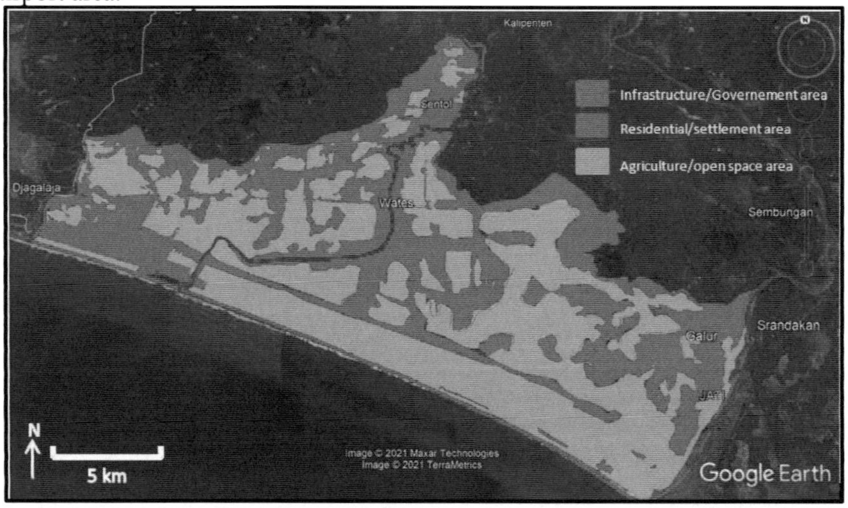

Figure 5. Recent simplified land use map of Wates Groundwater Basin.

4.2. Estimated Liquefaction Susceptibility of the area

The map of each parameter was carried out and can be seen in figure 6. Using the totally eight parameters of liquefaction susceptibility, all of the parameters then be overlayed and resulting the estimated liquefaction susceptibility map of the area as shown in figure 7.

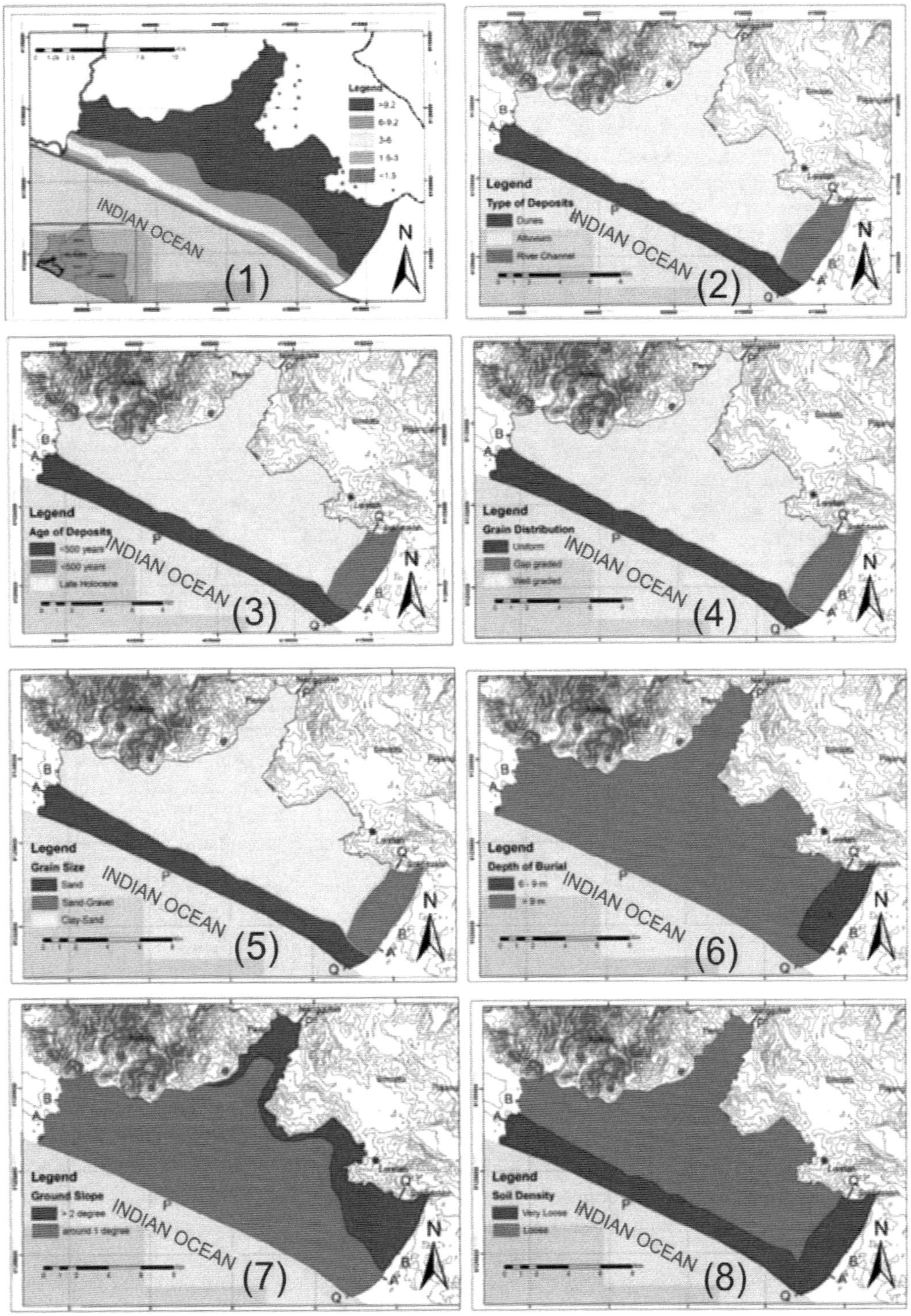

Figure 6. Maps of eight parameters (1) Groundwater depth, (2)Type of deposits, (3) Age of Deposits, (4) Grain Distribution, (5) Grain Size, (6) Depth of burial, (7) Ground Slope, and (8) Soil Density.

The groundwater depth data, the type of deposit data, age of the deposit, grain distribution, grain size, depth of burial, ground slope, and soil density data are interpreted and estimated from geological maps and geological cross-section data gained from the previous research that done by Maulana (2017)[7]. All of the data was then reclassified so it can fit the Elton et al (1995)[11] classification. After all of the results of the map, the polygon data is then converted into the raster data so each pixel can have the value of each parameter. All of the raster data was then overlayed using GIS software by implementing the Weighted Sum feature. The result of the weighted sum feature will be the total score of each pixel. The resulting raster was then reclassified to be the susceptibility value of the area.

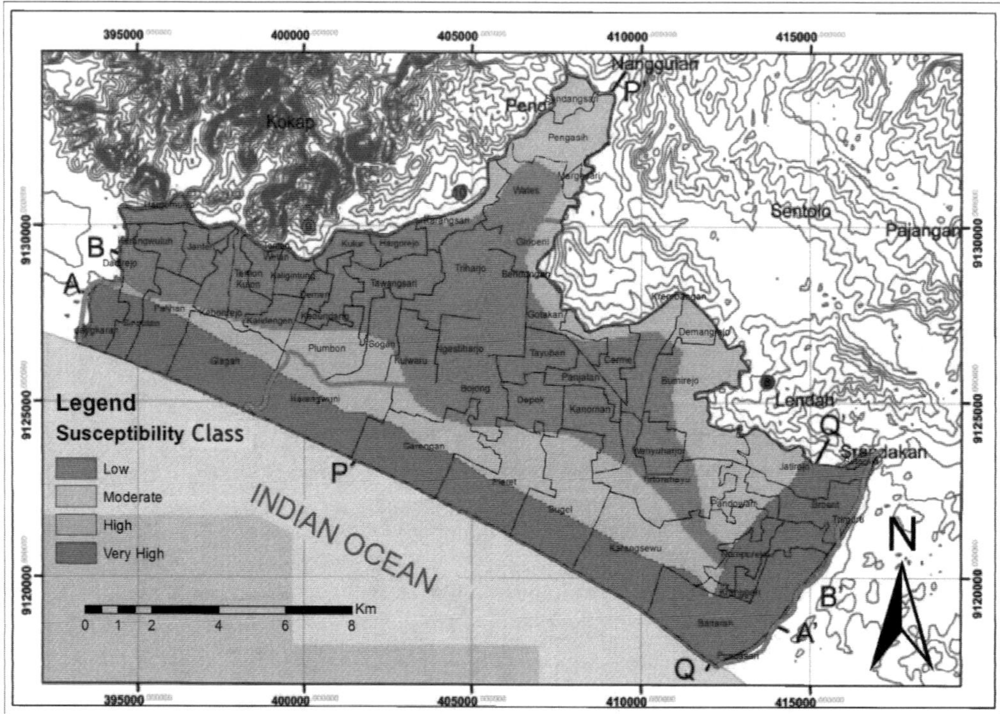

Figure 7. Estimated liquefaction susceptibility map of the research area using weighted sum tool.

The four susceptibility values/index shown in figure 7 revealed from the reclassification of the sum value of the parameters. A total score lower than 22 will be classified as low, 22-25 classified as moderate, 25-27 classified as high, and 28-31 classified as very high susceptibility. The result shows that the low susceptibility area is distributed in the middle part of the basin while the very high susceptibility is ranged in the south and east parts near the edge of the basin. The result is aligned with the regional liquefaction susceptibility map of the Yogyakarta area made by Buana et al (2019)[1] with the addition of a more detailed classification in the basin area.

The AHP method was also done in this research. Table 2 shows the comparison matrix of the parameters, while the normalization and weighting results are shown in Table 3. The final weight of all parameters involved in liquefaction susceptibility estimation is shown in Table 4. The result of the calculation for this weight in terms of Consistency Index (CI) is 0.013, while the Random Index (RI) is 1.41 because of 8 total parameters, so the resulted Consistency Ratio (CI/RI) is 0.0097 or 0.9%. The CR is classified as acceptable, due to the value being <0.1 or <10%, so the parameters' weight is can be used. Figure 8 shows the liquefaction susceptibility estimation of the research area by the AHP method.

Two susceptibility classes were produced from the AHP method, apart from the equal weight method that produced 4 classes. The south and east parts are classified as high susceptibility class, while the other part is classified as a moderate class. The result also confirmed the regional planning, however, the detailed mapping needed to be done in the east part of the research area where the high and moderate class is shown. The east part needs to be more concerned due to the planning of the Southern Causeway (JJLS) development but the area has low availability of detailed data.

Table 2. Comparison matrix of the liquefaction susceptibility estimation parameters.

Parameters	Type of Deposit	Age of Deposit	Depth of WT	Grain Size Distrib	Grain Size	Depth of Burial	Ground Slope	Soil Density	Total
Type of Deposit	1.00	4.00	1.00	2.00	2.00	4.00	3.00	1.00	18.00
Age of Deposit	0.25	1.00	0.25	0.33	0.33	1.00	1.00	0.33	4.50
Depth of WT	1.00	4.00	1.00	2.00	2.00	4.00	3.00	1.00	18.00
Grain Size Distrib	0.50	3.00	0.50	1.00	1.00	3.00	2.00	0.50	11.50
Grain Size	0.50	3.00	0.50	1.00	1.00	3.00	2.00	0.50	11.50
Depth of Burial	0.25	1.00	0.25	0.33	0.33	1.00	0.50	0.25	3.92
Ground Slope	0.33	1.00	0.33	0.50	0.50	2.00	1.00	0.25	5.92
Soil Density	1.00	3.00	1.00	2.00	2.00	4.00	4.00	1.00	18.00
Total	4.83	20.00	4.83	9.17	9.17	22.00	16.50	4.83	
	R	**S**	**T**	**U**	**V**	**W**	**X**	**Y**	

Table 3. Normalized of comparison matrix.

Parameters	ToD	AoD	DoGW	GSD	GS	DoB	GS	SD	Total	Parameter Weight	
Type of Deposit	0.21	0.20	0.21	0.22	0.22	0.18	0.18	0.21	1.62	0.20	**H**
Age of Deposit	0.05	0.05	0.05	0.04	0.04	0.05	0.06	0.07	0.40	0.05	**I**
Depth of WT	0.21	0.20	0.21	0.22	0.22	0.18	0.18	0.21	1.62	0.20	**J**
Grain Size Distrib	0.10	0.15	0.10	0.11	0.11	0.14	0.12	0.10	0.94	0.12	**K**
Grain Size	0.10	0.15	0.10	0.11	0.11	0.14	0.12	0.10	0.94	0.12	**L**
Depth of Burial	0.05	0.05	0.05	0.04	0.04	0.05	0.03	0.05	0.35	0.04	**M**
Ground Slope	0.07	0.05	0.07	0.05	0.05	0.09	0.06	0.05	0.50	0.06	**O**
Soil Density	0.21	0.15	0.21	0.22	0.22	0.18	0.24	0.21	1.63	0.20	**P**
Total	1.00	1.00	1.00	1.00	1.00	1.00	1.00	1.00	1.00	1	

Table 4. Final Weight Parameters.

No	Parameters	Weight	Percentage
1	Soil Density	0.203912	20.4%
2	Type of Deposit	0.2025862	20.3%
2	Depth of WT	0.2025862	20.3%
3	Grain Size Distrib	0.1170128	11.7%
3	Grain Size	0.1170128	11.7%
4	Ground Slope	0.0625326	6.3%
5	Age of Deposit	0.0501502	5.0%
6	Depth of Burial	0.0442072	4.4%
	Total	1	100%

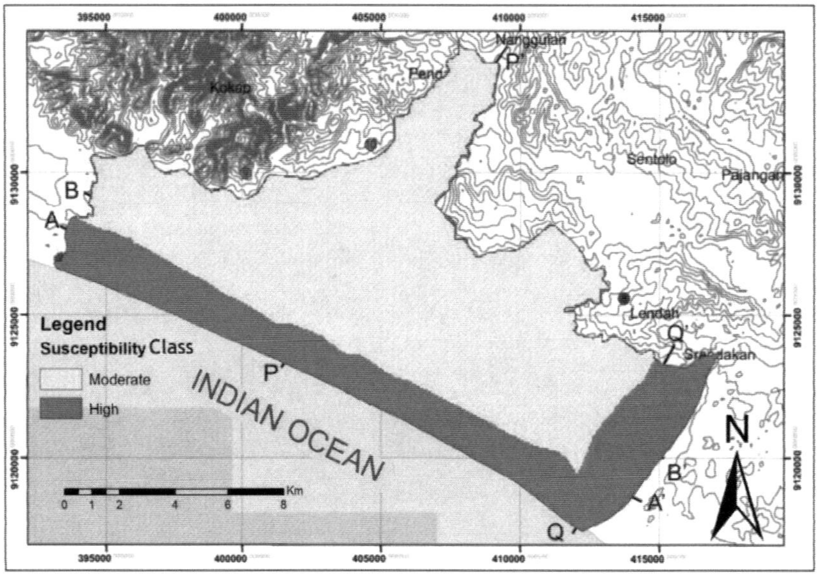

Figure 8. Liquefaction Susceptibility Estimation Map using AHP Method.

5. Conclusion

The estimated liquefaction susceptibility map was carried out in this research. The usage of the Geographic Information System (GIS) is one of the keys to producing the tentative map before the complete mapping and field data sampling are meant to be done. The southern part of Wates Groundwater Basin is confirmed to be a very high susceptibility class of liquefaction, which will be very easy to be triggered by the earthquake. The others area of the basin classified as low-high susceptibility. The latest land use map was also carried out in this research. Those maps will be the reference to determine the sampling area for geophysics and SPT/CPT data acquisitions. The detailed and integrated mapping will be done to obtain a more accurate map so the specific development of the area can be safer from the undesirable liquefaction potential event.

6. References

[1] Buana, T.W., Hermawan, W., Rahdiana R.N., Wahyudin, R.W., Hasibuan, G., Wiyono., and Sollu, W.P. 2019. Atlas Zona Kerentanan Likuefaksi Indonesia. Bandung: Badan Geologi Kementerian Energi dan Sumber Daya Mineral (ESDM) Indonesia

[2] Jalil, A., Fathani, T.F., Satyarno, I. and Wilopo, W., 2021. Liquefaction in Palu: The Cause of Massive Mudflows. Geoenviron Disasters. 8:21. https://doi.org/10.1186/s40677-021-00194-y
[3] Kusumawardani, R., Chang, M., Upomo, T.C., Huang, R.C., Fansuri, M.H. and Prayitno, G.A., 2021. Understanding of Petobo liquefaction flowslide by 2018.09. 28 Palu-Donggala Indonesia earthquake based on site reconnaissance. Landslides, pp.1-20. https://doi.org/10.1007/s10346-021-01700-x
[4] Dwianta, A.S. 2016. Geologi Dan Potensi Likuifaksi Berdasarkan Kondisi Hidrogeologi Dan Sifat Fisik Tanah Pada Lokasi Pembangunan Bandara Internasional Daerah Temon Dan Sekitarnya, Kecamatan Temon, Kabupaten Kulon Progo, Daerah Istimewa Yogyakarta (Undergraduate Thesis, UPN" Veteran" Yogyakarta).
[5] Anggara, V., Fatnanta, F. and Yusa, M., 2020. Analisis Likuefaksi Berdasarkan Data CPT Untuk Perancangan Pondasi Di Pesisir Pantai Kabupaten Kulon Progo DIY. Jurnal Rekayasa Sipil (JRS Unand), 16(2), pp.77-90.
[6] Artati, H.K., Pawirodikromo, W. and Purwanto, E., 2020. Analisis Potensi Likuifaksi Pada Pasir Vulkanik Di Pantai Glagah Kulonprogo Berdasarkan Data N-SPT. Teknisia, 25(2), pp.108-120.
[7] Maulana, D. 2017. Simulasi Dampak Pengambilan Air Tanah Yang Berlebih Pada Cekungan Air Tanah Wates (Undergraduate Thesis, Universitas Gadjah Mada).
[8] Rahardjo, W., Sukandarrumidi, dan Rosidi, HMD, 1995. Peta Geologi Lembar Yogyakarta, Jawa, skala, 1:100.000. Bandung: Ministry of Energy and Mineral Resources of Indonesia.
[9] Pusat Studi Gempa Nasional (PusGen). 2017. Peta Sumber dan Bahaya Gempa Indonesia. Kementerian PUPR
[10] Google. Google Earth 2021. Accessed 12 December 2021. https://earth.google.com/web
[11] Elton, D.J., Juang, C.H. and Sukumaran, B., 1995. Liquefaction susceptibility evaluation using fuzzy sets. Soils and foundations, 35(2), pp.49-60.
[12] Badan Informasi Geospasial. Ina-Geoportal 2021. Accessed 10 December 2021. https://tanahair.indonesia.go.id/portal-web
[13] Saaty, T.L., 2008. Decision making with the analytic hierarchy process. International journal of services sciences, 1(1), pp.83-98.
[14] Green, R. A and Bommer, J. J. 2019. "What is the smallest earthquake magnitude that needs to be considered in assessing liquefaction hazard?," Earthq. Spectra, vol. 35, no. 3, pp. 1441–1464

Evaluation of The Effect of Rock Joints on the Stability of Underground Tunnels

A Rahim[1*], T M Nasae[2], Z Hareyani[2], T F K Ngoroyemoto[3], I Hasim[1], M F Sapawie[1], H Hussin[4]

[1]Department of Geotechnics & Transportation, School of Civil Engineering, Faculty of Engineering, Universiti Teknologi Malaysia, Malaysia
[2]School of Material and Mineral Resources, Universiti Sains Malaysia, Penang, Malaysia
[3]Head of Department, Mining and Mineral Processing Engineering, Manicaland State University of Applied Sciences, Mutare, Zimbabwe
[4]Department of Geoscience, Faculty of Earth Science, Universiti Malaysia Kelantan, Jeli Kelantan
*Corresponding author's e-mail: afikahrahim@utm.my

Abstract. To facilitate the water supply to the main cities in Malaysia such as Kuala Lumpur, Putrajaya and Cyberjaya, Pahang Selangor Raw Water Transfer Tunnel project is constructed in 2015. Facility tunnel of 44.6 km length with 5.2 m diameter are being constructed to transfer water from Karak to Hulu Langat. Granitic rock with dominant intrusive zones in various range of I until V were observed in the studied tunnel length. This tunnel has crossed sheared zone that categorized into fair, poor, and very poor rock classes with moderate to heavily jointed rock. At NATM-1, the discontinuities like joints that is close and far from the fault are one of the most important which caused to instability of the tunnel and reduced the performance of tunnel performance. In this paper, the joint orientation, overburden and distributions of discontinuities (joints and faults) into the tunnel were evaluated. These data were then simulated using discrete fracture networks with the direction of tunnelling excavation in the tunnel. Movement of blocks in the tunnel roof and wall is possible due to the creation of more intersection points in the critical zone. It also highlights how minor features, such as step-over joints in rock mass, can have a significant impact on instability.

1. Introduction

Main problems in infrastructure development are the geological and topographical conditions. The development and progress of a region in fact cannot be separated from the ability of the region to interact with other areas on a larger scale. The intended interactions are economic, social and resource activities. To support this, the means of supply of raw water in this case becomes very vital. During the constructing of the Pahang Selangor Raw Water Transfer Tunnel (PSRWT), the tunnel was built in metasedimentary and igneous rock, which upset the in-situ stress field and caused large ground movements, making the tunnel unstable.

Hayashi [7], Brown [8, 9], Ladanyi and Archambault [10], Einstein and Hirshfeld [11], and Reik and Zacas [12] have studied the strength and deformation response of linked rocks, joined masses of modelling materials, and laminated rocks. They looked into how joint sets and angles affected the strength of a rock mass. Chenevert and Gatline [13], Attewell and Sandford [14], and Brown et al. [15]

all looked at shale and slate. Deklotz et al. [16], Akai et al. [17], McCabe and Koerner [18], Nasseri et al. [19,20], and Singh [21] looked at how gneisses and schists reacted, while Ramamurth Rao et al. [23], Pomeroy et al. [24], and Allirote and Boehler [25] looked at how sandstone Tien and Tsao [26] did their research with things that didn't exist. When all of their work is looked at and analysed, the greatest strength at failure is either b1401 or 901, where b is the angle of the break with respect to s1, and the least strength is usually around b14301, or 45(j/2), where j is the friction angle along the plane of weakness.

Most of the time, the Karak, Pahang area is made up of many joints close to faults, which causes complex changes in stress and strain when tunnelling is done. During the construction of PSRWT, this project used both methods, the New Austrian Tunnelling Method (NATM) and Tunnel Boring Machine (TBM). While tunnelling through a variety of structures such as joints and faults, the rock mass condition and overburden will encourage tunnel instability and collapse. The primary goals of this research have been to investigate the influence of joint orientations, overburden, and modelling on tunnel face stability for varying dips. Then, for a complete range of 0–90 degrees, the influence of variation in dip direction difference on rock strength is explored.

1.1. Pahang Selangor Raw Water Transfer Tunnel Project.

The Malaysian government has designated Shimizu – Nishimatsu – UEMB – IJM Joint Venture (SNUI JV) as the project's contractor, through the Ministry of Energy, Green Technology and Water Malaysia (KeTTHA) since 90s and the tunnel is completely constructed in March 2015. This project is known as the Pahang – Selangor Water Transfer Project (PSRWT). The tunnel is crossing from Karak (inlet) to Selangor , Hulu Langat (Outlet). The tunnel is 5.2 meters in diameter and 44.6 kilometers long (Figure 1)(KETHA, 2000). To build a large and lengthy tunnel through variable meta-sedimentary and granite rock beneath the Titiangsa Mountains without incident, a good safety precaution is required. A site inspection should be carried out as a precautionary step. Proper site inspections are required before any additional development can begin in order to learn about the unseen subsurface state.

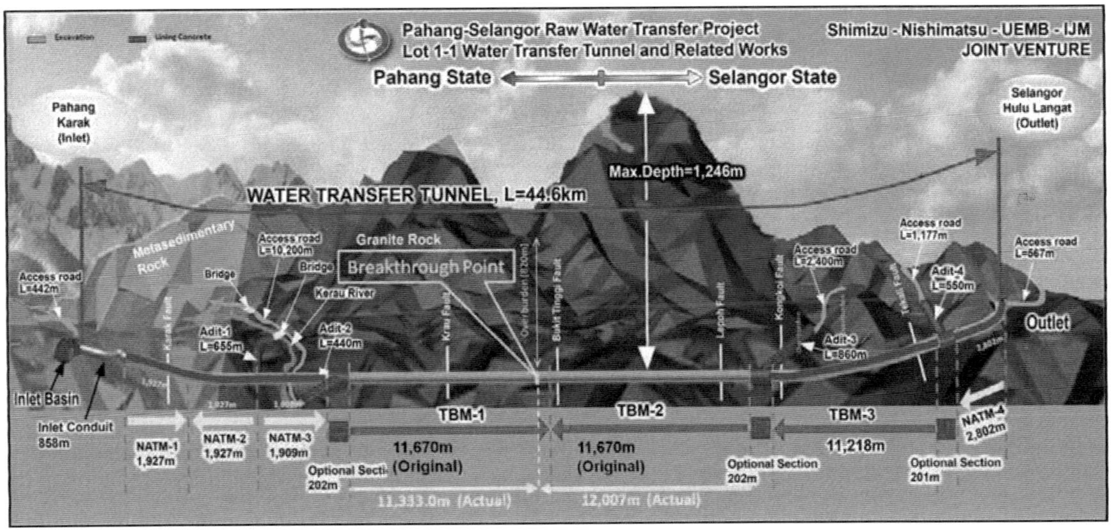

Figure 1. Study area NATM-1 (yellow box) allinged with the Pahang – Selangor Raw Water Transfer Tunnel (modifed after KETHA, 2000).

1.2. Meta-sedimentary rock formation.

The research area (NATM-1) is located within the Karak Formation, a meta-sedimentary rock that stretches beneath Sungai Karak (Figure 2). The Karak Formation contains metamorphic rock from the Devonian period. Metamorphic refers to the changes in mineral composition and texture that occur when

IOP Conf. Series: Earth and Environmental Science 1071 (2022) 012006 doi:10.1088/1755-1315/1071/1/012006

a rock is exposed to pressures and temperatures that are different from those that created it. Magmatic rocks invade meta-sedimentary rocks often, which are enhanced by tectonic processes, indicating post-depositional events in the Sector's evolution. While little is known about the cellar where they gathered, they usually represent the first identifiable occurrence (Padget, 2004). The research area is located within the Karak Formation, a meta-sedimentary rock that stretches beneath Sungai Karak. The Karak Formation contains metamorphic rock from the Devonian period. Metamorphic refers to the changes in mineral composition and texture that occur when a rock is exposed to pressures and temperatures that are different from those that created it. Magmatic rocks invade meta-sedimentary rocks often, which are enhanced by tectonic processes, indicating post-depositional events in the Sector's evolution. While little is known about the cellar where they gathered, they usually represent the first identifiable occurrence (Padget, 2004). Meta-sedimentary rock is not as hard as igneous rock, such as granite. Even though it is hard, sedimentary rock breaks it up into many cracks and joints. The stability of a meta-sedimentary rock formation will change because of this crack and joint. Meta-sedimentary rocks often re-crystallize as they age and become more metamorphic, but their sedimentary nature is clear. Most rocks are made up of quartzite, micaceous quartzites, and pelitic (argillaceous) gneisses (Padget,2004).

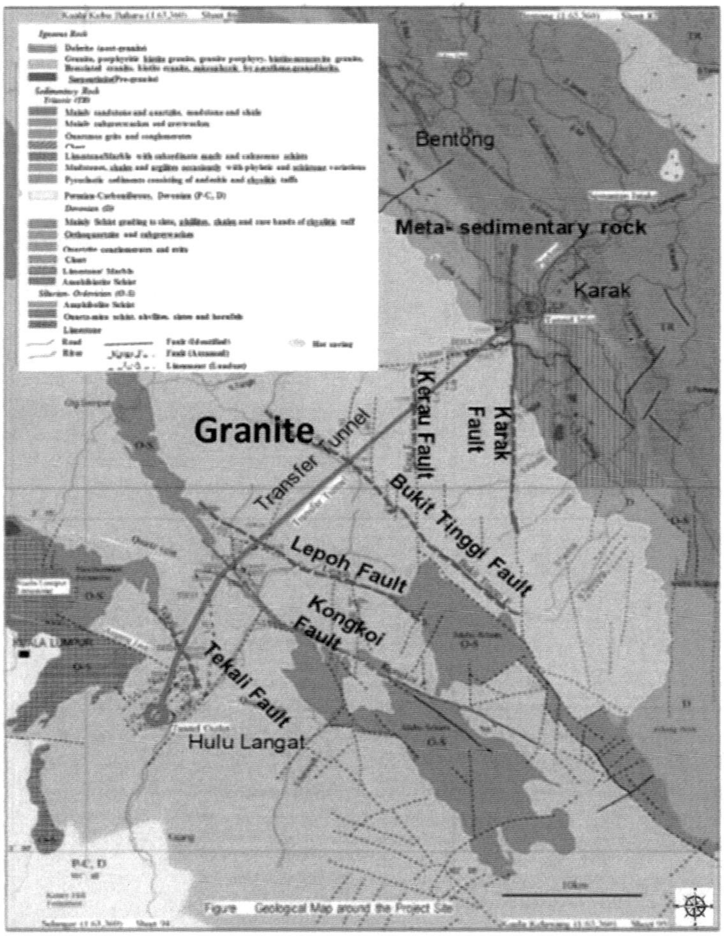

Figure 2. Geological Map of the study area (yellow box) and the 44km PSRWT (blue line) (modifed after KETHA, 2000).

1.3. Joint Pattern Development.

Geometric function of joint pattern in rock is crucial (Olson, 1993). The function of geologists in assessing common dangers and the stability of a site for engineering works has a long history that may be traced back to our ancestors' lore. In this study, researchers tried to do a thorough investigation to find out how rocks react to different confining pressures when the dip and dip direction differences of induced discontinuities change. In reality, there are a lot of times when discontinuities with different dip directions cross each other, especially when tunnelling in this shear zone. Interference of these breaks with the path of the load would change the way the rock broke, with a different amount of intact rock parts working together. In this study, artificial samples made of plaster and with clear breaks were used to show how the described combined effects work. Joint patterns at the study area (Figure 3) is most likely due to the influence of the surrounding environment. The surrounding environment, such as overburden and rivers, can result in a different joint pattern in meta-sedimentary rock than the Malaysian regional joint pattern (Figure 4). Uplift and volume expansion are the most common causes of intragranular joints in metamorphic rocks. But these seams happen in volcanic rocks because the rocks shrink in size in different ways as they cool. Most of the time, thermal stress is what causes joints to form in rocks that are cooling (Samanta, 2001). A shear or stress rupture can result in a rock fracture. Shear breakage causes faults, while rupture causes joint propagation. Joints are generally systematic and occur in sets over broad areas of more than 1500 km2, whereas faults are rarely systemic and are generally clustered in limited zones linked with a master fault or fold (Engelder and Lash, 2009). Fracture mechanics can be used to describe fracture start, propagation rate, and incremental propagation path (Olsen, 1993). Olsen (1993) says that cracks in rocks often start in a non-linear way because there is a lot of inelastic deformation on the grain scale before cracks form on the larger scale. The intragnular joints are not the result of tectonic processes because they don't have a preferred orientation in terms of an outside coordinate system. Instead, they might be caused by thermal tensions that form as a flow of lava cools. Textural features like the way crystals grow and the presence of phenocrysts in a fine-grained matrix also point to a sudden cooling of magma with early-formed crystals (Samanta, 2001).

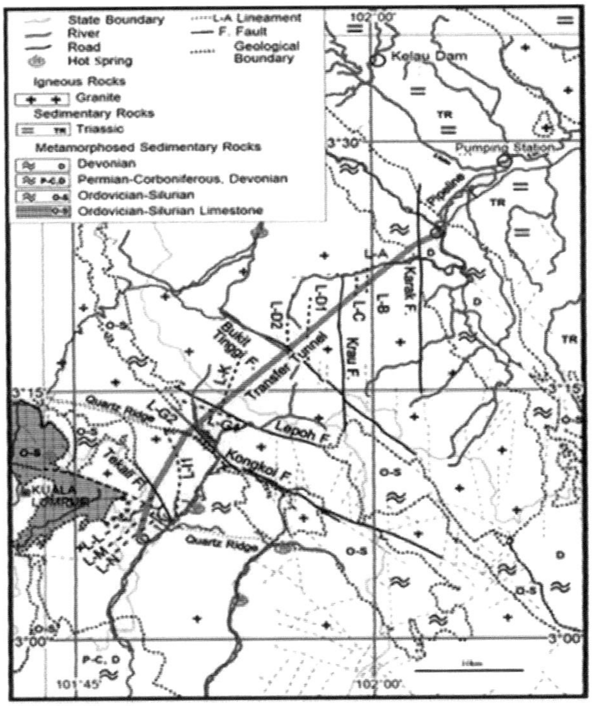

Figure 3. Cross section of tunnel with the Lineament (L-) and fault line (F.). Study area is marked by the yellow box.

2. Methodology

2.1. Tunnel Mapping.

The study site has been selected as NATM-1, Karak. Tunnel mapping / fieldwork provides a good opportunity to learn more about the geology of the research area as well as perform duties including tunnel mapping, digging and charging, and probing drilling. Apart from that, during this examination, the other tunnels in Karak, namely NATM-2, NATM-3, and TBM-1, were visited as a side search. This enables a comparison of tunneling properties between the NATM-1 and other tunnel components, such as geological composition, rock type, and tunnel development.

A tunnel mapping face (Figure 4 left). exercise is a critical operation carried out by geologists on site in order to record the tunnel face (Figure 4 right). It's finished after the mucking but before the support system is installed. The tunnel face was sketched after each blast to determine the stability, joint collection, and presence of groundwater. It must be evaluated and registered for future use as a guide.

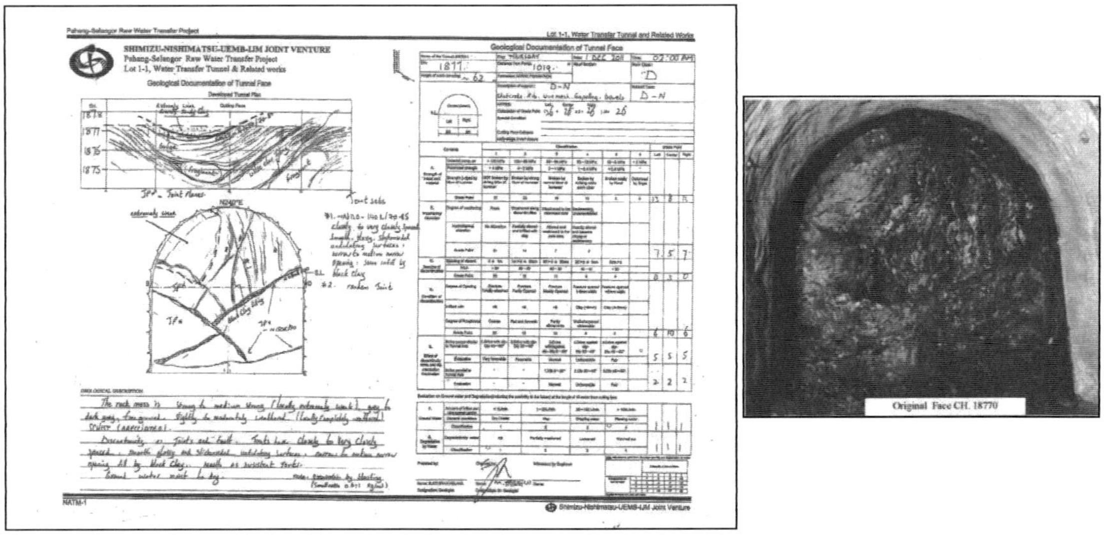

Figure 4. Tunnel Face Mapping (left), Tunnel Face (right).

3. Data Interpretation and Analysis

After the raw data has been extracted. The dip will be input into the program, along with its location, for study. The data will be analyzed using the Rocscience Dips tool. Based on plots from the dip and dip direction, this program generates a Fisher Concentration (Figure 5 a) and Rose Diagram (Figure 5 b) based on stereonet in rock mechanics, with the Rose Diagram illustrating the number of joints that go in each direction. By observing the shift in join pattern along the tunnel, these items can be studied. The joint pattern can change along the tunnel due to factors such as regional faults, overburden height, and climate. The findings of this study will be utilized to identify the source of specific joint pattern behavior and predict the possibility of the joint collapsing. To assess these data, it will take a lot of consultation with the supervisor who has enough knowledge and expertise in this subject. The interpretation process could go awry if there is no consultation, resulting in erroneous outcomes.

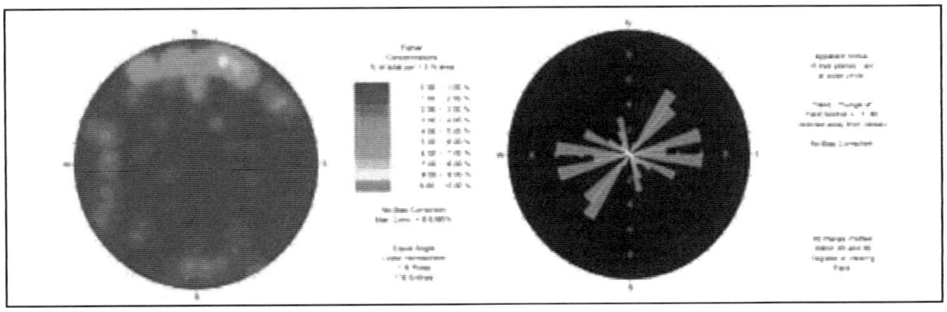

Figure 5. (a) Joints concentration (b) Direction of joint density (Rose Diagram).

4. Results and Discussion

4.1. Relationship between Joint Orientation and Overburden.

To develop the model, the extracted data was digitalized and entered into software. The pattern changes depending on the environment around the tunnel faces. The patterns begin with TD-0 and end with TD-909. The rock type existing at the tunnel face, the stability of the rock basis as determined by the Rock Mass Classification (Japan Highway Public Corporation; JH), the elevation of overburden, and the groundwater in the vicinity have all been considered. The joint pattern and the overburden can be seen in Figure 6.

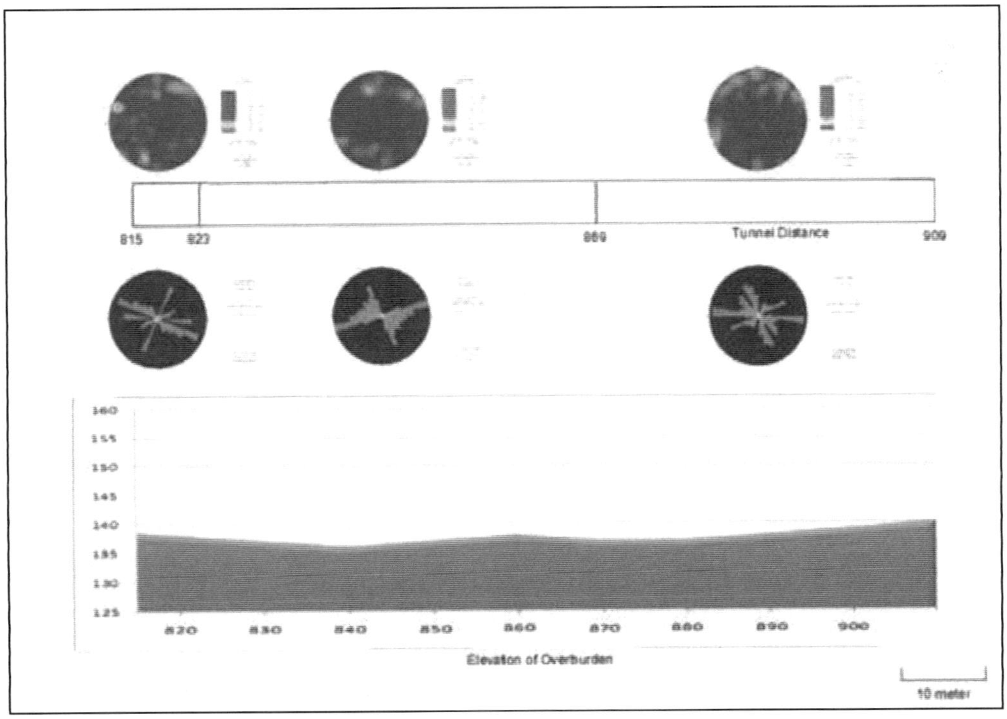

Figure 6. Joint Pattern of TD-815 to TD-909 with its elevation of overburden.

From Figure 6 , no significant relationship can be correlated. The rock stability in the NATM-1 area is threatened by numerous fractures and joints at the shear zone of Karak with the composition of the shale, homfels, phyllite, siltstone, sandstone, and slate as well as clay, breccla, quartzite, and chert were all found too. 1767 joints were recorded from the top of the tunnel face to the bottom. There are 430 faults, 278 smooth planar joints, 94 foliation joints, 9 bedding joints, and 956 random joints among the total joints. The majority of the joints were centered between N90° and 130°E / 60° and 80°. Except for TD-0 to TD-35, TD-81 to TD-96, TD-158 to TD-212, TD-229 to TD-268, TD-467 to TD-548, TD-759 to TD-813, and TD-823 to TD-909, the directional of joint pattem followed the regional fault pattem.

Throughout the tunnel, the elevation of overburden fluctuates. It reaches to a height of 205 meters above sea level from sea level. Then it dips to 150 meters above sea level and a river runs across a section of the overburden, affecting the joint pattem.

The seepage or groundwater can be dry, highly wet, or high inflow. It can occasionally produce flooding in the tunnel, posing a barrier to tunnel building. The largest groundwater flow rate is 300 L/min, which occurs when the river meets the overburden. From fair rock with weathered rock mass and some clay in joints to extremely poor rock with soil, crushing, and squeezing zones, rock stability ranges from class C-I to E.

4.1.1. Cause of Joint Development. As mentioned, the joint pattems have certain features in specific spots. These qualities are influenced by the climate in that place. After evaluating the features of joint pattem in each set of joints, there are some major environmental influences on joint pattems in an area. In comparison to the rest of the tunnel, the overburden at the inlet is very thin (thickness). As a result, the weathering of the rock formation in that location has begun. As a result of this process, the joint pattern was altered, and the rock structure became uneven. So, the regional fault pattern was somewhat distorted by the lateral dominant joint. It reveals that the weathering mechanism is one of the causes of joint development.

Between TD-550 and TD-654, the Karak River can be found at the top of the overburden. According to the Rose Diagram (Figure 5a and 5b) at the time, the joint's orientation is following the regional fault pattern. The orientation of the joints is unaffected by the river's position on top of the overburden. Regardless, this portion has much too many joints, with 192 joints compared to an average of 60-80 joints in other sections.

The joint pattern along the tunnel shows several improvements as the level of the overburden changes. There is no apparent link between these two traits and the modifications, however. As a result, no conclusive association can be inferred from this research. The relationship between these two objects will require more investigation. According to the paper, the Joint pattems have certain features in specific places. These qualities are influenced by the climate in that place. After evaluating the features of joint pattem in each set of joints, there are some major environmental influences on joint pattems in an area. In comparison to the rest of the tunnel, the overburden at the inlet is thin. As a result, the weathering of the rock formation in that location has begun. As a result of this process, the joint pattern was altered, and the rock structure became uneven. As a result, the regional fault pattem was somewhat distorted by the lateral dominant joint. It reveals that the weathering mechanism is one of the causes of joint development.

Between TD-550 and TD-654, the Karak River can be found at the top of the overburden. According to the Rose Diagram at the time, the joint's orientation is following the regional fault pattern. The orientation of the joints is unaffected by the river's position on top of the overburden. Regardless, this portion has much too many joints, with 192 joints compared to an average of 60-80 joints in other sections. According to this study, a river on the overburden caused several fractures and joints to form within a rock mass in a specific location, while the orientation of the joints remained constant.

The joint pattern along the tunnel shows several improvements as the level of the overburden changes. There is no apparent link between these two traits and the modifications, however. As a result, no conclusive association can be inferred from this research. The relationship between these two objects will require more investigation.

4.1.2. Potential Failure in Tunnel Process with the Joint/ Fault Distance. Several incidences of tunnel falure during tunnel processing have occurred in NATM-1. This occurrence occurred at two locations in the research area: TD-220 and TD-816. Due to the joints and defects that emerge on the tunnel's façade, the tunnel has collapsed at these points. Fortunately, no SNUI -JV employees were wounded in the incident. It is critical to anticipate potential failures prior to the tunneling phase in order to avoid future accidents.

When TD-220 and TD-816 are compared, N320 E/45°-50° at TD-220 and N320 E/45°-50° at TD-816 are the sheared joints that can fail. While both TDs are in the same location, the sort of rock present in these sites is not. Based on this, there is a link between these two instances. Tunnel failure could theoretically be caused by the type of rock in a region, but based on this observation, no definite associations exist as to which type of rock causes tunnel failure. As a result, the directional of sheared at both TD has been chosen as the guideline for predicting the region in the tunnel phase that has possible failure for the evaluation of prospective failure in this tunnel.

Based on the shear zone of TD-220, certain groups have dominant joints that adopt the sheared joints direction. These classes have been marked as having a high chance of failing tunnel processing. The four classes are TD-229 to TD-256, TD-256 to TD-268, TD-815 to TD-823, and TD-823 to TD-869. In these groups' dominant and almost dominant joints, the shear zone at the failure location is close. These classes contain both TD and TD with failure incidents. As a result, these groups are at risk of failing, and they must proceed with prudence throughout the tunneling phase in these locations. If this type of joint is present in that site, a thorough examination of the tunnel face is required to avoid any failures.

4.1.3. Numerical Simulations: Model Establishment and Parameters. After collecting the joint data, three-dimensional discrete element software is used to look more closely at the numerical results (3DEC -discrete element method). Often, parts of the physical medium are different or don't fit together. Derived in terms can be seen as breaks in a material's properties or in its meso and macro structures. According to the concept of discrete composition, the medium has mechanical properties that vary from place to place. This means that when the medium is stressed, it changes in an unexpected way. In continuous media, things that aren't always the same interact with one another. This means that discrete media can be thought of as the sum of continuous media. The most basic parts of a rock mass, as an example of a general interpretation, are rock blocks with different lithological properties (Continuum) and geological discontinuities (discontinuous features). When pushed from the outside, rock blocks can function as a continuous medium, with "discontinuities" that allow them to interact with one another (discontinuous features). Rock blocks can behave like a continuous medium when the tension between the breaks is too great. The process of moving or separating rock blocks is known as shearing.

There is a clear asymmetric distribution phenomenon due to the clear geological features of the dominant joint angle of the surrounding rock in the main deformation area of the tunnel. This is evident from the tunnel's changing shape, the pressure on the rocks surrounding it, and the tension on the steel. The stress-causing failure features of a soft rock tunnel in a high geostress joint/fault fracture zone are identified, and the appropriate numerical model is developed to fully mimic the properties of the surrounding rock. The first stress field of a rock mass in the computer model is simply the stress field caused by gravity. The model is buried 500 metres underground, and geostress is applied to the top of it to simulate the stress caused by the weight of the strata above it. An elastic-plastic constitutive model is used for the rock surrounding it. A Coulomb slip constitutive model is used for the joints. The calculating model is 50 metres long and 50 metres wide. The tunnel is 50 metres long as well. The tunnel model is depicted in Figure 7. The joint is shown as a fault by default.

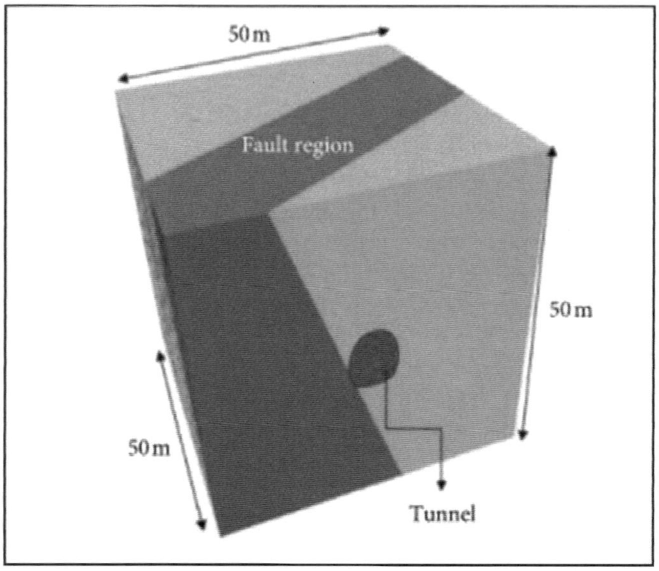

Figure 7. Numerical simulation of tunnel model.

4.2. Mechanism Analysis of Tunnel Collapse in Fault and Highly Jointed Zone

The joint/fault has an 18-20 metre range of effect on the tunnel ahead. During this procedure, the tunnel penetrates the fault zone for the first time. The surrounding rock has been disturbed as the tunnel has been plotted in both directions, left and right. The largest displacement, as seen, is 50 cm. The rock around the fault fracture zone is shattered, and its mechanical characteristics are not as good as those found elsewhere. During the excavation phase, the fault fracture zone exhibits more vault settlement, arch bottom uplift, and horizontal convergence of the tunnel body than the rest of the zone (Figure 8 a,b,c).

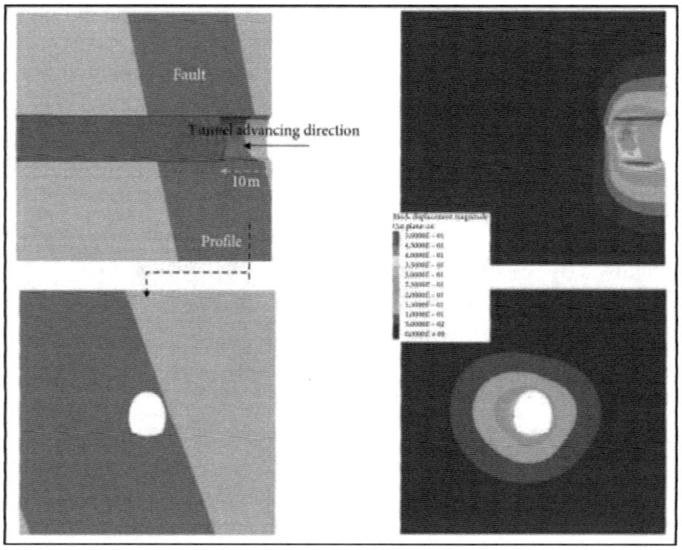

Figure 8 (a). The range of tunnel entering fault is 18-20 m.

The tunnel gradually penetrates the fault area during this process, and the fault regions surrounding the tunnel eventually converge. The surrounding rock of the tunnel has been displaced significantly in the fault area, and the deformed area on left or right sides gradually shifts from the considerable high stress zone deformation to their basic symmetrical deformation. There is the most movement between 21 and 30m at the point where the second and third phases of the left elliptical tunnel intersect (Figure 8b).

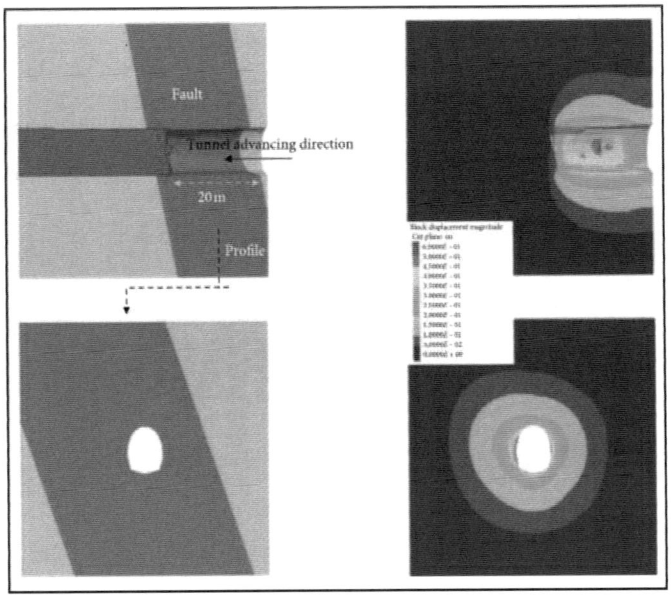

Figure 8 (b). The range of tunnel entering fault is 21-30 m in distance.

From where the fault is at the beginning of the tunnel, it moves more than 30 m forward. The tunnel gradually moves away from the fault line during this process, with the left side of the tunnel moving quicker than the right. The rock on both the left and right sides of the tunnel has shifted significantly in the fault area, and the deformation on both sides gradually shifts from the massive deformation on the left side to the basic symmetrical deformation on both sides. The most movement occurs at the junction of steps 2 and 3 of the left elliptical tunnel (Figure 8 c).

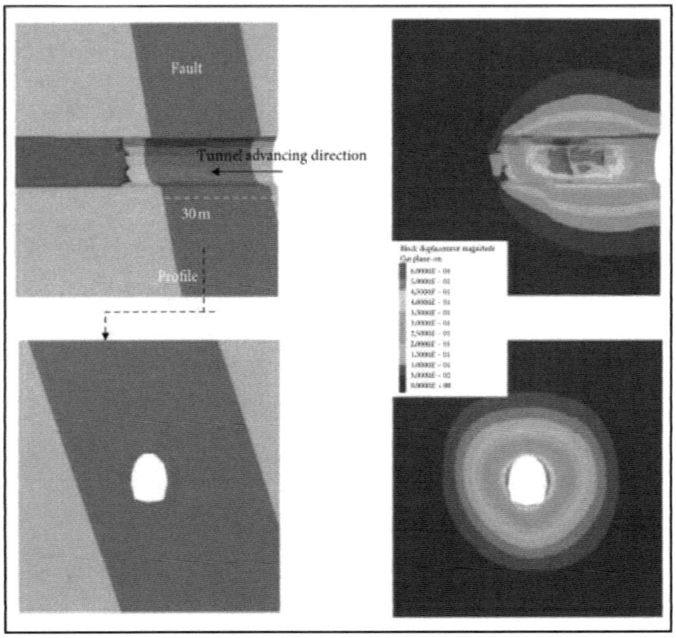

Figure 8 (c). The range of tunnel entering fault is at more than 30 m in distance.

5. Conclusion

Using field observation, theoretical analysis, numerical modelling, and other methodologies, this study investigates tunnel deformation during fault fracture zone crossing and the advance reinforcement effect under varied rock support conditions.

The following are the primary conclusions:

1. The joints present in meta-sedimentary rock influence the tunnel design in NATM-1. In the tunneling phase, it has caused considerable difficulties, which has slower the excavation pace in compared to the targeted performance about 20-30% recorded by KETTHA report in 2000. Several procedures and attempts have been made to deal with the current situation of excessive joints, especially by geologists and tunnel engineers, in order to complete the tunnel construction without incident or casualties. Weathering and the presence of a river in an area cause meta- sedimentary rock to create joints. According to the findings of this study, these two reasons have resulted in a shift in joint pattern. Overburden elevations appear to alter joint pattern as well, however there are no definitive relationship between joint pattern and overburden. As a result, more research is required to prove a solid relation between these two reasons. There are several points in the tunneling phase that have the potential for failure that all workers should be aware of in order to prevent incidents that can raise production costs, take longer, and result in personnel casualties.
2. The strength of surrounding rock in tunnel fault fracture zone is low, and the tunnel excavation leads to stress concentration, which aggravates the instability and collapse of the tunnel. Due to the change of rock dip angle and the position of soft and hard rock discontinuity, the deformation is obviously asymmetric.
3. Because advance support, advance bolt, and no grouting advance small pipe are unable to control rock deformation, advance grouting small pipe must be utilised to strengthen rock.
4. As the grouting reinforcement zone thickens, fault fracture zone rock deformation decreases considerably. A 2-meter grouting radius limits economic and construction distortion.

References

[1] Engelder T and Lash G 2009 Systematic joints in Devonian black shale: A target for horizontal drilling in the Appalachian Basin.
[2] EPU Ninth Malaysia Palan 2006-2010
[3] Girmscheid G and Schexnayder C 2002 Drill and Blast Tunneling practices. Practice Periodical on Structural Design and Construction vol **7**(3), pp 125-133.
[4] Hamblin W K and Christiansen E H 2004 Earth's Dynamic Systems, New Jersey: Prentice Hall, Pearson Education pp 166-187.
[5] Hoek E 2001 Big Tunnel in Bad Rock *Journal of Geotechnical and Geoenvironmental Engineering* The Thirty-Sixth Karl Terzaghi Lecture pp 726-740
[6] Hutchison C S and Tan D N K 2009 Geology of Peninsular Malaysia. Kuala Lumpur: The University of Malaya and the Geological Society of Malaysia
[7] Karakus M and Fowell R J 2004 An Insight into the New Austrian Tunneling Method (NATM) ROCKMEC'2004-VIIth Regional Rock Mechanics Symposium, Sivas, Turkey
[8] KeTTHA ('Kementerian Tenaga, Teknologi Hijau dan Air') 2000 Site Investigation Report.
[9] Madon M Abu Bakar Z A and Ismail H H 2010 Jurassic-Cretaceous Fluvial Channel and Floodplain Deposits along the Karak-Kuantan Highway, Central Pahang (Peninsular Malaysia) *Geological Society of Malaysia* Bulletin vol **56** pp 9-14.
[10] Metcafle I 2000 The Bentong-Raub Suture Zone *Journal of Asian Earth Science* vol **18** pp 691-721.
[11] Mughieda O S 2012 Stress Analysis of Non-persistent Rock Joints under Biaxial Loading *World Academy of Science, Engineering and Technology* vol **62** pp 834-839.
[12] Olsen J K 1993 Joint Pattern Development: Effect of Subcritical Crack Growth and Mechanical Crack Interaction *Journal of Geophysical Research* vol **98** (B7) pp 12,251-12,265.
[13] Padget P 2004 Meta-Sedimentary Rocks, Associated Intrusions and Tectonic Features of the Precambrian in Eastern Bamble, South Norway: An Interpretative Study. *Geological Survey of Norway Bulletin* vol **422** pp 39-51.
[14] Rabcewicz L 1964 The New Austrian Tunneling Method Waterpower part One pp 453-473 pp 511-515
[15] Rabcewicz L and Golser J 1973 Principles of Dimensioning the Supporting System for the New Austrian Tunneling Method Waterpower pp 88-93.
[16] Regnier J L Ring U Passchier C W Gessner K and Gungor T 2003 Contrasting Metamorphic Evolution of Meta-Sedimentary Rocks from the Cine and Selimiye Nappes in the Anatolide Belt, Western Turkey *Journal of Metamorphic Geology* vol **21** pp 699-721.
[17] Romana M (ITA-AITES World Tunnel Congress) 2009 Could the New Austrian Tunneling Method be Used Safely in Cities? Budapest, Hungary
[18] Samata S K 2001 Development of Micro-Scale Joints in Volcanic Rocks under Thermal Stress *Proc. Indian Acad. Sci. (earth planet. Sci.)* vol **101**(3) pp 191-203
[19] Sapari N Tipol F H Rahamat Noor N f and Mohamed Zaid S N 2011 Joint Pattern in Granite and Its Relationship with Its Slope Failure: Bukit Lanjan Rockslide Revisited.
[20] Singh B and Goel R k 1999 Rock Mass Classification: A practical Approach in Civil Engineering Amsterdam: Elsevier pp 267.
[21] Tan K W and Mazlin M 2009 An Appropriate Intuitional Framework Towards Integrated Water Resources Management in Pahang River Basin, Malaysia.
[22] Terzaghi k 1946 "Rock defects and loads in tunnel supports" Rock tunneling with steel supports, Youngstown, Ohio: R V. Proctor and T L White, eds Commercial Shearing and Stamping CO.
[23] Hayashi M 1996 Strength and dilatancy of brittle jointed mass—the extreme value stochastics and anisotropic failure mechanism. *Proc 1st Int Cong Rock Mech,* Lisbon vol **1** pp 295– 302.
[24] Brown E T 1970 Strength of models of rock with intermittent joints *J Soil Mech Found Div Am Soc Civ Eng*; 96:1935–49.

[25] Ladanyi B Archambault G 1972 Evolution de la resistance an cisaillement d'um massif rocheux fragmente. *In: Proceedings of the 24th international geological congress, Montreal* p 249–60.
[26] Einstein H H Hirschfeld R C 1973 Model studies in mechanics of jointed rocks J Soil Mech Found Div Am Soc Civ Eng; 90:229–48.
[27] Lama R D Vutukuri V S 1978 Handbook on mechanical properties of rocks *Clausthal Trans Tech*
[28] Reik G Zacas M. 1978 Strength and deformation characteristics of jointed media in true triaxial compression. *Int J Rock Mech Min Sci Geomech Abstr*; 15:295–303.
[29] Chenevert M E Gatline C 1965 Mechanical anisotropies of laminated sedimentary rocks Soc Pet Eng J:67–77.
[30] Attewell P B Sandford M R 1974 Intrinsic shear strength of brittle anisotropic rock-I: experimental and mechanical interpretation. *Int J Rock Mech Min Sci*;11: 423–30.
[31] Brown E T Richard L R Barr M V 1977 Shear strength characteristics of Delabole slate *Proceedings of the conference on rock engineering, Newcastle* p 31–51.
[32] Deklotz E J Brown J W Stemler O A 1966 Anisotropy of schistose gneiss *Proceedings of the 1st international congress on rock mechanics* vol **I** Lisbon; p 465–70.
[33] Akai K Yammamoto K Ariola M 1970 Experimentele forschung Uber anisotro-pische eigenschaften von kristallinen schierfern *Proceedings of the 2nd international congress on rock mechanics, Belgrade* p 181–6.
[34] McCabe W M Koerner R M 1996 High pressure shear strength of and anisotropic mica schist rock *Int J Rock Mech Min Sci;* 12:219–28.
[35] Nasseri M H Seshagiri Rao K Ramamurthy T 1996 Engineering geological and geotechnical responses of schistose rocks from dam project areas *Ind J Eng Geol;* 44:183–201.
[36] Nasseri M H Rao K S Ramamurthy T 1997 Failure mechanism in schistose rocks *Int J Rock Mech Min Sci* vol **34**(3–4):219.
[37] Singh V K Singh D Singh T N 2001 Prediction of strength properties of some schistose rocks from petro graphic properties using artificial neural networks Int J Rock Mech Min Sci vol **38**(2) pp 269–84.
[38] Ramamurthy T Rao G V Singh J A 1988 strength criterion for anisotropic rocks Proceedings of the 5th Australia–New Zealand conference on geomechanics, Sydney p 253–7.
[39] Rao K S Rao G V Ramamurthy T A 1986 strength criterion for anisotropic rocks *Ind Geotech J* vol **16**(4):317–33.
[40] Pomeroy C D Hobbs D W Mahmoud A 1971 The effect of weakness plane orientation on the fracture of Barnsley hard coal by triaxial compression. *Int J Rock Mech Min Sci*; 8:227–38.
[41] Allirote D Boehler J P 1978 Evaluation of mechanical properties of a stratified rock under confiningpressure *Proceedings of the 4th international congress on rock mechanics* vol **1** Montreux p 15–22.

Acknowledgments

This work was supported by the Fundamental Research Grant Scheme (FRGS) awarded by the Ministry of Education of Malaysia entitle 'Enhanced Coalbed Methane Recovery Processes by Coupling Fluid Flow of Mukah-Balingian Coal Deposits' with the referral no. R.J130000.7851.5F371.

This work was supported by the Universiti Teknologi Malaysia Encouragement Research Grant (UTMER) awarded by Universiti Teknologi Malaysia entitle 'Diagnosis of Alkali-Aggregate Reaction – Polarizing Microscopy and SEMEDS Analysis' with the referral no. Q.J130000.3851.19J79.

Preliminary Study of Landslide Hazard in Kutai Kartanegara Regency, East Kalimantan using Digital Elevation Model

S M Polawan[1*] and N A Raharjanti[1].

[1]Undergraduate Program of Geological Engineering, Faculty of Science and Engineering, Universitas Muhammadiyah Kalimantan Timur, Samarinda, Indonesia.
*Corresponding author's e-mail : ssm340@umkt.ac.id

Abstract. The Kalimantan Island is part of the Sundaland crust, namely the Eurasian Continental Plate. The plate is moving to the southeast colliding with the Indo-Australian plate which is moving north. Whereas for Kutai Kartanegara Regency in the Kalimantan Island position is far from the collision zone, so it is relatively stable tectonically. This is important to research that is, due to tectonic processes that occurred earlier, resulting in the formation of geological structures, especially faults. The purpose of the study was to determine the morphotectonic and landslide hazards in the Kutai Kartanegara Regency, where this research was carried out quantitatively with data collection techniques, then analyse landslide hazards based on data; DEM (Digital Elevation Model) including the slope, slope direction, and slope length for vulnerability analysis, geological data from Regional Geological Maps, which include rock formations, distances from faults data and administrative boundary spatial data in the form of vector GIS for the preparation of landslide hazard maps. The result showing Kutai Kartanegara predominantly categorized as moderate to high hazard zone. The low hazard zone covering 9.88% area, moderate hazard zone covering 35.81% area, and 54.31% area is high hazard zone of landslide. The analysis showing that the hazard of the research area consist predominantly sedimentary rocks and controlled by structural geology identified as thrust fault, strike slip fault, and fold which include in Anticlinorium Samarinda. Those lithology and geological structure features along with the slope are identified as controlling factor to the landslide hazard in the research area.

1. Introduction

Indonesia is a country with multiple disaster threats, but Indonesia is also a place for disaster research or (disaster laboratory) so that Indonesia has a very large disaster potential, because it's traversed by a row ring of fire. The background in this study is a need for an analysis related to spatial geomorphological aspects as an early stage in disaster mitigation, identifying the potential and vulnerability disasters caused at the research focus location, namely Kutai Kartanegara regency.

Kutai Kartanegara regency based on geographic and administrative is included in East Kalimantan Province. Kutai Kartanegara Regency is located at 115°26'28" East Longitude – 117°36'43" East Longitude and 1°28'21" North Latitude–1°08'06" South Latitude. Kutai Kartanegara is surrounded by Malinau Regency in the north, Kutai Timur Regency and Makassar Strait in the east, North Penajam Pasir and Balikpapan City in the south, and North Kutai Regency in the West. The land area of the Kutai Kartanegara Regency is around 27.263.10km^2 or 2.726.310 Ha around 12% from the total area of East Kalimantan Province, meanwhile the water area is 4.097 km [1].

Kutai Kartanegara Regency was divided into 18 districts and 237 villages. The districts in Kutai Kartanegara Regency are Kecamatan; Samboja, Muara Jawa, Sanga-Sanga, Anggana, Muara Badak, Marang Kayu, Tenggarong Seberang, Loa Janan, Loa Kulu, Tenggarong, Sebulu, Muara Kaman, Kota Bangun, Muara Muntai, Muara Wis, Kenohan, Kembang Janggut, dan Tabang. Kutai Kartanegara Regency has many rivers that spread in all district and become the main transportation facilities along with land transportation. The biggest river is Mahakam River about 920 km long.

The disaster potential that can happen ini Kutai Kartanegara Regency can be divided into geological disaster including landslide, and hydrometeorology disaster including, flood, drought, forest fires, and extreme weather [2]. From the description above, the problem formulation that becomes the limitation of the research is disaster geomorphology one of the approaches in disaster studies. Disaster geomorphology examines aspects of landforms, processes, and results of physical processes that have the potential and can cause disasters. Sustainable development requires spatial data and information about the potential for landslides and information dissemination to the public about disasters.

2. Regional Geology

Kalimantan is a part of Sundaland and a part of Eurasia [3]. Eurasian plate moves to the southeast and collide with Indo-Australian plate which moves to the north. The direction of Sulawesi movement reversed causing the direction from north – south become northwest – southeast and now have the direction west – east. The subduction zone located in western Sumatra, southern Java extends to Bali and Nusa Tenggara. The Position of Kalimantan is far from these subduction zone, therefore Kalimantan relatively stable in tectonic, but due to the tectonic activity which happened beforehand faults were formed in the Island [4].

Physiographic condition of Kutai Kartanegara Regency varies and can be categorized into ten (10) physiographic units namely: (1) Beach Sand Sediment, (2) Tidal Swamp, (3) Alluvial Plain, (4) Meander Belt, (5) Swamp, (6) Alluvial Valley, (7) Terrain, (8) Plain, (9) Hill, and (10) Mountain. Based on [5, 6] physiographic classification Kalimantan divided into five zones, namely Kutai Basin Zone, Kuching Highland Zone, Schwanner Block, Pasir Basin Zone, and Pathenosphere Block (Figure 1).

The physiographic characteristic of East Kalimantan is part of Tertiary Kutai Basin located in southern of Kuching Highland. The Basin was divided from Barito Basin by tectonic called paternoster cross high. The stratigraphic characteristic of Kutai Basin consists of sediments that deposited from early Tertiary and filling the basin to the east [7]. The earlier sediment deposited in the west side of the basin about 1000 – 2000 m in depth, than moved toward east continuously and increase in depth (Figure 2).

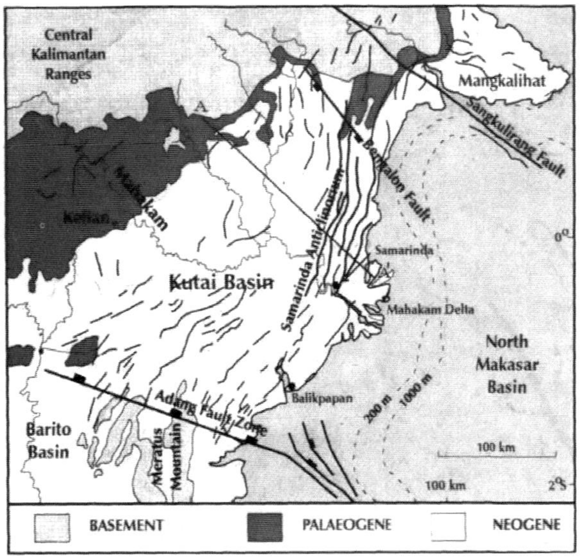

Figure 1. Regional geology of the Kutai Basin in [7].

3. Methodology

The aim of the research is to identify morphology-tectonic and the hazard of landslide in Kutai Kartanegara Regency. The data was collected from study literature and analysed with quantitative method to identify the zone of the landslide hazard. Through quantitative predictions using statistical combination factors (Probabilistic) or means predicting events that have occurred in the past in an area that has not experienced an event but has similar conditions to an area that has experienced it before, a probabilistic method is used as the approach formula. The location of the research was Kutai Kartanegara Regency and research was done in November 2020 to February 2021. The process of the research that done presented in Figure 2 explains the sequence of the hazard index analysis process from two data, namely the landslide potential zone (slope >15% or 8.51° and the runout potential zone (angle threshold min. 8.51°) for result area delineation and class hazard and then hazard index. The final result of the landslide hazard in the research area then presented as landslide hazard map to identify the area distribution of the hazard.

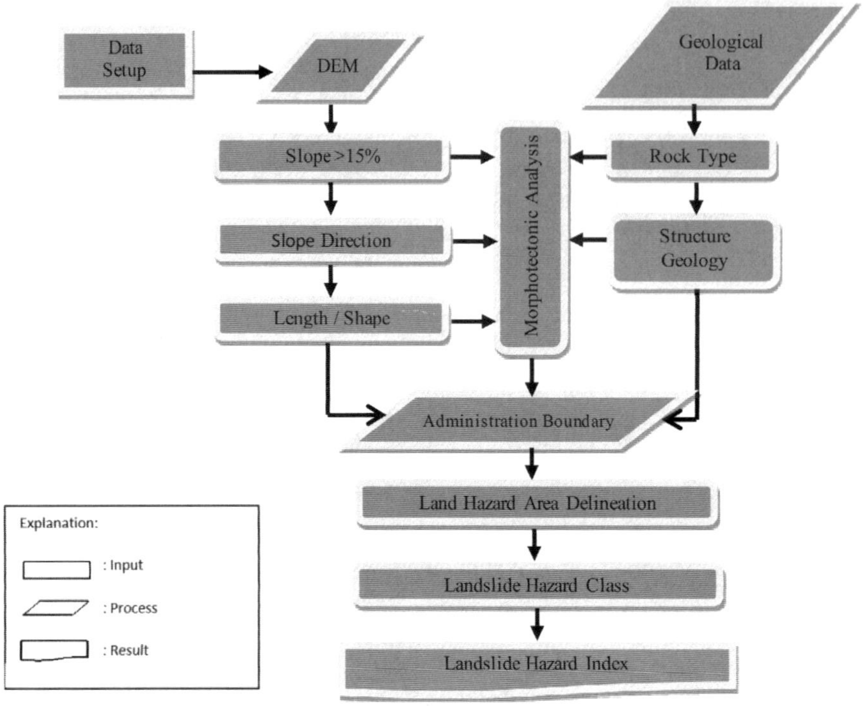

Figure 2. The process design to produce landslide hazard map based on DEM and Geological Data.

3.1. *Quantitative data collection*

The data collected in this research categorized as primary and secondary data. The secondary data used in the research to analyse the landslide hazard are (1) DEM to identify the slope, the direction of the slope, and the dimension of the slope to analyze the vulnerability to the landslide. The data was collected from the official website DEMNAS [8]. (2) Geology Data taken from Regional Geology Map [9] including rock formation and the structure geology, and (3) spatial data of the administrative border to delineate the area of the research. The primary data are strike dip and lithology from regional data and confirmed by field observation of landslide affected area.

3.2. *Hazard analysis*

Hazard analysis method is done by identifying the areas that have potential in slope failure, calculating the probability of the event, and interpretation of the landslide that happen. All of these process was analyze using software ArcGIS Dekstop-ArcMap.

The analyses of the data that done are:

- Morphotectonic analyisis including historical geology,
- Calculation of the parameters, including determining the score and weight of the data that already collected. The classification, score determination, and weighting using parameter for producing landslide hazard map with deterministic method based on Technical Module to Arrange the Analysis of the Landslide Disaster Risk (*BNPB*) [10].
- Delineating the border of morphology unit and administrative border to identify the potential area of landslide.
- Calculate the landslide index. The formula that used based on *BNPB* [10]:

$$\text{Landslide hazard zone} = (\text{weight} \times \text{score})\ \text{landslide potential zone} + (\text{weight} \times \text{score})\ \text{runout potential zone} \quad (1)$$

The values of the indexes from formula (1) then divided in to three: (1) the lowest hazard have the value of the index ≤ 0.333, (2) the moderate index have value $0.333 < H \leq 0.666$ and (3) the high index of hazard have value $H > 0.666$ [10].

4. Results

The geomorphological analysis divided the Kutai Kartanegara regency into four category of the slope from the flattest to steep respectively: <15% (gentle slope), 15 – 30% (moderate steep slope), 30 – 50% (steep slope), and 50 – 70% (extremely steep slope). The geomorphological map (Figure 3) showing the research area predominantly categorized in moderately steep slope area that occurs in the northwest and southeast part. Relatively more plain area which categorized as gentle slope area concentrated in the middle of the Kutai Kartanegara Regency.

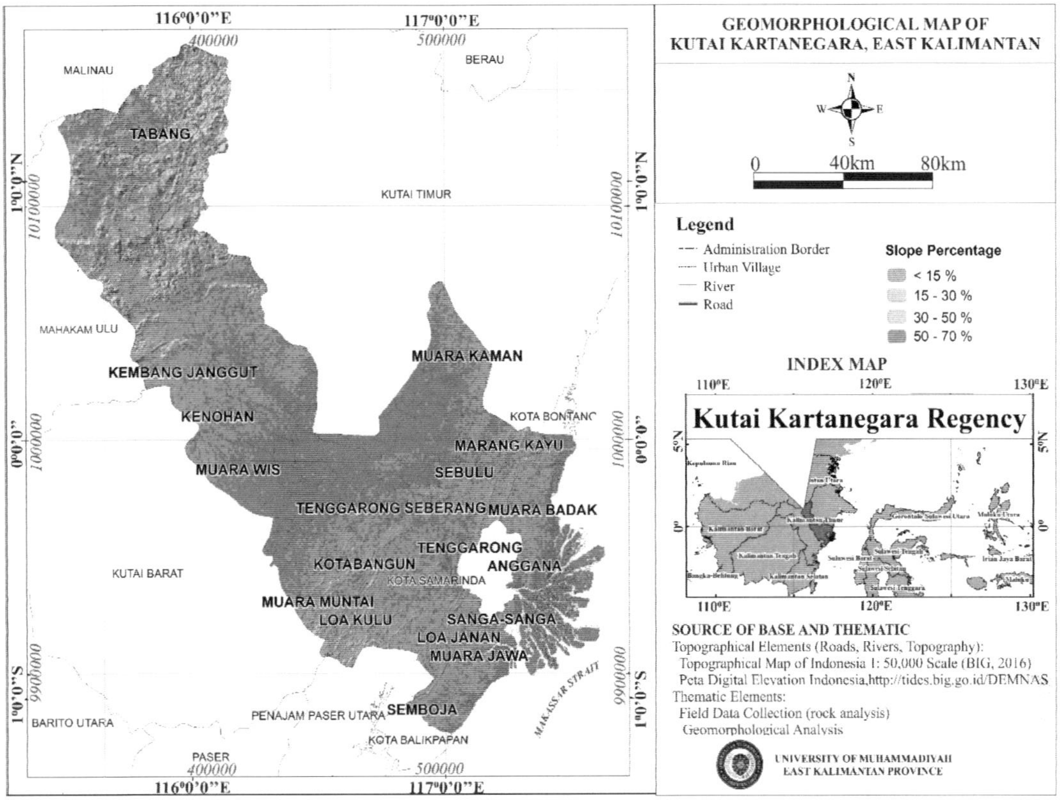

Figure 3. The geomorphological map in Kutai Kartanegara Regency

The landslide hazard analysis conducted combining the geomorphological (DEM) and geological features. Based on the analysis that was conducted the landslide in the Kutai Kartanegara Regency divided into 3 categories namely low, moderate, and high (Table 1). The area that categories as low hazard class identified only Anggana districts which the area of this low hazard zone is ±76.469Ha. The moderate hazard zones occupied in most of the districts including Kembang Janggut, Kenohan, Kotabangun, Loa Janan, Marang Kayu, Muara Badak, Muara Jawa, Muara Kaman, Muara Muntai, Muara Wis, Sanga-Sanga, Semboja, and Kota Tenggarong with total area ± 901.841Ha. There are only four districts categorized as high landslide hazard class, which are Loa Kulu, Sebulu, Tabang, and Tenggarong Seberang with the total area is ± 932.696Ha. Even though there are only four districts categorized as high landslide hazard, these class also occupied the other hazard class area. The result of the calculation showing the high landslide hazard class in the Kartanegara Regency have the largest distribution with the total area is ±1,367.936Ha.

IOP Conf. Series: Earth and Environmental Science 1071 (2022) 012007 doi:10.1088/1755-1315/1071/1/012007

Table 1. The result of landslide hazard base on districts in Kutai Kartanegara.

No	District	Area of Each Hazard Class (Ha)			Hazard Class
		Low	Moderate	High	
1	Anggana	76,469	23,373	6,121	Low
2	Kembang Janggut	6,367	79,582	102,111	Moderate
3	Kenohan	4,450	82,231	35,195	Moderate
4	Kotabangun	50	46,245	44,405	Moderate
5	Loa Janan		34,219	28,652	Moderate
6	Loa Kulu		45,004	124,448	High
7	Marang Kayu	1,564	66,989	44,415	Moderate
8	Muara Badak	17,479	28,989	14,840	Moderate
9	Muara Jawa	18,480	18,689	8,232	Moderate
10	Muara Kaman	63,503	225,175	45,272	Moderate
11	Muara Muntai	22,428	38,382	24,890	Moderate
12	Muara Wis	31,027	77,078	40,633	Moderate
13	Sanga-sanga	1,395	8,511	2,082	Moderate
14	Sebulu		22,778	44,320	High
15	Semboja	5,576	41,824	23,686	Moderate
16	Tabang		35,963	726,832	High
17	Tenggarong		12,416	14,671	Moderate
18	Tenggarong Seberang		14,406	37,096	High
	Kutai Kartanegara	**248,787**	**901,841**	**1,367,936**	**Moderate**

The comparison between the slope class and hazard class is made in the Table 2. From the comparison, the area that has more plain slope (<15%) or gentle sloping the Kutai Kartanegara Regency also has lowest hazard class which identified in Anggana sub-district. The area with moderately steep and steep slope also associated with moderate hazard occupied most of the sub-district in the research area. The extremely steep slope (50 – 70%) associated with high hazard class which has distribution in Loa Kulu, Sebulu, Tabang, and Tenggarong Seberang.

The geological analysis was done by collecting regional data and confirmed with field observation. The geological map in the Figure 4 showing Kutai Kartanegara Regency consists of three types lithology namely volcanic rocks, sedimentary rocks, and alluvial deposit. The sedimentary rocks have the largest distribution which spread throughout the research area. The volcanic rocks occur concentrated in the northwest research area. The youngest lithology unit is alluvium deposit concentrated in the middle of the research area and delta at the southeasternmost of Kutai Kartanegara Regency.

Table 2. The result slope class and hazard class.

Slope Class		Hazard Class	Sub-district
<15%	Gentle Sloping	Low	Anggana
15-30%	Moderately Steep	Moderate	Kembang Janggut, Kenohan, Kotabangun, Loa Janan, Marang Kayu, Muara Badak, Muara Jawa, Muara Kaman, Muara Muntai,
30-50%	Steep		Muara Wis, Sanga-Sanga, Semboja, Kota Tenggarong.
50-70%	extremely Steep	High	Loa Kulu, Sebulu, Tabang, Tenggarong Seberang

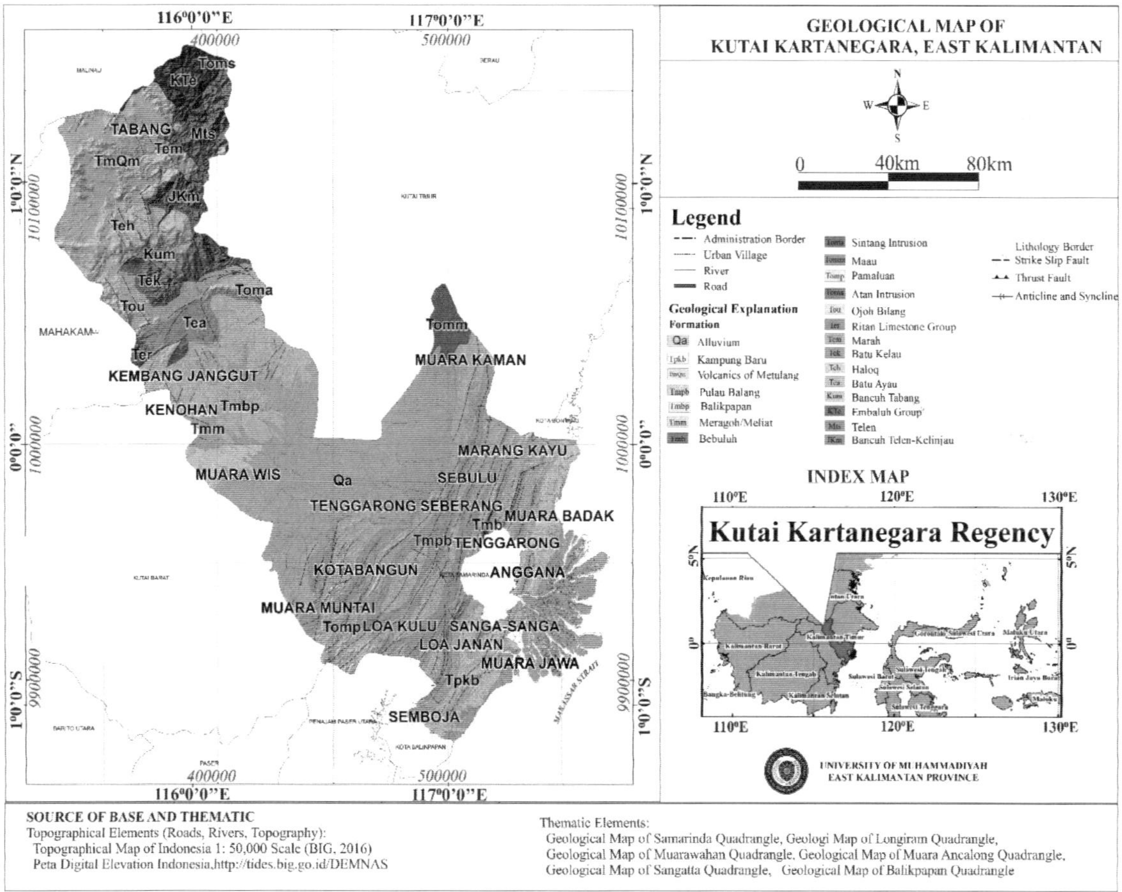

Figure 4. The geological map of Kutai Kartanegara Regency.

IOP Conf. Series: Earth and Environmental Science 1071 (2022) 012007 doi:10.1088/1755-1315/1071/1/012007

The analysis of geological factors of the landslide including lithological characteristics and geological structure is made. The resumed analysis in the Table 3 showing the landslide hazard classification calculated from the lithology, the distance from the fault as the discontinued plane, and the length or the dimension of the slope.

Table 3. Data analysis of the distance of faults and length of slope.

Districts	Landslide Hazard Classification	Description Area				
		Lithology	Fault Distance (m)	Slope Direction	Slope Dimension (length) (m)	Total Hazard Area (Ha)
Anggana	Low - High	Alluvial (F: Qa) Sediment (F: Tpkb, Tmbp, Tomp)	> 400	Plain- Southeast	< 200 - 500	105.963
Kembang Janggut	Low - High	Alluvial (F: Qa) Volcanic (F: Tea, Tmm) Sediment (F: Ter, Tmbp)	> 400	Plain - Northeast	< 200 - 1000	188.060
Kenohan	Low - High	Alluvial (F: Qa) Sediment (F: Tmbp, Tpkb) Volcanic (F: Tmm)	> 400	Plain - Southwest	< 200 - 1000	121.876
Kota Bangun	Low - High	Alluvial (F: Qa) Sedimen (F: Tmpb, Tomp)	> 400	Plain - Northwest	< 200 - 1000	90.700
Loa Janan	Moderate - High	Sediment (F: Tmbp, Tmpb, Tomp, Tpkb)	> 400	Plain - Northwest	< 200 - 500	62.871
Loa Kulu	Moderate - High	Sediment (F: Tmbp, Tmpb, Tomp)	> 400	Strike slip - Southeast	< 200 - 1000	169452
Marang Kayu	Moderate - High	Alluvial (F: Qa) Sediment (F: Tmb, Tmpb, Tomp, Tpkb)	> 400	Strike slip - North-west	< 200 - 1000	113.004
Muara Badak	Low - High	Alluvial (F: Qa) Sediment (F: Tmbp, Tmpb, Tomp, Tpkb)	> 400	East - North	< 200 - 1000	61.294
Muara Jawa	Low - High	Alluvial (F: Qa) Sediment (F: Tmbp, Tmpb, Tomp, Tpkb)	> 400	East - North	< 200	45.401
Muara Kaman	Moderate - High	Alluvial (F: Qa) Sediment (F: Tmpb, Tomp)	> 400	East - North	< 200 -> 1000	333.950
Muara Muntai	Low - High	Alluvial (F: Qa) Sediment (F: Tmpb, Tomp)	0 -> 400	East - North	< 200 -> 1000	85.700
Muara Wis	Low - High	Alluvial (F: Qa) Sediment (F: Tmbp, Tmpb, Tpkb, Tomp) Volcanic (F: Tmm)	0 -> 400	East - North	< 200 -> 100	148.737
Sanga-Sanga	Low - High	Alluvial (F: Qa) Sediment (F: Tmbp, Tmpb, Tpkb)	0 -> 400	Strike slip - North	< 200	11.988
Sebulu	Moderate - High	Alluvial (F: Qa) Sediment (F: Tmbp, Tmpb, Tomp)	> 400	Datar - Utara	200 -> 1000	67.097
Semboja	Moderate - High	Alluvial (F: Qa) Sediment (F: Tmbp, Tmpb, Tomp, Tpkb)	> 400	Strike slip - North	< 200 - 500	71.086
Tabang	Moderate - High	Alluvial (F: Qa) Sediment (F: Tmbp, Toma, Tou, Tek, Jkm, Kte, The, Tem, Ter, Kum, Mts, TmQm, Toms) Volcanic (F: Tea)	0 -> 400	Strike slip - North	< 200 -> 1000	762.795
Tenggarong	Moderate - High	Alluvial (F: Qa) Sediment (F: Tmbp, Tmpb, Tomp)	0 -> 400	Strike slip - North	< 200 - 1000	27.086
Tenggarong Seberang	Moderate - High	Alluvial (F: Qa) Sediment (F: Tmbp, Tmpb, Tomp)	0 -> 400	Strike slip - North	< 200 - 100	51.502

F = Formation

The analysis of the landslide hazard supported by the stream flow map (Figure 5) of the Kutai Kartanegara Regency showing that the research area predominantly controlled by rectangular river.

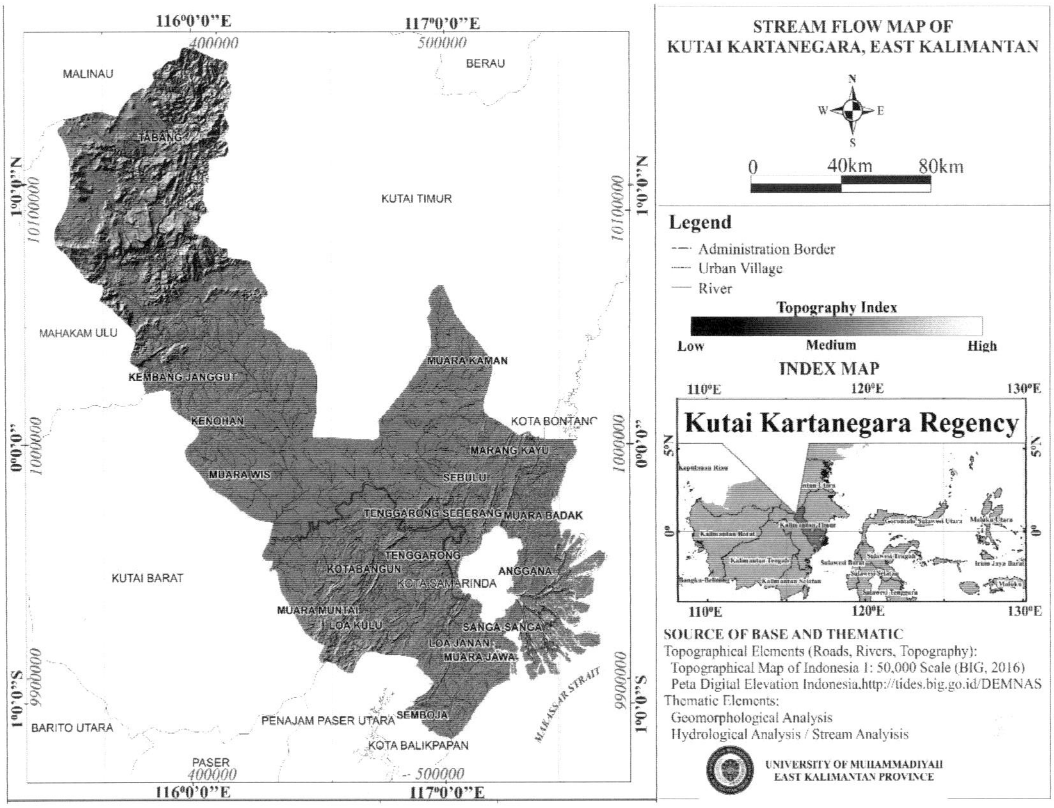

Figure 5. The stream flow map of the Kutai Kartanegara Regency.

The landslide hazard map is made to identify the distribution of the hazard in the research area. The landslide hazard index is analyzed from geomorphological and geological parameters which the steps explained in the methodology section. The landslide hazard map shown in Figure 6 the divided into three categories associated with the slope percentage that divided into four class namely:

- The green color, has slope class <15% (gentle slope) categorized as low hazard class only occurs in Anggana District.
- The yellow color, has slope class 15 – 30% (moderate steep slope) categorized as moderate hazard class, which occupied almost all the area in Kutai Kartanegara Regency.
- The orange color, has slope class about 30 – 50% (steep slope) and red color has slope class about 50 – 70% (extremely steep) categorized as high hazard index.

5. Discussion

Based on the result that shown on the geomorphological and geological map the geological structure has important role to the morphology of the Kutai Kartanegara Regency. The evidence of structural controls showed in the topographic lineament which has characteristic high topography, elongated, than cut suddenly. Another morphological feature that shows structural control is the rectangular type drainage pattern which have characteristic the smaller drainage make relatively perpendicular to the main drainage (Van Zuidam) [11] (Figure 5). Based on classification of slope (Van Zuidam) [11], Kutai Kartanegara covered with 35.41% of gently sloping area, 44.71% area is categorized as moderately steep area, 4.94% categorized as steep area, and 0.14% as extremely steep area. Kutai Kartanegara area is divided into 3 zones of landslide hazard based on BNPB classification [10] which are low hazard zone covering 9.88% area, moderate hazard zone covering 35.81% area, and 54.31% area is high hazard zone of landslide (Figure 6).

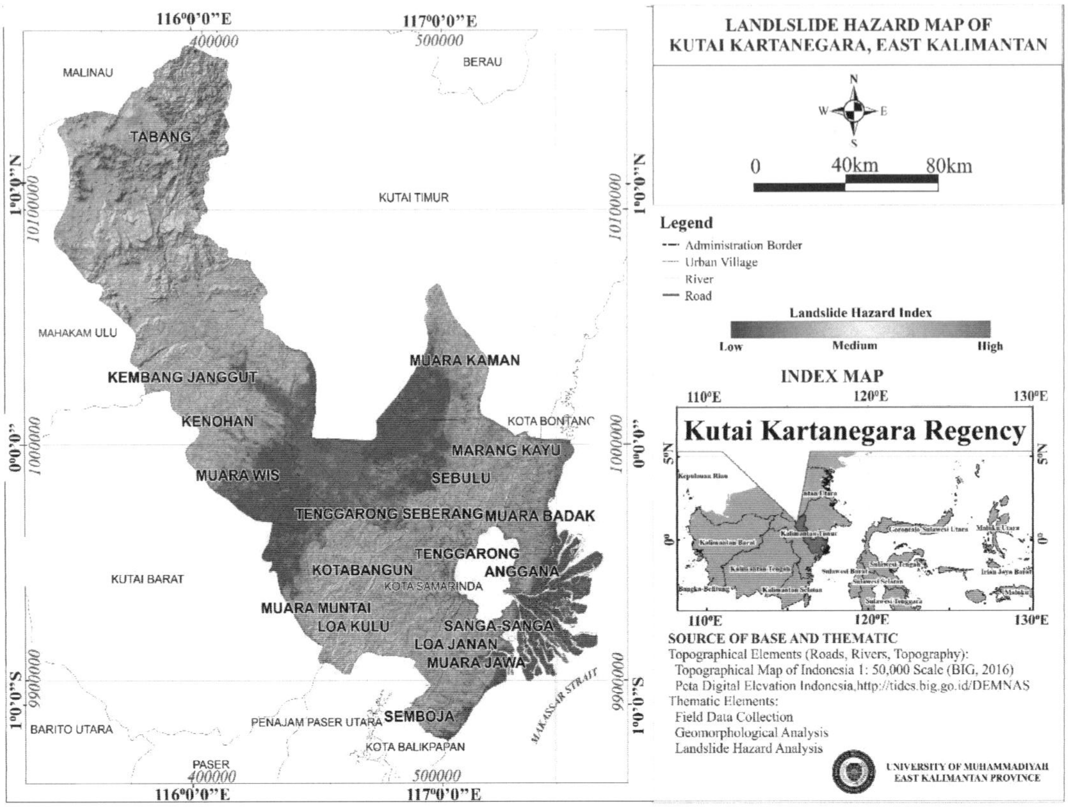

Figure 6. The landslide hazard map of the Kutai Kartanegara Regency.

From the result, it can be identified that the landslide in the research area is not just triggered by the slope, but also geological characteristics including lithology and structural geology. It corresponds to two factors that influence the landslide stated by Karnawati (2005) which are conditioning factors and triggering factors [12]. The conditioning factors are the components in the slope that can cause landslide or soil movement namely morphology, lithology type and stratigraphy, geological structure, geohydrology, and land use [12]. BNPB divided lithology into three types: alluvial, sedimentary rocks, and volcanic, and each lithology has different strength or resistivity so it will influence their resistance to other factors which are structural geology as internal factor and climate as external factor.

The area presentations of each lithological type in the Kutai Kartanegara Regency are alluvial occupied 25.29% area, sedimentary rocks occupied 60.04% area, and volcanic rocks occupied 14.61% area. From those data, it can be identified that the research area predominantly consists of sedimentary rock. The sedimentary rocks considered have low resistance to the pressure from geological structure which has destructive characteristic. The evidence showed in the geological map of the research area in the Figure 4 most of area of the Kutai Kartanegara effected by geological structures, only at the middle of Kutai Kartanegara the geological structure cannot be detected due to covered by alluvial deposit. The geological structures that developed in research area are thrust fault, strike slip fault, and fold which included in Anticlinorium Samarinda [9].

6. Conclusion & Recommendation

Kutai Kartanegara Regency is area that relatively prone to landslide. The analysis of landslide hazard, the research area divided into three landslide hazard class namely low, moderate, and high hazard class. The low hazard zone only identified in Anggana district. The moderate hazard class has the widest spread in the research area including in district Kembang Janggut, Kenohan, Kotabangun, Loa Janan,

Marang Kayu, Muara Badak, Muara Jawa, Muara Kaman, Muara Muntai, Muara Wis, Sanga-Sanga, Semboja, and Kota Tenggarong. The high hazard class has the largest area, but only four districts categorized in this class namely Loa Kulu, Sebulu, Tabang, and Tenggarong Seberang. The factors control the landslide hazard are the slope along with geological features such as geological structures and lithology. The area with steeper slope associated with higher hazard class and vice versa. Kutai Kartanegara dominated consists of sedimentary rocks which has relatively low resistance to the pressure from geological structure that destructive and make the landslide hazard class higher.

The mitigation of the potential landslide disaster that may happen in the research area is recommended to be done. The more data analysis such as vulnerability and capacity to calculate the landslide risk of the research area is suggested. The risk analysis will identify the potential loss that may happen due to landslide disaster. The risk management is one of methods in disaster mitigation.

References

[1] RPIJM 2018 *RPIJM Kabupaten Kutai Kartanegara*. Kutai Kartanegara.
[2] Syamsidar, S M and Alam F 2019 *Memahami Bencana Banjir dan Longsor* (Samarinda: RV Pustaka Horizon)
[3] Moss S, Chambers J, Cloke I R, Nas D S 1997 *New observations on the sedimentary and tectonic evolution of the Tertiary Kutai Basin, East Kalimantan*
[4] Satyana A H 1999 *Kalimantan An Outline of The Geology of Indonesia* (Indonesian Association of Geologist) p.69-89
[5] Van Bemmelen R W 1949 The Geology of Indonesia Vol IA (The Hague: Netherlands, Govt. Printing Office) p.732.
[6] Nuay 1985 Kerangka Tektonik Pulau Kalimantan: *Peta Geologi Lembar Kotabaru*
[7] Allen G P and Chambers J L 1998 Sedimentation In The Modern and Miocene Mahakam Delta (Indonesia: IPA)
[8] https://tanahair.indonesia.go.id/demnas/#/ accessed 19 April 2021
[9] Supriatna, S. Sukardi, R, and R. Rustandi. (1995). Peta Geologi Bersistem, Lembar Samarinda, Kalimantan sekala 1:250.000, Pusat Penelitian dan Pengembangan Geologi Bandung.
[10] Badan Nasional Penanggulangan Bencana 2019 *Petunjuk Teknis Kawasan Rawan Bencana Tanah Longsor* Jakarta
[11] Van Zuidam (1985). A*erial Photo – Interpretation in Terrain Analysis and Geomorphologic Mapping*, The Hague: Netherlands, Smith Publisher
[12] Karnawati, D. (2005). Bencana Alam Gerakan Massa Tanah di Indonesia dan Upaya Penanggulangannya, Jurusan Teknik Geologi Fakultas Teknik Universitas Gadjah Mada, Yogyakarta

Acknowledgments

Spirit of Muhammadiyah for science, gratefulness for UMKT Rector, research partner, and especially experts team on the Samarinda City disaster have provided input and case samples so that they can be applied in the Tenggarong city area, finally hopefully we can fight together again in more challenging research.

IOP Conf. Series: Earth and Environmental Science 1071 (2022) 012008 doi:10.1088/1755-1315/1071/1/012008

Flood assessment after Cempaka Cyclone 2019 in Imogiri District, Bantul Regency of Yogyakarta, Indonesia

Lornn Odm[1], **Egy Erzagian**[1,2], **and Wahyu Wilopo**[1,2*]

[1]Department of Geological Engineering, Faculty of Engineering, Universitas Gadjah Mada, Yogyakarta, Indonesia.
[2]Center for Disaster Mitigation and Technological Innovation (GAMA-InaTEK) Universitas Gadjah Mada, Yogyakarta, Indonesia
*Corresponding author's e-mail: wilopo_w@ugm.ac.id

Abstract. Imogiri District is one of the areas located at the foot of the southern mountains, with many rivers crossing the area. The Cempaka Cyclone in March 2019 caused a big flood in the Imogiri district and its surrounding area. Flood may affect various sectors of the socioeconomics life of a society, such as sectors of the economy, and agriculture, especially a severe effect on environmental and surface/groundwater systems in areas. The research objective is to assess flood hazards and their impact in the Imogiri District, Bantul Regency, Yogyakarta Special Region. The research was conducted by field survey and interview with an impacted community from the flood. The outcome of this study revealed that the area with high flood-prone potentials is distributed in the eastern and central part of the study area, along the river flood zone. The flood disaster caused property and agricultural losses but did not cause any casualties. The community was unprepared for the flood disaster, which had not occurred as big as in 2017.

1. Introduction

Flood damages are consistently rising, and flooding is a significant and worsening disaster for many societies. Climate change predictions forecast an increase in extreme events in the future. In Indonesia, floods take the second-largest share in disaster events after tornadoes. From 2015 to 2019, the National Disaster Management Agency (BNPB) recorded 3,388 floods throughout the country or an average of 678 events per year [1]. With 978 flood events, 2017 is marked as the most intense year. Flood commonly take place around but are not limited to rivers, coastal areas, or even plains [2-3]. Floods sometimes are the products of the interaction between social and environmental processes [4]. Natural factors which may cause floods to include heavy rainfall, high tide, and high flood intensity, on the other hand, human factors that may cause include improper land use, heavy deforestation, and blocked or damaged drainage channels [5]. The flood can cause massive losses, such as property damage and disruption of the infraction and economic activities [6]. Flood has many negative impacts on human health because floods sometimes pose a high risk of injury to the victims, environment, and plants. There are many methods to develop a flood hazard map by applying GIS software [7-9].

In March 2019, a flood occurred in Bantul Regency, Yogyakarta Special Region, Indonesia. Thirteen of 17 districts in Bantul Regency were flooded, with more than 4,400 people evacuated, two deaths, and many others destroyed, such as infrastructures, agriculture, and environments. Therefore, this study aims to develop a flood hazard map and assess the impact on the community.

2. Methodology

The research intends to find out the factors that cause flooding and flood hazard assessment in an area affected by the Cempaka Cyclone in 2019 in Imogiri District, Bantul Regency, as shown in Figure 2. The research was conducted by field survey and interviewing an impacted community from the flood disaster. The survey was conducted around one week after the flood occurrence; the key questions focus on the affected property, the elevation of water during the flood, the time of water rise, an area affected by the flood, the cause of the flood, and make select criteria. We used interviews and direct observation to collect the data in this study. The information related to the flood occurrence process, community knowledge, and the impact of floods on the community were asked to 25 people directly both the respondents from affected and non-affected an area by the flood. The total number of respondents is 25 people representing the impacted community from the flood disaster on March 17, 2019, in Imogiri District.

The flood peak discharge can be calculated based on the Nasional Standard of Indonesia (SNI) 2415:2016 [10], as follow:

$$Q_p = 0.00278 \times C \times I \times A \quad (1)$$

Where,
Qp is the peak flood discharge (m^3/s);
C is the runoff coefficient (0.25-0.40) in residential suburban
I is the intensity of rain during the time of concentration (mm/hour), and
A is the area of the flow area (Ha).

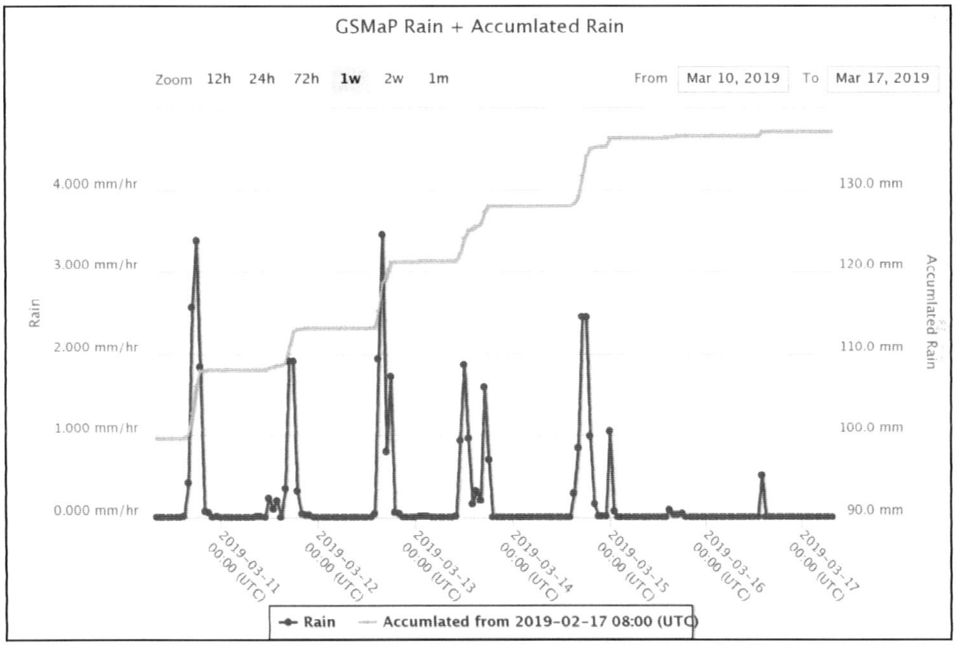

Figure 1. Rainfall intensity during the flood in an area based on the JAXA satellite.

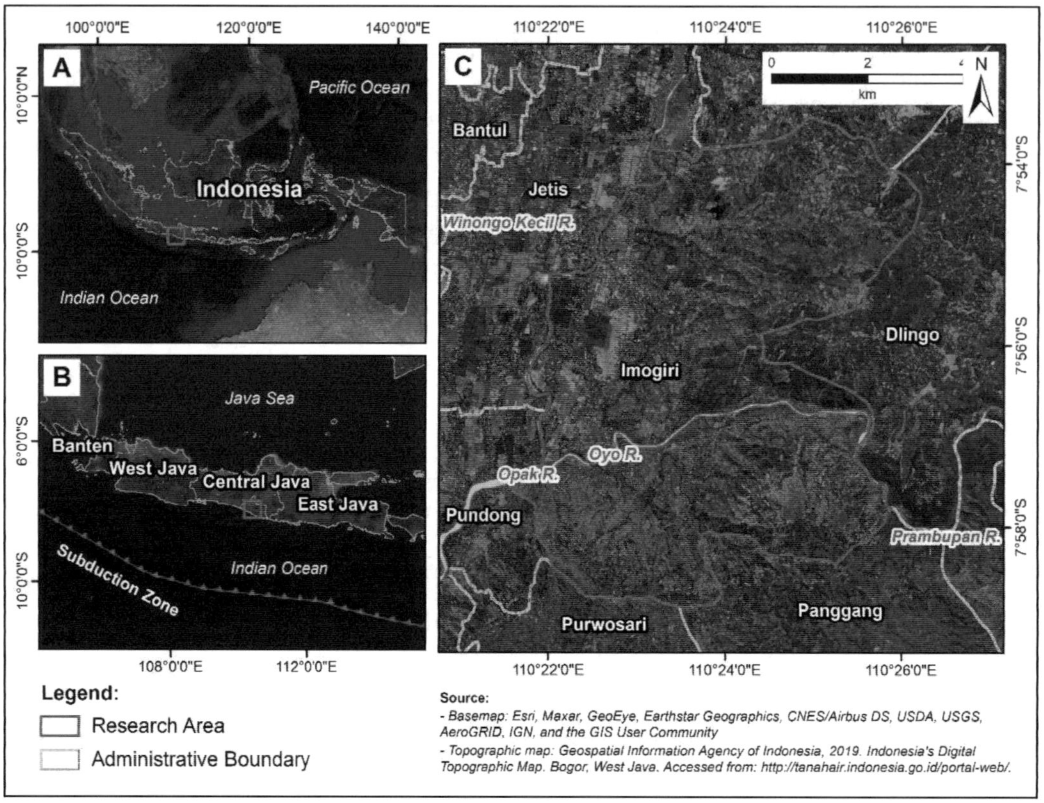

Figure 2. Research area in Imogiri district.

3. Result and Discussion

3.1. Flooding hazard zone

This study successfully produced a flood hazard map about flooding in the Imogiri district, Bantul Regency of Indonesia, where the flood hazard occurred on March 17, 2019. The information about the flooding zone area was collected by direct survey and interviews with local people. The results of this study can be conducted to analyze and estimate the flood hazard in the study area based on the data collected and carried out results. Figure 3 shows the flood zone distribution, the highest area in the western part of the Imogiri district (dark blue color). The distribution of the flooding zone was controlled by the river system, both main and tributary rivers. There are two big rivers in the research area, the Opak River and the Oyo River. In addition, it has tributary rivers, such Celeng river in the area.

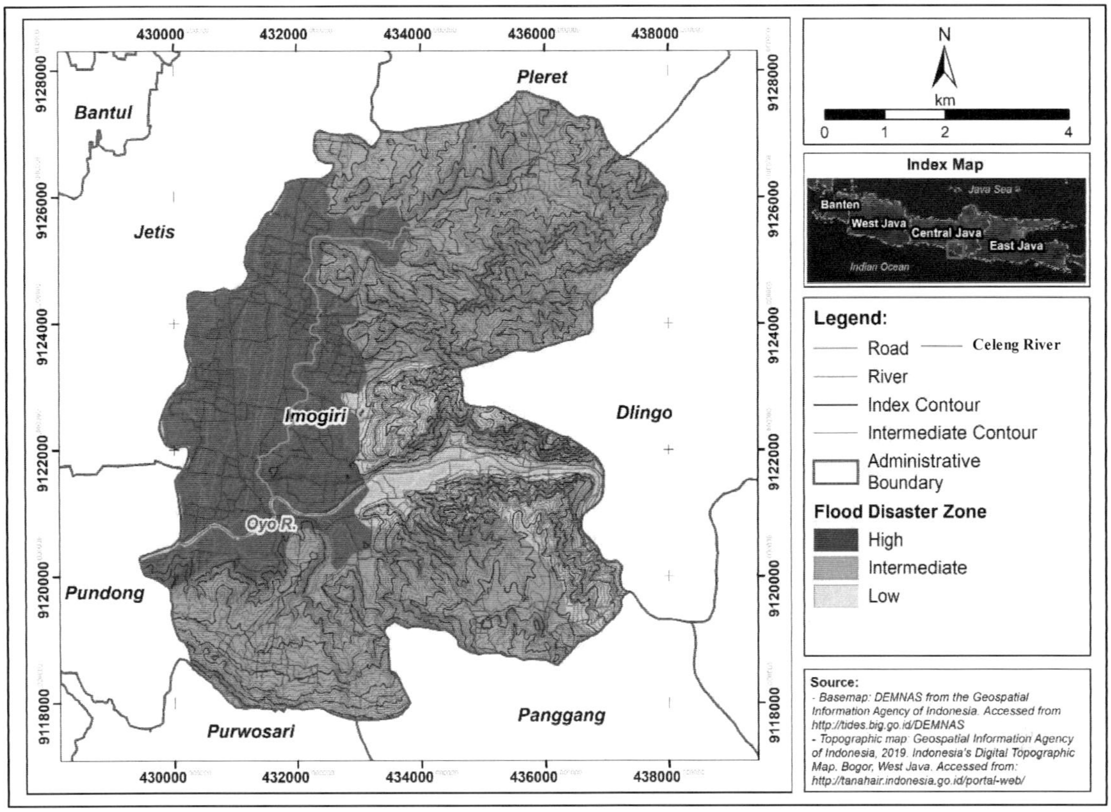

Figure 3. Flood hazard map in Imogiri district

3.2. *Cause of flood*

Generally, most flood is caused by heavy rainfall. It is also similar to the flood occurrence in the Imogiri Flood. According to the data from JAXA Satellite Japan [11], it can be identified that rainfall intensity before and during the day when the flood occurred is high. On March 17, 2019, the rainfall intensity reached 127 mm, while the previous day was 11.6 mm/day, 13.6 mm/day, and 22 mm/day.

Floods occur in the main river in the Opak and Oyo rivers and the Oyo river's tributaries. The tributary that experienced more severe flooding was the Celeng river in an area. This river divides the Imogiri district into western and eastern regions, flowing north to south. This tributary has a width of approximately 5 meters and a depth of 1,5 meters with a catchment area of 15.66 km^2, as shown in Figure 4. According to equation 1, the flood peak discharge reaches around 12.67 m^3/sec with a runoff coefficient of 0.55 (30 houses/ha) [10]. Based on direct observation, the volume of the river body cannot accommodate surface runoff from high-intensity rainfall. Therefore, that water is overflowing along the river channel. Subsequent rain events will affect the ability of the soil to infiltrate rainwater so that most of the rainwater becomes surface runoff. Based on information from residents, the process of rising river water levels is rapid and reaches its peak in less than 1 hour.

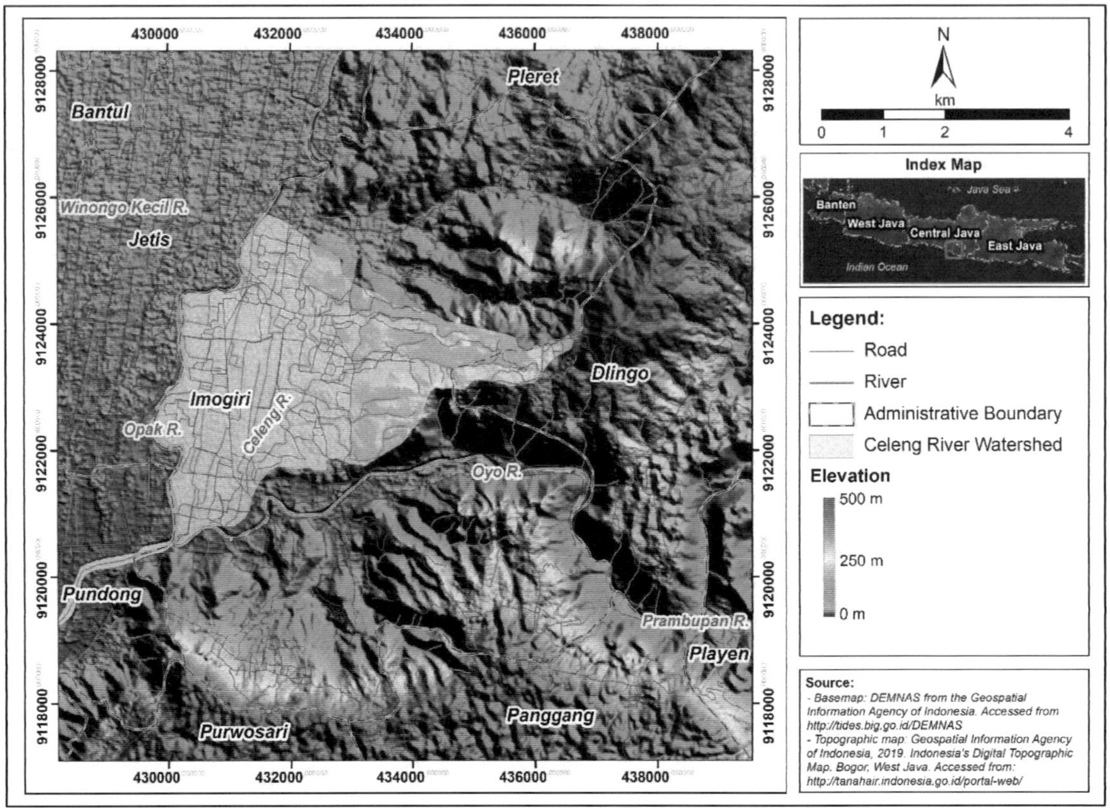

Figure 4. Celeng River watershed

3.3. Impact of Flooding

Every natural disaster will undoubtedly impact the affected area, not least the flood disaster that hits an area. Areas affected by flood disasters will receive risks both in the form of material and non-material losses. The risks are personal and public losses, such as some household furniture items being washed away by floods, houses becoming dirty, and garden and agricultural land collapsing. While public damage means the damage to some residents' facilities, such as asphalt roads damaged by some asphalt material washed away due to the impact of flood currents, some guardrails collapsed, and some tombs were destroyed. After several surveys of people in affected locations, in the case of recent flood disasters in the Bantul Regency, Yogyakarta, especially in the Imogiri sub-district. Here is some documentation while at the affected location.

The data collected from the people living in the location and its surrounding area can be glued to identify the type of property affected by the flood, the cause of the flood, and the impact of losses due to flood hazards. Based on interview data, the residents experienced flooding in stages previously, namely, the increase in flood water levels rose slowly. Then in several places, based on data was interviewed and a survey is, around 40% of the whole an area that it sudden rise in floodwater level or less than 1 hour is estimated to be influenced by locations closer to the source of the overflowing flood.

Based on the survey results, it was shown that the dominant property damaged by flooding was residents' houses which were 72%. Then agriculture is the second most damaged by flooding, at around 16%. The industry is a property type that is not dominantly experiencing flood damage in an area. The mapping area is densely populated with small industrial types and minimal large-scale industry, with only 12% damaged by flood. It is known that residents around the dominant flood area have understood that their area is included in an area with the potential for flooding, which was around 84%. This knowledge is obtained from several experiences of floods before they hit the Imogiri area in 2017. Another 16% of respondents stated that they did not understand that their area was a flood swamp area.

Based on the interviews, the most dominant impact of losses caused by floods is economic losses due to damage during floods. Other impacts felt by residents are property loss, agriculture, and health.

4. Conclusion

In general, areas highly susceptible to flooding in the Imogiri sub-district are located along rivers in large rivers such as the Oyo River and the Opak River and small rivers such as the Boar River. The denser settlements along the floodplain will increase the incidence of flooding due to narrowing the river channel. Floods in this area were triggered by the high intensity of rains and the small cross-section of rivers or drainage channels that could not accommodate surface water. Flood events occur very quickly where the peak is reached in less than 1 hour. This flood caused damage to residential areas and a small part of agricultural areas. In general, the community already understands that they live in flood-prone areas but have not made any efforts to prepare and mitigate. Therefore, it is necessary to carry out socialization activities to increase flood preparedness and response capabilities. In addition, an effort to improve drainage channels or deepen Celeng riverbeds in the study area needs to be carried out. It is undeniable that this case study still has many shortcomings, so further research is expected by adding more variables data to support the processing and improving the accuracy of the prediction of flood-prone and quality of flood hazard assessment in the Imogiri District, Bantul Regency.

References

[1] National Disaster Management Authority (BNPB), 2020, *Indonesian Disaster Risk Index*, BNPB.

[2] Altan, O; Backkhau, R; Boccardo, P; Zlatanova, S; Geoinformation for Disaster and Risk Management; *Joint Board of Geospatial Information Societies*: Copenhagen, Denmark, 2010.

[3] Merz, B.; Kreibich, H.; Schwarze, R.; Thieken, A. Assessment of economic flood damage. *Nat. Hazards Earth Syst. Sci.* 2010, **10**, 1679–1724.

[4] Rahmati, O.; Zeinivand, H.; Besharat, M. Flood hazard zoning in Yasooj region, Iran, using GIS and multi-criteria decision analysis. *Geomat. Nat. Hazards Risk* 2016, **7,** 1000–1017

[5] Harikli D, Skilodimuo.; George D, Bathrellos.; Dimitrios E, Alexakis.; *Flood Hazard Assessment mapping in Burned and Urban Areas. Sustainability 2021*, **13**, 4455. https://doi.org/10.3390/su13084455

[6] Herath, HMM; Wijesekera, NTS.; A State-of-the-Art Review of Flood Risk Assessment in Urban Areas. *IOP Conf. Series: Earth and Environmental Science* 281 (2019)

[7] Armenakis, C.; Du, EX.; Natesan, S.; Persad, RA.; Zhang, Y.; Flood Risk Assessment in Urban Areas Based on Spatial Analytics and Social Factors. *Geosciences* 2017, **7**, 123.

[8] Tate, E.C.; Maidment, D.R.; Olivera, F.; Anderson, D.J. Creating a terrain model for floodplain mapping. *J. Hydrol. Eng.* 2002, **7**, 100–108.

[9] Ahamad, M. S., Ali, R., Zakaria, N. A., Ab. Ghani, A. & Chang, C. K. 2011. Flood loss assessment of the 2003 Muda River flood. In: *3rd International Conference on Managing Rivers in the 21st Century: Sustainable Solutions for Global Crisis of Flooding, Pollution and Water Scarcity*, pp. 601–606.

[10] National Standardization Agency (BSN), 2016, SNI 2415:2016: The procedure for calculating the planned flood discharge, *National Standardization Agency*

[11] Earth Observation Research Center (EORC), Japan Aerospace Exploration Agency (JAXA). JAXA Global Rainfall Watch (GSMaP_NRT), http://sharaku.eorc.jaxa.jp/GSMaP/index_e.htm, Accessed on 25 June 2019.

Acknowledgments
The authors would like to thank the Department of Geological Engineering, the Faculty of Engineering, University of Gadjah Mada, and the CUBE ASEAN Project 2021 sponsored by JICA AUN/SEED-Net for the support.

IOP Conf. Series: Earth and Environmental Science 1071 (2022) 012009 doi:10.1088/1755-1315/1071/1/012009

Predicting the volume of slope failure material using the 2-dimensional limit equilibrium method

Hendra Santoso and I Gde Budi Indrawan*

Department of Geological Engineering, *Faculty of Engineering, Gadjah Mada University, Jl. Grafika No. 2, Yogyakarta 55281*, Indonesia
*Corresponding author's e-mail: igbindrawan@ugm.ac.id

Abstract. Mining operations in areas with the potential for landslides require careful analysis of the risks that will occur so that the consequences of losses can be anticipated early with the approach to predicting the volume of landslide material. This study aims to evaluate and predict materials that are at risk of slope failures on the slopes of the PT. TMR mining located in Bengalon East Kutai District of East Kalimantan Province. The method used is to collect geological data, soil shear strength parameters and earthquake factors and the influence of groundwater on the body of the slope. Estimation of the predicted volume of slope failures based on the area of each slip surface is calculated using the prismoidal cross section method and multiplied by the probability of slope failures. The results show estimated volume of 122,298 m^3 if the slope failure is caused by saturated slope conditions and 2,438 m^3 if the slope failure is caused by a semi-saturated slope.

1. Introduction

The safety aspect in mining is highly important, especially safety related to mining operations. The most significant risk from mining activities is the risk of death and material loss caused by the slope failure of a mine slope. The level of risk will increase with increasing volume of slope failure material and objects that become victims of the slope failure, as has been done by [1], which assesses the risk criteria for each slope and its implications for excavation optimization. The method that has been commonly used to analyze slope stability is the limit equilibrium method and as an indicator of slope stability is the slope safety factor (FoS), often referred to as the deterministic method, and the Probability of Slope failure (PoF), which uses a probabilistic method approach. Many studies related to slope stability in Indonesia using the limit equilibrium method have been carried out, including [2][3][4]. Research to predict the volume of slope failures that will occur in soft materials using the 3-dimensional limit equilibrium method has been previously carried out by [5], those study used the Rocscience Slide 3D software limit equilibrium method (LEM). Another method that can be used to predict landslide materials is a combination of volume predicted by landslides based on the area on a 2-dimensional cross-section and the probability of failures (PoF).

The research area is in one of the mining pits of PT. Tawabu Mineral Resource (PT. TMR), located in Bengalon District, East Kutai Regency, East Kalimantan Province (**Figure 1**). There have been slope failures in several sections which resulted in material losses considering the costs incurred to move the slope failure material even though there were no casualties at the time of the incident. For this reason, the purpose of this study is to determine the stability of the existing slope based on the slope failure probability method and be able to predict the volume of material that will undergo slope failure based on the slope failure probability of the existing slope.

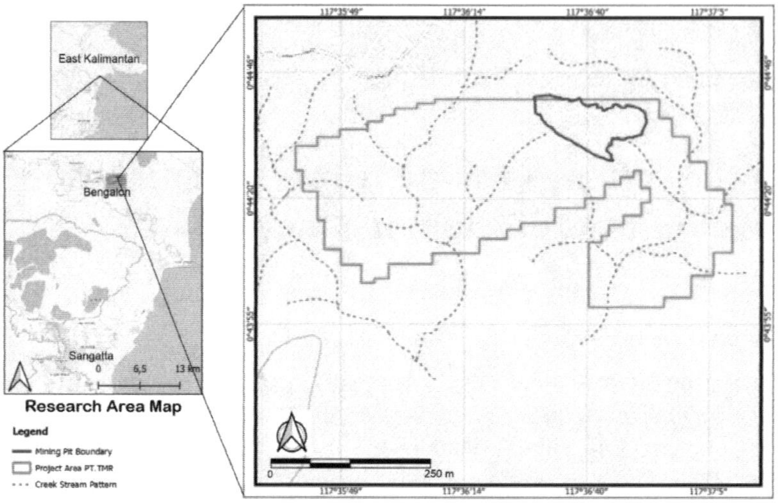

Figure 1. Location map of Research Area.

2. Geological Setting and engineering characteristics

2.1. Geomorphology

Based on the geomorphological classification [6], the research area is located in the swamp landform unit (F4) which occupies 60% of the entire research area, while the rest of the research area is the plain landform unit (D5).

2.2. Stratigraphy

Regionally, the study area is part of the physiographic zone of the Kutai Basin and is stratigraphically located in the Menumbar Formation. This formation has the same age as the Balikpapan Formation [7], this formation thick coal deposits are often found, widely mined in East Kalimantan and one of them by PT. TMR. The lithology in the study area consist of 2 rock units, namely the older claystone unit and the younger alluvial deposit unit. Most of the mining area is comprised of the claystone unit. This unit comprises alternating claystone, coal, sandstone, siltstone, and limestone. The youngest unit that was deposited later was the alluvial deposit unit. This unit was deposited from quaternary to recent. The typical composition of alluvial deposits are loose sand, mud and clay, and soft claystone.

2.3. Lithology

The following will describe the characteristics of each lithology found at the research site, which includes:

- Soil

Based on the color characteristics, the soil is divided into 2, namely soil layers at a depth of 0-4 meters and 4-7.5 meters.

- The surface soil layer has a characteristic of light brown color consisting of clay and sandy clay material. The characteristics of the soil layer are firm (S3)– very stiff (S5)[8], with hardness ranging from 0.06 to 0.31 MPa. This layer is a weathering zone which forms a nearly horizontal slope following the topographic contour.
- The soil layer found at the drill points GT-TMR01 (highwall) and GT-TMR-02 (low wall) did not contain much organic material, and the water content and plasticity were quite low. Based on the position and thickness of the material, the soil for GT-TMR02 is 3 meters below the elevation of GT-TMR01, and the soil material is 1 meter thicker than the soil in GT-TMR01.

Based on the physical and mechanical properties testing, the soil in GT-TMR01 is low plasticity silt (ML), while the soil in GT-TMR02 is a high plasticity clay (CH).

- The second soil layer show similar characteristics to the surface soil layer but is dark brown and light gray to dark in color. This second soil layer has formed a stratum alongside the rock layer. Based on Unconfined Compressive Strength.. (UCS) testing in the laboratory, this soil hardness ranges from 0.14 to 0.21 MPa, classified as rigid clay (S4).

- Claystone

Claystone layers generally have a light gray to dark gray color with a low to high level of plasticity. Layers of clay are found from a depth of 5 - 100 meters,alternating with layers of sandstone or siltstone. These layers are always found between the floor and roof layer of coal. This unit has thicknesses ranging from 0.2 to more than 10 meters with a general direction of N59°E and relatively uniform slopes ranging from 25-26°. The hardness of this rock ranges from 0.25 to 1.43 Mpa. From the drilling results in GT-TMR01 and GT-TMR02, this rock is quite dense with a high RQD value(>80%).

- Siltstone

The siltstone layers aregenerally light to dark gray in color. These layers are always found between the sandstone and claystone layers. The siltstone from the drilled sample has a low plasticity index intercalating with claystone.

- Coal

Coal in the study area is brownish-black in color with luster ranging from dull to brilliant. When scratched, the coal will give a brownish-black streak. The direction of the cleavage is in the direction of the layer. The hardness of the coal is soft to medium. In the research area, the thickness of the mined coal ranges from 1 to 6.5 meters, divided into five main layers. The roof and floor layers are generally claystone or carbonaceous claystone layers. Thick coal seams usually have a fairly thick interburden spacing that can reach 39 meters. The general direction and slope of coal follow the general direction of the other rock units (N59°E), with a slope of 25 – 26°.

- Sandstones

Sandstones are generally bright yellow in color with very fine to fine sand grain sizes with a dominant mineral composition of quartz. Generally, this sandstone layer is cemented by clay material, so that in the drilling description, it is referred to as sandy clay. The sand material is quite stable and can be formed on a bench with a slope of 50° and a height of 10 meters. The thickness of the sandstone ranges from 1 to 10 meters.

3. Research Methods

This study aims to determine the volume of material in the event of a slope failure by performing several stages. The stages used to achieve the objectives of this research can be seen in the flow chart **Figure 4**, and can be explained as follow:

3.1. Data Collecting

- Geology and Hydrogeology
  The condition of geomorphology, stratigraphy, geological structure and the depth of the groundwater are important parameters to determine the character of the area. Stratigraphy was determined based on two drilling data located at the highwall and lowwall positions with a depth of 100 meters.
- Physical and mechanical properties of rock
  According to the ISRM classification, 2007 [8], the soil and rock composition in the study area includes soil with a soft to very hard hardness (S5) and very weak rock (R1), so the criterion used is MohrCoulomb. The total number of drilling samples tested was 29 samples. The physical properties parameter required for slope stability analysis is the unit weight (γ) while the main mechanical properties parameters are cohesion (c) and angle of friction (φ). The limestone strength type is Generalized-Hoek Brown. UCS parameters, Geological Strength Index (GSI),

Intact Rock constant (mi) and Disturbance Factor (D) were obtained from field description assumptions and GSI calculator calculations.

- Topographic map
  The map used in the slope stability analysis is a topographic map of the mine situation in dxf format. This map is intended to form a cross-sectional profile in a 2-dimensional slope stability analysis.

3.2. The determination of the parameters

- Statistical parameters
  The statistical parameters of γ, c and include the mean, standard deviation, and the maximum and minimum values of the data, respectively.
- Earthquake Parameters
  Seismic data is collected from the *Pusat Studi Gempa Nasional* [9] accessed through the website (Design Spectra Indonesia) [10]. The seismic load of the earthquake in the study area is 0.0650 g. Refer to SNI 8460-2017[11] in the section on earthquake requirements for slopes, the horizontal earthquake coefficient (kh) is determined to be 0.5 from the peak velocity of the earthquake, result as kh in the calculation of slope stability is 0.03.

3.3. Slope Stability Analysis

Principally, the slope stability analysis method used in this study is the principle of limit equilibrium. A slope is classified as stable if the shear strength of the slope (s) is greater than the shear stress (τ) required to reach equilibrium. The relationship between shear strength and shear stress is what determines the slope's Factor of Safety(FoS) as can be written in equation 1 below:

$$\mathrm{FoS} = \frac{s}{\tau} \quad (1)$$

Simplified Bishop method, carried out using the Rocscience Slide software, is used to calculate the shear strength and shear stress acting on a slope. In addition, the Limit Equilibrium Method (LEM) Derivative is a slope stability analysis method, based on a mathematical approach related to the stabilizing and destabilizing forces acting on the slope, determine slope behavior and stability conditions, they can be grouped into 2 [12]. Deterministic Methods: this method shows whether the slope is stable or not. After the conditions are known or assumed.

- Probabilistic method: this method uses a statistical approach from the analyzed shear strength parameters. Many studies on the slope failure probability method have been carried out, for example, by [13][14] [15]. On the basis that the number of uncertainty factors in a slope from the aspect of geometry, characteristics, loads, and slope failure predictions originating from topographic conditions, structural geology, groundwater level, soil shear strength. [16], the slope safety factor (FoS) obtained from the deterministic analysis cannot ensure the possibility of slope failures at certain FoS values so that the slope failure probability approach is needed to determine the prediction of slope failures that will occur with the PoF and FoS parameters which can be obtained at the same time in one calculation step. In this study, the slope failure probability method is very easy to apply using rocscience slide software by Monte Carlo simulation. The analysis is divided into 13 sections shown in **Figure 2**

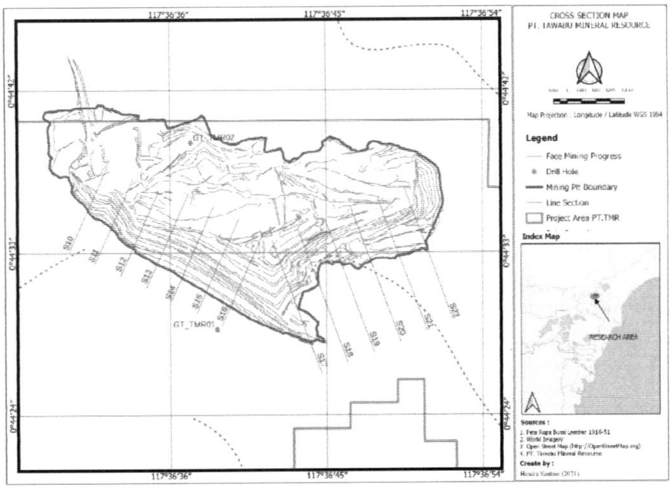

Figure 2. Cross section Map of slope stabilty analysis.

The analysis is focused on section S12 and sections S18-19 considering that in this area signs of slope failures have occurred. Further analysis was carried out by matching the field observation and the results of the slope stability analysis.

Acceptance criteria for the level of slope stability refers to the Decree of the Minister of Energy and Mineral Resources No. 1827K/30/MEM 2018 which regulates the standard set for the general mining sector in Indonesia.

Table 1. Consequences of Failure CoF and Acceptance Criteria of Mining Industries in Indonesia based on Law of ESDM ministry No. 1827K/30/MEM 2018*[17]*.

Type of slope	Consequences of Failure (CoF)	Acceptance criteria Static Factor of Safety (FoS) (min)	Dynamic Factor of Safety (FoS)(min)	Probability of failure (max) PoF (FK≤ 1)
Single slope	Low - High	1,1	None	25 – 50%
Inter-Ramp	Low	1,15 – 1,2	1,0	25%
	Medium	1,2 – 1,3	1,0	20%
	High	1,2 – 1,3	1,1	10%
Overall Slope	Low	1,2 – 1,3	1,0	15 – 20%
	Medium	1,3	1,05	10%
	High	1,3 – 1,5	1,1	5%

3.4. Estimation of volume of predicted slope failure material

Slope failure volume is the amount of slope failure material caused by the excavation process or other external factors and the amount of rock or soil mass that experiences displacement due to the stress release process and accumulates to find a new balance of rock or soil mass. The volume of slope failure can indicate the level of impact parameters that may be caused by slope failure events, which can then be used to determine the level of slope failure risk in areas that have a high to low level of potential risk [5].

By knowing the area of the slip failure, it can be calculated the volume of the slope failure that will occur based on the area of the slope failure area shown by software based on the limit equilibrium method. The way to calculate the slip surface area of each cross-section is to take measurements in Autocad software or other similar software by digitizing the image format file (.*jpg), which is the result of the analysis obtained from the rocscience slide software. After knowing the cross-sectional area of each slip surface, the slope failure material volume can be directly calculated using the prismoidal method [18] as shown in equation 2 and **Figure 3** below.

$$V = (S1 + 4M + S2)\frac{L}{6} \qquad (2)$$

Where:

S1, S2 = area of the first and last cross section (m^2)
M = area of midsection (m^2)
L = distance between S1 and S2 (m)
V = Volume of slope failure material (m^3)

If PoF is the probability of slope failures (%), then the volume of material predicted for slope failures ($V_{prediction}$) is the product of the total slope failure volume multiplied by PoF, or can be written into the following equation:

$$V_{prediction} = PoF \times V \qquad (3)$$

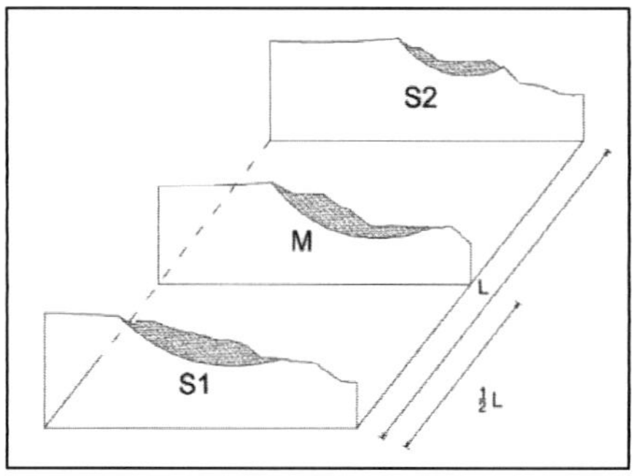

Figure 3. Slope failure volume calculation illustration using the prismoidal cross-section method *[18]*.

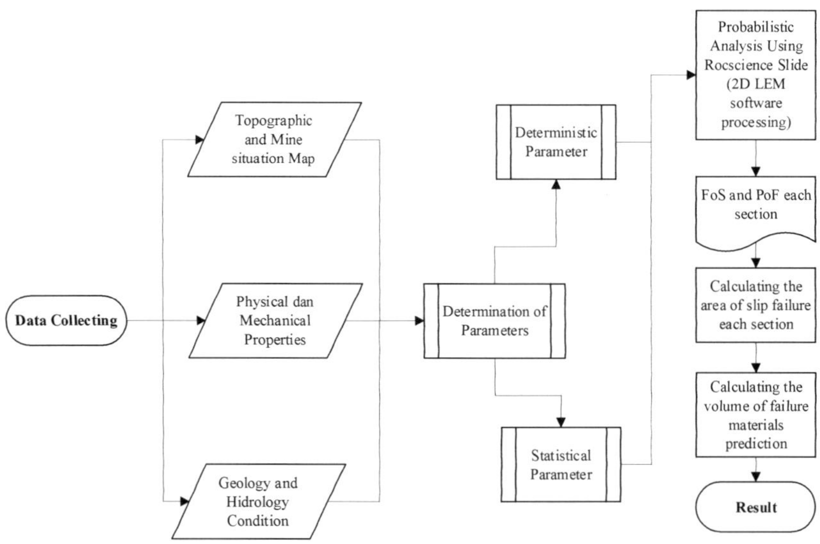

Figure 4. Research flow chart.

4. Results of analysis and discussion

Based on the results of physical and mechanical properties testing (*Appendix A*), the characteristics of the material are divided into three, namely claystone 1 (depth of 0- 7.8 meters), claystone 2 (depth of 7.8-16.14 meters), and siltstone (depth of 16.44-99 meters). The values for coal and sandstone matersials are determined based on the reference study [19]. Statistical parameters of each lithology used in the probability analysis is listed in ***Table 2***.

Table 2. Statistical parameters of slope stability analysis.

No.	Material Name	Property	Distribution	Mean	Std. Dev.	Rel. Min	Rel. Max
1	Claystone 1	Cohesion	Normal	37,57	11,33	3,1	47,92
2	Claystone 1	Phi	Normal	15,38	0,15	15,27	15,48
3	Claystone 1	Unit Weight	Normal	16,22	3,01	14,09	18,35
4	Claystone 2	Cohesion	Normal	37,57	42,99	34,95	46,18
5	Claystone 2	Phi	Normal	17,72	1,01	17,01	18,44
6	Claystone 2	Unit Weight	Normal	18,41	0,09	18,35	18,47
7	Siltstone	Cohesion	Normal	37,57	54,51	8,18	302,22
8	Siltstone	Phi	Normal	34,36	4,22	14,04	40,74
9	Siltstone	Unit Weight	Normal	17,87	1,73	10,82	20,75
10	Batubara	Cohesion	Normal	45	0	0	0
11	Batubara	Phi	Normal	35	0	0	0
12	Batubara	Unit Weight	Normal	15	0	0	0
13	Sandstone	Cohesion	Normal	320	0	0	0
14	Sandstone	Phi	Normal	47	0	0	0
15	Sandstone	Unit Weight	Normal	24	0	0	0

Cross-sectional profile from the lowwall to the highwall (**Figure 5**) shows the dominant lithology that composes the slopes of the PT. TMR is claystone. The slope of the rock has a varied pattern with an angle of 20-25°. Based on the results of the UCS test, most of the overburden materials have UCS that range from 0.06-1.43 MPa, which means that most of the materials were very soft soil (S0) to stiff (S5) and very soft rock (R1) according to the ISRM 1978 classification. The type of slip surface that will occur in soft materials like this is generally a circular failure. **Figure 6** shows the slope failures that have occurred in sections S18-S19 and slope failures in alluvial materials caused by groundwater pressure at shallow depths. The destabilization process naturally leaves fractures which can trigger further slope failures.

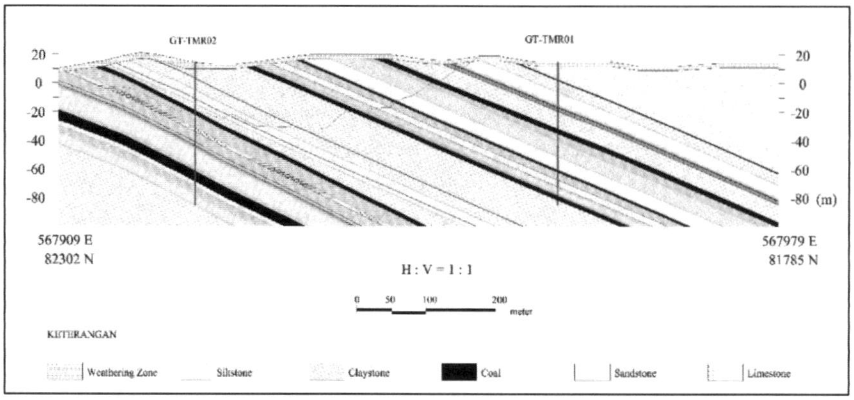

Figure 5. Stratigraphic section from lowwall to highwall direction.

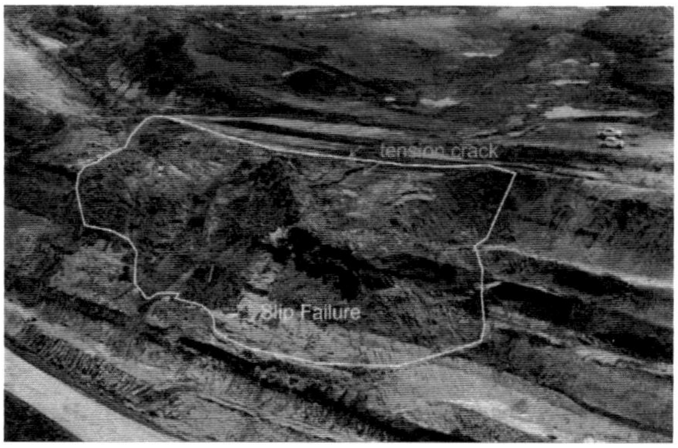

Figure 6. Slope failure that has occured in sections S18 and S19.

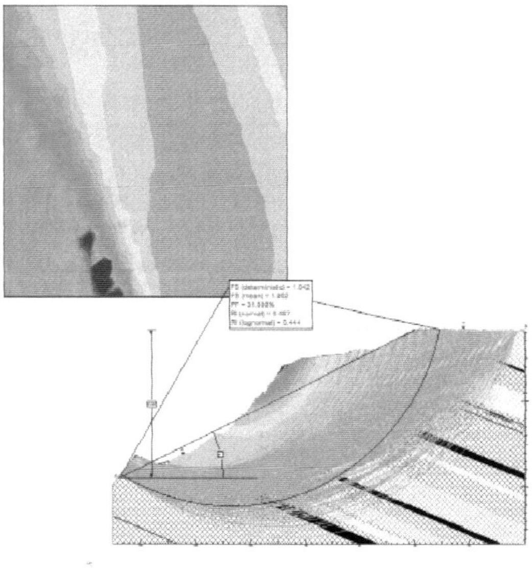

Figure 7. LEM analysis in Section S18.

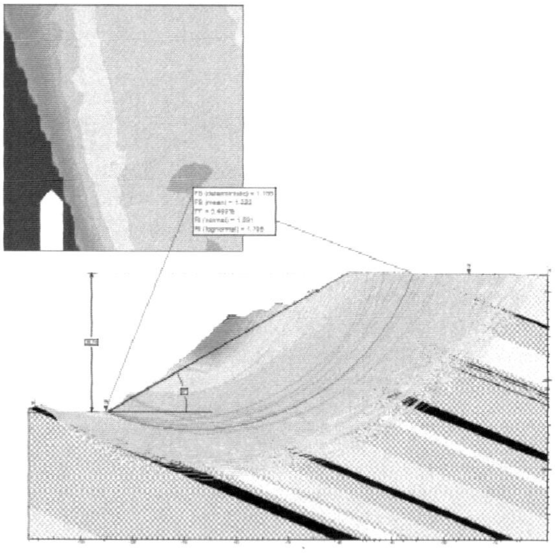

Figure 8. LEM Analysis inSection S19.

Slope stability analysis in section S18 and section S19 showed that the difference of FoS values was not significant. Based on the analysis results shown in **Table 3**, FoS section S18 = 1.06 with PoF= 31.8% (**Figure 7**) while FoS section S19 = 1.22 and PoF = 5.4% (**Figure 8**). Meanwhile in the semi-saturated condition, section S18 has FoS=1.3 and PoF=1.3% while section S19 has FoS=1.48 and PoF 0.1%. From these two conditions, it is explained that the presence of groundwater flow to the body of the slope in section S18 has more influence on the increase in stress and the decrease in shear strength caused by the increase in pore water pressure due to the presence of groundwater. In contrast, in section S19 the effect of groundwaterwater is not too significant. The difference in shear strength and shear stress on a slope with a distance of 50 meters is sufficient to prove that the condition of soil stability can differ depending on the geometry of the slope and the shape of the groundwater table, which affects the stress and pressure in the soil.

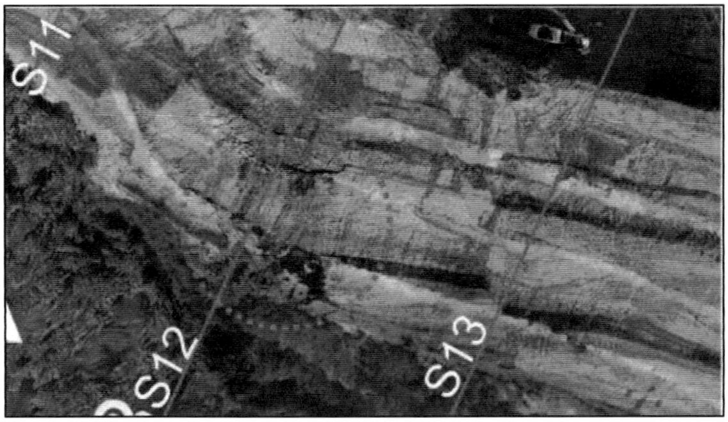

Figure 9. Slope failure in section S12 (circled area).

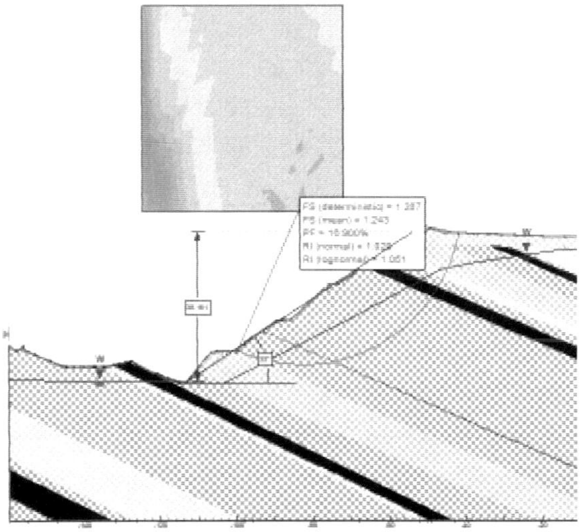

Figure 10. LEM analysis in Section S12.

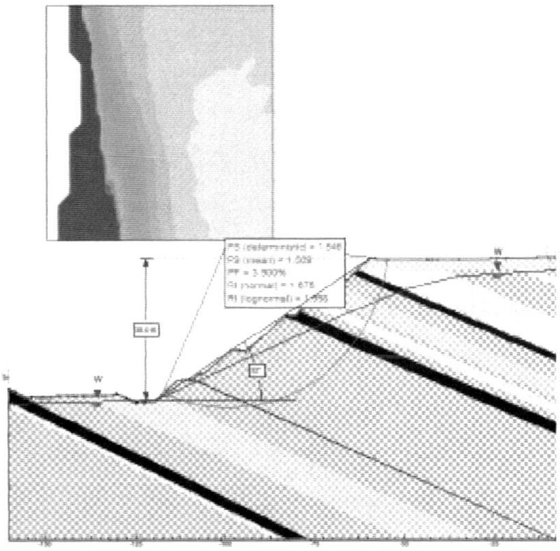

Figure 11. LEM analysis in Section S13.

Figure 9 shows that the slope failure occurred in section S12, but in section S13 the slope looks quite stable. It is assumed that the slopes that have experienced slope failures, namely the S12 slope, use residual shear strength parameters, while the slopes that have never experienced slope failures use the peak shear strength parameters. In saturated slope conditions, the results of the analysis of section S12 produce FoS values = 1.24 and PoF = 16.9% (**Figure 10**), while in section S13 the values of FoS = 1.57 and PoF = 3.9% (**Figure 11**) . In the semi-saturated slope condition, section S12 has FoS=1.6 and

PoF=0.2%, while section S13 has FoS=2.01 and PoF=0%. Both sections have the same chance of slope failures and decrease when the slope is semi-saturated.

Analysis of all highwall slopes in saturated conditions, using the slope failure probability method, shows PoF values ranging from 0 to 31.8% and in semi-saturated conditions ranging from 0-17.2%. The slopes that have the highest PoF in saturated conditions are section S18 slopes, while the lowest slopes are S10, S11, S20, and S21 slopes.

The result of the calculation of probability of failure ***Table 3*** and ***Table 4*** further by measuring the area of the slip surface multiplied by the probability of slope failure, obtained estimates of material predicted a slope failure on the condition of the slopes saturated amounted to 122 298 m^3, while the half-saturated predict in occurrence of slope failure has a slope failure volume of 2438 m^3. From these calculations, it can be seen that the level of groundwater saturation of the slope greatly affects the magnitude of the risk that will occur.

Table 3. Estimated results of highwall slope failure material volume prediction for saturated slope conditions.

Section	Saturated Condition			Slip failure area (m^2)	Area at risk of failure (m^2)	Material volume at risk of failure (m^3)
	FoS (mean)	FoS (Deterministic)	PoF (%)			
S10	2,767	2,721	0	781,69	-	1.209,62
S11	2,015	1,99	0	1.890,96	-	5.159,67
S12	1,243	1,207	16,9	858,90	145,15	2.891,12
S13	1,569	1,546	3,90	988,23	38,54	4.546,64
S14	1,387	1,373	2,50	1.904,61	47,62	11.000,09
S15	1,219	1,207	12,0	2.638,29	316,59	8.368,08
S16	1,443	1,428	0,30	2.005,73	6,02	30.912,91
S17	1,110	1,091	21,3	3.115,06	663,51	41.366,03
S18	1,062	1,042	31,8	3.300,33	1.049,50	12.159,12
S19	1,221	1,199	5,40	1.896,25	102,40	853,31
S20	1,912	1,872	0	1.214,59	-	638,59
S21	1,524	1,552	0	2.225,95	-	2.554,35
S22	1,286	1,257	12,39	618,49	76,63	638,59
			Total			**122.298,12**

IOP Conf. Series: Earth and Environmental Science 1071 (2022) 012009 doi:10.1088/1755-1315/1071/1/012009

Table 4. Estimated results of highwall slope failure material volume prediction for semi-saturated slope conditions.

Section	Semi Saturated Condition			Slip failure area (m^2)	Area at risk of failure (m^2)	Material volume at risk of failure (m^3)
	FoS (mean)	FoS (Deterministic)	PoF (%)			
S10	2,978	2,926	0,2	552,48	1,10	364,15
S11	2,351	2,32	1,2	858,31	10,30	132,31
S12	1,600	1,567	0,2	697,25	1,39	11,62
S13	2,015	1,99	0	854,83	-	-
S14	1,782	1,768	0	1.515,37	-	-
S15	1,548	1,534	0	1.756,87	-	53,38
S16	1,688	1,677	0	2.005,73	-	398,78
S17	1,415	1,396	0,2	3.202,82	6,41	810,22
S18	1,343	1,318	1,3	1.710,08	22,23	248,47
S19	1,484	1,460	0,1	1.896,25	1,90	15,80
S20	2,253	2,227	0	2.046,31	-	67,18
S21	1,865	1,835	0	1.963,34	-	268,72
S22	1,576	1,499	17,2	46,87	8,06	67,18
			Total			**2.438**

5. Conclusion

Based on the above analysis and discussion, the results can be concluded that highwall slope stability showed the value of FoS under saturated conditions is in the range of 1.06 – 2.76 and PoF is in the range of 0-31.8%. The slopes of S17 and S18 are very risky and are above the predetermined threshold, but if they are in a semi-saturated condition, almost all of them are below the specified threshold. It is recommended to manage mine groundwater preventively and curatively to mitigate the saturation of slope that can lead to slope failures. Estimation of the volume of slope failure material that is predicted by the extent of the slip surface and the degree of probability of slope failure, the estimated volume is122,298 m^3 when a slope failure caused by the condition of the slopes are saturated and 2,438 m^3 when slope failures caused by a semi-saturated slopes.

Appendix A

Results of physical and mechanical properties tests

Sample	UNIFIED	Depth	Water Content	Unit Weight			Internal friction angle (residua	Cohesion (residual)	Internal friction angle	Cohesion (peak)	UCS
	Material	h	Wn	γn	γsat	γd	φ_r	C_r	φ_p	C_p	
		(m)	(%)	kPa	kPa	kPa	o	kPa	o	kPa	MPa
GT-TMR01											
S01	ML	2.08-2.38	22.11	16,16	17,87	13,24	22,16	29,11	27,7	36,38	0,06
S02	CH	5,2-5,5	27,32	17,3	17,82	13,58	21,26	36,32	26,57	45,4	0,21
S03	CL(Zst)	7,8-8,1	19,69	17,7	18,75	14,79	15,22	48,33	24,23	70,61	0,26
S04	CL(Zst)?	12.37-12.67	19,80	16,42	18,29	13,71	18,02	42,25	27,92	61,49	0,25
S05	CH(Cst)	16.14-16.44	18.03	17,65	19,2	14,95	23,49	89,22	32,75	158,87	0,37
S06	ML(Zst)	23.55-23.85	19,69	18,2	19,24	15,21	24,07	87,06	39,35	123,56	0,47
S07	CH(Cst)	27,64-27,94	18,28	19,06	19,62	16,12	31,34	183,82	40,44	326,56	0,3
S8	CH(Cst)	30.0-30.29	18.55	18,31	19,14	15,44	26,62	172,25	38,48	297,14	0,83
S9	CH(Cst)	37.08-.37.38	17.35	18,83	19,78	16,02	29,25	89,22	40,03	137,29	0,52
S10	CH(Cst)	44,6-44,9	20,39	18,61	19,35	15,46	23,41	92,55	38,83	131,41	0,62
S11	CH(Cst)	54.97-55.27	16,31	18,47	19,67	15,89	23,14	91,96	38,48	130,92	0,47
S12	CH(Cst)	72.33-72.63	15,52	18,87	19,94	16,34	26,85	163,43	38,84	281,84	1,01
S13	MH(Zst)	80,7-81,0	13,17	18,59	19,74	16,42	28,38	158,24	40,78	272,62	1,43
S14	CH(Cst)	84.43-84.73	13,82	19,37	20,43	17,01	28,64	146,57	37,92	259,88	1,1
S15	CH(Cst)	99,6-99,9	17,37	18,43	19,47	15,67	27,02	94,12	37,42	144,16	0,5

IOP Conf. Series: Earth and Environmental Science 1071 (2022) 012009 doi:10.1088/1755-1315/1071/1/012009

Appendix B

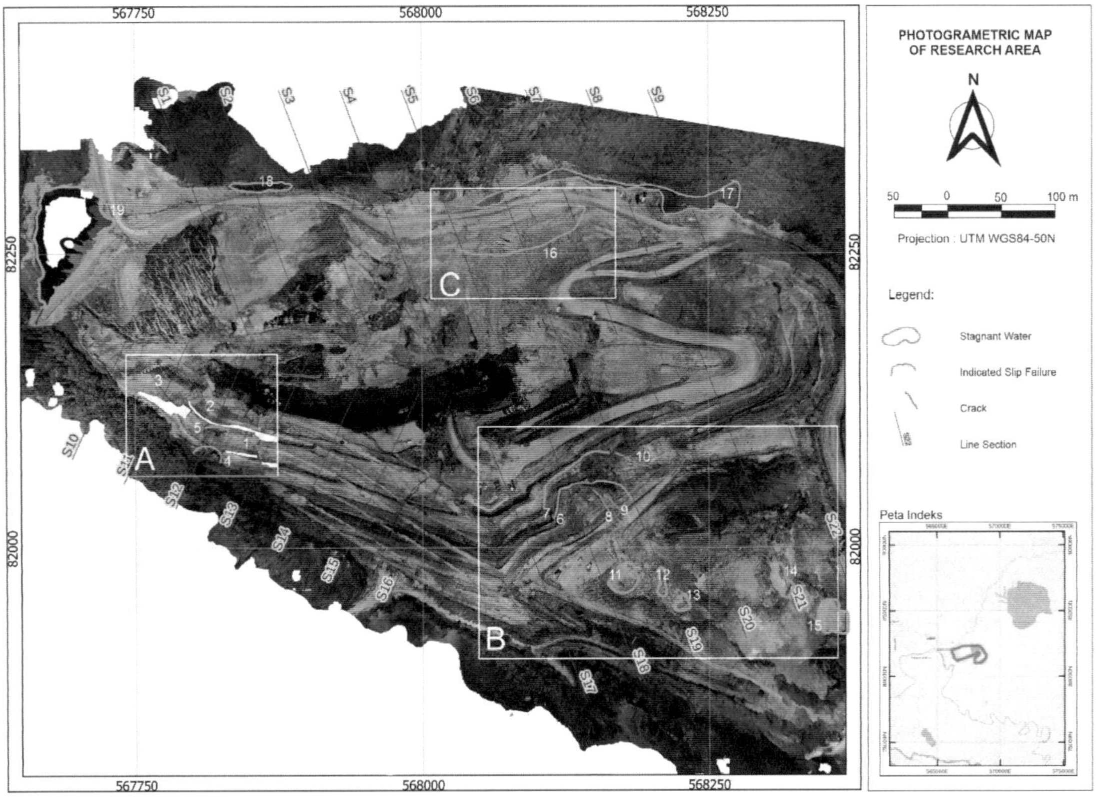

References

[1] L.-F. Contreras, E. Hormazabal, R. Ledezma, and M. Arellano, "Geotechnical risk analysis for the closure alternatives of the Chuquicamata open pit," p. 16, 2019.
[2] F. Aprilia and I. G. B. Indrawan, "An Evaluation Of Rock Slope Stability Using Limit Equilibrium Analyses," *Journal of Applied Geology*, vol. 6, no. 2, Art. no. 2, Sep. 2015, doi: 10.22146/jag.7220.
[3] T. J. Lalitya, I. G. B. Indrawan, and A. Bassmantra, "Analisis kestabilan lereng tambang terbuka batubara dengan metode probabilitas pada highwall dan lowwall Pit Tania Panel 2, PT. Kaltim Prima Coal, Kalimantan Timur," p. 18, 2018.
[4] H. Trihatmanto, M. A. Zuhri, and I. Saleh, "Kajian probabilitas kemantapan lereng pada highwall, untuk rekomendasi optimalisasi Pit Pelikan PT.Kaltim Prima Coal Site Sangatta, Kabupaten Kutai Timur, Provinsi Kalimantan Timur," p. 23, 2017.
[5] M. A. Azizi, R. Karim, I. Marwanza, and M. K. Ghifari, "Prediction of Material Volume of Slope Failure in Nickel Surface Mine Using Limit Equilibrium Method 3D," *ind. mining, prof. jurnal*, vol. 1, no. 1, pp. 43–48, Dec. 2019, doi: 10.36986/impj.v1i1.13.
[6] R. A. Van Zuidam, *Guide to Geomorphologic Aerial photographic interpretation and mapping View node catalog data*. ITC: Enschede The Netherlands, 1983.
[7] Sukardi, N. Sikumbang, I. Umar, and R. Sunaryo, "Peta Geologi Lembar Sangatta, Kalimantan 1:250.000 = Geological Map of the Sangatta Quadrangle, Kalimantan," Pusat Penelitian dan Pengembangan Geologi, Bandung, Indonesia, 1995.
[8] ISRM, *The Complete ISRM Suggested Methods for Rock Characterization, Testing and Monitoring: 1974-2006*. ISRM Turkish National Group and the ISRM, 2007.
[9] Pusat Studi Gempa Nasional (Indonesia), Ed., *Peta sumber dan bahaya gempa Indonesia tahun 2017*, Cetakan pertama. Bandung: Pusat Penelitian dan Pengembangan Perumahan dan Permukiman, Badan Penelitian dan Pengembangan, Kementerian Pekerjaan Umum, 2017.
[10] "Desain Spektra Indonesia." http://rsa.ciptakarya.pu.go.id/2021/index.php?pga=0.0650&ss=0.1240&s1=0.0980&tl=16&kelas=2#grafik (accessed Nov. 15, 2021).
[11] Badan Standarisasi Nasional, *SNI 8460-2017 Persyaratan Perancanagan Geoteknik*. 2017.
[12] L. I. González de Vallejo and M. Ferrer, *Geological Engineering*. CRC Press Taylor & Francis Group, 2011.
[13] J. M. Duncan, S. G. Wright, and T. L. Brandon, *Soil Strength and Slope Stability*. Hoboken,New Jersey: John Wiley & Sons, Inc., 2014.
[14] Y. Wang, Z. Cao, and S.-K. Au, "Efficient Monte Carlo Simulation of parameter sensitivity in probabilistic slope stability analysis," *Computers and Geotechnics*, vol. 37, no. 7, pp. 1015–1022, Nov. 2010, doi: 10.1016/j.compgeo.2010.08.010.
[15] H. El-Ramly, N. R. Morgenstern, and D. Cruden, "Probabilistic slope stability analysis for practice," *Canadian Geotechnical Journal*, vol. 39, pp. 665–683, Jun. 2002, doi: 10.1139/t02-034.
[16] O. K. H. Steffen, P. J. Terbrugge, J. Wesseloo, and J. Venter, "A Risk Consequence Approach To Open Pit Slope Design," *The South African Institute of Mining and Metallurgy International Symposium on Stability of Rock Slopes in Open Pit Mining and Civil Engineering*, 2008.
[17] Menteri ESDM, *Keputusan Menteri Energi Dan Sumber Daya dan Mineral Nomor 1827/K/30/MEM/2018 Tentang Pedoman Pelaksanaan Kaidah Teknik Pertambangan Yang Baik*. 2018, pp. 57–59.
[18] B. E. Meserve and R. E. Pingry, "Some Notes on the Prismoidal Formula," *The Mathematics Teacher*, vol. 45, no. 4, pp. 257–263, 1952.
[19] D. Martin and P. Stacey, *Guidelines for open pit slope design in weak rocks*. CSIRO Publishing, 2018.

IOP Conf. Series: Earth and Environmental Science 1071 (2022) 012010 doi:10.1088/1755-1315/1071/1/012010

Application of stone column method for liquefaction mitigation in Palu Bay area after the 2018 Palu earthquake

Fahrizal Patriaman[1,2], Teuku Faisal Fathani[1,3*]

[1]Department of Civil and Environmental Engineering, Faculty of Engineering, Universitas Gadjah Mada, Yogyakarta 55281, Indonesia; fahrizal.p@mail.ugm.ac.id, wilopo_w@ugm.ac.id

[2]Directorate of Strategic Infrastructure, Directorate General of Human Settlement, Ministry of Public Works and Housing, Republic of Indonesia

[3]Center for Disaster Mitigation and Technological Innovation (GAMA-InaTEK), Universitas Gadjah Mada, Yogyakarta 55281, Indonesia

*Corresponding author's e-mail : tfathani@ugm.ac.id

Abstract. On 28 September 2018, an earthquake occurred on the north-south Palu Koro shear fault, extending through Palu City and other cities in Central Sulawesi Province, Indonesia. In addition to the tsunami, the event triggered liquefaction not only on the land but also in the coastal area. Thus, mitigation efforts to prevent potential disasters in Palu Bay coastal area are required. This research covered the Palu Bay coastal area, Central Sulawesi Province and conducted through the core drillings and Standard Penetration Tests on the area. Low liquefaction potential was achieved through soil improvement by applying stone column method based on the specified criteria. Mitigation effort on the liquefaction hazards comprised of stone column method which applied the Priebe's calculation method. The mapping result indicated the need for stone columns with deeper and closer spacing distance on the western and southern areas than those in the eastern areas. Due to the wide variety of stone column designs resulted from the mitigation analysis, simplification stone column designs was made to facilitate the construction implementation.

1. Introduction

Located on the Pacific Ring of Fire, Indonesia is common with frequent earthquakes. On 28 September 2018, an earthquake occurred on the north-south Palu Koro shear fault, extending through Palu City and other cities in Central Sulawesi Province, Indonesia [1]. The event also triggered other disasters: liquefaction and tsunami. As revealed by several studies, the tsunami event occurred around Palu Bay coastal area were caused by landslide [2–4]. Moreover, liquefaction events on the mainland areas were also identified [5–7]. However, according to Sassa and Takagawa [8], liquefaction disasters occurred not only on the land but also on the coastal areas. The evidence is shown in **Figure 1** according to Sassa and Takagawa [8] and documentation by the Ministry of Public Works and the Housing Republic of Indonesia.

Based on the results of soil profile investigation analysis by Patriaman *et al.* [9] Palu Bay area was dominated by non-cohesive soil (sand), which indicated high liquefaction potential. In addition, the groundwater level position varied between 0.09 m and 4.48 m below the ground level.

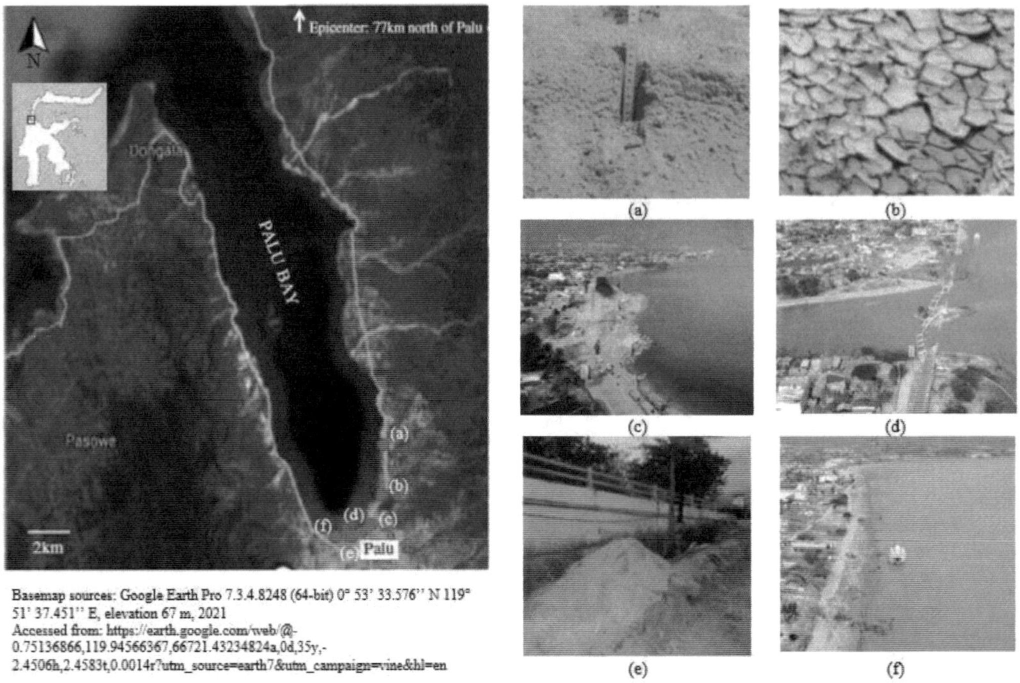

Figure 1. Evidence of Liquefaction Potential in the Palu Bay coastal area based on Google Earth image interpretation [8,10], and documentation by the Ministry of Public Works and the Housing Republic of Indonesia: (a), (b), (e) sand boil (c), (d), (f) shoreline reduction.

Analysis of liquefaction potential was carried out using drill log data along the Palu Bay coast. Patriaman *et al.* [9] conducted the analysis for 0-20 m depth from the ground surface. Due to the seismic activity of the Palu Koro Fault and the geological conditions of the subsoil, Palu Bay area is vulnerable to liquefaction. Thus, mitigation efforts are essentials.

The objective of this study was to analyze liquefaction mitigation efforts for area with high liquefaction potential. The research covered the Palu Bay coastal area, Central Sulawesi Province as presented in **Figure 2**. This study used 22 points of core drillings and Standard Penetration Test (SPT) data along the Palu Bay coastal areas as seen in **Figure 2**. The final result of this research was a mapping of liquefaction mitigation designs for the Palu Bay coastal area.

2. Selecting the liquefaction mitigation methods for Palu Bay area

The essential idea in preventing the occurrence of liquefaction is to perform one of the following conditions: (1) the densification of loose sand in order to enable the soil to resist the liquefaction, (2) the lowering of groundwater level in order to prevent the soil from saturated, or (3) the drainage of groundwater in order to reduce excessive pore water pressure [11]. In this study, soil improvement techniques were selected because the subsidence and drainage of groundwater was not possible for the Palu Bay coastal area. The basic principle of improving poor soil capacity (in various aspects) is improving soil properties according to the preferred improvement objectives [12]. Priority in selecting the soil improvement methods to prevent liquefaction is usually given to the construction efficiency, reliability, and cost-effectiveness [13].

Traditional liquefaction mitigating methods consist of soil replacement, sand compaction column, vibration compaction, dynamic compaction, blast compaction, grouting, deep mixing column method, gravel column method, dissipation using screen pipes, and lowering of the groundwater table [14]. These methods have been widely applied in engineering practice and construction technology using proven effective and reliable instruments. The stone column method was selected in this study because it involved the replacement of poor soil with sand or a combination of sand and crushed stone and build a vertically resistant system that allowed penetration to soft layer by extending depth of 20 to 30 m. The stone column could also be considered as reinforcement component. Moreover, due to the high permeability of rock material, rock columns enabled the increasing consolidation rate in soft soils that were useful in drainage and reduced liquefaction potential [15–18]. Accordingly, this method was considered suitable for Palu Bay coastal area because this area was dominated by sandy soil.

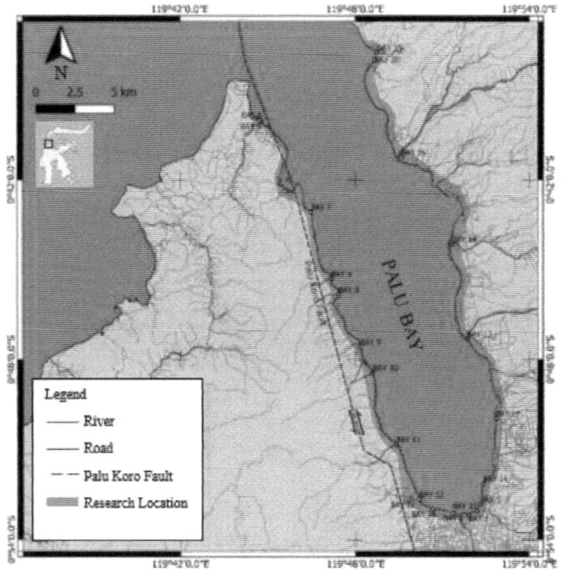

Basemap sources: Geospatial Information Agency of Indonesia, 2021
Topographic map: Geospatial Information Agency of Indonesia, 2021
Accessed from: https://tanahair.indonesia.go.id/portal-web/download/perwilayah

Figure 2. Research Area (Indonesia's digital topographic map from Geospatial Information Agency of Indonesia [19]).

3. Research method

At the preliminary stage of this research, the determination of the liquefaction mitigation method considered the environmental conditions of the research object to obtain appropriate mitigation design. By applying calculation of the liquefaction potential by Patriaman *et al.* [9], the calculation of the stone column design was done by using the Priebe's method [20]. According to Ng and Tan [21] and Sexton, *et al.* [22], the Priebe's method is the most popular design method for estimating the base settlement of stone column reinforcement. It is established on the concept of a unit cell where the column is considered in plastic state while the surrounding soil behaves elastically. The settlement given as reduction improvement factor n_1 is defined as the final settlement's ratio with and without the column.

According to Priebe [20], the value of the safety factor (SF) will determine the value of stone column diameter (D), the distance between the centre of the stone columns (s), depth of the stone column, and a new safety factor (SF') as a form of liquefaction mitigation. The value of the distance of the stone column design is obtained by performing a back analysis of the predetermined SF' value. This calculation is assisted by the goal seek function in Microsoft Excel software. Furthermore, the calculation results of the distance and depth of the stone columns are mapped using QGIS 3.16.4 software.

3.1. Unit cell stone column

In this study, all stone columns were assumed to have the same behaviour. Thus, the calculation was required only to one unit [23]. In general, stone columns were uniformly distributed in a triangular or rectilinear configuration. The loading area for each column could be either a hexagon or a square. The loading region was converted into a circle (cylindrical) with the same cross-sectional area to allow axial symmetry conditions. The unit cell used in this study is a triangular pattern because the equilateral triangle pattern provides the densest arrangement of stone columns in a particular area [15]. **Figure 3** shows a Priebe's simplified unit cell modelling [20]. The cross-section of the unit cell applied was a triangular pattern so that the equation of the equivalent diameter value obtained was:

$$d_e = 1.05s \tag{1}$$

where d_e is the unit cell diameter (m), and s is the distance between the centre of the stone columns (m).

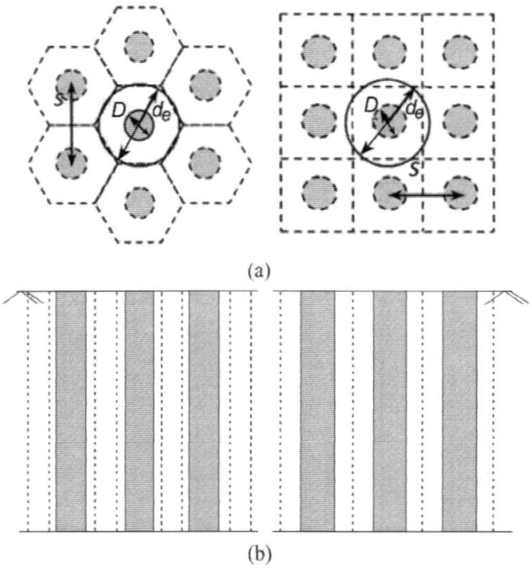

Figure 3. Simplification of unit cells under axial symmetry conditions: (a) top view [23] (b) front view.

3.2. Basic Improvement Factor (n_0)

The initial step improvement factor is denoted by n_0. The stone column improves the mechanical characteristics of the initial soil compared to its pre-treatment properties (without columns) [16]. Stone column planning assumes that bulging is constant over the length of the column, the column material is incompressible, and the bulk densities of the soil and column are neglected [16,22]. Priebe [20] proposes that the evaluation results as expressed in the following equation:

$$n_0 = 1 + \frac{A_C}{A}\left[\frac{0.5 + f\left(\mu_s, \frac{A_C}{A}\right)}{K_{aC} \cdot f\left(\mu_s, \frac{A_C}{A}\right)} - 1\right] \tag{2}$$

where n_0 is the basic improvement factor, A_C is the total area of the unit cell (m^2), A is the area of the stone column (m^2), μ_s is Poisson's ratio, and K_{aC} is the active lateral pressure (kN). The value of $f\left(\mu_s, \frac{A_C}{A}\right)$ is obtained from the equation below:

$$f\left(\mu_s, \frac{A_C}{A}\right) = \frac{(1-\mu_s)\cdot\left(1-\frac{A_C}{A}\right)}{1 - 2\mu_s + \frac{A_C}{A}} \tag{3}$$

Furthermore, the value of K_{aC} is obtained under the following equation:

$$K_{aC} = tan^2\left(45^\circ - \frac{\varphi_C}{2}\right) \tag{4}$$

where φ_c is the shear angle of the stone column filling material. The A_C value is obtained from the following equation:

$$A_C = 0.25\pi D^2 \tag{5}$$

where D is the diameter of the stone column (m). The value of A is obtained from:

$$A = 0.25\pi {d_e}^2 \tag{6}$$

In general, conditions that occur with the value of Poisson's ratio $\mu_s = \frac{1}{3}$ is under this equation:

$$n_0 = 1 + \frac{A_C}{A}\left[\frac{5 - \frac{A_C}{A}}{4 \cdot K_{aC} \cdot \left(1 - \frac{A_C}{A}\right)} - 1\right] \tag{7}$$

3.3. Column compressibility considerations

The embankment material in the compacted stone column can still be compressed. Therefore, any loading that causes settlement is not related to column protrusion [20]. The results of the compression will increase the area of the stone column and its influence. Therefore, it is necessary to find the value of $\Delta\left(\frac{A}{A_c}\right)$ as an addition to the ratio of $\left(\frac{A_C}{A}\right)$. By identifying the value of the shear angle of the stone column filling material (φ_c) and the ratio of the constraint modulus of the column to the surrounding soil $\left(\frac{D_C}{D_S}\right)$, Priebe [20] proposes the value of $\Delta\left(\frac{A}{A_c}\right)$ can be obtained using the following equation:

$$\Delta\left(\frac{A}{A_c}\right) = \frac{1}{\left(\frac{A_C}{A}\right)_1} - 1 \tag{8}$$

where $\Delta\left(\frac{A}{A_c}\right)$ is an addition to the ratio of $\left(\frac{A_C}{A}\right)$, $\left(\frac{A_C}{A}\right)_1$ is the value of the ratio of the cross-sectional area of the column and the grid size at $\mu_s = \frac{1}{3}$. $\left(\frac{A_C}{A}\right)_1$ taken from the lowest positive value from the following equation:

$$\left(\frac{A_C}{A}\right)_1 = -\frac{4 \cdot K_{aC} \cdot (n_0 - 2) + 5}{2 \cdot (4 \cdot K_{aC} - 1)} \pm \frac{1}{2} \cdot \left(\left[\frac{4 \cdot K_{aC} \cdot (n_0 - 2) + 5}{4 \cdot K_{aC} - 1}\right]^2 + \frac{16 \cdot K_{aC} \cdot (n_0 - 1)}{4 \cdot K_{aC} - 1}\right)^{\frac{1}{2}} \tag{9}$$

The incompressible stone column material has a reducing effect on the value of the basic improvement factor n_0, so the value of the reduced improvement factor n_1 needs to be calculated by using the equation below:

$$n_1 = 1 + \frac{\overline{A_C}}{A}\left[\frac{5 - \frac{\overline{A_C}}{A}}{4 \cdot K_{aC} \cdot \left(1 - \frac{\overline{A_C}}{A}\right)} - 1\right] \tag{10}$$

where $\frac{\overline{A_C}}{A}$ is modified area ratio. The value of $\frac{\overline{A_C}}{A}$ is obtained from the equation below:

$$\frac{\overline{A_C}}{A} = \frac{1}{\left(\frac{A}{A_c}\right) + \Delta\left(\frac{A}{A_c}\right)} \tag{11}$$

3.4. Analysis of liquefaction potential in the repaired soil

According to Priebe [20], the stone column method is suitable for soil improvement, especially in seismic areas. The stone column has a certain flexibility and prevents liquefaction potential. The stability effect is resulted from the frictional resistance of the column bearing external loads, a large amount of soil weight, and the ability to reduce excess pore water pressure in the soil. A sharp drop in pore water pressure towards the column is substantial because it creates a filter cake effect that maintains the lateral support required for the bearing capacity of the column. Moreover, it also prevents a higher soil infiltration rate into the column even if the column material does not meet the specified filter criteria.

Complex conditions in seismic events are often investigated for more or less homogeneous soils. Nevertheless, practical criteria for evaluating the liquefaction potential were developed empirically. Priebe [20] proposed reducing the Cyclic Stress Ratio (*CSR*) that may occur in seismic events to be the same as the reducing load on the soil between the columns by replacing Vibro under the following equation:

$$\propto = \frac{p_s}{p} = \frac{1}{n} \tag{12}$$

where $\propto$ is the reduction factor for *CSR* value, p_s is the soil stress between the stone columns in kN/m^2, p is the total overburden stress of the soil without the stone column in kN/m^2, and n is an improvement factor. The value of safety factor (SF') after soil improvement using the stone column method can be calculated by the following equation:

$$SF' = \frac{CRR}{CSR \cdot \propto}; \propto < 1 \tag{13}$$

where $\propto$ is the reduction factor for *CSR* value, CRR is the Cyclic Resistance Ratio, and *CSR* is the Cyclic Stress Ratio.

4. Results and discussion

Stone column performance in this study was measured only based on the increased safety factor (SF) value of the depth of liquefaction. The SF value based on the calculations result by Patriaman *et al.* [9] using SPT data that showed high liquefaction potential, especially in Palu Bay's western and southern coastal areas.

The stone column design requires the value of the diameter and the distance between the center of the stone columns to be determined in order to obtain the basic improvement factor (n_0) value. In this study, the designed diameter of the stone column was 1 m. Then, the n_0 value could be determined using Equation (7). Next, the n_0 value should be reduced because the compacted material was still compressible. Therefore, it was required to calculate the reduced improvement factor n_1 using the equation (10).

The new safety factor value (*SF'*) is the final result of the Priebe's calculation Method. The value of *SF'* was obtained from Equation (13). However, according to Seed and Idriss [24], non-liquefiable classification requires the soil layer to be over a value of 1. Therefore, back analysis is done to ensure wether the *SF'* value is classified into non-liquefiable classification. In this study, the value of *SF'* was determined as 1.2. According to Sonmez [25], a safety factor of 1.2 is assumed as the lowest potential value of non-liquefied category. The goal and seek function provided by the Microsoft Excel software supported the calculation of *s* value by using predetermined *SF'* and *D* values. **Table 1** shows the analysis results of the Priebe's method.

As presented in **Table 1**, the basic improvement factor (n_0) calculation results show that the n_0 value will increase along with the decrease of the SF value. The low safety factor value represents the poor soil conditions. Hence shortened spacing between stone columns (*s*) is required to increase the safety factor value to meet the expected criteria. Furthermore, the value of n_0 will affect n_1 due to stone column compression, which will decrease the potential liquefaction value in accordance with the resulted $\propto$ value. **Table 1** shows the *s* value as the minimum allowable spacing and the minimum allowable depth of the stone columns. The calculation result shows that the spacing (s) and depth ranged between 1.03 – 5 m and 5 – 20 m, respectively.

Most of the Palu Bay SF values [9] were below 1 and indicated high liquefaction potential. They signified the importance of mitigation efforts through soil improvement. Following the soil improvement by using the stone column method, the SF' value of the area increased to exceed 1 and reached low liquefaction potential as in the expected criteria. The difference of the values of safety factor before and after the stone column construction is presented in **Figure 4**.

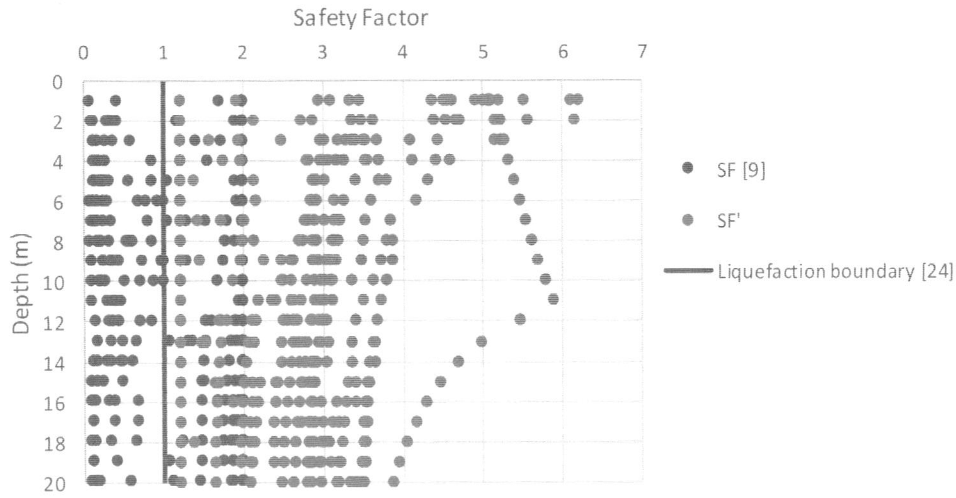

Figure 4. Safety factor values before (SF-blue colour) and after stone columns construction (SF'-orange colour).

QGIS software was applied to map the spacing and depth of the stone columns at the project site. The width of the research location is 50 m from the shoreline to the mainland. Further research is needed to determine mitigation with the stone column method in the water (sea). **Figure 5a** shows the mapping results of spacing between the stone column points and **Figure 5b** shows the mapping results of the installation depth of stone columns. As presented in **Figure 5a**, Palu Bay's western and southern areas required closer column spacing than the eastern area due to the Palu Koro Fault crossing the west region of Palu Bay, which resulted in higher significant earthquake potential. Similarly, western and southern areas required deeper installation of stone columns than the eastern areas.

Figure 5a and Figure 5b show a wide variety of stone column mitigation designs. Thus, simplification of the designs was required by dividing the Palu Bay coastal area into 5 zones. The zone division was adjusted to the uniformity of the spacing and depth values of the stone column design in each area in order to achieve optimum design.

Figure 5c shows the simplified map of the stone column construction design. Zone A included 2 locations: northwest of Palu Bay and south of Palu Bay. Zone A in the northwest of Palu Bay represented the calculation results at points BAY-4, BAY-5, BAY-6, BAY-7, and BAY-8. Meanwhile, Zone A in the south of Palu Bay represents the calculation results at BAY-2, BAY-11, BAY-12, BAY-18, and BAY-22. In this zone, according to the calculation results presented in **Table 1**, the minimum required depth is 20 m. Furthermore, the minimum stone column distance required in this zone was 1.03 m according to the value of s in BAY-7. However, to facilitate the field work, the the stone column spacing was shortened to 1 m. Furthermore, Zone B represented the calculation results at point BAY-9 and its surroundings with required stone column spacing and depth of 1.09 m and 14 m, respectively. However, to facilitate the field work, the spacing was shortened to 1 m.

Zone C represented the calculation results of points BAY-1, BAY-3, BAY-17, BAY-19, and BAY-23. The required depth and minimum spacing of the stone columns in this zone according to the *s* value of BAY-17 were 14 m and 1.14 m, respectively. However, to facilitate the field work, the spacing was shortened to 1 m. Zone D represented the calculation results of point BAY-13 and its surroundings. The required depth and minimum spacing distance for the stone column in this zone were 13 m and 1.8 m, respectively.

The spacing distance and depth of the stone columns in these designs resulted from the minimum design values of the area, that it was highly possible for a quite far location to require deep installation, and vice versa. This condition applied in Zone E, which represented the calculation results of points BAY-15 and BAY-16. The required depth and spacing distance of stone columns in this zone were 10 m and 1.2 m, respectively, according to the *s* value in BAY-17. The calculation results show that the spacing distance in this zone was denser than in Zone D. The zoning, as mentioned earlier, was expected to facilitate the implementation of the stone column construction.

Table 1. Soil profile and calculation results of stone column design.

Drilling Point	Ground Water Level (m)	Dominant Soil [9]	Lowest SF [9]	Stone column method			
				Minimum Depth (m)	Minimum *s* value (m)	*D* (m)	*SF'*
BAY-1	0.80	Sand	0.39	14	1.42	1	1.2
BAY-2	0.67	Silt	0.37	12	1.38	1	1.2
BAY-3	0.87	Gravely Sand	0.49	10	1.57	1	1.2
BAY-4	2.99	Gravely Sand	0.35	20	1.44	1	1.2
BAY-5	1.23	Silty Sand	0.08	20	1.05	1	1.2
BAY-6	0.90	Silty Sand	0.10	20	1.07	1	1.2
BAY-7	1.12	Gravely Sand	0.07	11	1.03	1	1.2
BAY-8	4.48	Gravely Sand	0.28	14	1.32	1	1.2
BAY-9	0.09	Sandy Gravel	0.12	14	1.09	1	1.2
BAY-10	2.89	Gravely Sand	2.00	-[a]	-	-	-
BAY-11	1.35	Sand and Gravel	0.30	5	1.36	1	1.2
BAY-12	0.75	Silty Sand	0.08	20	1.05	1	1.2
BAY-13	1.90	Sandy Gravel	0.84	12	2.76	1	1.2
BAY-14	2.11	Sandy Gravel	2.00	-[a]	-	-	-
BAY-15	1.25	Silty Sand	0.23	6	1.25	1	1.2
BAY-16	2.76	Gravely Sand	0.84	6	2.74	1	1.2
BAY-17	2.75	Silty Sand	0.29	16	1.34	1	1.2
BAY-18	1.22	Sandy Silt	0.19	18	1.20	1	1.2
BAY-19	1.05	Clayey Silt	0.15	8	1.14	1	1.2
BAY-20	2.30	Silty Sand	0.88	-[a]	-	-	-
BAY-22	1.57	Gravely Sand	0.17	20	1.16	1	1.2
BAY-23	1.35	Silt	0.85	9	2.82	1	1.2

[a] Not susceptible to liquefaction

5. Conclusion

Based on the potential liquefaction analysis, Palu Bay coastal area was indicated with liquefaction susceptibility that signified the importance of mitigations efforts, such as soil improvement, to prevent the area from liquefaction disasters hazard. Stone columns was selected as the mitigation method applied in this study. This method was considered suitable for sandy-soil dominated areas such as in Palu Bay coastal area.

This study applied the Priebe's method for the stone column calculation. Soil improvement through stone column method was followed with increasing SF value that exceeded 1 and resulted in low liquefaction potential as expected in the criteria.

The mapping results showed that Palu Bay's western and southern areas required stone columns with closer spacing than the eastern area. It was due to the Palu Koro Fault crossing the west area containing more significant earthquake potential. Consequently, it required deeper stone column on the western and southern areas than the eastern areas.

Due to the variety of mitigation designs exhibited by the analysis results, the stone column mitigation design was simplified by dividing the Palu Bay coastal area into 5 zones. Such simplification was expected to facilitate the implementation of stone column construction. Waste material from the ruins of the building can be utilized for the stone column construction. However, further studies are required in order to ensure whether such material meets the applicable requirements.

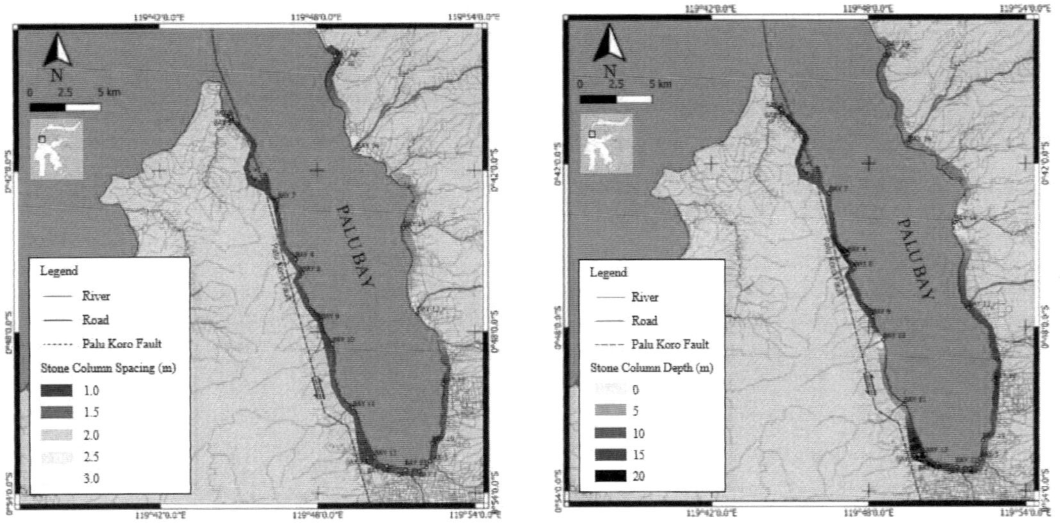

(a) (b)

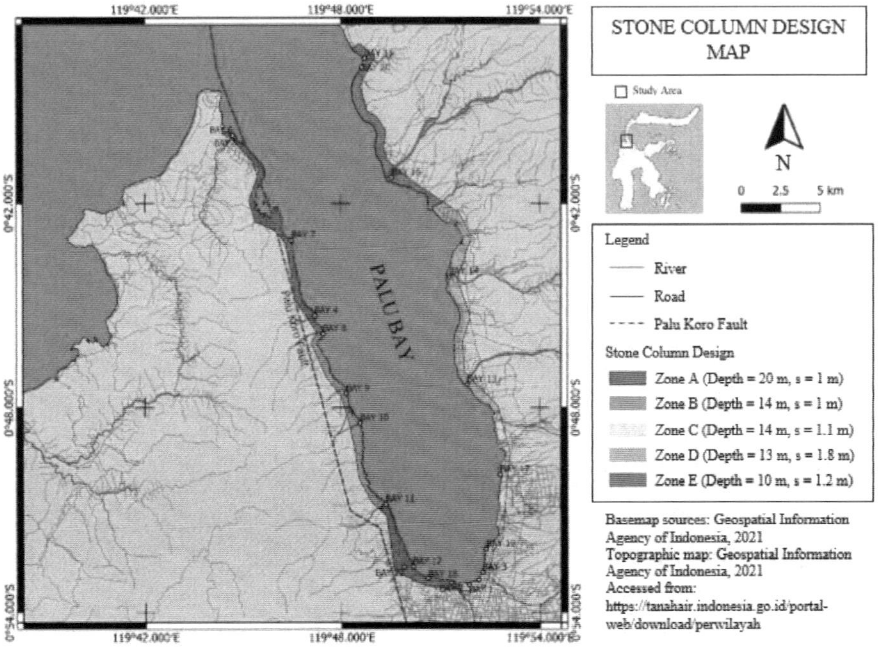

(c)

Figure 5. Map of the stone column design: (a) spacing distance (b) depth (c) simplified map (Indonesia's digital topographic map from Geospatial Infomation Agency of Indonesia [19]).

References

[1] Meteorology Climatology and Geophysics Council Republic of Indonesia 2018 *PRESS RELEASE NO: UM.505/9/D3/IX/2018* (Jakarta)

[2] Liu P L F, Higuera P, Husrin S, Prasetya G S, Prihantono J, Diastomo H, Pryambodo D G and Susmoro H 2020 Coastal landslides in Palu Bay during 2018 Sulawesi earthquake and tsunami *Landslides* **17** 2085–98

[3] Husrin S, Azahra F Y, Prihantono J, Tussadiah A and Abida R 2020 Hydrodynamics of Palu Bay during the Event of 2018 Palu Earthquake and Tsunami *Bull. Mar. Geol.* **35** 23–34

[4] Nakata K, Katsumata A and Muhari A 2020 Submarine landslide source models consistent with multiple tsunami records of the 2018 Palu tsunami, Sulawesi, Indonesia *Earth, Planets Sp.* **72**

[5] Hidayat R F, Kiyota T, Tada N, Hayakawa J and Nawir H 2020 Reconnaissance on liquefaction-induced flow failure caused by the 2018 Mw 7.5 sulawesi earthquake, palu, Indonesia *J. Eng. Technol. Sci.* **52** 51–65

[6] Kusumawardani R, Chang M, Upomo T C, Huang R C, Fansuri M H and Prayitno G A 2021 Understanding of Petobo liquefaction flowslide by 2018.09.28 Palu-Donggala Indonesia earthquake based on site reconnaissance *Landslides* **18** 3163–82

[7] Jalil A, Fathani T F, Satyarno I and Wilopo W 2021 Liquefaction in Palu: the cause of massive mudflows *Geoenvironmental Disasters* **8**

[8] Sassa S and Takagawa T 2019 Liquefied gravity flow-induced tsunami: first evidence and comparison from the 2018 Indonesia Sulawesi earthquake and tsunami disasters *Landslides* **16** 195–200

[9] Patriaman F, Fathani T F and Wilopo W 2021 Liquefaction potential analysis in Palu Bay area *IOP Conf. Ser. Earth Environ. Sci.* **930**

[10] Google Earth Pro 7.3.4.8248 (64-bit) 2021 Palu Bay Map 0° 53' 33.576'' N 119° 51' 37.451'' E, elevation 67 m *[online] Available at: https://earth.google.com/web/@-0.75136866,119.94566367,66721.43234824a,0d,35y,-*

2.4506h,2.4583t,0.0014r?utm_source=earth7&utm_campaign=vine&hl=en
[11] Hamada M 2014 Soil liquefaction and countermeasures *Springer Ser. Geomech. Geoengin.* 125–52
[12] Darwis 2017 *Dasar-Dasar Teknik Perbaikan Tanah* (Yogyakarta)
[13] Porbaha A, Zen K and Kobayashi M 1999 Deep Mixing Technology for Liquefaction Mitigation *J. Infrastruct. Syst.* **5** 21–34
[14] Huang Y and Wen Z 2015 Recent developments of soil improvement methods for seismic liquefaction mitigation *Nat. Hazards* **76** 1927–38
[15] Barksdale R D and Bachus R C 1983 *Design and Construction Stone Columns Vol 1* vol 1 (Virginia)
[16] Ellouze S, Bouassida M, Hazzar L and Mroueh H 2010 On settlement of stone column foundation by Priebe's method *Proc. Inst. Civ. Eng. Gr. Improv.* **163** 101–7
[17] Nicholson P G 2015 *Soil Improvement and Ground Modification Methods* (Waltham)
[18] Nazariafshar J, Mehrannia N, Kalantary F and Ganjian N 2019 Bearing Capacity of Group of Stone Columns with Granular Blankets *Int. J. Civ. Eng.* **17** 253–63
[19] Badan Informasi Geospasial 2021 Central Sulawesi Province Map *[online] Available at: https://tanahair.indonesia.go.id/portal-web/download/perwilayah*
[20] Priebe H J 1995 The design of vibro replacement *Gr. Eng.* **28** 31–7
[21] Ng K S and Tan S A 2014 Design and analyses of floating stone columns *Soils Found.* **54** 478–87
[22] Sexton B G, McCabe B A and Castro J 2014 Appraising stone column settlement prediction methods using finite element analyses *Acta Geotech.* **9** 993–1011
[23] Castro J 2017 Modeling stone columns *Materials (Basel).* **10**
[24] Seed H B and Idriss I M 1970 *A Simplified Procedure for Evaluating Soil Liquefaction Potential* (National Technical Information Service)
[25] Sonmez H 2003 Modification of the liquefaction potential index and liquefaction susceptibility mapping for a liquefaction-prone area (Inegol, Turkey) *Environ. Geol.* **44** 862–71

Acknowledgement
The authors would like to express their gratitude for the support provided by the Office of River Area of Sulawesi III, Directorate General of Water Resources, Ministry of Public Works, and Public Housing, Republic of Indonesia.

Distribution, petrographic analysis, and preliminary source identification of Hindu-Buddhist artifacts in Karangsambung-Karangbolong National Geopark and surrounding areas, Kebumen, Central Java, Indonesia

J Primizakaria[1], N I Setiawan[1*], C Ansori [1,2]

[1]Geological Engineering Department, Engineering Faculty, Universitas Gadjah Mada, Yogyakarta, Indonesia
[2]Geotechnology Research Center, National Research and Innovation Agency, Bandung, Indonesia.
*Corresponding author's e-mail: nugroho.setiawan@ugm.ac.id

Abstrak. Hindu-Buddhist artifacts (e.g., linga, yoni, statues, and craft objects such as pottery) were discovered in Karangsambung-Karangbolong National Geopark (KKNG) and surrounding areas, Kebumen, Central Java, Indonesia. The information related to the distribution, composition, and source of the Hindu-Buddhist artifacts is essential to support the development and management of the KKNG, particularly on the relation of lithology units with cultural diversity. A geological field survey was conducted to produce a distribution map of the artifacts, then overlaid with the geological map of Banyumas and Kebumen sheets scale of 1:100,000. Totally of 10 Hindu-Buddhist artifacts were distributed in the research area, grouped into 3 areas, namely Karangbolong Hills, Kebumen, and southern Kebumen areas. Furthermore, 10 samples of artifacts were collected and petrographically analyzed to acquire their compositions and to determine their source identification. Based on their distribution and petrographical analyses, the artifacts were made of andesite (linga of Tugu, linga, and yoni of Sumberadi, tomb of Trasidi, Celeng Stone, yoni of Rowo), tuffaceous sandstone (tomb of Kuwu Panjer), and sandy micrite (Ganesha statue of Kejawang) sourced from Halang Formation. Meanwhile, andesite rocks used to construct the Kalbut Stone and linga of Pejarakan were sourced from Gabon Formation.

1. Introduction

The ancient Kebumen area was home to several older stratum of societies that lived in the Megalithic era, marked by the stepped pyramid and andesite structures [1]. In addition, the existence of the Ancient Kebumen area was closer to the influence and territory of Medang or Ancient Mataram Kingdom (1016–732 AD) and the Pasir Luhur or Ancient Galuh Kingdoms (1482–1475 DC) [1,2]. The artifacts and analysis of religious symbols provide information that various socio-religious stratum have contributed to the socio-cultural life of the present-day Kebumen society [1].

Karangsambung and Karangbolong are formed by complex tectonic settings. It is shown by the geological diversity of morphology, lithology, geological structures, and geological history (Figure 1) [3,4]. This area is located in the South Serayu Physiographic Zone, composed of igneous, metamorphic, and sedimentary rocks of the Cenozoic to Mesozoic ages [3,4]. Humans widely used the outcrops of the stones to support their life and culture. One of the cultures, which is an artifact, is scattered in Karangsambung and Karangbolong areas. The Kebumen and Karangbolong area artifacts were made of volcanic materials such as andesite and tuffaceous sandstone. The materials were selected based on their physical condition, which can be sculpted, compact, robust, and easily found in their living area [1]. Research on the source of the material was carried out based on the petrographic character approach to understand the origin of the artifacts.

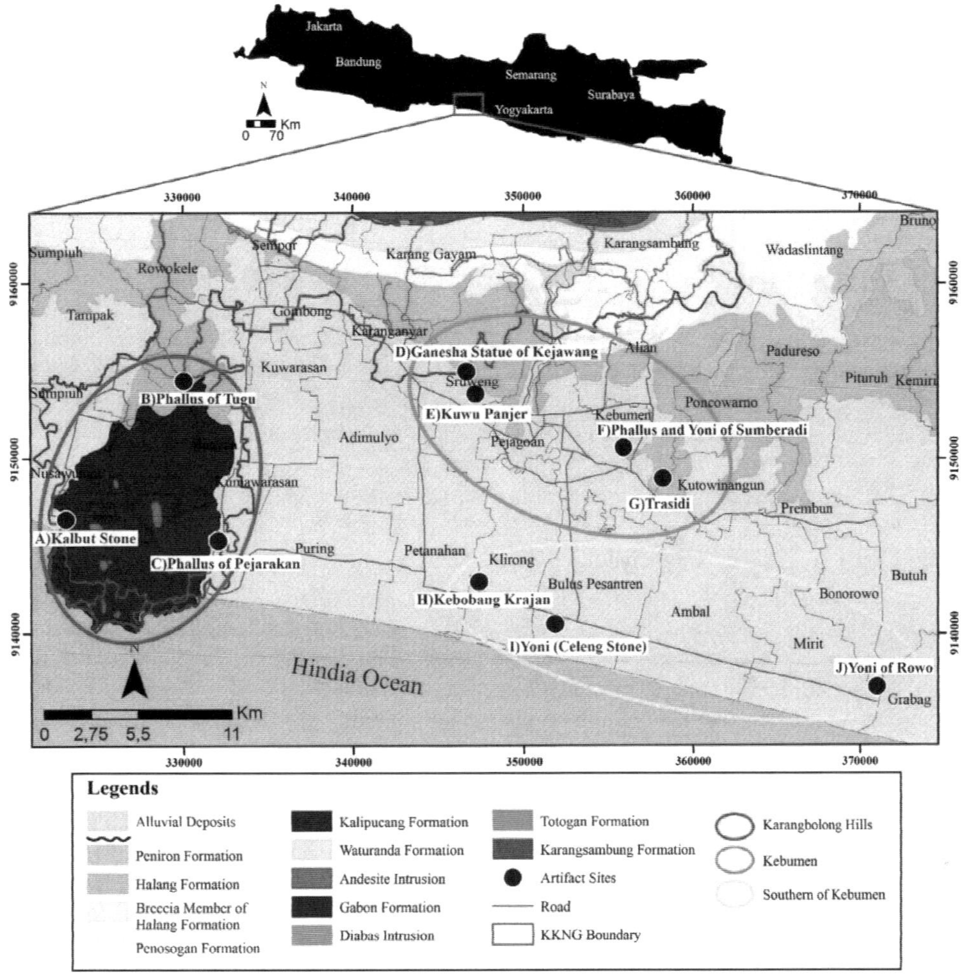

Figure 1. Distribution map of Hindu-Buddhist artifacts in the Karangsambung-Karangbolong National Geopark and its surrounding areas overlaid on the geological map of Kebumen and Banyumas Sheets [3,4]. Road maps and administrative regions of Central Java were provided by the Indonesian Geospatial Information Agency [5].

Karangsambung and Karangbolong have been designated as national geopark since 2018. It covers an area of 543,599 km^2 and encompasses 12 subdistricts and 118 villages (Figure 2). The presence of the Hindu-Buddhist artifacts in the Karangsambung-Karangbolong National Geopark (KKNG) and surrounding areas provide more value to the cultural sites of the national geopark. However, the sources of the material artifacts still. Hence, this study aims to identify the Hindu-Buddhist artifact's material, including the characteristics and distribution in the KKNG area. This research supports the development and management of the KKNG, which aims to protect the earth's diversity and conservation of the environment and to support broad earth science education.

2. Simplified Regional Geology of the KKNG area

A simplified regional geology map can be seen in Figure 1. The oldest rock formation in the Karangsambung is Luk-Ulo Mélange Complex (e.g., high-pressure metamorphic rocks, ophiolites), with the ages of the Cretaceous [6]. This complex is formed by subduction and accretionary complexes [6]. The mélange rocks are unconformably overlain by the Middle to Late Eocene of the Karangsambung Formation. Based on Asikin et al. [3], the formation comprises sedimentary rock such as limestone, conglomerate, sandstone, and basalt fragments embedded in the scaly clays matrix. Oligocene to Early Miocene of Totogan Formation overlies this formation, composed of clay breccia with claystone, sandstone, limestone, and basalt fragments. Above this formation is Early Miocene of Waturanda Formation, consisting of coarse sandstone and breccia with andesite, basalt, sandstone, and tuff fragments. Early to Middle Miocene of Penosogan Formation overlies this formation, composed of alternating calcareous sandstones, claystone, tuff, marl, and calcarenite. Above this formation is Halang Formation with the ages of Late Miocene to Pliocene, composed of alternating sandstone, limestone, marl, and tuff, with breccia interlayers. The Halang Formation is unconformably overlain by Pleistocene to Recent ages of volcanic and sedimentary deposits. The deposits consist of alluvium, young Mount Sumbing volcanic, and dan coastal deposits.

Meanwhile, the Karangbolong area consists of Gabon, Kalipucang, Halang Formations, and alluvial deposits [4]. Based on Asikin et al. [4], the Late Oligocene to Early Miocene of Gabon Formation comprises breccia with andesite fragments, tuff, coarse sandstone, lapilli, and lava deposits. Furthermore, andesite intruded the Gabon Formation in the Early to Middle Miocene. Above the Gabon Formation, is Kalipucang Formation, composed of reef and clastic limestones with bituminous shale at the bottom. Halang Formation, deposited in the Karangsambung and the Karabolong area, overlaid the Kalipucang Formation in the Middle Miocene to Pliocene.

3. Methodology

Mapping activities were carried out to determine the distribution of the artifacts. The primary distribution of the Hindu-Buddhist artifacts in the research area is based on Kebumen Regency cultural heritage objects and local communities' information. Overlaid the distribution of the artifacts with the geological map is one of the steps to acquiring information about the artifact's material sources from rock characteristics [7,8]. The artifact's distribution map is overlaid with the geological map scale 1:100,000 of Banyumas and Kebumen sheets [3,4] (Figure 1). The approach that can be used is mapping the distribution of the artifacts, collecting rock samples from the artifacts, and conducting petrographical analysis [9].

Representative samples are taken from the artifact rocks, which are later analyzed petrographically to acquire the textures and compositions of the rocks. The size of the rock samples is approximately ~10 cm^3 from the artifacts with the legalized permission from Kebumen Regency National Geopark Management Agency and local communities. In all cases, unaltered and unweathered parts of artifacts were selected for thin sections analyses. The rock descriptions of the artifacts are compared with the geological formation's lithology, where the artifacts are seated.

4. Site Location

The distribution of Hindu-Buddhist artifacts in the Karangsambung-Karangbolong Geopark area can be divided into 3 regions, namely Karangbolong Hills, Kebumen, and the southern part of Kebumen (Figure 1). Photographs of the artifacts can be seen in Figure 2, which are consistently in order with the location in the Figure 1.

Figure 2. Artifacts are scattered in the Karangsambung-Karangbolong Geopark and surrounding area. The location of artifacts can be seen in Figure 1. (a) Kalbut Stone; (b) Linga of Tugu; (c) Linga of Pejarakan; (d) Ganesha Statue of Kejawang; (e) tomb of Kuwu Panjer; (f) Linga and Yoni of Sumberadi; (g) tomb of Trasidi; (h) well of Kebobang Krajan; (i) Celeng Stone; (j) Yoni of Rowo.

4.1. Kalbut Stone
This site consists of a linga and 2 stone water tanks (Figure 2a). It is located in Candirenggo Village, Ayah, Kebumen, Central Java. One stone water tank has 0.6 m length, 0.8 m width, and 0.5 m height. The other stone water tank has 0.5 m length, 1.5 m width, and 0.8 m height. Based on the petrographical observation, it shows characteristics of andesite with porphyritic texture. The phenocryst consists of plagioclase, clinopyroxene, and opaque minerals. The alteration of pyroxene into goethite indicates that the rock has undergone an oxidation process (Figure 3a).

4.2. Linga of Tugu
The linga of Tugu is located in Tugu Village, Buayan, Kebumen, Central Java (Figure 2b). The linga has 0.3 m in diameter and 1 m in height. The thin section shows the characteristics of the andesite. The rock has porphyritic and trachytic textures, indicating that the rock originally was lava. The phenocryst consists of plagioclase, clinopyroxene, and opaque minerals, which are embedded in the glassy groundmass (Figure 3b).

4.3. Linga of Pejarakan
Linga of Pejarakan is located in Pejarakan Village, Buayan, Kebumen, Central Java (Figure 2c). The artifacts in this area are 2 linga, which have different sizes. The enormous linga has 0.2 m length, 0.2 m width, and 0.4 m height, while the smaller linga has 0.2 m length, 0.1 m width, and 0.3 m height. The thin section shows the characteristics of the andesite. The rock has porphyritic and trachytic textures, indicating that the rock originally was lava. The phenocryst consists of plagioclase, clinopyroxene, and opaque minerals embedded in the glassy groundmass (Figure 3c).

4.4. Ganesha Statue of Kejawang

The artifact is a god Ganesha statue located in Kejawang Village, Sruweng, Kebumen, Central Java (Figure 2d). The statue has 0.3 m length, 0.3 m width, and 0.6 m height. The thin section shows the characteristics of sandy micrite. The rock has a texture of poor to moderate sorting. The fragments are microfossils and volcanic materials (plagioclase and clinopyroxene), which are embedded in the micritic matrix (Figure 3d). Some microfossils have been dissolved, which indicates the rock has undergone intensive weathering.

4.5. Tomb of Kuwu Panjer

The site is a rock tomb located in Karangsari Village, Sruweng, Kebumen. Central Java (Figure 2e). The rock tomb has a size of approximately 3 m long and 2 m wide. The thin section shows the characteristics of tuffaceous sandstone. The rock has a fine-grain size with good sorting. It is dominantly composed of glass, opaque minerals, and microfossils (Figure 3e). Furthermore, it indicates that the rock was resedimented volcanic rock.

4.6. Linga and Yoni of Sumberadi

The site is located in Sumberadi Village, Kebumen, Central Java. The artifacts in this area are linga and yoni (Figure 2f). The linga has 1 m long, 1 m wide, and 1 m in height, while the yoni has 0.2 m in diameter and 0.2 m in height. The thin section shows the characteristics of andesite rock. The rock has porphyritic and trachytic textures, which indicate that the rock originally was lava. It has a phenocryst composition of plagioclase, clinopyroxene, and opaque minerals embedded in the glassy groundmass. The alteration of plagioclase into clay minerals might be indicated that the rock has undergone an intense weathering process (Figure 3f).

4.7. Tomb of Trasidi

The site is a rock tomb located in Argopeni Village, Kebumen, Central Java (Figure 2g). The site is 3 m long and 4 m wide. The thin section shows the characteristics of the andesite. The rock has porphyritic and trachytic textures, which indicate that the rock was lava. The phenocryst consists of plagioclase, clinopyroxene, and opaque minerals embedded in glassy groundmass (Figure 3g).

4.8. Well of Kebobang Krajan

The site is a well with an inner wall made of pottery (Figure 2h). It is located in Gebangsari Village, Klirong, Kebumen, Central Java. The well has 0.6 m in diameter and 5 m in depth. The samples taken at this site were earthenware. The observation of a thin section shows dominantly composed of plagioclase and clinopyroxene. It is interpreted that the soil used to manufacture the pottery results from weathering volcanic rocks (Figure 3h).

4.9. Celeng Stone

The yoni artifact is located in Ayamputih Village, Buluspesantren, Kebumen, Central Java. The yoni has 0.5 m length, 0.4 m width, and 0.5 m height. The yoni's condition is half-destroyed (Figure 2i). The thin section shows the characteristics of andesite rock. The rock has porphyritic and trachytic textures, which indicate that the rock originally was lava. It has a phenocryst composition of plagioclase, clinopyroxene, and opaque minerals embedded in glassy groundmass (Figure 3i).

4.10. Yoni of Rowo

The yoni artifact is located in Rowo Village, Bonorowo, Kebumen, Central Java (Figure 2j). The yoni is 0.4 m long, 0.4 m wide, and 0.5 m in height. The thin section shows characteristics of andesite. The rock has porphyritic and trachytic textures, which indicate that the rock originally was lava. It has a phenocryst composition of plagioclase, clinopyroxene, and opaque minerals embedded in glassy groundmass (Figure 3j).

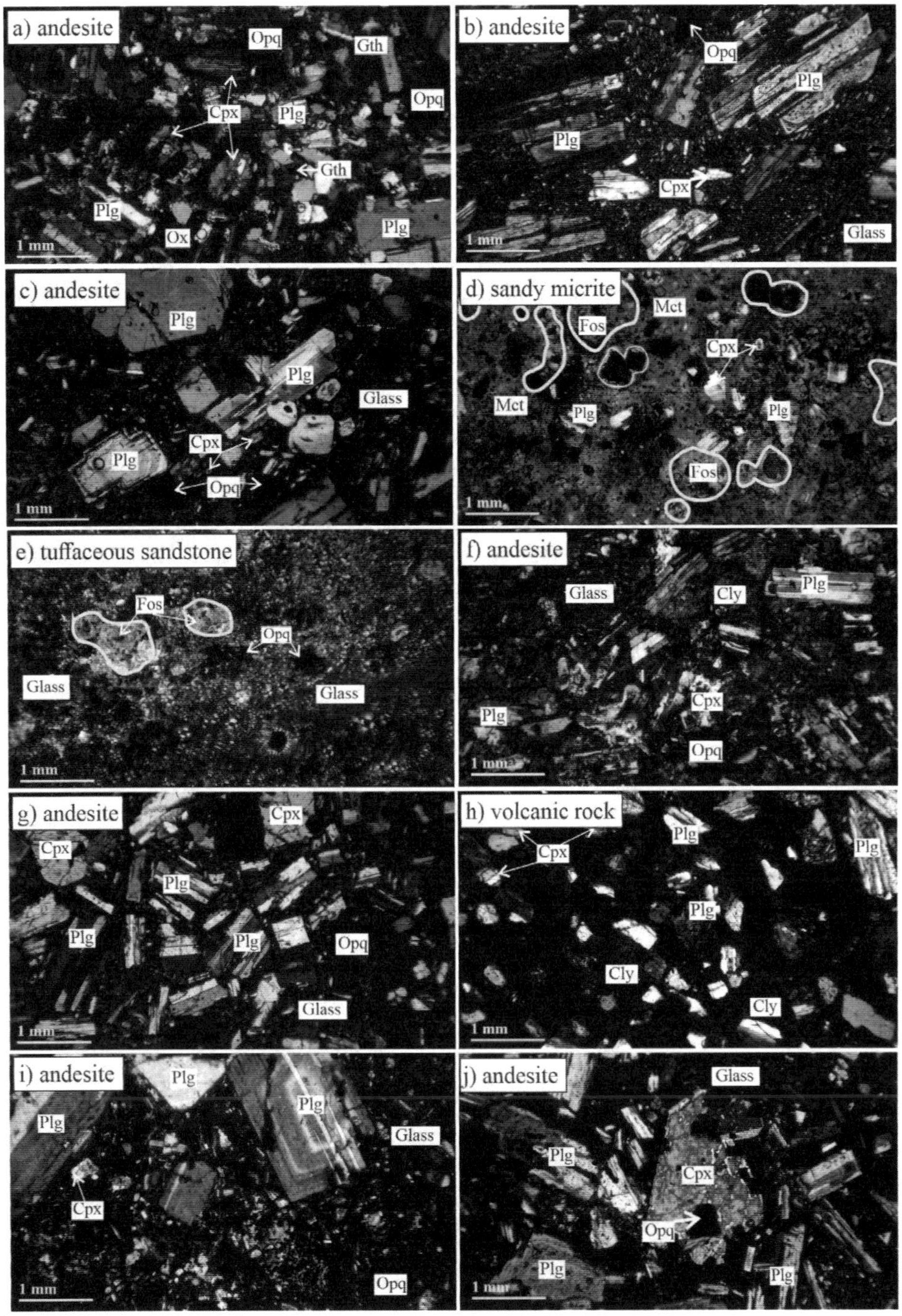

Figure 3. Cross-polarized thin-section images of representative artifact samples from the studied area. (a) Andesite of Kalbut Stone; (b) andesite (linga of Tugu); (c) andesite (linga of Pejarakan); (d) Sandy micrite (ganesha statue of Kejawang); (e) tuffaceous sandstone (tomb of Kuwu Panjer); (f) andesite (linga and yoni of Sumberadi); (g) andesite (tomb of Trasidi); (h) volcanic rock (well of Kebobang Krajan); (i) andesite (Celeng Stone); (j) andesite (yoni of Rowo). Abbreviation of the minerals as follows: Plg = plagioclase; Cpx = clinopyroxene; Opq = opaque minerals; Fos= fossil; Mct = micrite; Cly = clay.

5. Discussion

Based on a field survey conducted in the KKNG area, the distribution of Hindu-Buddhist artifacts can be divided into 3 regions, namely the Karangbolong Hills, Kebumen, and the southern Kebumen (Figure 1). Artifacts in the Karangbolong Hills area are sites of Kalbut Stone, Linga of Pejarakan, and Linga of Tugu. Artifacts in the Kebumen area are ganesha statue of Kejawang, tomb of Kuwu Panjer, yoni of Sumberadi, and tomb of Trasidi. Meanwhile, artifacts in the southern part of Kebumen are the well of Kebobang Krajan, Celeng Stone, and yoni of Rowo. Artifact sites that have been used as cultural heritage objects by the Education Department of Kebumen Regency are the sites of tomb of Kuwu Panjer, Kalbut Stone, tomb of Trasidi, well of Kebobang Krajan, and Linga and Yoni of Sumberadi. The summary of artifact name, location, formation, rock type, and composition can be seen in Table 1.

Table 1. Summary: artifact name, location, formation, rock type, and composition.

Artifact Name	Location Region	Seated on Formation [3,4]	Rock Type	Composition	Possibly Sources Formation
Kalbut Stone	Karangbolong Hills	Kalipucang	Andesite	Plg, Cpx, and Opq	Gabon
Linga of Tugu	Karangbolong Hills	Halang	Andesite	Plg, Cpx, Opq, and glass	Halang
Linga of Pejarakan	Karangbolong Hills	Alluvial Deposits	Andesite	Plg, Cpx, Opq, and glass	Gabon
Ganesha Statue of Kejawang	Kebumen	Halang	Sandy micrite	Altered microfossils, Plg, Cpx, and micrite	Halang
Tomb of Kuwu Panjer	Kebumen	Halang	Tuffaceous sandstone	Glass, Opq, and altered microfossils	Halang
Linga and Yoni of Sumberadi	Kebumen	Halang	Andesite	Plg, Cpx, Opq, and glass	Halang
Tomb of Trasidi	Kebumen	Halang	Andesite	Plg, Cpx, Opq, and glass	Halang
Well of Kebobang Krajan	Southern Kebumen	Alluvial Deposits	Volcanic rock	Plg, Cpx, Opq, and glass	Halang
Celeng Stone	Southern Kebumen	Alluvial Deposits	Andesite	Plg, Cpx, Opq, and glass	Halang
Yoni of Rowo	Southern Kebumen	Alluvial Deposits	Andesite	Plg, Cpx, Opq, and glass	Halang

Mineral abbreviation : Plg (plagioclase); Cpx (clinopyroxene); Opq (opaque minerals)

Geologically, Kalbut Stone is located in the Kalipucang Formation; linga of Pejarakan is situated in the alluvial deposit, while linga of Tugu, Ganesha statue of Kejawang, tomb of Kuwu Panjer, yoni of Sumberadi, and tomb of Trasidi are located in the Halang Formation. Kebobang Krajan, Celeng Stone, and Yoni of Rowo are situated in the alluvial deposit (Figure 1).

The Kalbut Stone is seated in the Kalipucang Formation. However, the lithology of Kalbut Stone was not sourced from the Kalipucang Formation because the lithology of the Kalbut Stone is andesite, while the formation consists of reef limestone [4]. The seated of linga of Pejarakan, well Kebobang Krajan, Celeng Stone, and yoni of Rowo are in the Alluvial Deposit. However, the artifacts were not sourced from the Alluvial Deposits. The lithology of the artifacts are andesite and volcanic rock, while the Alluvial Deposit is composed of clay, silt, sand, and gravel [4].

The Halang Formation was deposited in the Late Miocene to Early Pliocene ages with a turbidite flow mechanism in a marine environment [10]. This formation is deposited conformably above the Penosogan Formation and has a total thickness of approximately 2.4 km [3,11]. This formation has volcanic facies with lithological characteristics of andesite. It is dominantly composed of plagioclase and pyroxene phenocrysts embedded in the glassy groundmass [10]. Furthermore, this formation also consists of pyroclastic rock, tuff with dominant glass and crystal compositions. The occurrences of microfossils indicate that these pyroclastic rocks were resedimented in marine environments [10]. The formation also consists of sedimentary sandy micrite rocks dominantly composed of micrite composition [10]. These characteristics are similar to the characteristics of the rocks that comprise the artifacts in the study area. The similarity is also supported by the artifact's proximity to the Halang Formation. Therefore, it can be interpreted that the artifacts in the KKNG area are made from andesite (linga of Tugu, linga and yoni of Sumberadi, tomb of Trasidi, Celeng Stone, yoni of Rowo), tuffaceous sandstone (tomb of Kuwu Panjer), and sandy micrite (Ganesha statue of Kejawang) were sourced from the Halang Formation (Table 1).

The Gabon Formation was deposited in the Early Miocene - Late Oligocene, composed of volcanic breccia, andesite lava, tuff, lapilli, and lava material [12]. The characteristics of andesite are dominantly composed of plagioclase and pyroxene phenocrysts. The phenocrysts are embedded in the glassy groundmass [12]. This formation comprises the Karangbolong Hills and is close to the Kalbut Stone and linga of Pejarakan. It is interpreted that the andesite rocks that compose the Kalbut Stone and linga of Pejarakan were sources from the Gabon Formation (Table 1). However, the possible quarry locations of Halang and Gabon Formations have not been determined. Detailed investigations with geochemical analyses are needed to give solid suggestions and conclusions of the artifact's sources.

6. Conclusion

The distribution of Hindu-Buddhist artifacts in the Karangsambung-Karangbolong National Geopark can be divided into 3 regions: Karangbolong Hill, Kebumen, and South Kebumen, with a total of three regions of 10 artifacts. The artifacts comprise linga, yoni, statue, well, and water stone tank. Based on the petrographic analysis, the artifacts were made from andesite (linga of Tugu, linga and yoni of Sumberadi, tomb of Trasidi, Celeng Stone, and yoni of Rowo), tuffaceous sandstone (tomb of Kuwu Panjer), and sandy micrite (Ganesha statue of Kejawang) were sourced from the Halang Formation. In contrast, the andesite rocks that compose the Kalbut Stone and linga of Pejarakan were sources from the Gabon Formation.

References

[1] Hindarto T and Ansori C 2020 Sistem Sosial dan Keagamaan Masyarakat Megalitik dan Hindu Kuno di Lima Wilayah Kecamatan di Kabupaten Kebumen *Jurnal Analisa Sosiologi* **9** 224–66

[2] Nastiti T S 2014 Jejak-jejak Peradaban Hindu-Buddha di Nusantara *Kalpataru: Majalah Arkeologi* **23** 35–49

[3] Asikin S, Handoyo A, Busono H, and Gafoer S 1992 Geologic map of Kebumen quadrangle, Java, scale 1: 100.000 *Geological Research and Development Center, Bandung*

[4] Asikin S, Handoyo A, Busono H and Gafoer S 1992 Geologic Map of Banyumas Quadrangle, Java. scale 1: 100 000 *Geological Research and Development Center, Bandung*

[5] Badan Informasi Geospasial *https://tanahair.indonesia.go.id/*

[6] Alfing J, Bröcker M and Setiawan N I 2021 Rb–Sr geochronology of metamorphic rocks from the Central Indonesian Accretionary Collision Complex: Additional age constraints for the Meratus and Luk Ulo complexes (South Kalimantan and Central Java) *Lithos* **388–389** 1–11

[7] Malyk-Selivanova N, Ashley G M, Gal R, Glascock M D, and Neff H 1998 Geological–geochemical approach to "sourcing" of prehistoric chert artifacts, northwestern Alaska *Geoarchaeology: An international journal* **13** 673–708

[8] Shipley III W E and Graham E 1987 Petrographic analysis and preliminary source identification of selected stone artifacts from the Maya sites of Seibal and Uaxactun, Guatemala *J Archaeol Sci* **14** 367–83

[9] Harjanto A 2011 Vulkanostratigrafi di Daerah Kulon Progo dan Sekitarnya, Daerah Istimewa Yogyakarta *Jurnal Ilmiah Magister Teknik Geologi* **4** 30–45

[10] Hendratno A and Monica S 2018 Studi fasies vulkanik pada Formasi Halang di Kecamatan Somagede, Kabupaten Banyumas *Proceeding Seminar Nasional Kebumian ke-11* pp 964–81

[11] Ibrahim H R and Mardiana U 2014 Studi stratigrafi pada Formasi Halang berdasarkan data measured section pada Daerah Ciwaru, Kecamatan Ciwaru, Kabupaten Kuningan, Provinsi Jawa Barat *Universitas Padjajaran*
[12] Ansori C and Wardhani F A 2019 Tipe Magmatik Batuan Beku Formasi Gabon di Tinggian Karangbolong, Kebumen *Jurnal Geologi dan Sumberdaya Mineral* **20** 63–74

Acknowledgments
We want to thank the Geological Engineering Department, Engineering Faculty, Universitas Gadjah Mada, and the RTA Program for providing the research grant to facilitate this research.

Morphometric Characteristic and Distribution of Hummocky Hills in Debris Avalanche Deposit of Galunggung Volcano, West Java, Indonesia

A S Alfin[1], **H E Wibowo**[1*], **A Harijoko**[1,2]

[1]Departement of Geological Engineering, Faculty of Engineering, Universitas Gadjah Mada. Jalan Grafika no 2, Yogyakarta 55281, Indonesia
[2]Center of Disaster Studies, Universitas Gadjah Mada, Bulaksumur, Yogyakarta 55281, Indonesia
*Corresponding author's e-mail: haryo.edi.w@ugm.ac.id

Abstract. Sector collapse of a volcanic body produces topography of horseshoe-shaped caldera incisions and a widespread hummocky hill resembling debris avalanche deposit. The study of morphometry, distribution, and the alignment of the hummocky hills is necessary to understand the source and deposition process of the debris-avalanche deposit. Galunggung volcano is, one of the volcanoes in Java Island, attributed with 7 km wide of caldera opens towards the east direction (N 135° E) where hummocky hills cover an area of ~32 km2. Morphometry of the hummocky hills is 500-700,000 m2 in size, elongation ratio of 1.3-3.9. Hummocky hills are distributed 0.5 to 20 km distance from the summit of the Galunggung volcano. The elongation orientation of the hummocky hills is deviated from the main flow direction forming various angles. We divide the deviation angle into parallel (0-45°) and perpendicular orientation (45-90°). The maximum size of the hummocky hills generally decreases towards increasing distance from the summit. These hills also show changes in alignment from perpendicular to parallel, suggesting a typical free flow pattern of debris-avalanche deposits.

1 Introduction

Sector collapse is a volcano body collapse that causes rock slides [1]. Several Quaternary volcanoes in Java Island, Indonesia, have undergone sector collapses, like Raung Volcano [2], Galunggung Volcano [3], and Anak Krakatau Volcano [4]. The tsunami of Banten Beach in 2018 was generated from the collapse of the southwest slope of Mount Anak Krakatau [4]. The Sector collapse resulted in a debris avalanche and debris flow deposits. Sector collapse is hazardous because it can transport voluminous debris-avalanche deposits and debris flows deposits, covering area tens of kilometers from the source of the collapse [5]. With a high level of danger, it will be very vulnerable for residents around the slopes of the volcano [5].

Debris avalanche could be a product of large-scale sector collapse triggered by magma intrusion, hydrothermal variations, high rainfall, tectonic activity, or melting ice [6]. This transported rock mass will produce debris avalanche deposits. Debris avalanche deposit will produce a wavy topography spreading on the slopes of the volcano called hummocky hills. The existence of those hills is evidence of a sector collapse.

Galunggung Volcano has evidence of sector collapses from the presence of horseshoe-shaped caldera in the summit area and hummocky hills on its southeastern flank up to the distal zone. Previous work

[3] stated that debris avalanche deposits covered 170 km2 of the south-eastern part of Galunggung Volcano and reached 23 km distance from the crater (Fig.1).

Mitigation of volcanic disasters requires a thorough understanding of the respective volcanic processes. We reported the morphometric characteristics and distribution of the hummocky hills in the eastern flank of the Galunggung volcano. We aim to understand the depositional process of the debris avalanche deposit from the sector collapse of this volcano.

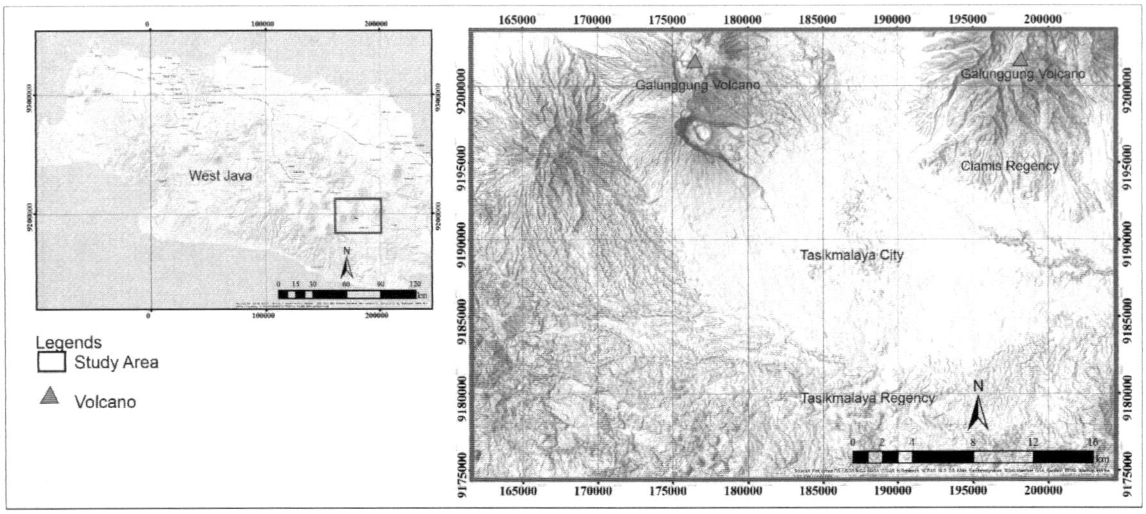

Figure 1. Research area on the east flank of Galunggung volcano represented by hummocky topography around Tasikmalaya City. (base map from World Hillshade ESRI images is freely downloaded from (https://www.arcgis.com/index.html).

2 Geology of Galunggung Volcano

Previous work [3] divides the Stratigraphy of the Galunglcano into 3 formations: the Old Galunggung Formation, Tasikmalaya Formation, and Cibanjaran Formation. The lithology of the Old Galunggung Formation consists of basaltic lava and basaltic pyroclastic flow deposit. The Tasikmalaya Formation was formed by rocks associated with the formation of the Galunggung amphitheater. Cibanjaran Formation resembles recent volcanic activities such as 1822, 1894, 1918, and 1982 – 1983 eruptive events producing pyroclastic fall, pyroclastic flow deposit, and lava dome.. This research focuses on the Tasikmalaya Formation. This formation was formed between 4200 to 150 years ago. The massive debris avalanche deposits formed hummocky morphology and are associated with pyroclastic-flow deposits and lahar deposits.

3 Methods

This research combines DEM analysis and field observations. DEM analysis is used to measure morphometric parameters, including basal area and longest axis, and distance from summit. In addition, this analysis is also used to measure displacement angle which represents deviation angle of the individual flow direction to the orientation of the longest axis (Fig. 2) following method of [7,8,9,10]. DEM data, freely downloadable from (http://tides.big.go.id/DEMNAS/), is converted into hill-shade with shadows orientation of 0°, 45°, 90°, 135°, 180°, 225°, 270°, 315°, and angle of enlightenment 30°. Field observation was conducted to obtain morphological, lithological, and structural data of the debris avalanche deposit.

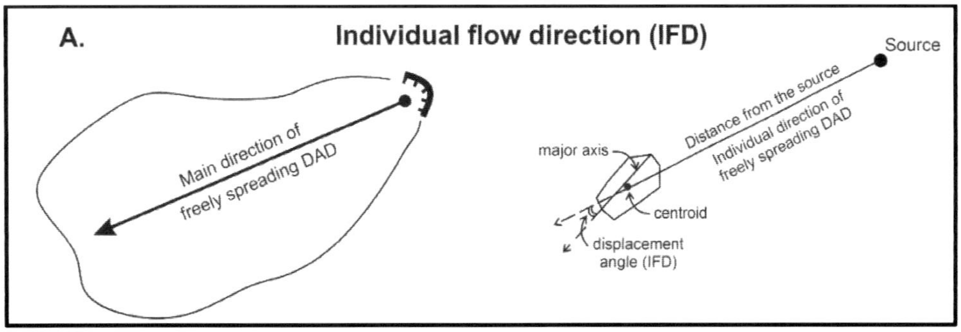

Figure 2. Morphometric parameter of debris avalanche hummock for freely spreading, adapted from [7]. DAD refers to Debris Avalanche Deposits.

4 Result

4.1. Lithology and structural geology of hummocky hills

We observed lithology in 74-observation points. The lithology variations of the hummock hills are lava, tuff breccias, and lapilli tuff. Some lava shows jigsaw crack structures typical of debris avalanche products (Figure 3). These lavas mostly form the isolated hummocky hills morphology. The lava block has a grey colour, solid, compact, and shows porphyritic texture with plagioclase and pyroxene phenocrysts embedded in the aphanitic groundmass.

Tuff breccias are found in hummock morphology and are sometimes together with andesitic lava. Volcanic breccias are poorly sorted, contain sub-angular – sub-rounded andesite fragments in a sand-silt-sized matrix. Lapilli tuff is generally found in the uppermost layer of hummock hills and is thought to be deposited after the sector collapse event of the Galunggung volcano. The lapilli tuff is white, relatively well sorted, and is composed of scoria and andesite lithics.

The distribution of lithology in the hummocky hills is dominated by two facies, block and mixed facies. Block facies are intact units of volcanic bodies that are transported and deposited in intact form. Mixed facies are rocks that have very intense orderly cracking which the rocks experienced brecciation and are then filled with sand and pyroclastic deposits [6]. In the Galunggung debris avalanche deposit, block facies are distributed in the southern, western, and eastern parts while mixed facies are in the central and northern parts.

The geological structure found in the study area is a normal fault with N300°E orientation. The occurrence of this fault is localized in the hummocky hill of STA 20 and not associated with any major lineament in the area.

Figure 3. Photography of the outcrop of. a) hummocky hills taken from Situgede lake. Symbol "H" resembles location of individual hill morphology. b) block facies composed of jigsaw crack lava, c) block facies composed of pyroclastic flow deposit mixed with matrix composed of mud.

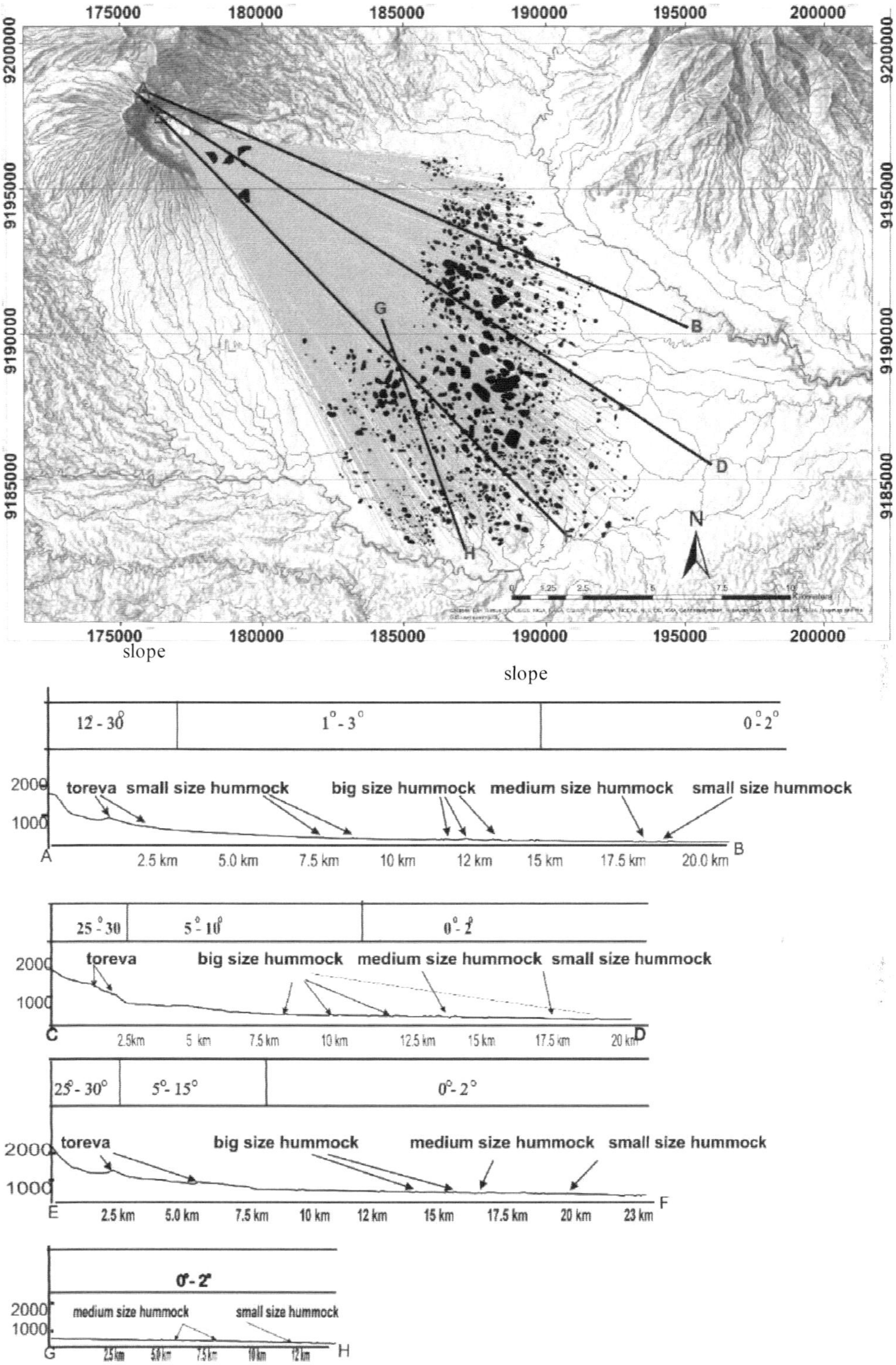

Figure 4. Delineated hummocky hills (black polygon) each with the individual flow direction (grey line). Topographic section lines (black lines) cover both parallel and perpendicular patterns to the flow direction. Base map from World Hillshade ESRI images is freely downloadede from (https://www.arcgis.com/index.html).

4.2. Morphometric of the hummocky hills

We identified about 754 hummocky hills morphology from 0.5 to 20 km distance from the summit. The hummocky hills have wavy morphology with conical to elongated shape hills and slopes between 30°-60° (Figure 3). Basal size of the hummocky hills ranges from 500 to 700,000 m^2. The size of the hummocky hills is classified into three classes: big (> 60,000 m^2), medium (60,000-2,500 m^2), and small (<2,500 m^2) following model-based objective on probability plot of [11]. Occurrence of the hummocky hills varied downslope. Topographic profiles show that the hummocky hills start to appear in the medial zone (elevation 1213 masl, 17° slope), absence in elevation 646 masl at 3° slope, and reappear since elevation 475 masl, 2° slope to the last occurrence in elevation 320 masl at relatively flat terrain (Figure 4).

The longest and shortest elongation ratios range from range from 1.3 to 3.5. Elongation ratio represent morphology roundness. We carefully selected hummocky hills with elongation >1.3 to avoid discrepancy of deviation angle due to basal roundness of the hummocky hills. Individual flow direction (IFD) ranges from N115°E to N174°E with displacement angle of the hummocky hills ranging from 0.1 to 89°.

5 Discussion

5.1 Morphology of Galunggung Debris Avalanche Deposit

Generally, morphology of sector collapse's debris-avalanche are characterized by the occurrence of horseshoe-shaped craters on volcanic cones, hummocky hills in the medial and distal areas, natural embankments (toreva), marginal cliffs, and distal cliffs. The wavy hills produced by debris flow deposits have a smoother appearance and a regular pattern than hummocky hills produced by debris-avalanche deposits [6]. DEM analysis and field observation suggest that Galunggung volcano has characteristics of debris avalanche. It has a horseshoe-shaped crater on the summit of the Galunggung volcano, toreva block on the proximal zone, and hummocky hills spread in the distal zone. The hummocky hills on the eastern distal zone of the Galunggung volcano show wavy morphology with conical to elongated shape hills and have slopes between 30°-60° (Fig. 3 a).

5.2 Lithological distribution and geological structure of the hummock hills

Hummocky hills are composed of dominated lava blocks and pyroclastic breccia blocks. The lava blocks experienced brecciation and jointing which were then filled with volcanic material. In some places, we found intact lava blocks with varying degree of facture intensity. These deposits are generally composed of basalt and andesite lava blocks. Lahar deposits are also found in low topography. This deposit is composed of igneous rock with a rounded shape, poorly sorted, and has different size fragments.

The hummocky hills spread in 500 m to 25 km distance from the summit. The long-distance supports the confinement process which occurs when the body of the Galunggung volcano collapses. During the process, the transported blocks disintegrates creating small fragments. Rocks that are less consolidated disintegrate and become a filler matrix in fractured lava blocks. Field observation shows that component of hummocky hills in Galunggung volcano, from near to far of the source, is varied from intact rock to brecciated rock.

The geological structure of hummocky hills is influenced by the flow regime. The compressional flow regime forms folds and reverse faults. The extensional flow regime forms a normal fault. An isolated normal fault in hummocky hills is observed in the distal zone of the Galunggung Formation. The extensional flow regime is formed due to the pull of gravity so that the volcanic blocks spreading. The spreading block countinous faulting and transported from a grabben. This graben forms an inter-hummocky area that separates the hummocky hill horst.

5.3 Distribution of Hummocky Hills Product of Galunggung Volcano Sector Collapse

Bivariate data analysis comparing the dimensions of the hummock hills with the distance from the source shows a decrease in hummock size as they moved away from the source (Figure 5). The closest hummocky hills from the source have the largest size of 390,000 m² at a distance of 0.6 km, indicating that this is a toreva block. It stands on the flank of Galunggung with a topographic slope of 25°. Hummocky hills do not occur at a distance of 4.5 – 8.5 km slope 2 -5°. The hummocky hills started to occur at a distance of 8.5 – 21 km in topographic slope 0 – 2°. In this zone, the size of hummocky hills ranges from 25,000 m² to 4,000 m² (Figure 6).

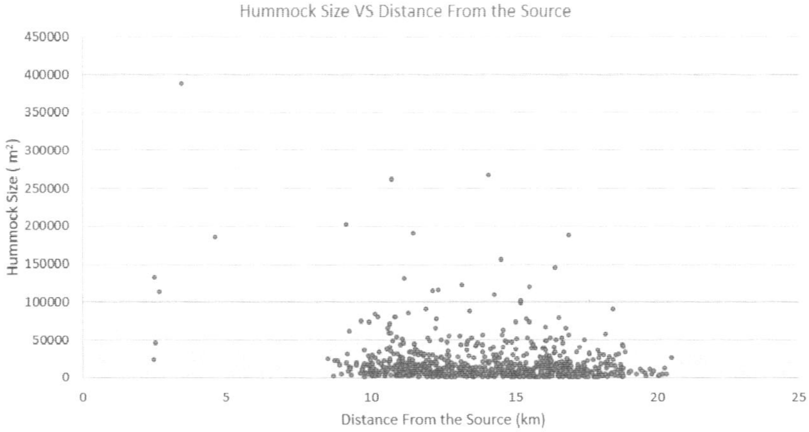

Figure 5. Bivariate plot of hummock area and distance from the source. Maximum size of the hummocky hills generally shows descending trend to increasing distance from the source.

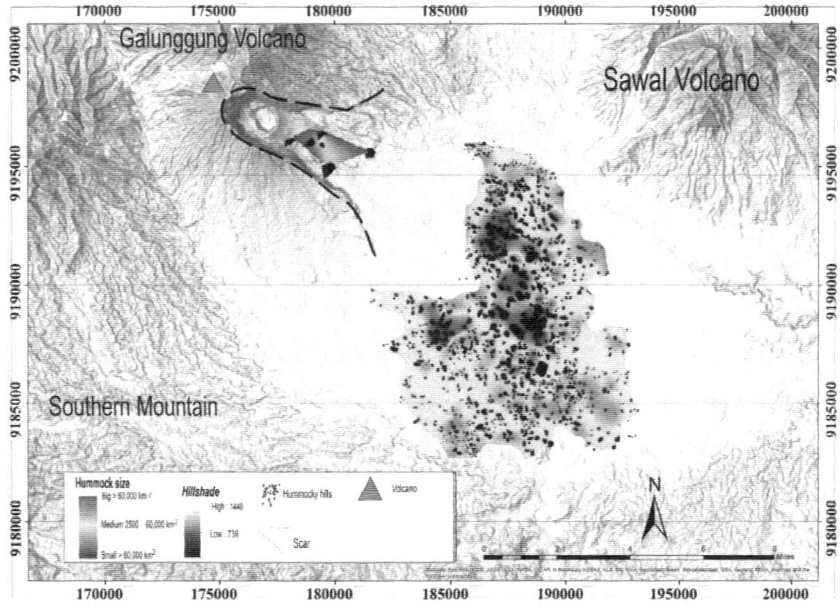

Figure 6. Hummock size distribution maps of Galunggung debris avalanche deposits. Base map from World Hillshade ESRI images is freely downloadable from (https://www.arcgis.com/index.html).

Distribution map of the size of the hummocky hills following the three classes of big (> 60,000 km²), medium, and small (<2,500 km²) hummocky hills are presented in Figure 6. Generally, big hummocky hills are distributed in a relatively narrow area at a distance of 0.5 – 4.5 km with topography slopes of 5

– 25° typical of toreva domains. At a distance of 8.5 – 20 km, big hummocky hills appear in the proximity of the main flow direction. Medium and small hills are relatively dispersed in the hummocky hills areas.

Generally, there is a pattern of hummock size decreasing trends as it moves away from the source of the deposits. This is closely related to the debris avalanche transport mechanism. In the early stage of the sector collapse, the block was breakdown into various sizes. The long and fast transportation of debris avalanche deposits resulting in further disintegration creating smaller size fragments as the transportation progresses. The fission continues until the final stage of the breakdown so that the size of the volcanic block gets smaller as it moves away from the source [8]. From these data, big size hummocky hills becomes representative data in the interpretation of debris avalanche distribution.

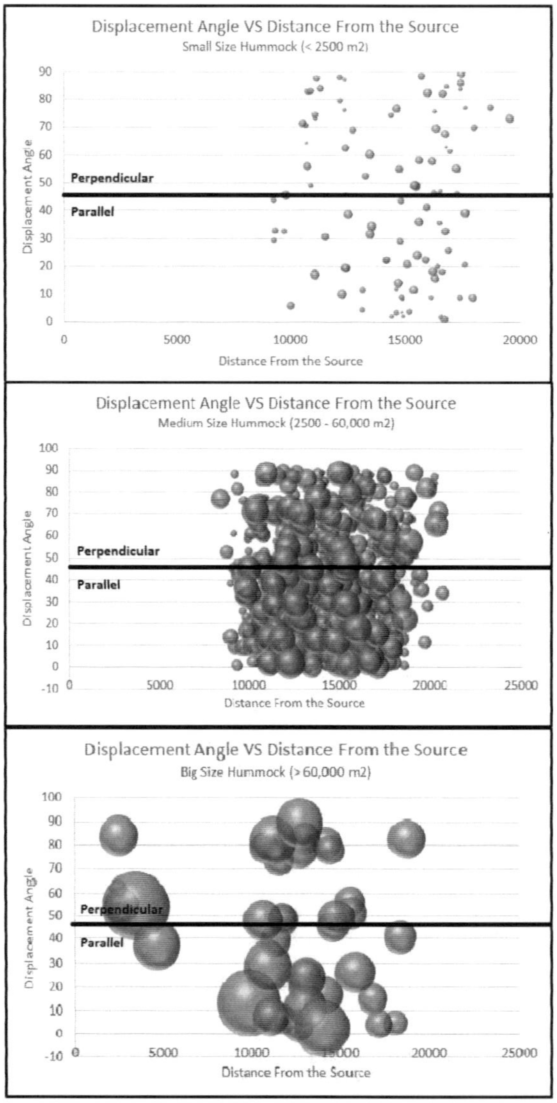

Figure 7. Bivariate data of hummock displacement angle and hummock size against distance from the source for three classes of hummocky hills size of small, medium, and big size. Displacement angle is divided into parallel (<45°) and perpendicular (>45°). Big size hummock shows that dominant population is shifted from perpendicular to parallel trend with increasing distance from the source.

Other than various size, debris avalanche tends to have various displacement angle which may resemble initial topography and flow regime [8]. Figure 7 shows the variation of displacement angle, divided into parallel (<45) and perpendicular (>45), versus distance from source for three classes of hummocky hills. At a distance of 500 – 4500 m with a slope of 5 – 25°, big hummocky hills are dominated by perpendicular oriented hills. At a distance of 8500 m on a slope of 0 – 2°, big hummocky hills are dominated by parallel-oriented hills. Medium size hills show similar pattern to that of big ones. However, the pattern in the medium size is not as clear as that of big ones. The change of displacement angle from the perpendicular, downhill in the topographic slope of 5 – 25°, to parallel dominant in flat terrain with a topographic slope of 0 – 2° aligned with the pattern suggested by [8].

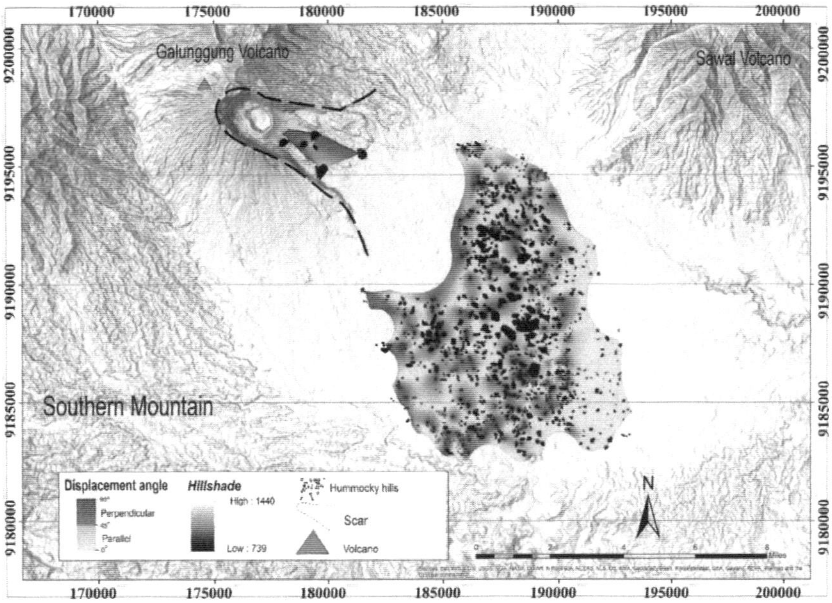

Figure 8. Distribution maps of hummocky hills displacement angle of Galunggung debris avalanche deposits. Base map from World Hillshade ESRI images is freely downloadable from (https://www.arcgis.com/index.html).

The spatial distribution of the displacement angle of the hummocky hills is presented in Figure 8. Big size hummocky hills with perpendicular displacement angles are concentrated in the proximal zone at a distance of 0.5 – 4.5 km from the summit and downhill slopes reaching 5 – 25°. These hummocky hills likely represent compressional flow regime zone. In contrast, big size hummocky hills at a distance of 8.5 – 20 km from the summit and relatively flat terrain slopes range 0 – 2° of the distal zone that is dominated by perpendicular displacement angle likely represent the extensional flow regime zone. This extensional flow regime is supported by the occurrence of a normal fault, localized, in hummocky hills STA 20 with an orientation of N300°E. This normal fault orientation is well aligned with the main flow direction of the debris avalanche deposit. The normal fault in the hummocky hills might be formed due to the extensional flow regime due to gravity load creating rifting on the block so that it is pulled into the volcanic depression area [12].

6 *Conclusion*

The sector collapse of the Galunggung volcano produced a unique morphology such as horseshoe-shaped crater open eastwards on the summit area, toreva blocks on the proximal zone, and hummocky hills on the distal zone. The hummocky hills size ranges from 500 to 700,000 m² and are classified into three classes of big (> 60,000 km^2), medium, and small (<2,500 km^2). Ratios of the longest and shortest axis, elongation ratio, of the hummocky hills range from 1.3 to 3.5. Individual flow direction (IFD) ranges from N115°E to N174°E with displacement angle of the hummocky hills ranging from 0.1 to 89°. The lithology of the hummocky hills of Galunggung Volcano is composed of andesitic lava with jigsaw crack structure and pyroclastic flow deposit.

The big-sized hummocky hills show decreasing size towards the distance from the summit. In addition, the displacement angle of the big size hummocky hills changed from the perpendicular, in the proximal zone at a distance of 0.5 – 4.5 km from the summit and downhill slopes reaching 5 – 25°, to parallel, a distance of 8.5 – 20 km from the summit and relatively flat terrain slopes range 0 – 2° of the distal zone. These morphometric and distribution trends, especially those of big size hummocky hills, characterize debris avalanche deposits of the Galunggung volcano as a free release spreading type.

References

[1] Yoshida H 2014 Hummock alignment in Japanese volcanic debris avalanches controlled by preavalanche slope of depositional area Geomorphology **223** 67–80

[2] Moktikanana, MLA, Wibowo HE, Harijoko A 2021 Hummock size and alignment in Gadung debris avalanche deposit, Raung Volcanic Complex, East Java, Indonesia. IOP Conference Series: Earth and Environmental Science **851** 1 - 2

[3] Bronto S 1989 Volcanic Geology of Galunggung, West Java, Indonesia. University of Caterbury. Science: Theses and Dissertations [4255] **38** – 101

[4] Walter TR, Haghshenas Haghighi M, Schneider FM 2019 Complex hazard cascade culminating in the Anak Krakatau sector collapse. *Nat Commun* **10**-39

[5] Siebert L, Roverato M 2020 A Historical Perspective on Lateral Collapse and Volcanic Debris Avalanches **15** 209–35

[6] Ui T, Takarada S, Yoshimoto M 2000 Debris avalanches in Sigurdsson H, Houghton B F, McNutt S R, Rymer H, Stix J (eds) Encyclopedia of Volcanoes **617**–26

[7] Yoshida H, Sugai T, Ohmori H 2010 Longitudinal downsizing of hummocks by the freel spreading volcanic debris avalanches in Japan the Quaternary Research (Daiyonki-Kenkyu) **49 (2)** 55–67.

[8] Yoshida H, Sugai T, Ohmori H 2012 Size–distance relationships for hummocks on volcanic rockslide-debris avalanche deposits in Japan. Geomorphology **136 (1)** 76–87

[9] Yoshida H 2016 Magnitude-frequency distribution of hummocks on rockslide-debris avalanche deposits and its geomorphological significance Geosciences **6**-7

[10] Yoshida H and Sugai T 2010 Quantitative examination of hummock alignment in the debris avalanche deposits: Zenkoji debris avalanche, Usu volcano, Japan. *Geographical Review of Japan Series B* **18 429**–44

[11] Sinclair 1991 A fundamental approach to threshold estimation in exploration geochemistry: probability plots revisited, Journal of Geochemical Exploration, Volume 41 **1**–2

[12] van Wyk de Vries B and Davies T 2015 Landslides, debris avalanches, and volcanic gravitational deformation in Sigurdsson H, Houghton B F, McNutt S R, Rymer H, Stix J (eds) Encyclopedia of Volcanoes 665–8.

Acknowledgments

This reaserch is fully funded by the Geological Engineering Department Internal Funding program year 2021, entitled *"Karakteristik Morfologi dan Distribusi Produk Sector Collapse Gunung Galunggung".*

Geochemistry of arc alkaline magmatism of Java Island, Sunda Arc: a statistical review

Esti Handini[1*], I Wayan Warmada[1], Toshiaki Hasenaka[2], Nicholas D Barber[3], Tomoyuki Shibata[4]

[1]Geological Engineering Department, Universitas Gadjah Mada, Indonesia
[2]Center for Water Cycle, Marine Environment, and Disaster Mitigation, Kumamoto University, Japan
[3]Department of Earth Sciences, University of Cambridge, UK
[4]Graduate School of Science, Hiroshima University, Japan
*Corresponding author's e-mail: esti.h@ugm.ac.id

Abstract. Alkaline magmatism develops in the rear arc area of Java, Sunda Arc, at different range of slab depth; ~270-580 km in central section and ~150 km in the east. We collate published geochemical data of volcanic rocks from four alkaline volcanoes (Muria, Lasem, Bawean, and Ringgit-Beser) and perform statistical analysis to evaluate geochemical characteristics of each suite. A set of major and trace elements is scaled and transformed using principal component analysis (PCA) and then followed by implementation of k-means algorithm to cluster the data points based on Euclidian distances. K-means clustering of the dataset suggests that Central Java alkalines are most elevated in K_2O and total alkali. The algorithm further suggests that Muria samples can be clustered into two, owing to these components. These two clusters, however, are not well reflected on trace element-based clustering. Lasem volcanics show distinct cluster high in Na_2O/K_2O and SiO_2, while Bawean samples are mixed into both Muria clusters. Ringgit-Beser alkalines show two distinct clusters tied to MgO and enrichment in Ba, Rb, and Sr. Our findings suggest that the potassium and LILE enrichment in these alkaline rocks is independent of slab depth and is most likely regulated by tectonic-related arc segmentation in Java subduction zone.

1. Introduction

Java Island is part of a long subduction zone (or "arc") system called the Sunda Arc which extends from the western tip of Sumatra to clusters of islands east of Flores. This island is home to 30+ Quaternary volcanic centres which mainly erupt calc-alkaline magmas. Four of these volcanic centres have distinctively erupted alkaline magmas [1, 2, 3], which all originate from the rear of central and eastern section of the island (Figure 1). This style of alkaline magmatism is however absent in Western Java. The spread of this arc alkaline volcanism corresponds to varying slab depth, a distance measured from the arc surface (base of the volcano) to the surface of the down going plate. Recent estimates of slab-top depth [4] indicate that the alkaline volcanism in Central Java corresponds to a greater slab depth (~270 km and beyond) than those in the rear of East Java (~150 km).

In Central Java, arc volcanoes Muria and Lasem are formed at around ~270 km above the slab depth on mainland Java. This depth-to-slab is greatest at Bawean (~580 km), a Quaternary volcanic island in the north of Java with extreme alkaline enrichment [3]. The lateral section of this arc segment has distinctively wider and deeper range of slab depth compared to other sections of the island. In the eastern end of Java Island, the volcano distribution is much narrower, spreading from ~90 km beneath the volcanic front to ~150 km beneath the rear-arc. Ringgit-Beser and Lurus are the two alkaline volcanic centers from this section, the other being Lurus. From the same slab depth as these two East Java alkaline volcanoes, the volcanic centers from Central Java section mostly erupt less alkali-enriched high-K calc-alkaline magmas (i.e., Dieng, Telomoyo, and Ungaran). This discrepancy in along-arc alkaline patterns as a function of slab depth may hint at different source processes in generating these alkaline magmas from Central and East Java. While advanced petrological and mineralogical analysis is key to further give an accurate petrogenetic model, we consider that statistical review of the available geochemical datasets will shed a light on comparability of the magmas generated from these alkaline volcanic centers. Such a review will form a robust basis for a holistic petrogenetic study of alkaline magmatism in Java.

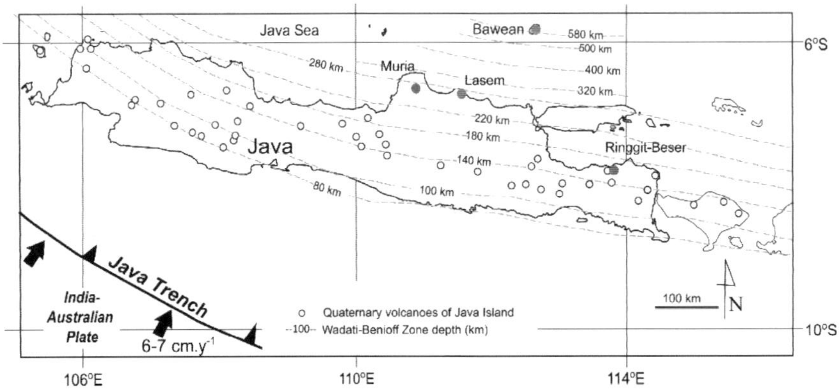

Figure 1. Volcano distribution of Java Island. The colored circles represent the alkaline volcanic centers in the rear of Java arc. The depth of Wadati Benioff zone represents the slab depth, a measured distance from the slab surface to the base of the volcano, following the *Slab 2.0* models [4]. Java Island outline and volcano distribution are delineated from SRTM map provided by https://srtm.csi.cgiar.org/.

This paper uses well-known statistical analyses (i.e., principal component analyses and k-means clustering) to review and revisit the clustering patterns of these alkaline volcanic samples based on published whole rock geochemistry analysis. The principal component analysis (PCA) is used to remove the noise originated from raw correlated variables into a smaller number of new variables (components). This model is useful to identify correlated variables and subtle patterns underlying such large datasets with many variables. This review aims to test whether the degree of potassium and large ion lithophile elements (LILE) enrichment is a function of slab depth variability or tectonic-related arc segmentation.

2. Alkaline magmatism of Java, Sunda Arc

This section briefly discusses (1) the geological setting that delivers Java arc and (2) geochemistry of the arc alkaline magmas from previous published works.

2.1. Geology of Java Island, Sunda Arc

Java Island, a part of Sunda Arc, is derived from northward subduction of Indo-Australian plate beneath the south-eastern margin of Eurasian plate. Subduction in Java initiated in the Late Oligocene to Early Miocene [5] at the rate of 6-7 cmy^{-1} [6]. Compressional force is dominant in the Quaternary Java arc, which may explain the back-arc spreading in the region of the Java Sea (e.g., Bawean) [6]. The basement material beneath Central Java is thought to be an amalgamation of an earlier Cretaceous subduction systems i.e. accreted terranes, and has been found to be more oceanic than continental in character beneath East Java as compared to Central Java [5]. The genetic link between Java's alkaline magma's and current subduction system is still debated [1, 2, 3], and is not considered in this paper.

Previous work on Muria [1] reported younger highly potassic (HK) series and older potassic (K) series magmas. The HK series is higher in K_2O and is more enriched in LILE and light rare earth elements (LREE) compared with the older K-series. The authors [1] further suggest that the generation of the two magma series indicates a petrogenetic transition from intraplate to a subduction processes. Other studies of alkalines from Java also find similar characteristics of elevated K_2O, LILE and LREE in volcanic rocks from Ringgit-Beser [2] and Lasem [3, 7]. High K_2O content in these magmas is commonly accompanied by increase in LILE and LREE (i.e., Rb, Sr, Ba, La).

2.2. Geochemical review of Javan alkalines, Sunda Arc

This subsection provides a brief review of previous works to give an understanding of the alkalinity of volcanoes studied here. The volcanic products from Muria, Lasem, Bawean, and Ringgit-Beser are all enriched in K_2O and total alkali (K_2O+Na_2O) as they are all plotted above the subalkaline dividing line (Figure 2). Note that Lurus, another alkaline volcano in East Java, is not considered here due to lack of data availability. Data points from all these volcanoes all show somewhat similar level of total alkali enrichment, peaking at 7-9 wt%, with some of those from Bawean and Muria show higher values of total alkali (12-13 wt%). Lasem's sample population peaks at 2.5 wt% K_2O, noticeably lower than populations from other volcanoes (5-6 wt%) and is plotted in the boundary line of alkaline and sub-alkaline division (Figure 2). Among the data points from these four volcanic centres, Lasem samples (at ~270 km of slab depth) are the most evolved suite as shown by high SiO_2 and slightly lower MgO. By contrast, Muria's sample population, which is thought to source from similar slab depth as Lasem, shifts toward lower SiO_2 at the same level of total alkali enrichment. Lasem's sample population tends to show higher range of Na_2O/K_2O ratios than those from other volcanic centers. This is mainly owing to elevated Na_2O and lower K_2O at Lasem compared to other centres. This would also put Lasem samples in the field between potassic rocks and high-K calc-alkaline.

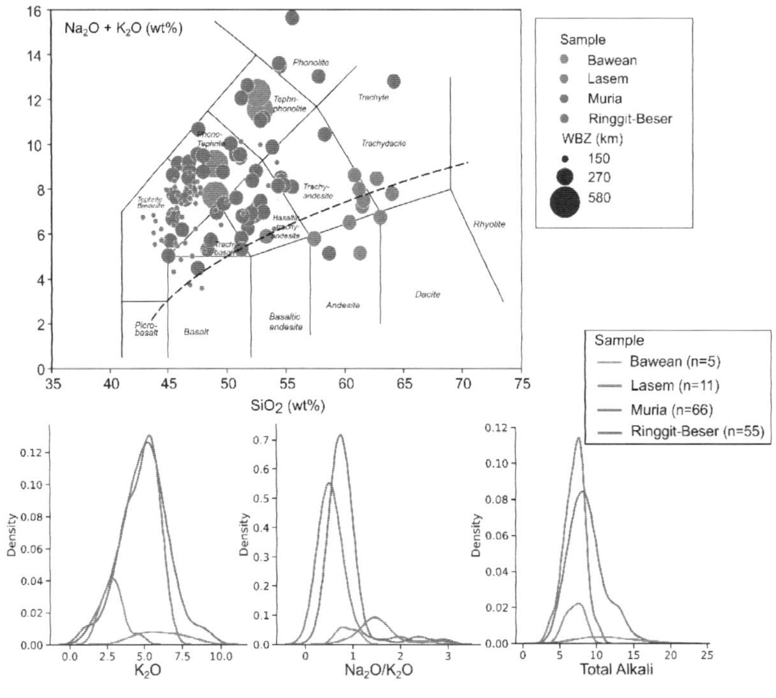

Figure 2. A summarized review of K_2O, Na_2O/K_2O and K_2O+Na_2O of the Javan arc alkalines [1, 2, 3, 7]. The total alkali silica (TAS) silica diagram is after [8], dotted line in TAS diagram is dividing line between alkaline and sub-alkaline field after [9]. The number of available data from Bawean and Lasem is very low, compared with those from Muria and Ringgit-Beser. Kernel density estimate (KDE) plot uses Gaussian kernel, a default provided by seaborn [10] library.

Figure 2 shows that regardless of the distinct slab depth difference between the alkaline magmas sourced from Central vs. East Java, the alkalinity of the volcanics is comparable between the two regions. Muria and Ringgit-Beser, which greatly contrast in slab depth (270 km vs. 160), even show almost the same degree of K_2O enrichment. Only Lasem shows a distinctly different evolutionary trend, owing to the nature of a more evolved suite.

3. Data availability and method

The alkaline magmas reviewed in this paper is sourced from precompiled dataset downloaded from GeoRoc database [11] under the "Sunda Arc" category [12; https://georoc.eu/ accessed June 2021]. The original precompiled dataset contains 2722 data points of volcanic, metamorphic, and sedimentary rocks from across Sunda arc. Among these data points, all the calc-alkaline, non-Javan alkaline, and non-volcanic rocks were omitted. The final datasets contain 137 geochemical data points, collated from published works on Java arc alkaline products [1, 2, 3, 7]. The final datasets were standardized based on z score values using the following formula

$$z = (X - \mu) / \sigma \tag{1}$$

where X is a single data value, μ is the population mean, and σ is the population standard deviation. Data points with z score > 3 are considered outlier and omitted, which means the final dataset covers 99% of the sample. This is also expected to reduce the uncertainties due to inter-laboratory analysis.

Principal Component Analysis (PCA) was done on two different datasets of (1) major elements and (2) trace element ratios, generated from the final datasets to address discrepancy in data completeness of each data point. The major elements PCA is generated from 137 data points, while trace element ratios PCA is from 127 data points, of which Bawean is absent. The dataset is split into training- and test-set at 80:20 ratio prior to PCA. Accuracy score and confusion matrix from the scikit-learn metrics library were employed to evaluate the performance of the model. PCA and k-means algorithms were run using scikit-learn library [13]. Principal components represent the calculated eigenvectors of the data covariance matrix. The number of components were determined at more than 80% of the total variance explained for both of PCA models in this paper. The variance retained in the new generated variable (principal components) is measured by the eigenvalue. The k-means clustering method is a technique used to identify clusters of data points in a dataset. In this paper, the k-means clustering is built upon major and trace element PCA models. K-means algorithms select *k*-centroids through an iterative process, with *k* equals to the number of original clusters. The resulting k-means clustering is evaluated using Silhouette coefficient (between -1 to 1) (i.e., larger number indicates cohesiveness of the samples in their cluster and each cluster is well separated). The Silhouette coefficient is calculated using the following formula

$$\text{Silhouette coefficient} = (b-a)/\max(a,b) \tag{2}$$

where a is the average intra-cluster distance, and b is the average inter-cluster distance. We provide Kernel density estimation (KDE) plots alongside the PCA models to help the reader visualize the distribution of data underlying principal components. The KDE plot uses the Gaussian kernel function provided by seaborn library, [10] and gives an estimated distribution of the variables observed in intra-cluster sample population.

4. Results and discussion

This section discusses the result of PCA models using major elements datasets, trace element datasets, and k-means clustering using both PCA models. The two PCA models use different sample sizes as a response to variability in data source materials. Mainly, this is because Bawean has insufficient trace element data to be modelled properly in the trace element PCA. The threshold used to determine the number of components in PCA model is above 80% of total variance. This threshold is satisfied by three principal components in both PCA models. The loading vectors of each original variables are represented by the blue lines on the plots below.

4.1. Major elements

PCA models of the major elements shown in Figure 3 make use of 10 species of main major elements (i.e., SiO_2, TiO_2, Al_2O_3, FeOt, CaO, MgO, MnO, K_2O, Na_2O, and P_2O_5) with the addition of total alkali (K_2O+ Na_2O) and ratio of Na_2O/K_2O. Figure 3 top left suggests that principal component 1 (PC1) accounts for the majority of variability in the data set (57.83%). The plot of PC1 vs. PC2 suggests that some Muria and Bawean magmas show prominent enrichment of total alkali, K_2O, and Na_2O. Despite this, the samples from Lasem show the highest ratio of Na_2O/K_2O, accompanied by higher range of SiO_2 and, to some degree, Al_2O_3. Ringgit-Beser and some Muria samples show weakly bimodal distribution in some of the major elements, as shown by KDE plot in Figure 3. However, this bimodality is buffered by our PCA model which allows evaluation of many variables at the same time.

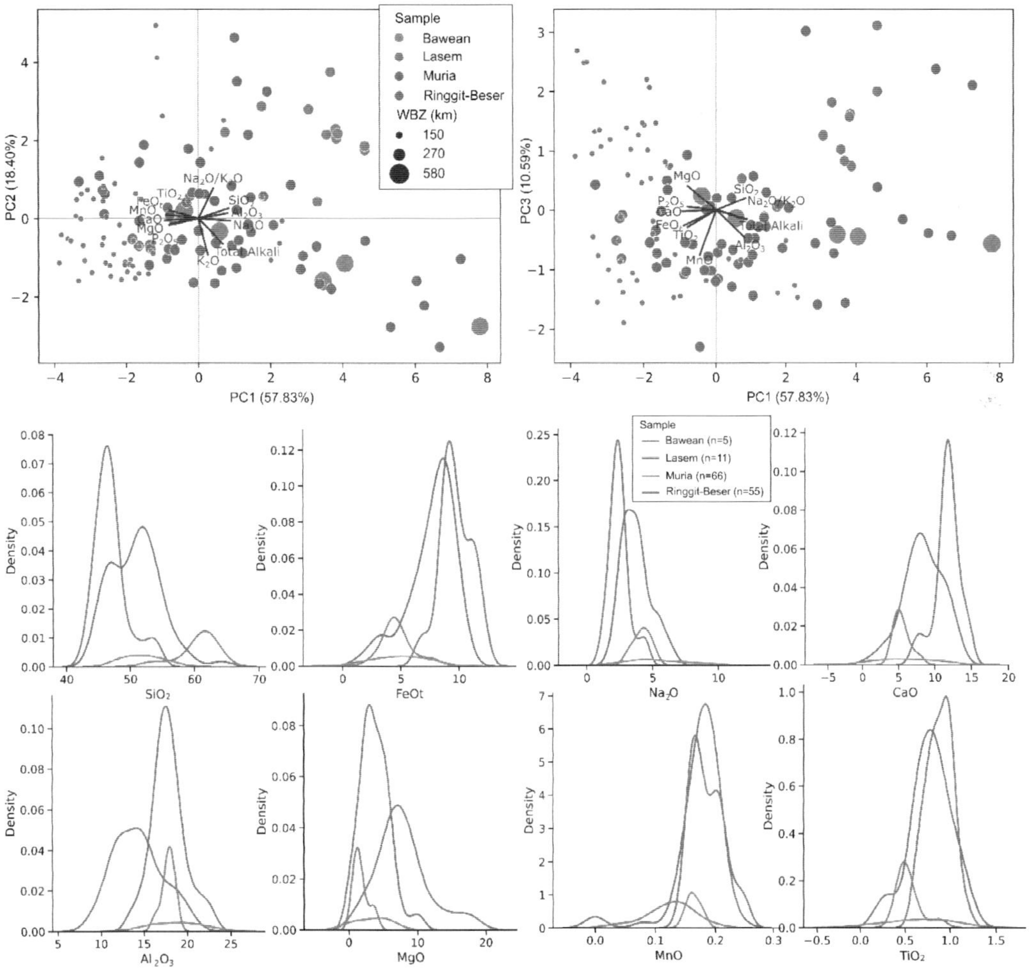

Figure 3. Major element PCA model of alkaline volcanic samples from Java. The three principal components make up for 87% of the total variance. Total data points used in the analysis is 137. PC = principal components. The model's accuracy for 3 components is at 79%; the model correctly predicts 22 data points out of 28 data test set.

The PC1 vs PC3 plot (top right, Figure 3) shows that Ringgit-Beser samples are characterized by high MgO, CaO, FeOt, and TiO_2. Furthermore, some of the Ringgit-Beser samples show very high MgO, accompanied by high CaO and P_2O_5. However, some of the Ringgit-Beser samples show a lower range of MgO accompanied by higher TiO_2, FeOt, and MnO. Furthermore, TiO_2 increases as MgO decreases in Ringgit-Beser magmas, indicating a clear signal of either crystal fractionation or differences in the source mantle. The high TiO_2 in lower range of alkaline rocks from Ringgit-Beser has been reported as the characteristics of K series [2]. This K series shows less enrichment of Ba, Nb, and K_2O compared with the EK (enriched potassic series), a characteristic that is attributed to different type of enriched mantle source (EMII and EMI, respectively) [2].

In summary, the PCA model of major elements shows that most alkaline volcanic samples from Central Java are characterized by variation in K_2O, Na_2O, total alkali, Na_2O/K_2O, and to some extent SiO_2 (particularly in Lasem). On the other hand, the alkalines from East Java section are characterized by variation in MgO, P_2O_5, CaO, FeOt, TiO_2, and MnO. Ringgit-Beser alkalines from East Java tend to show higher ranges of MgO at the same range of SiO_2 compared to the alkalines from Central Java, suggesting a lesser degree of magma differentiation. The PCA model also suggests that level of K_2O enrichment is independent of MgO content, a statement that is evidently true for Ringgit-Beser and Muria magmas. In addition to that, TiO_2 in Ringgit-Beser magmas behave differently as a response to decreasing MgO content. This characteristic is not present at a Central Java volcano like Muria. The extent of K_2O enrichment in Muria magmas fits the sample grouping proposed by previous work [1] which confirms that enriched potassic rocks (HK series) of Muria are also elevated in Rb, Sr, and Ba in comparison to the potassic series (K series).

4.2. Trace element

The trace element PCA model shown in Figure 4 uses ratios of Sr/Y, Ba/Nb, Rb/Sr, La/Nb, Ce/Nb, and Zr/Nb from Muria, Lasem, and Ringgit-Beser (excluding Bawean). The KDE plot of Muria samples shows weakly bimodal distribution in Sr/Y and Ba/Nb. It also shows that Ringgit-Beser samples are bimodal in Zr/Nb, La/Nb, and Ce/Nb distribution, while those from Muria tend to shift toward lower range of these three trace element ratios. Lasem samples show restricted distribution of all these trace element ratios, but are mostly overlapped with the Ringgit-Beser's peak.

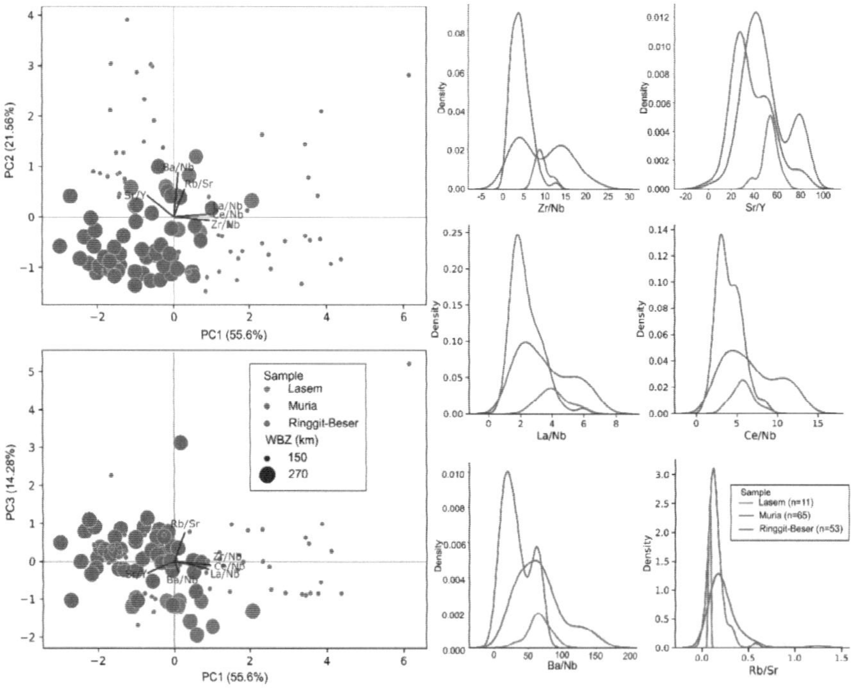

Figure 4. Trace element PCA model of alkaline volcanic samples from Java. The three principal components make up for 91.44% of the total variance. Total data points used in the analysis is 127. The

model's accuracy for 3 components is at 84%; the model correctly predicts 22 data points out of 26 data test set.

The plot of PC1 vs PC2 (top left, Figure 4) suggests that both PC1 and PC2 account for the variation of trace elements in the dataset (77.16%). This plot distinctly shows that Ringgit-Beser samples are clustered to two different loading vectors: (1) toward higher La/Nb, Ce/Nb, Zr/Nb, and (2) toward higher Ba/Nb and Sr/Y. Some of these samples from Ringgit-Beser show both characteristics and high Rb/Sr. Lasem samples are distributed in between those two apparent clusters of Ringgit-Beser. On the other hand, most of Muria samples are plotted on the lower range of these trace element ratios, but they do show particularly high Sr/Y.

The enriched potassic series from Muria and Ringgit-Beser (HK and EK series, respectively) is reported to have higher range of Ba, Rb, and Sr, compared with the less enriched potassic series (reported as K series in both volcanics) [1, 2]. This geochemical feature is true for the Ringgit-Beser samples but is not well defined in Muria samples. This further signifies that the grouping of Muria samples proposed in previous work [1], which is based on the alkalinity and select trace elements (Ba, Rb, and Sr), is not supported by more robust statistical methods like our trace element PCA model. At the same time, both of our major and trace element PCA models partially confirm the grouping proposed for Ringgit-Beser magmas in prior work [2]. Ringgit-Beser samples which are high in MgO are also elevated in Ba, Rb, and Sr.

4.3. K-Means clustering

The PCA models presented in previous subsections are further used as a basis for k-means clustering analysis shown in Figure 5. The k-means clustering for major element dataset is run on 4 original cluster from Muria, Lasem, Bawean, and Ringgit-Beser, and is on 3 (all but Bawean) for the trace element dataset. The overall predicted clusters from this method only slightly overlapped, indicating a good performance of the method as suggested by positive silhouette scores.

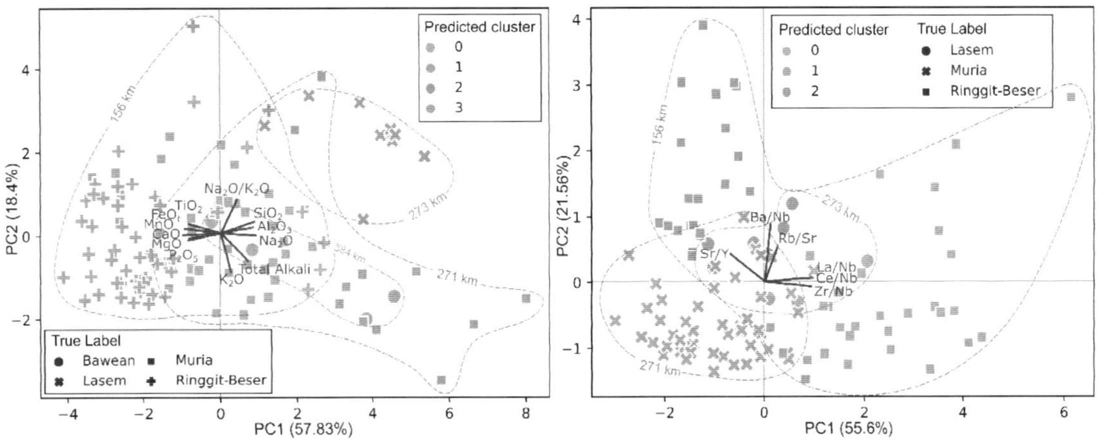

Figure 5. The result of k-means clustering analysis built from PCA models of major and trace element ratios. Major element n is 137, while trace element n is 127. The silhouette coefficient is 0.37 for major element clustering and is 0.43 for the trace element clustering. Coloured dotted lines delineate original sample clustering based on erupting volcanic centers. The number on the dotted line represents the suggested slab depth [4].

K-means clustering built from major elements PCA (Figure 5, left) suggests that most data points from Lasem are aggregated together. Most Ringgit-Beser data points are aggregated near the new determined k-centroid, but a few data points are spread out, overlap with Muria cluster. Muria data points show ubiquitous distribution, owing to the variation in K_2O and total alkali, allowing the formation of two clusters with different *k*-centroids. On the other hand, Bawean samples, which correspond to greatest slab depth and are limited in number, are fitted in the two Muria clusters. This k-mean clustering method suggests that variation in major elements of these alkaline magmas is independent of suggested slab depth. These results also confirm that the Ringgit-Beser magmas distinctively more primitive than the other alkaline magmas, confirming previous finding [2].K-means clustering built from trace element ratios PCA (Figure 5, right) shows that most Muria data points are aggregated together, suggesting the differences in LILE (Ba, Rb, Sr) between the proposed potassic and enriched potassic group (K and HK series, respectively) is not statistically robust. On the other hand, Ringgit-Beser data points are aggregated into two different clusters, one of which is more elevated in LILE such as Ba, Rb, and Sr. This predicted clustering of Ringgit-Beser magmas is accompanying the noticeable variability in MgO content. The same geochemical characteristics (e.g., high in MgO content and elevated LILE) is also observed in sub-alkaline magmas which correspond to various slab depth (<150 km) from eastern Java arc section [14]. The same geochemical feature, however, is not observed in sub-alkaline samples from volcanoes at the corresponding slab depth in the central section of Java arc. This finding further suggests that LILE enrichment in these alkaline arc magmas from Java is most likely related to arc segmentation and geometry, possibly a tectonic-related feature, rather than the slab depth.

5. Conclusion

We perform principal component analysis (PCA) and k-means clustering models on published major and trace element datasets of arc alkalines from the rear portion of Java Island, Sunda Arc. The formation of these arc alkaline rocks corresponds to varying slab depth, which is much greater in the Central Java section as compared to the East. Our PCA and k-means clustering models suggest that the level of potassium enrichment in these alkaline magmas from Java Island, Sunda Arc, is independent of slab depth. Muria volcanics show the greatest potassium and total alkali enrichment among the studied alkaline volcanoes. From the same level of slab depth or even greater, samples from Lasem and Bawean do not show similar levels of K enrichment in their magmas. The eastern Java potassic rocks from Ringgit-Beser, which correspond to shallowest slab depth in Javan alkalines, show much more primitive geochemical characteristics than those from Central Java. In terms of trace element ratios, the extent of the K-enrichment in Muria magmas does not translate into LILE enrichment (Ba, Rb, and Sr), contesting the validity of the proposed K and HK framework for magmas at Muria. This finding strengthens claims that Muria represents a more typical subduction zone volcano than previously thought. Our model identifies that variability in the LILE enrichment is more prominent in Ringgit-Beser magmas and is strongly correlated with the MgO content. Our findings from this statistical review suggest that the potassium and LILE enrichment is independent of the corresponding slab depth and is probably more affected by tectonic-related arc segmentation in Java subduction zone.

References

[1] Edwards CMH, Menzies MA and Thirlwall MF 1991 *J. Pet.* **32** 555-592.
[2] Edwards CMH, Menzies MA, Thirlwall MF, Morris JD, Leeman WP and Harmon RS 1994 *J. Pet.* **35** 1557-1595.
[3] Leterrier J, Yuwono YS, Soeria-atmadja R and Maury RC 1990 *J. Southeast. Asian. Earth. Sci.* **4** 171-187, doi: 10.1016/S0743-9547(05)80011-X.
[4] Hayes GP 2018 U.S. Geological Survey data release, doi:10.5066/F7PV6JNV.
[5] Hall R 2002 *J. Asian Earth Sci* **20** 353-431.
[6] Widiyantoro S and Van der Hilrst R 1996 *Geophys. J. International* **130** 167-182.
[7] Calanchi N, Lucchini F and Rossi PL 1983 *Mineral. Petrogr. Acta.* **27** 15-34.
[8] Le Bas JM, Le Maitre RV, Streckeisen A and Zanettin B 1986 *J. Pet.* **27** 745-750
[9] Irvine TN and Baragar WRA 1971 Can. J. Earth. Sci. **8** 523-548.
[10] Waskom M et al 2017 *mwaskom/seaborn: v0.8.1 (September 2017)*, Zenodo, doi doi.org/10.5281/zenodo.883859.

[11] Lehnert K, Su Y, Langmuir CH, Sarbas B and Nohl U 2000 *Geochem. Geophys. Geosyst.* 1, 1012.
[12] DIGIS Team 2021 *Goettingen Research Online/Data*. V1.
[13] Pedregosa *et al 2011* J. Machine. Learning. Res. **12**, 2825-2830.
[14] Handley HK, Blichert-Toft J, Gertisser R, Macpherson CG, Turner SP, Zaenuddin A and Abdurrachman M 2014 *Geochim. Cosmochim. Acta.* **139** 205-226.

Acknowledgements
The authors thank the two anonymous reviewers who have provided constructive comments to improve the manuscript. This study is part of a collaborative research between the authors. Nicholas D. Barber has his PhD funded by the Bill and Melinda Gates Foundation, and the Gates-Cambridge Trust.

Characteristics of Mengger Formation's volcaniclastic in Cisaat River, Brebes, Central Java, Indonesia

H N Aini[1], N I Setiawan[1*], M N Haq[1], C A Mahardika[1], S Nugroho[2], and R Setiawan[3]

[1]Department of Geological Engineering, Engineering Faculty, Universitas Gadjah Mada. Grafika street Number 2, Yogyakarta 55281, Indonesia.
[2]Conservation Office of Sangiran Early Man Site, Sragen, Central Java, Indonesia
[3]Centre for Geological Survey, Geological Agency, Bandung, Indonesia
*Corresponding author: nugroho.setiawan@ugm.ac.id

Abstract. The Cisaat River in Brebes Regency, Central Java, is one of the sites on Java Island where vertebrate fossils were found. Mengger Formation, composed of volcaniclastics material such as tuff and pumice fragments, contains these fossils. The presence of vertebrate fossils along with volcanic deposits has remained unclear of its deposition history. This research characterizes the volcaniclastic rocks in Mengger Formation based on geological field observation and petrography analysis. The study aims to identify and distinguish its primary and secondary volcaniclastic depositional processes. The measured stratigraphic sections reveal that the formation has 7 lithofacies from older to younger, namely grain-supported conglomerate, sandstone, siltstone, sandy mudstone, matrix-supported conglomerate, tuffaceous sandstone, and tuffaceous conglomerate. The primary volcaniclastic materials comprise pumice clast, plagioclase, pyroxene, hornblende, quartz, biotite, lithic fragments, and volcanic glass. In contrast, secondary volcaniclastic or epiclastic-sedimentary materials contain lithic fragments, quartz, plagioclase, pumice clast, and clay matrix. The origin of the resedimented volcaniclastic material was probably derived from a large volume of ignimbrite near the volcano. From the deposit and measurement section, it can be concluded that the type of the Mengger volcaniclastic.

1. Introduction

Bumiayu Site is located in Central Java Province, Brebes Regency, approximately ~30 km northwestern from Mount Slamet [1] (Figure 1). This site is known as one of the fossil-finding locations of vertebrates, invertebrates, hominids, and artifacts thought to have ages of Late to Early Pleistocene [2]. The archaeological contents in this area could be found in several locations, including Kaliglagah, Satir, and Cisaat Rivers [1]. A faunal fossil in Java Island, based on the biostratigraphy context, is divided into 2 groups of fauna, namely Satir Fauna (2–1.5 Ma) and Cisaat Fauna (1.2–1 Ma) [3]. Satir and Cisaat faunas are the first mammal groups in the western part of Java Island that lived during the Late Pliocene about 2 million years ago [4]. Fragments of fossils such as *Elephas sp.* and *Cornu Bubalus palaeokarabau* were discovered in tuffaceous sand gravel deposits, member of the Mengger Formation, the upstream of the Cisaat River [1,6]. The Mengger Formation, which carries these fossils, is composed of volcaniclastic material with dominantly tuff [5]. The distribution of the Mengger Formation is relatively limited, with a total thickness of approximately 150 m [5]. The age of the Mengger Formation is 1.53 ± 0.08 Ma (Pleistocene) based on K-Ar radiometric dating of the hornblende phenocryst extracted from the pumice fragments in the lahar deposit of the Mengger Formation [6]. While based on the occurrences of mammal fossils belonging to the Upper Vertebrate Zone, it is the Early Pleistocene [7]. It remains unclear whether the eruption occurred simultaneously as the fossil-bearing deposits in the Mengger Formation.

In this paper, we focus on the volcaniclastic characteristics of the Mengger Formation by identifying and distinguishing primary and secondary volcaniclastic depositional processes. The study of volcaniclastic deposits can directly or indirectly reflect primary volcanic processes such as their type, style, and history [8].

2. Geology of Cisaat River

The research area is part of the Physiographic Zone of the North Serayu Mountains, bordered by the South Serayu Zone and the Quaternary volcanic arc in the south and in the west and east with the Bogor Zone and Kendeng Zone, respectively [9]. The geological formation from old to young comprises Kalibiuk, Kaliglagah, Mengger, and Gintung Formations (Figure 1) [5]. The bottom of the Kalibiuk Formation comprises claystone and fossilized blue marl. The middle part contains green sandstone lenses with abundant mollusks. In comparison, the upper part has a thin interlayer of sandstone. The depositional environment was intertidal with ages of late Early Pliocene to early Late Pliocene. The thickness of this formation is approximately 2,500 m [5].

The upper part of the Kaliglagah Formation consists of cross-bedded coarse sandstone. In contrast, the lower part consists of black clay, lignite, freshwater mollusk fossils (*Sulcospira sp.*), intercalation with tuffaceous sandstones, and conglomerates. The total thickness of this formation is approximately 350 m. The depositional environment was estimated to be terrestrial to shallow marine with the age of the Late Pliocene [5]. The formation was conformably overlain with the Mengger Formation during the Pleistocene epoch. The Mengger Formation consists of gray tuff, tuffaceous sandstone, tuffaceous claystone, and conglomerate with a total thickness of approximately 150 m. Furthermore, the Gintung Formation was conformably superimposed on the Mengger Formation during the Middle to Late Pleistocene. This formation is characterized by the andesitic conglomerate, sandstone, calcareous sandstone, and occasional black claystone. The depositional environment was terrestrial to transitional, with a unit thickness of 800 m [5].

Fragments of *Elephas sp.* and *Cornu Bubalus palaeokarabau* fossils were found within the tuffaceous sand-gravel deposits of the Mengger Formation in upstream of the Cisaat River [1]. Fossil-bearing rock layers containing volcanic material indicate that the Central Java area's hominid and vertebrate life cycle was strongly influenced by volcanic activities [1].

3. Methods

A geological field survey was conducted along the Cisaat River (Figure 1). The lithological measured section was done approximately 500 m upstream to downstream of Cisaat River, particularly on the Mengger Formation (Figure 2). The geological features that are observed and measured are lithological variations, contacts between layers, sedimentary structures, rock textures, mineral compositions, and thickness. Two representative rock samples are collected and further analyzed petrographically to obtain rock textures and compositions of tuffaceous sandstone and tuffaceous conglomerate. The petrographic classification of mixed pyroclastic–epiclastic rocks and sandstone in this paper follows Schmidt (1981) and Pettijohn (1987) [10], respectively.

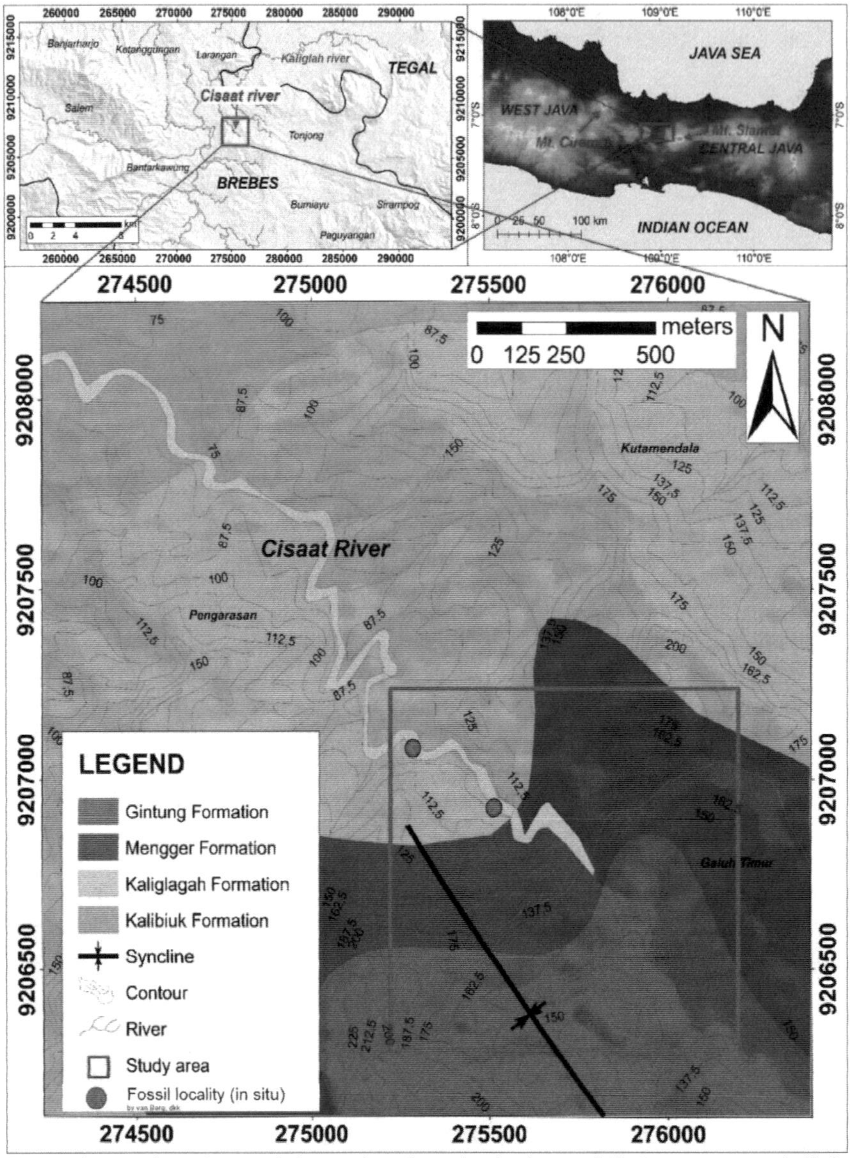

Figure 1. Research area location and the regional geological map [4]. Base-map and digital elevation model (DEM) acquired from the Indonesian Geospatial Information Agency (https://tanahair.indonesia.go.id/portal-web/).

IOP Conf. Series: Earth and Environmental Science 1071 (2022) 012014 doi:10.1088/1755-1315/1071/1/012014

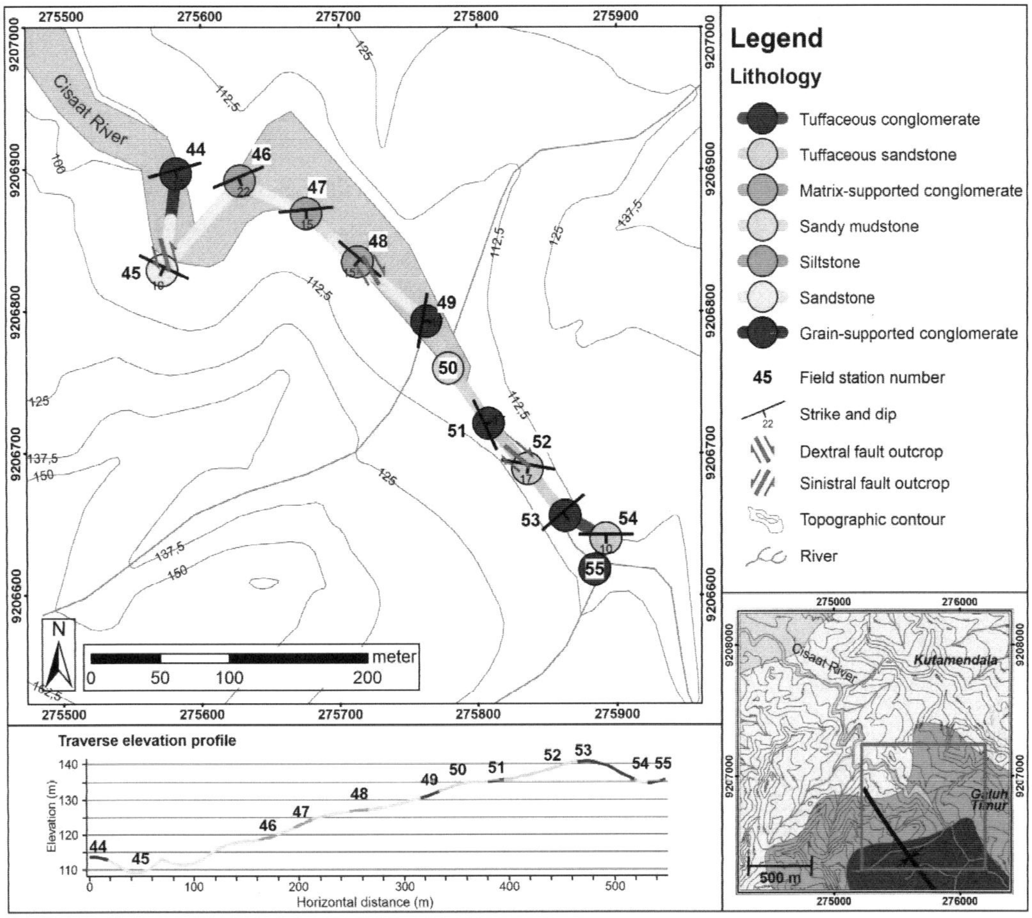

Figure 2. Traverse map of Mengger Formations along the Cisaat River. The measured section of the Mengger Formation is started from station number 44 to 55. The detailed measured section can be seen in Figure 3. Base map and digital elevation model acquired from Indonesian Geospatial Information Agency (https://tanahair.indonesia.go.id/portal-web/).

4. Results

Mengger Formation comprises 7 lithofacies, which were named based on structures, textures, and mineralogical compositions. From younger to older, the lithofacies consist of grain-supported conglomerate, sandstone, siltstone, sandy mudstone, matrix-supported conglomerate, tuffaceous sandstone, and tuffaceous conglomerate (Figure 3). Outcrop photographs are shown in figure 4, while thin section photomicrographs are shown in figure 5. The resume of petrographical analyses is presented in Table 1.

4.1. Grain-supported conglomerate

This lithofacies was observed at station 44 (Figure 4a), comprised of poorly sorted fine to cobble grained andesitic-basaltic and sandstone lithic fragments, with grain supported and sub- to well-rounded grains, which are embedded in a matrix (35%) of fine- to coarse-sand sized. It is characterized by massive and normal graded bedding structures. Granule to cobble-sized quartz fragments is also observed. This lithofacies is estimated to be the lower boundary of the Mengger Formation, which has erosional contact with the Kaliglagah Formation's sandstone (Figure 3).

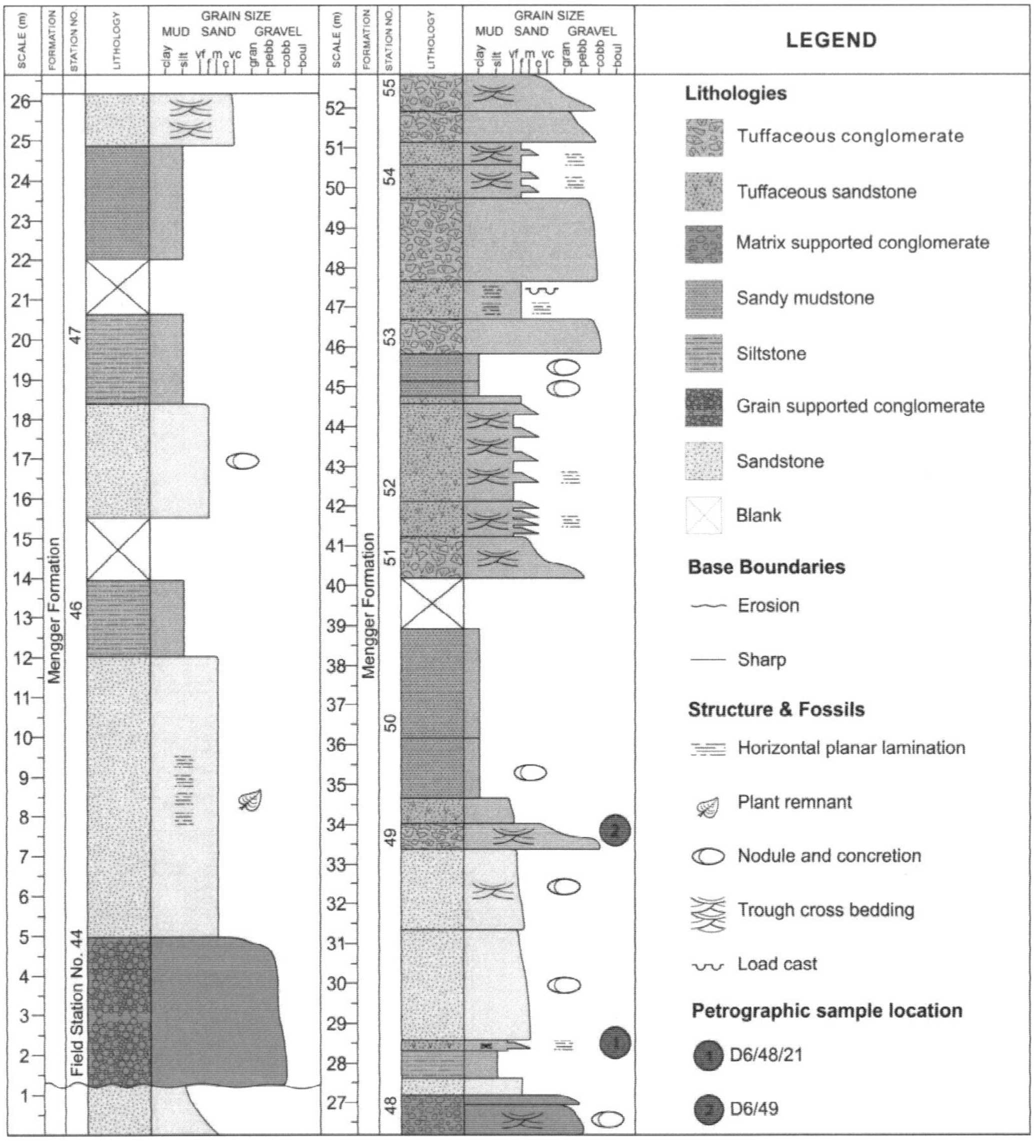

Figure 3. Measured stratigraphic section of Mengger Formation in Cisaat River. The traverse map of the measured section can be seen in Figure 2.

4.2. Sandstone

This lithofacies shows differences in grain size. It consists of well-sorted, very fine sand to clay with grain-supported and sub-angular to sub-rounded grains. The fragments consist of carbonate material, lithic fragments such as mudstone and sandstone, and minerals fragments including hornblende, feldspar, quartz, and fossils. This lithofacies was observed at station 44, 46, 47, and 48 (Figure 4b). Plant remnants such as leaves and wood fossils were found abundantly in station 44. The lithology is characterized by lamination, massive, horizontal stratification, and cross-stratification.

Figure 4. Photographs of Mengger Formation in Cisaat River. **(a)** Grain-supported conglomerate contact with sandstone at field station 44, **(b)** sandstone at field station 48, **(c)** siltstone at field station 47, **(d)** sandy mudstone at field station 48, **(e)** pumiceous conglomerate at field station 47, **(f)** tuffaceous sandstone at field station 48, and **(g)** pumiceous conglomerate at field station 54. Scale using red or white strips equal to 10 cm.

4.3. Siltstone

This lithofacies was observed at stations 46 and 47 (Figure 4c). The lithology is non-calcareous with grain sizes of clay to fine sand. It has massive and horizontal stratifications. The rock composes of silt-sized siliciclastic material.

4.4. *Sandy mudstone*

This lithofacies was observed at stations 48, 49, 50, and 52 (Figure 4d). The grain sizes are very fine-sand to pebble with moderately sorted, grain-supported, sub-angular to well-rounded textures. It has massive and horizontal stratifications. The rock is non-calcareous. The fragments (5%) consist of igneous rock lithic such as andesitic rock, which is embedded in a matrix (95%) comprised of lithic components such as sandstone and mudstone, hornblende, feldspar, and quartz.

4.5. Matrix-supported conglomerate

This lithofacies was observed at stations 47 and 50 (Figure 4e). The grain size are granule to cobble-sized with rounded to angular shapes. This rock is non-calcareous. The compositions are siliciclastic materials such as sandstone and mudstone, pumice, and quartz.

Figure 5. Plane-polarized light **(a)** and cross-polarized light **(b)** images of tuffaceous arkose arenite. The sample code is D6/48/21, taken from field station 48. Plane polarized light **(c)** and cross-polarized light **(d)** images of tuffaceous lithic greywacke. The sample code is D6/49, taken from field station 49. Mineral abbreviation as follow: Plg=plagioclase, Px=pyroxene, Lith=lithic (include pumice clast), Hbl=hornblende, Qz=quartz, Opq=opaque mineral.

4.6. Tuffaceous sandstone

This lithofacies was observed at stations 48, 49, and 51, 53 to 54 (Figure 4f), comprising moderately sorted fine to coarse sand and angular to sub-rounded grains. Occurrence of volcanic materials such as tuff, glass, feldspar, hornblende, lithic fragment, small to medium size pumice with grain supported, densely fabric with point and floating contact. The intensity of volcanic materials increases gradually upward, which shows the change in the composition of mixed volcanic and sedimentary rocks into volcaniclastic rocks. The lithofacies are characterized by lamination, horizontal stratification, and cross-stratification. A tuffaceous conglomerate interbedding can also be found in this lithology. The thin section of the sample taken at field station 48 is light brown to pale green in plane-polarized light and brownish-gray in cross-polarized light (Figures 5a, 5b). The thin section consists of fragments with the size of 0.05 – 1.45 mm and <0.01 mm matrix. The textures are poorly sorted, grain-supported, point, and long contact, sub-rounded to sub-angular grain roundness. The rock is composed of fragments (91%) comprised of plagioclase (42%), andesitic-basaltic lithic (21%), biotite (10%), pyroxene (7%), hornblende (5%), and opaque mineral (5%). The matrix (10%) is composed of clay minerals (5%) and volcanic glass (5%). This lithofacies is interpreted as the primary product of volcaniclastic, with a petrographic name of tuffaceous arkose arenite.

4.7. Tuffaceous conglomerate

This lithology was observed at station 54 (Figure 4g). The grains are sub to well-rounded that range from fine sand to pebble. It has a poorly-sorted, grain-supported texture. The rock is non-calcareous and has abundant conglomerate grains with horizontal stratification. The fragments (60%) consist of pumice, lithic andesite, lithic sandstone, and lithic mudstone embedded in a matrix. The matrix has a percentage of 40%. The thin section is colorless to brown in plane-polarized light and brownish-gray in cross-polarized light (Figures 5c, 5d). The thin section consists of fragments of 0.01 – 1 mm in size with <0.01 mm matrix. The textures are poorly sorted, grain-supported, point contact, and sub-rounded to rounded grain roundness. The fragment of the rock (70%) consists of lithic igneous and sedimentary rocks (40%), plagioclase (8%), pyroxene (7%), opaque mineral (6%), quartz (5%), and hornblende (4%). The matrix (30%) comprises volcanic glass (24%), clay minerals (5%), and plagioclase (5%). The petrographic name of this rock is tuffaceous lithic graywacke. In addition, the rock is considered an epiclastic-sedimentary product.

Table 1. Resume of petrographical analyses.

No	Sample No.	Rock Type	Rock Composition (%)										Tot
			Primary (Fragment)						Seconda-ry		Matriks		
			Plg	Px	Hbl	Bt	Frg. Lith.	Frg. Vit.	Qz	Min. Opq	Clay	Glass	
1	D6/48/21	Tuffaceous Arkosic Arenite	40	7	7	5	20	6		5	4	6	100
2	D6/54	Tuffaceous Lithic Graywacke	8	7	4		40		5	6	25	5	100

5. Discussions

Based on field observation and petrographic analyses, the volcaniclastic rocks of the Mengger Formation comprise of mixed between epiclastic-sedimentary products and primary volcaniclastic. Epiclastic-sedimentary products are marked by tuffaceous conglomerate lithofacies, composed of poorly sorted, fine- to pebble-grained with sub- to well-rounded grain shapes. It has abundant conglomerate grains, medium to large (pebble size) pumice clasts, siliciclastic fragments, and lithic mudstone embedded in a matrix with horizontal stratification. Epiclastic-sedimentary products have a mixed volcanic and non-volcanic clastic composition, with volcanic clastic consisting of different compositions and types, rounded grain shape, and moderately to well-sorted (based on clastic density) [11].

Epiclastic-sedimentary is marked by petrographic textures that distinguish the sedimentary materials with a more rounded grain shape and eroded appearance (Figure 5a). In contrast, the primary volcaniclastic grains are more angular in shape (Figure 5c). The primary volcaniclastic materials comprise pumice clast, plagioclase, pyroxene, hornblende, quartz, biotite, lithic fragments, and volcanic glass. The epiclastic-sedimentary materials comprise lithic fragments, quartz, plagioclase, and clay matrix.

The primary volcaniclastic is marked by tuffaceous sandstone lithofacies, which contain a lot of pumice. The middle to the upper of the formation increasingly shows the presence of primary volcaniclastic at several stations. The succession is well exposed in stations 48, 51, 53, and 54. Primary volcaniclastic comprises moderately sorted, fine- to pebble-grained with subangular to subrounded textures. It has the composition of angular glass shards, free crystals, vitric ash, small to medium-pumice clasts, and siliciclastic fragments. Furthermore, the primary volcaniclastic displays sedimentary structures, including trough cross-stratifications, parallel lamination, and horizontal stratification, which indicating fluvial processes.

This character (primary volcaniclastic) is similar to the study of the distal facies of the Mushono volcanic ash [12]. The Mushono volcanic ash layer is one of the widespread Pliocene tephra in central Japan, consisting of vitric ash, pumice clasts, free crystals, and epiclastic debris, showing a wide range in grain-size, i.e., from clay size (vitric particle) to cobble size (pumice). Mushono reworked volcaniclastic deposits at 150 to 200 km away from the source area. However, due to their aerodynamic property, the transportation distance of decimeter-sized pumice clasts by Plinian eruption is limited to 50 to 60 km from the source [12]. The origin of resedimented volcaniclastic material was probably derived from large-volume ignimbrite near the volcano. The presence of non–welded ignimbrite in the proximal area is key to generating the volcaniclastic re-sedimentation in distal fluvial. In the proximal area, especially unconsolidated loosely packed ignimbrite has a higher reworking potential than the fallout deposits on an alluvial plain [8]. However, other factors such as topography and drainage patterns affect the re-sedimentation styles in distal areas [8]. That might also happen in Mengger Formation with pebble-sized pumice. Pumice and scoria are produced in abundance by explosive eruptions [11]. Pebble- to cobble-sized large pumice clasts are often recognized in Plinian deposits and ignimbrite [13].

The eruptive center of the Mengger Formation has not been clarified in previous studies. The possibly closest ignimbrite source to this formation is Mount Slamet (~30 km distance). The pumice deposit in the Slamet Volcano mainly consists of pyroclastic-flow deposits dominated by pumice with a small amount of lithic and sandstone fragments. The characteristic is subangular – subrounded with poor sortation. Meanwhile, pyroclastic surge deposit is characterized by cross-stratification. This deposit can only be found on the west side of Mt. Slamet and is interpreted as part of the Old Slamet volcanism stage in the estimated age of the Late Pleistocene [14]. However, K-Ar radiometric dating of the hornblende phenocryst extracted from the pumice fragments in the lahar deposit of the Mengger Formation yields an age of 1.53 ± 0.08 Ma (Pleistocene) [6]. In addition, the occurrences of mammal fossils in the Mengger Formation indicate the period of the Upper Vertebrate Zone [7]. Thus, absolute dating analysis of the oldest volcanism in Mt. Slamet might be necessary for the source estimation of the pumiceous Mengger Formation.

Based on the primary volcaniclastic history, the Mengger Formation experienced at least three periods of eruptions. The stratigraphic measurement, lithofacies, and petrographical analyses reveal the first and second periods of eruption are marked by tuffaceous sandstone pumice-rich with sharp contact between tuffaceous sandstone and siltstone. The evidence can be seen from the section of 28,5 m at station 48, and 33.5 to 34.5 m at station 49 (Figure 3). The third period is marked by tuffaceous sandstone that overlays interbedded tuffaceous conglomerates from 40 to 52.5 m with extremely different contact from the second period (tuffaceous sandstone and sandy mudstone). The vertical changes of volcaniclastic succession reveal the sedimentary response to a large-volume eruption. After the eruption of the Mengger volcanic, a large volume of volcaniclastic debris including gravel-sized pumice (tuffaceous conglomerates lithofacies) were discharged into the fluvial system. This period produces a larger size of pumice deposits than the first period. From the deposit and measurement section, we conclude that type of source of Mengger Volcaniclastic eruption is a Plinian Type, with distinct deposits of pyroclastic flow and pyroclastic surge.

The previous researchers [1,6] reveal that the fossils found in the Mengger Formation are loose and partly in tuffaceous sand-gravel deposits. However, we did not find any vertebrate fossils in this study. Therefore, we could not correlate the eruption's effects on vertebrates and hominids in the study area. The previous research [1,6] mentioned that fossils were buried in the tuffaceous sand deposits. Thus we assume that these fossils are older Mengger Formation. Furthermore, the fossils are transported and deposited into this formation. The second assumption is that these fossils became extinct during the volcanic eruptions. Large volumes of volcanic material are transported (volcaniclastic) and buried these fossils. Further studies are needed regarding the correlation between eruptions and fossil extinctions in the research area.

6. Conclusion

Based on field observations and petrographic analysis, the volcaniclastic rocks in the Mengger Formation are a mixture of epiclastic-sedimentary products and primary volcaniclastic. Epiclastic-sedimentary products are characterized by tuffaceous conglomerate lithofacies, while primary volcaniclastic is characterized by tuffaceous sandstone lithofacies. Both contain a lot of pumice. The origin of the resedimented volcaniclastic material was probably derived from a large volume of ignimbrite near the volcano. Based on the primary volcaniclastic history of the Mengger Formation, there were at least three periods of eruption. Vertical changes of the volcaniclastic succession reveal the sedimentary response to a large-volume eruption. Post eruption of the Mengger volcanic source, a large amount of volcaniclastic debris including gravel-sized pumice (tuffaceous conglomerates lithofacies), were discharged into the fluvial system. From the deposit and measurement section, it can be concluded that the Mengger volcaniclastic eruption source is Plinian type, with distinct deposits of pyroclastic flow and pyroclastic surge.

References

[1] Nugraha S 2014 *Jurnal Sangiran* **3** p 28
[2] Setijadi R 2008 Perubahan Iklim Kala Pliosen - Plistosen Daerah Bumiayu Ditinjau dari Bukti Palinologi *Din. Rekayasa* **4** p 4
[3] Sondaar P Y 1984 Faunal Evolution and The Mammalian Biostratigraphy of Java. Cour. Forsch. Inst. Senckenberg
[4] Zaim Y 2010 Geological Evidence for the Earliest Appearance of Hominins in Indonesia. *Vertebr. Paleobiol. Paleoanthropology* p 97–110
[5] Kastowo and Suwarna 1996 Geological map of Majenang Quadrangle, Jawa, Scale 1:100,000 (Bandung: Geological Survey Institute)
[6] Bergh G D Van Den, Wibowo U P, Setiabudi E, Kurniawan I, Storey M, Suharyogi I Y P and Puspaningrum M R 2019 The Early Pleistocene Terrestrial Vertebrate Faunal Sequence of Java, Indonesia p 10–11
[7] Marks P 1957 Stratigraphic Lexicon of Indonesia (Bandung: The Geological Survey of Indonesia)
[8] Kataoka, K 2003 Volcaniclastic remobilization and resedimentation in distal terrestrial settings in response to large-volume rhyolitic eruptions: examples from the Plio-Pleistocene volcaniclastic sediments, central Japan. *Journal of Geosciences, Osaka City University* Vol. 46, Art. 4, p. 47-65
[9] van Bemmelen R W 1949 The Geology of Indonesia. General Geology of Indonesia and Adjacent Archipelagoes *Gov. Print. Off. Hague* p 545–7; 561–2
[10] Lucchi F 2013 Stratigraphic methodology for the geological mapping of volcanic areas: Insights from the aeolian archipelago (southern Italy) *Geol. Soc. Mem.* **37** p 37–53.
[11] McPhie, J, M. Doyle, R. Allen 1993 Volcanic Texture, Tasmania: *Tasmanian Government Printing Office*
[12] Walker, G.P.L 1980 The Taupo pumice: product of the most powerful known (ultraplinian) eruption? *Jour. Volcanol. Ceotherm.* Res., 8, 69-94
[13] Cas, RAF dan Wright, J.V 1988 Volcanic Successions: Modern and Ancient. London: Chapman dan Hall
[14] Sutawidjaja, I. dan Sukhyar, R 2009 Cinder cones of Mount Slamet, Central Java, Indonesia. *Jurnal Geologi Indonesia,* vol. 4 no.1 pp.57-75.

Acknowledgment

We would like to extend our gratitude to team that has made this study especially field observation and data acquisition possible. This research was partly funded by the Ministry of Research and Technology of Indonesia through *Penelitian Terapan Unggulan Perguruan Tinggi (PTUPT)* funding program for year 2021. This research is entitled "*Rekontruksi Sejarah Geologi untuk Memahami Bencana Geologi Purba dan Implikasinya Terhadap Kepunahan Manusia Purba Bumiayu, Jawa Tengah*" contract Number 2154/UN.1/DITLIT/DIT-LIT/PT/2021.

Estimation of Volume and Column Height from Pumiceous Tephra-Fall Deposits of Mt. Raung, East Java, Indonesia

S M Cahyani[1], H E Wibowo[1*], M L A Moktikanana[1], A Fajarwati[2], and Kristianto[3]

[1]Departement of Geological Engineering, Faculty of Engineering, Universitas Gadjah Mada. Jalan Grafika no 2, Yogyakarta 55281, Indonesia
[2]Departement of Development Geography, Faculty of Geography, Universitas Gadjah Mada. Jalan Kaliurang Bulaksumur, Yogyakarta 55281, Indonesia
[4]Center of Volcanology and Geological Hazard Mitigation, Ministry of Energy and Mineral Resources, Bandung 40122, Indonesia
*Corresponding author's e-mail: haryo.edi.w@ugm.ac.id

Abstract. Pumiceous tephra-fall deposits observed mainly in the western flank of Mt. Raung are indicating past explosive eruption event. The volume and the eruption column height of the eruptive event is estimated using isopach and isopleth maps. These maps are constructed from seven observation points. The pumiceous tephra-fall deposits can be divided into 9 layers from bottom to top, White Pumice Fall 1–7, Yellow Pumice Fall, and Lithic-Rich Scoria Fall. These layers not separated by any soil or weathered layer indicates no significant time gap between them. Isopach and isopleth maps are generated from White Pumice Fall 6, Yellow Pumice Fall, and Lithic-Rich Scoria Fall considering the widespread distribution of these deposits. White Pumice Fall, Yellow Pumice Fall, and Lithic-Rich Scoria Fall are resulted from eruptive event with ~25–33 km, ~35 km, and ~29 km column height and 1.3 km^3, 0.73 km^3, and 0.32 km^3 volume, respectively. Total volume of 2.35 km^3 and the column height of 25-35 km suggest that Plinian eruption type of VEI 5 magnitude produces this pumiceous tephra-fall deposit sequence.

1. Introduction

Volcanic eruption is the process of magma releasing from inside the earth to the surface and when magma comes out erupting and generating eruptive column then it is called an explosive eruption [1]. Pyroclastic fall deposits are formed by explosive eruptions characterized by the formation of convective columns or plumes that carry pyroclastic and volatile upwards into the atmosphere and then the ejecta is carried and distributed by the wind [2]. The distribution of explosive eruption products in the form of pyroclastic falls is influenced by the wind (speed and direction), the height of the eruption column, and the intensity of the eruption. The greater the eruption intensity, the more volume of product is released and the higher the eruption column is. The explosive characteristics of an eruption can be generally indicated from the volcanic explosivity index (VEI) [3]. From the information on the volume of ejecta and the column height, the VEI scale can be determined. Mt. Raung is an active volcano attributed with 2 km wide caldera morphology on the summit. This caldera morphology indicates that Mt. Raung has had a large-scale explosive eruption in the past. The geological map by [4] reported the occurrence of two massive pyroclastic fall deposits in Mt. Raung. Those are 1) Rjp1, pumiceous tephra-fall deposits with thickness up to 2 m and 2) Rjp2, scoriaceous tephra-fall deposit related to VEI 5 eruption event in 1586. This study reports first estimation of volume and column height of the eruption producing the pumiceous tephra-fall deposits. Both parameters are then used to identify the magnitude and type of the eruption.

2. Geology of Mt. Raung

Mt. Raung is one of the most active Quaternary stratovolcanoes in East Java, Indonesia. This volcano is part of Sunda Volcanic Arc resulted from northward subduction of Indo-Australian Plate beneath Eurasian Plate [5]. Mt. Raung has 2 km wide circular caldera on the summit area [6]. This caldera structure is likely related to major explosive eruption even though the age of this caldera-forming event is unknown.

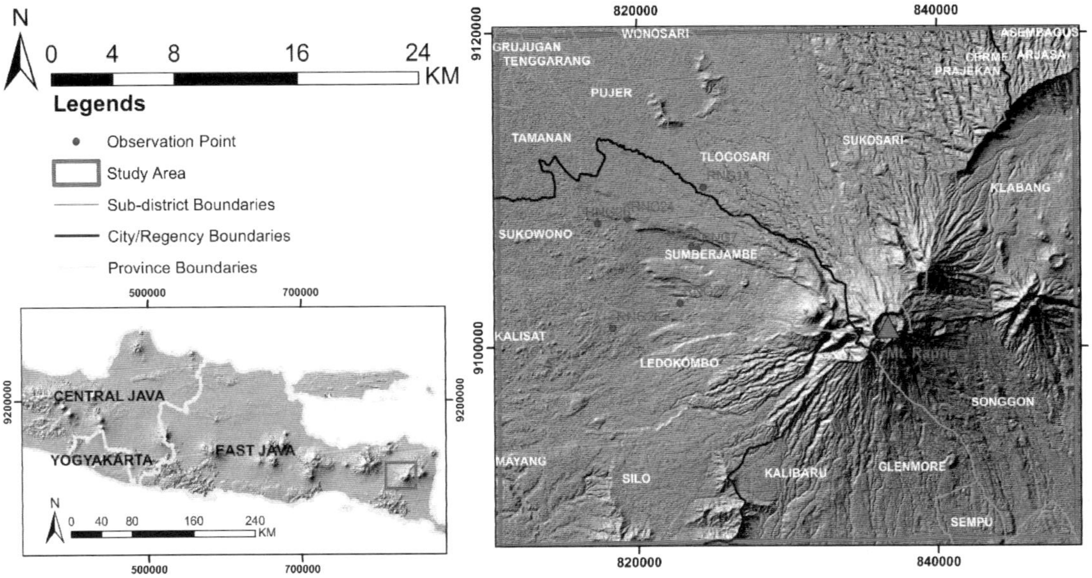

Figure 1. Location of Mt. Raung and distribution of observation point in this study. DEM data is downloaded from https://srtm.csi.cgiar.org/.

Geological map of Mt. Raung [4] divides the volcanic activity of Mt. Raung into three periods from the oldest to the youngest; they are Old Raung Volcanism, Gadung Volcanism, and Young Raung Volcanism (Figure 1). Pumiceous tephra-fall is reported as Rjp1 and is distributed on the western and southern flank of Mt. Raung. Historical eruption records show that Raung volcano is characterized by the explosive eruption mainly producing volcanic ash and pyroclastic flow. Early record of Mt. Raung activities comes from paroxysmal event in 1586. Other paroxysmal eruption was documented in 1597, 1638, 1890, 1953, and 1956 [6]. The 1956 eruption produced column height up to 12 km high, with ashfall distributed up to 200 km radius, covering Bali and Surabaya [6]. The last activity of Raung was recorded in January 2021.

3. Methods

Samples and field data are collected using spot sampling covering Mt. Raung in all direction, except northeast flank. Tephrostratigraphy is recorded in all of the observation points. Data of the pyroclastic fall deposits includes thickness, contact between each layer, structure of the deposit, composing materials, the average size of clasts, and dimension of five biggest clasts, both pumice and lithic. Correlation is mainly built from the tephrostratigraphy and the physical characteristics of the deposits. Isopach and isopleth maps of pumice fall deposits are generated considering the correlation result and the physical characteristics. Estimation of the eruption volume follows the method of [7] using isopach maps data. The thickness in the proximal and distal area cannot be assumed due to data limitation.

Volume estimation is limited using isopach data, thus uncertainty of estimated volume cannot be defined in absolute value. Estimation of the eruption column height follows the method of [8] using isopleth maps. Combination of both data are used in estimating the Volcanic Explosivity Index (VEI) and eruption type.

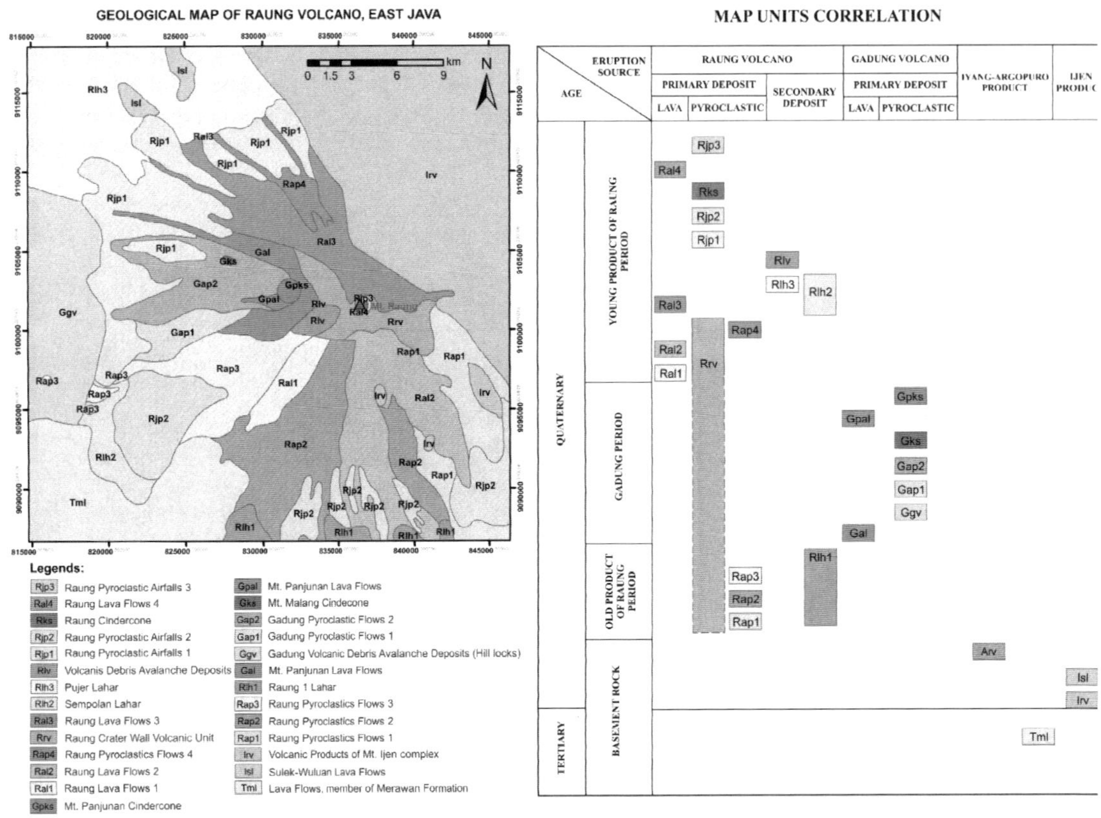

Figure 2. Geological map of Raung Volcano, East Java (Re-drawn from [4])

4. Results and Discussion

4.1. Tephrostratigraphy

Pumiceous tephra-fall deposits are observed in the western flank of Mt. Raung. Tephrostratigraphic records are made in seven observation points and the correlation is built accordingly. Pumiceous tephra-fall deposits of Mt. Raung can be divided into 9 units from bottom to top are White Pumice Fall (WPF) 1-7, Yellow Pumice Fall (YPF), and Lithic-Rich Scoria Fall (LSF) (Figure 3). These deposits are separated to other tephra fall sequence by thick soil above and below the sequence. Within the sequence, the tephra-fall deposits are considered fresh with minor weathering process and are not separated by any soil layer. The absence of weathering layer or soil within this sequence indicates no significant time gap between the deposition periods of each layers that can be considered as coming from the same eruptive period. White Pumice Fall (WPF) 1 can be found in RNG14, is about 9 cm thick, has massive structure, and is composed of white andesitic pumice fragments (1 mm to 40 mm in size). White Pumice Fall (WPF) 2 can be found in RNG14, is about 22 cm thick, has massive structure, and is composed of white andesitic pumice fragments (1 mm to 32 mm in size) and basaltic lithic (1 mm to 32 mm in size). White Pumice Fall (WPF) 3 can be found in RNG26, is about 45 cm thick, has normal gradation structure, and is composed of white andesitic pumice fragments (1 mm to 10 mm in size). White Pumice Fall (WPF)

4 can be found in RNG26, is about 5 cm thick, has massive structure, and is composed of white andesitic pumice fragments (<1 mm to 3 mm in size) and basaltic lithic (<1 mm to 2 mm in size). White Pumice Fall (WPF) 5 can be found in RNG26, is about 5 cm thick, has massive structure, and is composed of white andesitic pumice fragments (<1 mm to 3 mm in size) and basaltic lithic (<1 mm to 2 mm in size). White Pumice Fall (WPF) 6 can be found in all sites, except RNG07 and RNG24 (RNG06, 14, 23, 26, and 28), has thickness range of 20-45 cm, has normal gradation structure, and is composed of white andesitic pumice fragments (1 mm to 35 mm in size) and basaltic lithic (<1 mm to 17 mm in size). White Pumice Fall (WPF) 7 can be found in RNG26, is about 4 cm thick, has massive structure, and is composed of white andesitic pumice fragments (<1 mm to 2 mm in size) and basaltic lithic (<1 mm to 2 mm in size). Yellow Pumice Fall (YPF) can be found in all seven observation points (RNG06, 07, 14, 23, 24, 26, and 28), has thickness range from 15-46 cm, has normal gradation structure, and is composed of yellowish andesitic pumice fragments (<1 mm to 45 mm in size) and basaltic lithic (<1 mm to 15 mm in size). Lithic-Rich Scoria Fall (LSF) is the uppermost layer of the pumiceous tephra-fall deposits sequence with thickness range from 11-35 cm. This deposit can be found in all sites, except RNG24 (RNG06, 07, 14, 23, 26, and 28). This deposit has massive structure and is composed of andesitic lithic fragments (<1 mm to 15 mm in size) and scoria (<1 mm to 20 mm in size). These deposits' characteristics are aligned with the pumiceous pyroclastic fall deposit's characteristics (Rjp1) described by [6]. Estimation of volume and column height of the eruption process is focused on White Pumice Fall (WPF) 6, Yellow Pumice Fall (YPF), and Lithic-Rich Scoria Fall (LSF) layers considering their widely spread distribution. Representative outcrop picture of these deposits is presented in Figure 4.

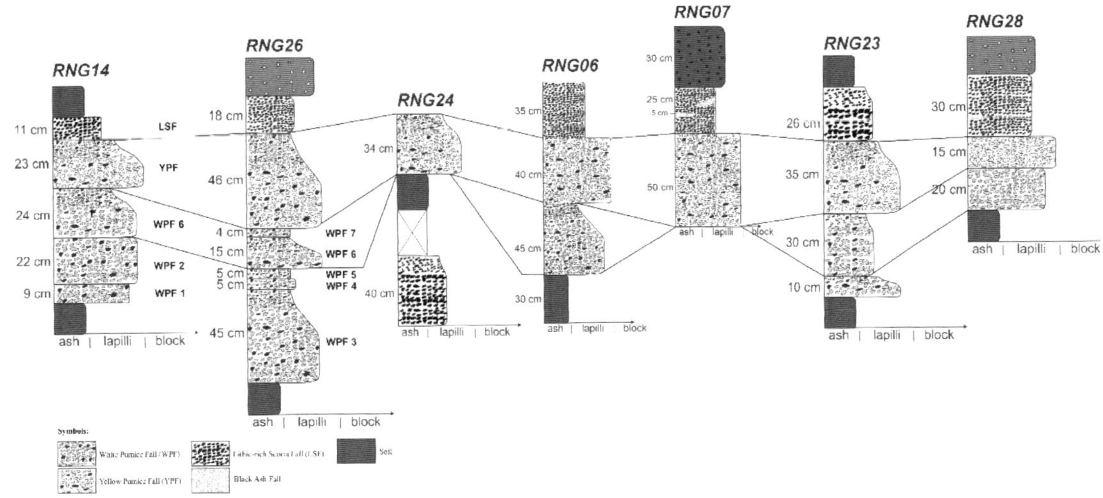

Figure 3. Stratigraphy of Mt. Raung pumiceous tephra-fall deposit divided into nine units of WPF1-7, YPL, and LSF. Stratigraphic correlation from the seven observation points are done at WPF6, YPL, and LSF layers.

IOP Conf. Series: Earth and Environmental Science 1071 (2022) 012015 doi:10.1088/1755-1315/1071/1/012015

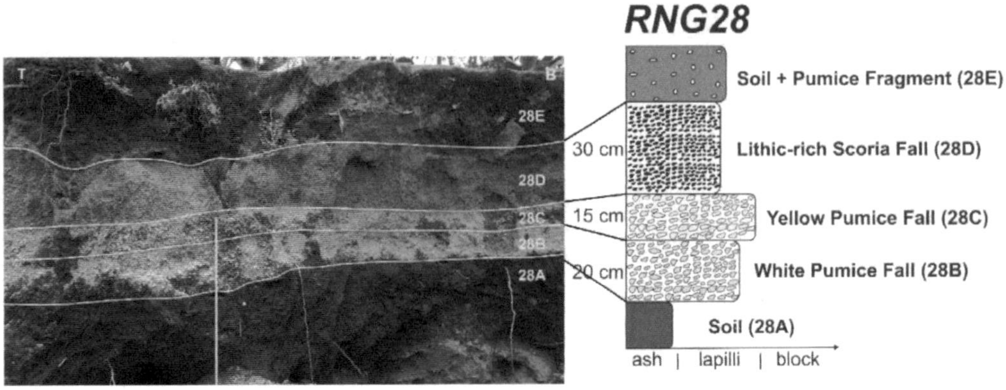

Figure 4. Outcrop and lithologic column of tephra-fall deposit at RNG28. Yellow stick scale resembles 1 m height.

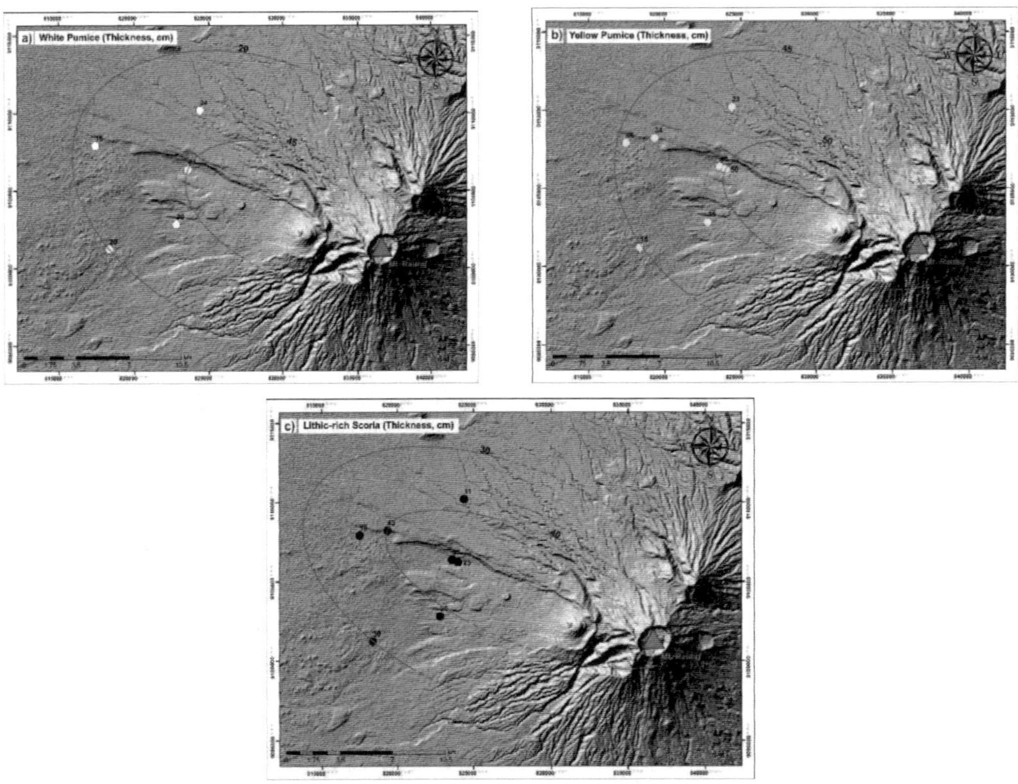

Figure 5. Isopach map of tephra fall deposits (a) White Pumice Fall (WPF) deposits, (b) Yellow Pumice Fall (YPF) deposits, (c) Lithic-Rich Scoria Fall (LSF) deposits. DEM data is downloaded from https://tanahair.indonesia.go.id/demnas/.

4.2. Volume estimation of Mt. Raung erupted materials

The isopach map is drawn using free-hand method following thickness distribution of the deposits. In an ideal condition, tephra deposits get thicker with larger fragments as they get closer to the eruption centre. The thickest deposit is prioritized as key data considering brittle characteristics of the tephra fall make it prone to thickness loss due to secondary process. At the location point with the largest thickness, an axis line is drawn towards the eruption center. This axis line presents the direction of tephra deposits distribution and wind direction that carries the pyroclasts. The isopach contour line shows approximate thickness of deposit. In addition, the isopach map is evaluated with the isopleth map assuring that the tephra-fall deposits can be correlated and are coming from Mt. Raung. The isopach maps are shown in Figure 5a-c.

The estimation of Mt. Raung erupted material's volume follows the Weibull-fit method of [7] from the thickness and isopach area value. There are Weibull parameters to describe important features of deposit thinning and to provide information about the eruption style and magnitude: 1) the characteristic decay length of deposit thinning in kilometer (λ), 2) a thickness scale in centimeter (θ), and 3) shape parameters (n) [7]. The estimated volume from the WPF 6, YPF, and LSF deposits are 1.3 km^3, 0.73 km^3, and 0.32 km^3, respectively. Minimum total volume of the eruptive event producing sequence of pumiceous tephra-fall deposit of Mt. Raung is 2.35 km^3 (VEI 5). This volume is considered as an underestimated number considering the limited thickness data availability.

4.3. Estimation of volcanic eruption column height

Estimation of column height is determined by following the method of [8] using isopleth map assuming density of lithic is 2500 kg/m^3, which is the basaltic lithic density (Figure 6a-c). The method of making the isopleth maps is almost similar to isopach maps, except maximum lithic size data is used in the isopleth maps. There are things to consider when plotting data on the Fallout and Dispersal of Tephra Diagram by [8]: value of maximum downwind range, crosswind range, and lithic diameter (Table 1). There are two isopleth lines for White Pumice Fall: 1) lithic 0.8 mm and 2) lithic 1.6 mm. Meanwhile, there is only one isopleth line for Yellow Pumice Fall and Lithic-Rich Scoria Fall, lithic 1.6 mm. The plotted data on the Fallout and Dispersal of Tephra Diagram by [8] from the value of maximum downwind range, crosswind range, and lithic diameter is presented in Figure 7.

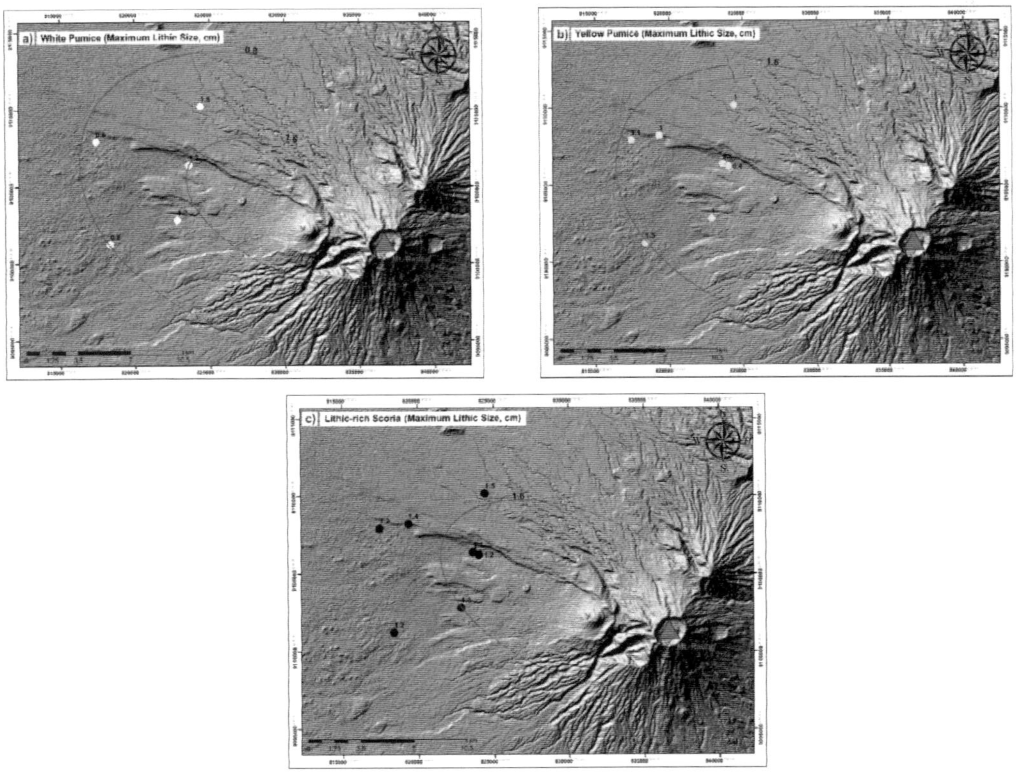

Figure 6. Isopleth map of tephra fall deposits (a) White Pumice Fall (WPF) deposits (b) Yellow Pumice Fall (YPF) deposits (c) Lithic-Rich Scoria Fall (LSF) deposits. DEM data is downloaded from https://tanahair.indonesia.go.id/demnas/.

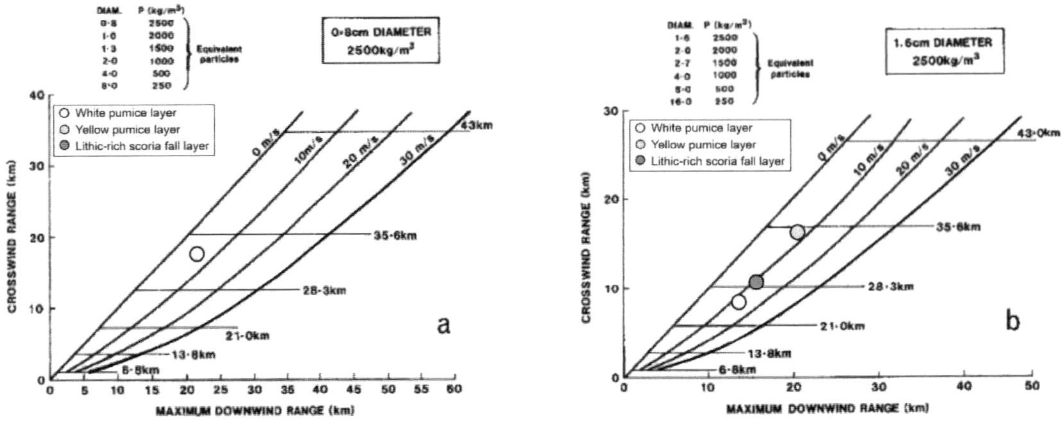

Figure 7. Plotting at Fallout and Dispersal of Tephra Diagram by [8] to estimate eruption column height (a) Lithic diameter 0,8 cm, (b) Lithic diameter 1,6 cm.

Table 1. Summary of isopach and isopleth parameters, and estimation results of volume and column height

Fall deposit	Thickness (cm)	Isopach area (km²)	λ (km) [7]	θ (cm) [7]	n [7]	Volume (km³) [7]	Maximum lithic size (cm)	Down-wind (km)	Cross-wind (km)	Column Height (km) [8]
White Pumice Fall (WPF) 6	45	112.585	4.88	586.87	0.2	1.3	0.8	21.2	15.3	33
	20	360.861					1.6	15.3	13.4	25
Yellow Pumice Fall (YPF)	50	103.82	36.84	53.94	2	0.73	1.6	21.2	15.3	35
	45	350.295								
Lithic-Rich Scoria Fall (LSF)	40	114.456	5.52	587.73	1.13	0.32	1.6	16.2	10.5	29
	30	316.468								

According to the Fallout and Dispersal of Tephra Diagram by [8], the eruption column of WPF6, YPF, and LSF deposit are ~25 – 33 km, ~35 km, and ~29 km, respectively. The value of crosswind and downwind connected to the maximum distribution of lithic size is proportional to the eruption column height and eruption intensity. As the bigger lithic is distributed further from the eruption centre, the value of crosswind and downwind increase. This indicates a high eruption column and great eruption intensity. Column height of the three thickest deposits in the sequence of pumiceous tephra-fall deposit of Mt. Raung show relatively similar height of 25–35 km suggesting explosive Plinian eruption type. This volume and column height estimation indicates that Mt. Raung has experienced VEI 5 magnitude eruptive event prior to the one which is recorded in 1586.

5. Conclusion

Sequence of pumiceous tephra-fall deposits of Mt. Raung composed of nine layers of WPF 1–7, YPF, and LSF. The volume estimation of WPF6 is 1.3 km^3, YPF is 0.73 km^3, and LSF is 0.32 km^3. Total volume is 2.35 km^3 (VEI 5) and is considered to be an underestimated number due to limited outcrop availability. In addition, eruption column height of WPF6 is ~25 – 33 km, YPF ~35 km, and LSF ~29 km (Plinian eruption type).

References

[1] Bronto S 2010 *Geologi Gunung Api Purba* ed U Hartono and N Suwarna (Bandung: Geological Agency of the Ministry of Energy and Mineral Resources) p 35

[2] Sparks R S J 1986 The dimensions and dynamics of volcanic eruption columns *Bull. Vulcanol.* **48** 3

[3] Newhall C G and Self S 1982 The Volcanic Explosivity Index (VEI): An Estimate of Explosive Magnitude for Historical Volcanism *Journal of Geophysical Research* **87** 1232–1234

[4] Sutawidjaja I S, Suparman, Sitorus K 1996 *Geological Map of Raung Volcano, East Java* (Bandung: Vulcanological Survey of Indonesia)

[5] Sabila F S N and Abdurrachman M 2018 Vulcanostratigraphy and Petrogenesis of Raung Vulcano, Jember and Bondowoso Area, East Java *Proceedings Pekan Ilmiah Tahunan IAGI 2018* (Pekanbaru: PIT IAGI)

[6] Center of Vulcanology and Geological Hazard Mitigation (CVGHM) 2014 *Data Dasar Gunung Api Raung* accessed 16 August 2019 from http://www.vsi.esdm.go.id/index.php/gunungapi/data-dasar-gunungapi/526-g-raung

[7] Bonadonna C and Costa A 2012 Estimating the volume of tephra deposits: A new simple strategy *Geology* **40** 415–418

[8] Carey S and Sparks RSJ 1986 Quantitative models of the fallout and dispersal of tephra from volcanic eruption columns *Bull. Vulcanol.* **48** 109–125

Acknowledgments

This reaserch is fully funded by the Ministry of Research and Technology of Indonesia through *Penelitian Terapan Unggulan Perguruan Tinggi (PTUPT),* funding program for year 2021, No. 8/E1/KPT/2021 and 3572/E4/AK.04/2021 and Contract No. 6/E1/KP.PTNBH/2021 and 006/E4/AK.04.PTNBH/2021 and 2140/UN1/DITLIT/DIT-LIT/PT/2021 and 6629/UN1/DITLIT/DIT-LIT/PT/2021, entitled *"Pengurangan Risiko Bencana Gunung Api Berbasis Integrasi Ilmu Kebumian: Studi Kasus Gunung Raung Jawa Timur"*. We also would like to thank Faiza Rahman Hakim for his assistance during fieldwork.

Reconnaissance study on the stratigraphy and characteristics of eruption products associated with basaltic caldera in Raung volcano, East Java, Indonesia

M L A Moktikanana[1] and A Harijoko[1,2*]

[1]Department of Geological Engineering, Faculty of Engineering, Universitas Gadjah Mada. Jalan Grafika no 2, Yogyakarta 55281, Indonesia
[2]Center of Disaster Studies, Universitas Gadjah Mada. Jalan Mahoni C 16, Yogyakarta 55281, Indonesia
*Corresponding author's e-mail: aharijoko@ugm.ac.id

Abstract. As one of the most active volcanoes with significant eruptive impact in Indonesia, the Raung volcano shows a trace of the pre-historical devastating eruption as evidenced by the morphology of caldera at its summit. Our fieldwork revealed the occurrence of four pyroclastic fall deposits, two lava flows, ignimbrite, block and ash flow deposit, and lahar as the products of Raung eruption. There is evidence of Plinian phase eruption represented by layers of pumice fall without any significant time gap, which is uncommon for an andesitic-basaltic stratovolcano. Thin section observation on lava and the lithic fragment of block and ash flow deposits reveals that the composition of Raung volcanic products is mainly basalt consisting of plagioclase, clinopyroxene, olivine, and orthopyroxene as the primary crystal phases. Hornblende and biotite also compose pumice as a minor constituent. This evidence implies that Raung magmatism may potentially evolve long enough to produce such massive pumice fall deposits.

1. Introduction

Situated in the Eastern Part of Java, only 50 km away from Bali, the Raung volcano is famous for its eruptive activity and broad impact in Indonesia. The eruption of this andesitic-basaltic volcano in the last few years generally drifted to the east. It forced the closure of I Gusti Ngurah Rai International Airport in Bali [1, 2], a gateway for Indonesia's tourism destinations. The current eruptive characteristic of the Raung volcano is Strombolian type ejecting widespread ashfall and crater-filling lava flows [3]. The caldera morphology on the summit indicates that a past major eruption occurred with much greater power than the recent modern eruptions.

Caldera forming eruption usually happens on a scale of VEI 5 [4] – that is equivalent to the Krakatau eruption in 1883 or hundred times greater than Merapi eruption in 2010. A record from [5] said that Raung volcano had experienced a VEI 5 scale eruption in 1593, which occurred over three days and killed thousands of lives. However, detailed investigations reporting the distribution and properties of the products associated with that large-scale explosive eruption were not yet available. The only evidence about explosive products of Raung volcano – particulary pyroclastic fall deposits – was reported by [6] on the Geological Map of Raung Volcano scale 1:100.000, revealing the presence of widely distributed pumice fall and scoria fall unit surrounding Raung volcano.

This study aims to present a reconnaissance study about several units of pyroclastic deposit and lava flow from Raung as the products related with basaltic caldera volcanism. We also report a detailed description of the presence of pumice fall deposit that represents a Plinian phase product that is uncommon in an andesite basaltic volcano. Furthermore, we hope this study may provide a new perspective and opportunity in the developing research related to the formation of basaltic caldera.

2. Geology of Raung Volcano

Raung volcano is an andesitic-basaltic stratovolcano with a 1.5 x 2 km caldera morphology on its summit (**Figure 1**). [6] separated the 3 phases of Raung volcanism into Old Raung Period, Gadung Period, and Young Raung Period. The products of Old Raung Period mainly consist of pyroclastic flows such as ignimbrite or block and ash flow deposit. Meanwhile, the Young Raung volcanism has a different eruptive behavior producing mostly basaltic lava, minor block and ash flow deposit, and widespread pyroclastic fall deposit, which are similar to the eruptive character of Raung in the modern era up until the present time.

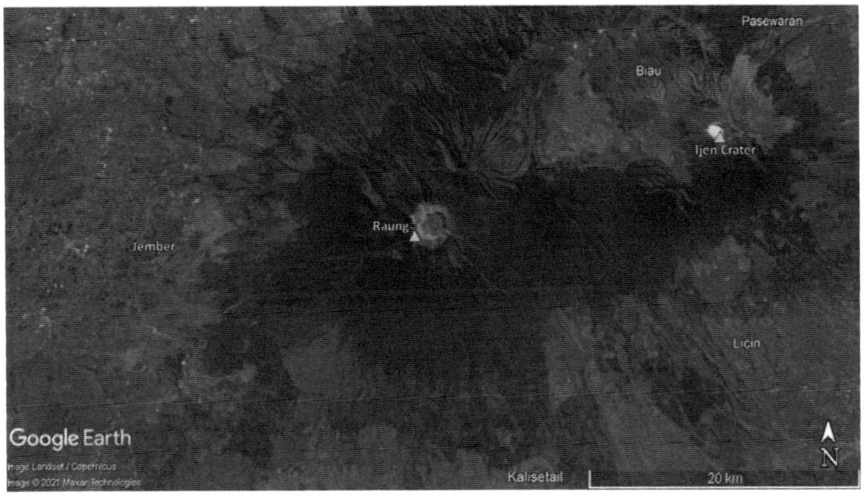

Figure 1. Morphology of caldera on the summit of Raung volcano. Image taken from Google Earth (2021).

The last magmatic eruption of Raung occurred on the period of February-April 2021 producing the explosion of ash plume up to 2500 m from the summit directed to the south and east [7]. Monitoring from 1900 to 2021 records 48 eruption events, with a recurrence of 1 to 5 years on an average VEI 1-3 [7]. Major eruption with a VEI scale of 5 was documented in 1593, while two VEI 4 eruption occurred in 1638 and 1817. Those three major eruptions are 50-180 years apart and may recur in the future.

3. Samples and Methodology

Field observation, stratigraphic measurement, and sample collection were conducted in June 2021, with a total of 9 observation sites including 5 points on the southern part (Banyuwangi) and 4 points on the western part (Jember) of Raung volcano (**Figure 2**). We recorded thickness, sedimentary structures (if any), components, grain size, and contact for pyroclastic deposits. Meanwhile, data recording for lava has the thickness and a general description of lithology.

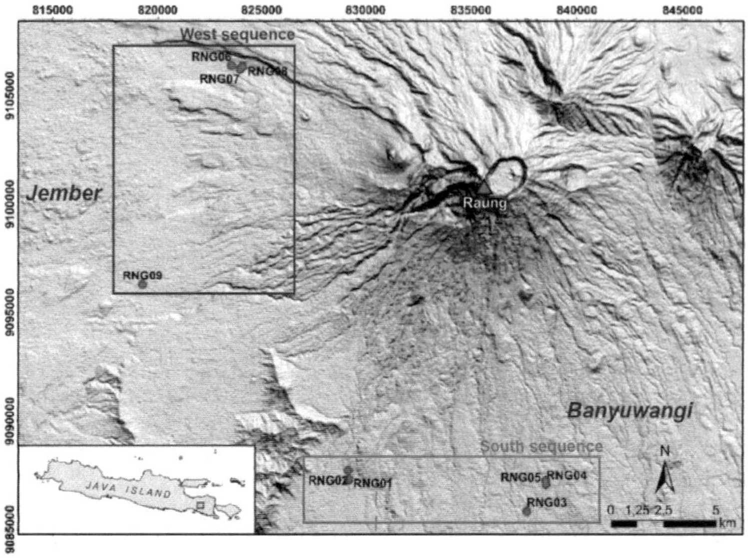

Figure 2. Field observation and sampling point location on the western and southern part of Raung volcano. Digital Elevation Model data can be accessed for free through (https://tanahair.indonesia.go.id/demnas/).

We conducted grain-size distribution and componentry analyses for six samples of pyroclastic fall deposits collected from 5 layers, from RNG04, RNG05, and RNG06 (**Figure 2**). Grain size distribution was performed using 1ϕ (5 μm) to -4ϕ (16 mm) sieve. Componentry analysis was determined by counting the number of grains for each lithology type, starting from 4 mm to >16 mm. The classification of grain types identified on the componentry analysis shown in **Figure 3** comprises pumice (white, grey, yellow, reddish-yellow, banded), scoria (dark grey, light brown, banded), dan igneous lithic (fresh and altered). Petrographic analysis was carried out on the most dominant juvenile type for each unit with 15 samples. Thin sections for pumice fall deposits are represented by white pumice, scoria fall deposits are characterized by dark grey scoria, and ignimbrite juveniles are represented by grey pumice.

4. Results

4.1 Stratigraphy

Our fieldwork discovered a different stratigraphic unit in the southern and western parts of Raung, which are difficult to correlate, so we separated the stratigraphy into two sections: South Sequence and West Sequence. Field documentations are presented in **Figure 4** while stratigraphic correlations for each sequence are shown in **Figure 5.**

Figure 3. Classification of grain types used in the componentry analysis.

4.1.1 South Sequence. Old Raung Ignimbrite. This unit was found in RNG02 and RNG03 at the very bottom as the base of the sequence, with total thickness of 0.5-2 m, covered by lahar deposit with erosional contact. It has a highly consolidated, lithic rich, light brown color, with lapilli-block fragment size (0.1 – 20 cm) and fine-coarse ash matrix (**Figure 4b**). It is composed of grey pumice as the dominant component, with reddish-yellow pumice and blocky lithics as other components. This unit might resemble the Raung pyroclastic flow 2 (Rap2) unit from Old Raung period as described by [6] and is estimated to be the oldest unit in the South sequence.

Lava. Two types of different lava were found in RNG02 and RNG05. In the RNG02 section, lava is present on top of the lahar deposits and forms the top of the waterfall. Lava RNG02 shows a columnar joint structure (**Figure 4c**), gray-colored with a porphyritic texture and contains plagioclase and pyroxene phenocrysts. In RNG05, lava presents as an intermittent river bed, dark gray with vesicular and aphyric texture (**Figure 4d**).

Milky white fine ash. This unit was widely distributed in the southern part, particularly at RNG01, RNG04, and RNG05. It has a cloudy white, slightly brownish color, very fine ash grain, and 5-10 cm thick (**Figure 4e**). This unit clearly stands out and differs from Raung's product which is generally darker in color (dark gray to black), so it is assumed to come from a volcano located quite far from Raung. Due to its wide distribution and distinctive character, this unit is used as a keybed for southern sequence correlation.

Scoria fall. The South sequence scoria fall was found at RNG04 and RNG05, 14 km from Raung summit. It was emplaced on top of paleosoil, younger than the milky white fine ash layer. This unit is 5-30 cm thick with normal grading structure, composed of medium ash – lapilli sized (0.5 - 4 cm) fragments and fine ash matrix (**Figure 4e**). The juveniles are dense, angular, crystal rich with spherical vesicle, dark grey (61%) and light brown (29%) scoria. The lithics are relatively finer than scoria and comprised of grey-colored fresh igneous lithic (7%) with occasionally reddish altered igneous lithic (3%).

Block and ash flow deposit. This unit is highly consolidated, 5-40 cm thick, occured at RNG04 and RNG05 intercalated with scoria fall (**Figure 4e**). It is composed of lapili - block (1-10 cm) fragments embedded in fine ash matrix. Juveniles are mainly consisting of basalt lithic with minor pumice (<5%, 0.1 - 0.2 cm size.)

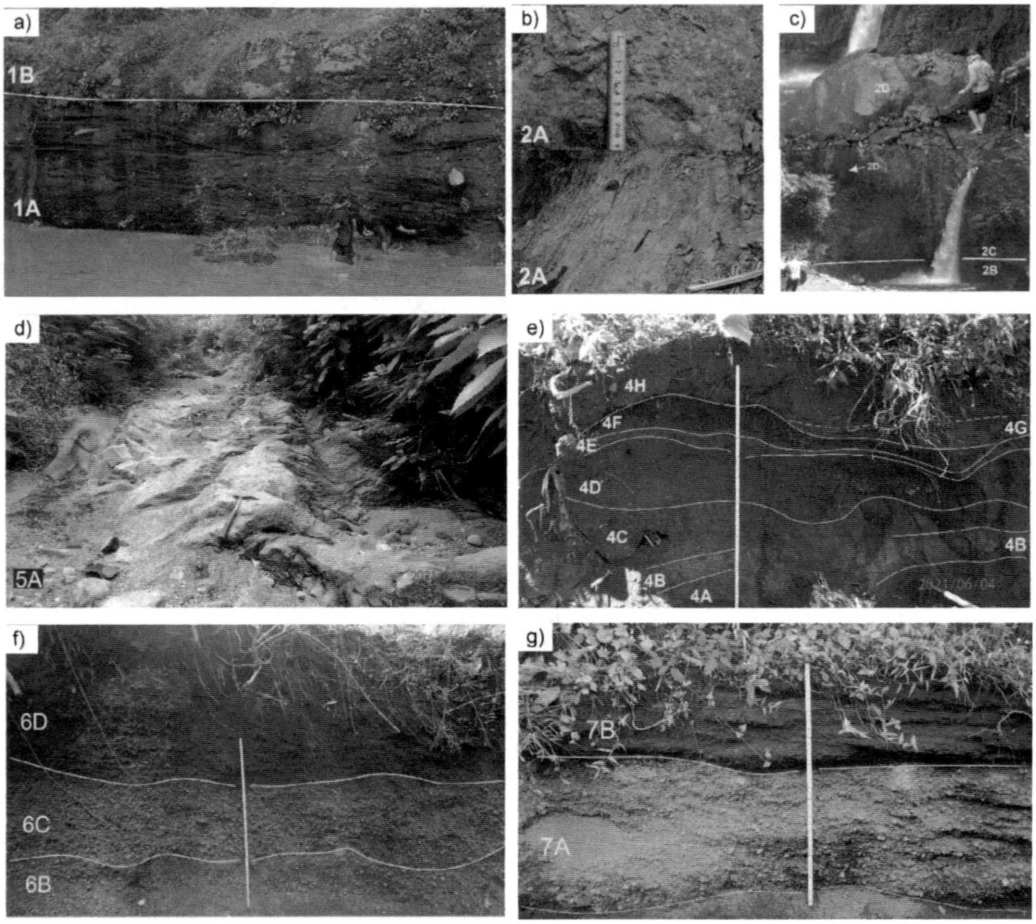

Figure 4. Outcrops of (a) lahar unit with cross bedding structure (1A), (b) Old Raung Ignimbrite (2A), (c) columnar jointed lava flow (2D) on top of lahar deposit, (d) Aphyric lava flow (5A) as river bedrock, (e) a sequence of scoria fall deposit (4D, 4F) intercalated with block and ash flow deposit (4E, 4G) on top of milky white fine ash layer (4B) found in the southern part of Raung volcano, (f-g) layers of white (6C) and yellow (6C, 7A) pumice fall covered by lithic rich scoria fall (6D, 7B) on the west sequence.

4.1.2 West Sequence. White pumice fall. White pumice fall is the oldest fall deposit of the West sequence. It was found at 14 km from the current eruption center (RNG06). It comprises 45 cm thick, coarse ash to lapilli sized (0.1-5 cm) fragments and medium-fine ash matrix (**Figure 4f and 4g**). The components are consisted of dominant white pumice (48%), banded pumice (34%), and grey pumice (14%), with very minor fresh igneous lithic (2%) and altered lithic (2%). Pumice has an average size of 0.4-1.6 cm and can reach up to 5 cm in size, while the maximum size of lithic is only 1.7 cm. Pumice fragments show angular shape, crystal poor character, rich in large spherical vesicles.

Yellow pumice fall. This 50-cm yellowish brown layer was found at RNG06 and RNG07, overlying the white pumice fall layer in gradational contact. This unit consists of coarse ash - lapilli (0.4-5 cm) sized fragments, relatively coarser than white pumice fall. The average pumice fragments size is 0.8-1.6 cm, with fragments larger than 1.6 cm composing 20 wt% of this unit. Meanwhile, the lithic components are relatively finer with maximum size of 0.8 cm. Pumice fragments are similar to those of white pumice layer which are angular, crystal poor and large-vesicle rich, but less dense. This unit is mainly composed of banded pumice (48%), followed by yellow pumice (24%), reddish-yellow pumice (24%), grey pumice (2%), and rare lithic (2%).

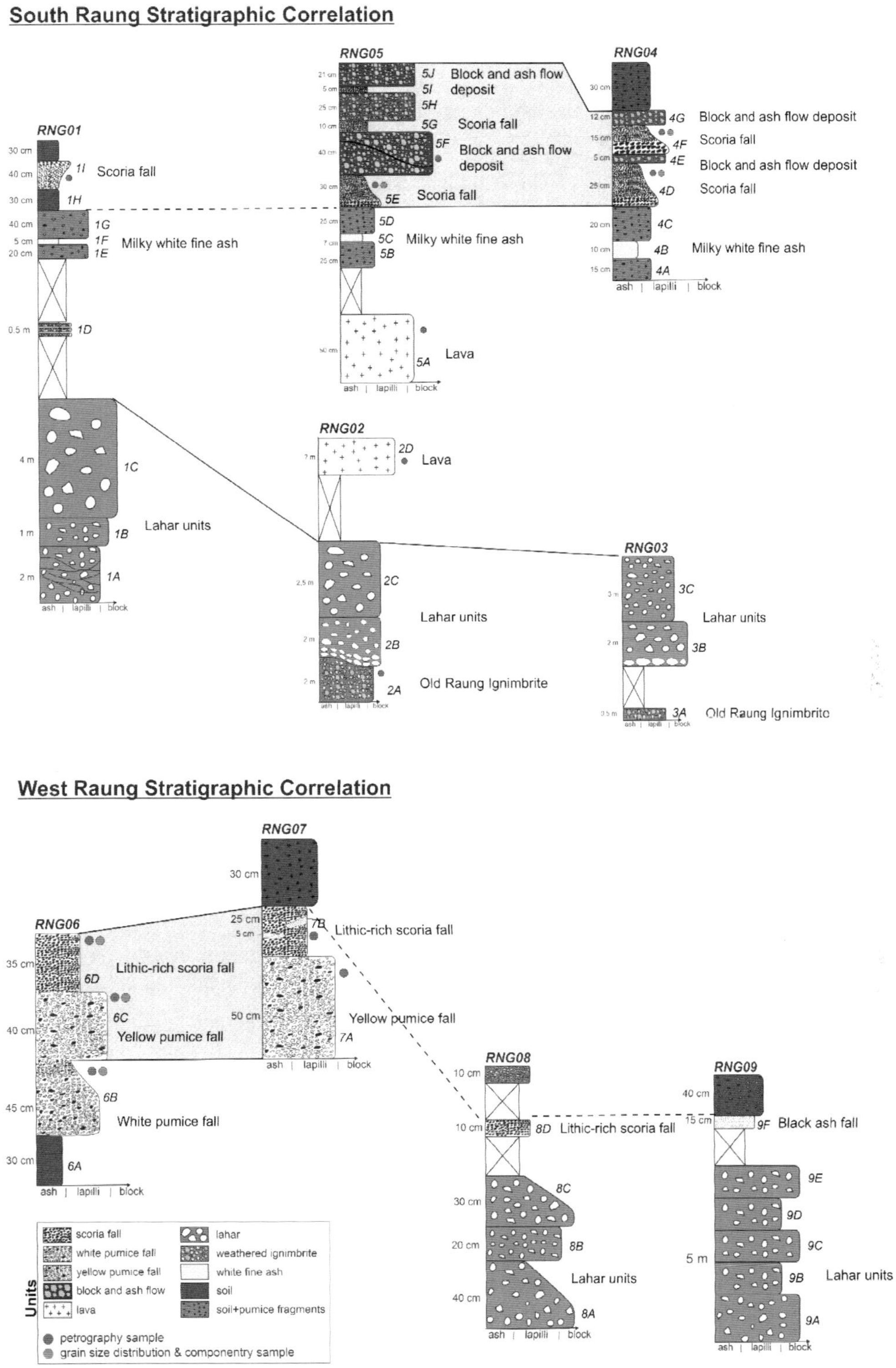

Figure 5. Stratigraphic correlation of south and west sequence of Raung volcano.

IOP Conf. Series: Earth and Environmental Science 1071 (2022) 012016 doi:10.1088/1755-1315/1071/1/012016

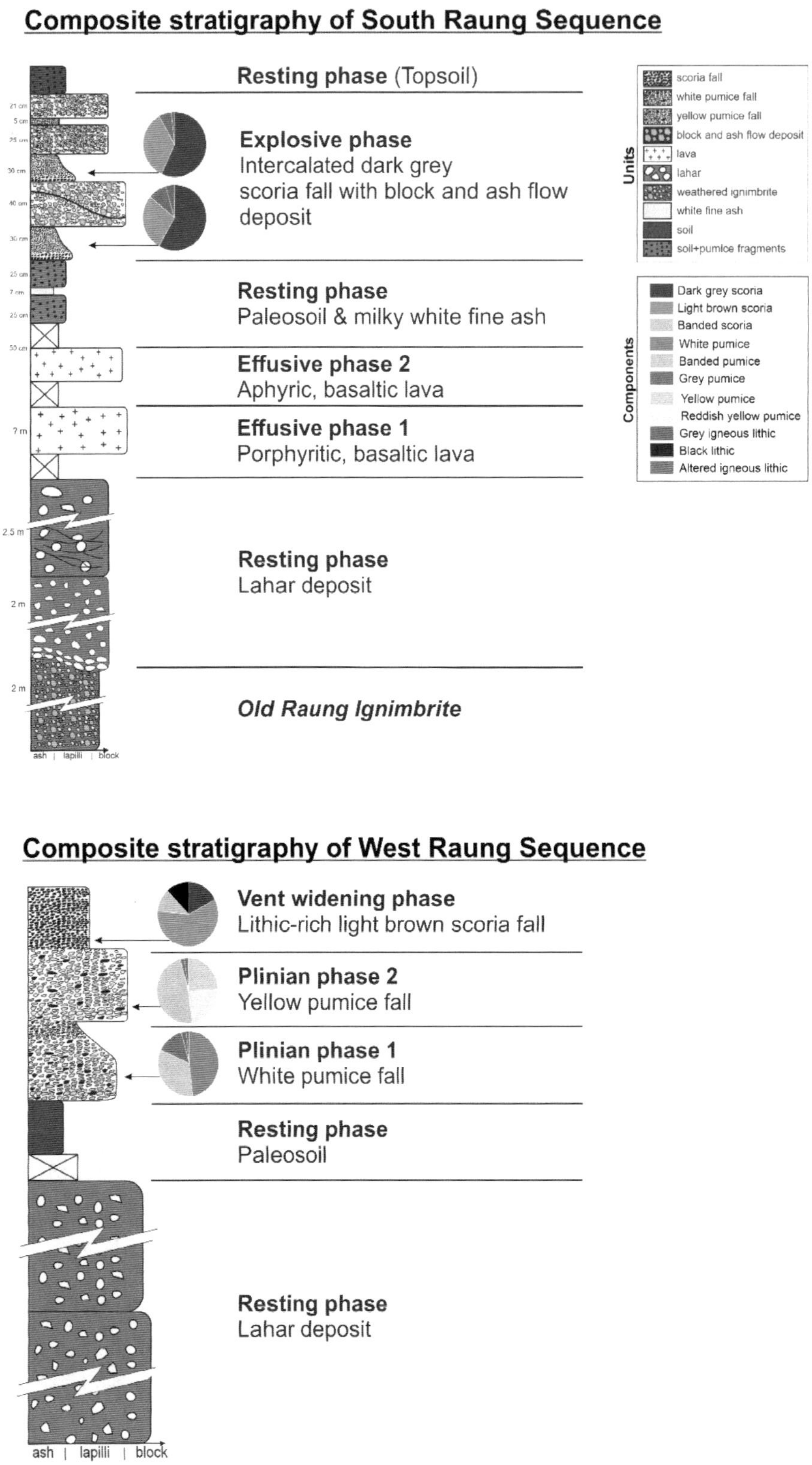

Figure 6. Composite stratigraphy of South and West Sequence compared with the portion of components from componentry analysis.

Scoria fall. The west sequence scoria fall is covering yellow pumice fall layer with sharp contact and has a relatively finer grain size than the white and yellow pumice layer. The scoria is angular and dense, with coarse ash - lapili (0.1-4 cm) fragments and average fragments size of 0.4-0.8 cm. This unit was found at RNG06 and RNG07 up to 35 cm thick. Slightly different from the one in the south, this unit appears the presence of banded scoria and absence of red altered lithic. West sequence scoria has a relatively more crystal poor character and richer lithic components than south sequence scoria. The components are composed of light brown scoria (59%), dark grey scoria (18%), banded scoria (12%), also a significantly increasing content of fresh igneous lithic (12%).

4.1.3 Secondary product. Lahar. Lahar deposits are highly consolidated and weathered, poorly sorted, with channel and lamination sedimentary structure. Lahar units are composed of various fresh to altered, rounded to subangular lithics, range from 5 – 100 cm size in coarse ash matrix with total thickness up to 5 m. Composite stratigraphy of South and West sequence is presented in **Figure 6**.

4.2 Petrography

Thin section observation revealed that lava flows have basaltic composition with crystal and microlite rich characteristic. Both aphyric and porphyritic lava have trachytic texture and similar mineralogical composition. Plagioclase, clinopyroxene, olivine and orthopyroxene are present as phenocryst, while the groundmass consists of plagioclase microlite, olivine microlite, and volcanic glass (**Figure 7a-b**).

Plagioclase crystals hava mainly clear and transparent feature with some sieve and zoning texture. Clinopyroxene and orthopyroxene also show clear feature with pale green color, while some of the olivine have altered to iddingsite on the crystal rim and fracture. The characteristics that distinguish those two lava types are the presence of vesicles and the size of microlite crystals. Aphyric lava has more vesicles and finer-size microlites, while porphyritic lava has very little to no vesicles with coarser microlites (**Figure 7a-b**).

Raung scoria generally has microlite rich characteristic but relatively more crystal poor compared with lava. There are slight differences between the southern and the western scoria observed in thin section. The southern scoria is more crystal rich and has irregular vesicles, whereas the western scoria is relatively more crystal poor with slightly more spherical vesicles. Crystal component in scoria consists of plagioclase as major phenocryst and microlite, also clinopyroxene as minor constituent (**Figure 7c**).

The pumice of fall deposit is highly vesicular and crystal poor with spherical-shaped vesicle. Plagioclase, hornblende, and biotite are present as crystal phase in pumice with very minor percentage (**Figure 7d**). On the other hand, pumice in ignimbrite shows highly weathered groundmass where the volcanic glass has replaced by clay minerals (**Figure 7e**). The mineralogical composition mainly composed of plagioclase and clinopyroxene with minor biotite.

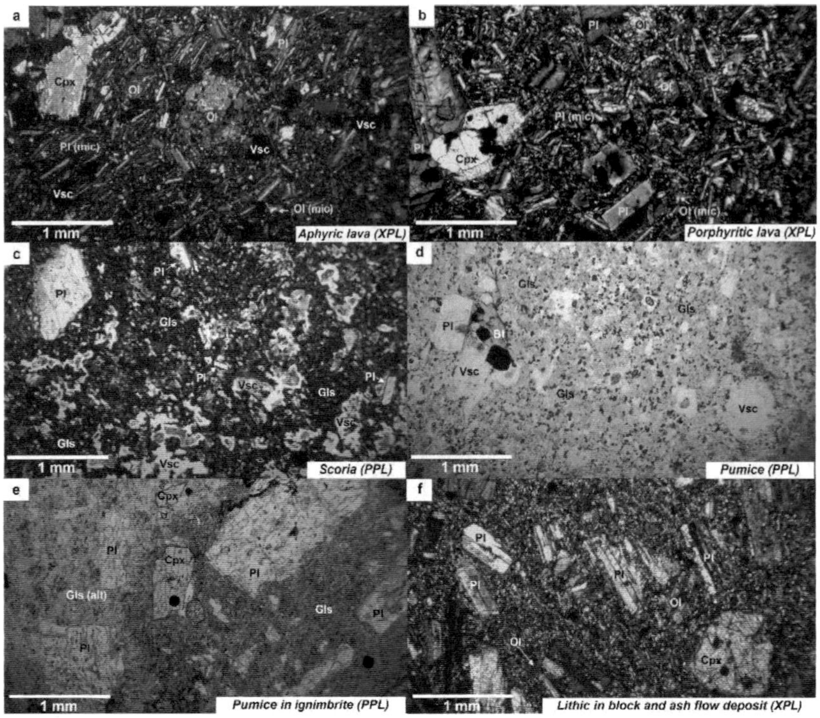

Figure 7. Photomicrograph of Raung volcanic products. Pl = plagioclase, Cpx = clinopyroxene, Ol = olivine, Bt = biotite, Gls = volcanic glass, Vsc = vesicle, (mic) = microlite, (alt) = altered.

The block and ash flow juvenile has a similar mineralogy with lava but it is more weathered. The presence of plagioclase, clinopyroxene, and olivine phenocryst are still recognizable. The groundmass of volcanic glass with plagioclase and olivine microlites are partially altered to clay minerals and Fe-oxide (**Figure 7f**).

5. Discussion

The south sequence was preceded with an explosive phase producing ignimbrite, followed by two effusive phases producing basaltic lava and one more explosive phase associated with scoria fall and block and ash flow deposit. The similar lava and block and ash flow composition indicates that the flow deposit was generated by lava dome collapse. Between the explosive and effusive phase recorded in the south sequence there was a quite long resting phase as indicated by deposition of lahar deposit and formation of two layers of paleosoil separated by the milky white fine ash layer. Meanwhile, there was no evidence of a time gap or pause in volcanic activity between the scoria fall deposit and the emplacement of block and ash flow deposits during the younger explosive phase. This indicates that multiple basaltic explosive eruption phases occurred in the past within a relatively short period of time. The west sequence shows a distinct character compared to the south sequence, remarked by the presence of two massive pumice fall layers with a total thickness reaching up to 85 cm. These pumice fall deposits found at a distance of 14 km from the eruption center with a maximum fragment size of 5 cm indicates an enormous Plinian phase eruption [8]. This Plinian phase eruption is likely to be related with the most differentiated magma of Raung volcano, supported by the presence of hornblende and biotite within pumice that represents a characteristic of more evolved magma than those producing lava or scoria. Lava flows have crystal-rich and vesicle-poor character while in contrast scoria and pumice are highly vesicular and crystal-poor. It might indicate that Raung explosive eruptions were associated with crystal-poor magma that supports bubble nucleation and vesicle development, so magma fragmentation occurred more intensively than crystal-rich magma.

The final stage of Plinian eruption was marked by deposition of scoria fall with a more abundant lithic component than the previous yellow and white pumice layer. Increasing lithic component in the scoria fall layer may indicate conduit widening as the last stage of Raung explosive eruption. The presence of scoria fall formed directly after pumice fall may reflect a condition of pre-eruptive magma chamber stratification [**9, 10**]. The magma differentiation process by fractional crystallization causes mafic magma to settle at the bottom and more evolved magma to be at the top of the magma chamber. When an explosive eruption occurs, the evolved magma came out first to form pumice fall, followed by scoria fall derived from the mafic magma. Sieve and zoning texture observed in lava along with the occurrence of banded pumice and banded scoria components may indicate magma mingling process within magma chamber. However, detailed stratigraphic and petrological studies with more data and analytical samples are needed to confirm the hypothesis of stratified magma chamber and magma mingling process. There was no paleosoil present between the pumice and scoria layers, so it can be interpreted that this Plinian eruption occurred continuously without a significant time gap. The one that remains unclear is whether this Plinian phase product is related with the caldera forming eruption in Raung, so further stratigraphic and petrological studies are still required to fill in the gap of our knowledge regarding Raung past behavior.

6. Conclusion

This preliminary study of southern and western sequences of Raung volcano presents evidence that there have been several explosive eruption events during the history of Raung activity. Our research shows that at least one explosive eruption event with ignimbrite product, one explosive eruption event with pumice fall products, two effusive eruption events producing lava flows, and one explosive eruption event resulting in alternating scoria fall and block and ash flow deposit ever happened in the past. The critical point to notice is that there is evidence of a Plinian phase represented by a very massive pumice fall deposits that must have been produced by an eruption scale much larger than the most recent volcanic activity. This shows that the volcanic activity in Raung is not restricted to producing lava, scoria, or block and ash flow deposit as a typical andesitic-basaltic volcanism, but also may potentially evolve long enough to be able to produce massive pumice fall deposits. Therefore, it is necessary to carry out further studies to answer whether the pumice fall deposits are associated with the formation of the Raung Caldera as a development to understanding the mechanism of basaltic caldera formation.

7. References

[1] Tempo.co 2015 *"Bali Airport Closed Again After Mt Raung Eruption"* accessed 17 November 2021 from https://en.tempo.co/read/683343/bali-airport-closed-again-after-mt-raung-eruption
[2] Republika 2015 *"Bali's Ngurah Rai Airport Closed Again on Sunday"* accessed 17 November 2021 from https://www.republika.co.id/berita/nrd1c4/balis-ngurah-rai-airport-closed-again-on-sunday
[3] Badan Geologi (Indonesian Geological Agency) 2011 *Data Dasar Gunungapi Indonesia. second ed.* (Bandung: Ministry of Energy and Mineral Resources Republic of Indonesia)
[4] Newhall C G and Self S 1982 The Volcanic Explosivity Index (VEI): an estimate of explosive magnitude for historical volcanism. *Journal of Geophysical Research* **87** 1231–1238. https://doi.org/10.1029/JC087iC02p01231
[5] Neuman van Padang M 1983 History of the volcanology in the former Netherland East Indies. Scripta Geol. **71** 1-76.
[6] Sutawidjaja I S, Suparman, Sitorus K 1996 Geological Map of Raung Volcano, East Java (Bandung: Volcanological Survey of Indonesia)
[7] Global Volcanism Program 2013 Raung (263340) in Volcanoes of the World, v. 4.9.3 (01 Feb 2021) Venzke E (ed.) (Washington D C: Smithsonian Institution)
[8] Cas R A F and Wright J V 1987 *Volcanic Successions Modern and Ancient*. (London: Allen and Unwin)

[9] Kennedy B M, Holohan E P, Stix J, Gravley D M, Davidson J R J, Cole, J W, Burchardt, S 2018 *Volcanic and Igneous Plumbing Systems of Caldera Volcanoes* in Burchardt, S., ed., *Volcanic and Igneous Plumbing System: Understanding Magma Transport, Storage, and Evolution in the Earth's Crust*. Elsevier p. 254-279

[10] Cioni R, Pistolesi M, and Rosi M 2015 *Plinian and Subplinian Eruptions* in H. Sigurdsson, ed., *Encyclopedia of Volcanoes*. Academic Press p. 519-535

Acknowledgments

This work is fully funded by the Ministry of Research and Technology of Indonesia through *Penelitian Terapan Unggulan Perguruan Tinggi (PTUPT)* funding program for year 2021, entitled "*Pengurangan Risiko Bencana Gunung Api Berbasis Integrasi Ilmu Kebumian: Studi Kasus Gunung Raung, Jawa Timur*" contract number 6/E1/KP.PTNBH/2021 and 2140/UN1/DITLIT/DIT-LIT/PT/2021. We also would like to thank Rahajeng Ardinni Noor for her assistance during fieldwork.

Grain size, mineralogical, and geochemistry of the 1996-2018 Volcanic Products of Anak Krakatau Volcano, Indonesia

D N Ardian[1*], H Darmawan[2], Wahyudi[2], B W Mutaqin[3], Suratman[3], N Haerani[4], Wikanti[5]

[1]Department of Geological Engineering, Faculty of Engineering, Universitas Gadjah Mada, Sekip Utara, Bulaksumur, Yogyakarta, Indonesia
[2]Laboratory of Geophysics, Department of Physics, Faculty of Mathematics and Natural Sciences, Universitas Gadjah Mada, Sekip Utara, Bulaksumur, Yogyakarta, Indonesia
[3]Department of Environmental Geography, Faculty of Geography, Universitas Gadjah Mada, Sekip Utara, Bulaksumur, Yogyakarta, Indonesia
[4]Center of Volcanology and Geological Hazard Mitigartion, Geological Agency, Ministry of Energy and Mineral Resources, Bandung, Indonesia
[5]Remote Sensing Application Center, National Institute of Aeronautics and Space of Indonesia, Pasar Rebo, Jakarta Timur, Indonesia
*Corresponding author's e-mail: dininuari@gmail.com

Abstract. Anak Krakatau Volcano is the only active volcano on Krakatau Volcanic Complex. It is located in the Sunda Strait as part of the Quaternary volcanic arc as a result of the Indo-Australian plate subduction under the Eurasian plate. The volcanic activity of the Anak Krakatau volcano since 1927 is considered to be very active with a combination of explosive and effusive eruptions. Lava flow tends to be concentrated in the southwestern part, except for the 1996 lava flow (north) and 1993 lava flow (northeast). In 2018 there was an eruption accompanied by flank collapse on the southwestern side, caused by the accumulation of the instability volcanic body due to volcanic and tectonic activity. The volcanic activity will be reflected in the resulting deposits. This study was conducted to determine the characteristics of the deposits, especially on the northern part for post-1996. The analysis carried out included the stratigraphic columns, granulometric, petrography, and geochemistry analysis.

1. Introduction

Krakatau Volcanic Complex is located in Sunda Strait as a part of Quaternary Volcanic Arc as a result of the subduction of Indo-Australia Plate under Eurasia Plate [1]. This volcanic complex consists of 4 parts, namely Sertung Island, Panjang Island, Rakata Island, which are the remnants of Krakatau Volcano, and the currently active Anak Krakatau volcano. The activity of Anak Krakatau was dominated by pyroclastic deposits until nearly 1960. Started in 1972 the activity was followed by andesite-basaltic lava flows deposited on almost all flank except the eastern part [2]. The highest intensity of Anak Krakatau lava flow occurred in 1990 and concentrated on the south, southwestern and west part. In 1996, lava flow flowed down to the north and northwestern side [3]. Afterwards, the lava flow concentrated back to the south, southwestern and west part [2].

By the end of 2018, a flank collapse eruption occured at the southwestern part of Anak Krakatau and initiated a tsunami on the west coast of Banten and south coast of Lampung [4]. Historically, several

IOP Conf. Series: Earth and Environmental Science 1071 (2022) 012017 doi:10.1088/1755-1315/1071/1/012017

flank collapses of Anak Krakatau had occured before in 1940 – 1950 [3] and 1963 [5] at the same sector. Referring to [6] the sector collapse of Anak Krakatau was triggered by accumulation of slope instability due to volcanic activity, tectonic activity, pore pressure changes, hydrothermal and mechanical weakening by alteration and erosion. High intensity of the volcanic activity increased slope instability due to cumulation of new materials. Hence, the materials will reflect the volcanic activity. We presented study of lithological measured-section combined with mineralogical and geochemical analyses of the post 1996 products of Anak Krakatau.

2. Geological Setting

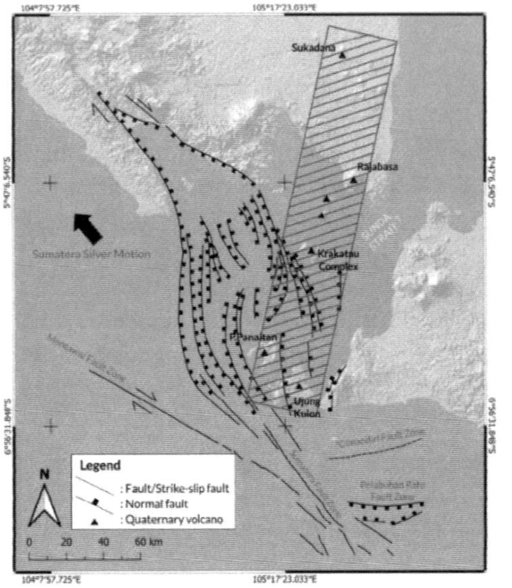

Fig 1. The map shows geological structures and a NE-SW volcanic line as a result of the evolution of Sunda Strait [7].

Sunda Strait is located between Java and Sumatra Island and acts as a transition zone where there is a change in the direction and speed of the subduction in Java and Sumatra as shown in **Fig 1** [7]. Refers to [8], the 40° change of the subduction in Java and Sumatra triggered an evolution of Sunda Strait.

Refers to [9] Krakatau Volcanic Complex is divided into 4 phases as shown in **Fig 2** including : (1) Pre-Historic Krakatau (2) Young Krakatau (Rakata, Danan, Perboewatan) (3) 1883 Phase (4) Anak Krakatau Phase. These four phases produce distinct products such as lava flow, scoria fall, and pumiceous ignimbrite, however, they are within the same liquid line decent trend of geochemical composition.

The products of Krakatau which represented by Rakata and Sertung by [2] shows basalt and dacite types. The catastrophic eruption of 1883 phase produced dacite – rhyolite, while the Anak Krakatau phase products shows variation of basaltic andesite – andesite – dacite – rhyolite. Basaltic andesite with SiO_2 less than 57% represent 1927 – 1980 eruptions. While the dacite - rhyolite with high amount of SiO_2 content reaches 75% was obtained from the mesostasis bomb as a result of 1981 vulcanian eruptions. This dacite – rhyolite products are unique as [10] explained that these products are resulted from a short differentiation period. After 1981, the products of Anak Krakatau, was represented by basaltic andesite type.

IOP Conf. Series: Earth and Environmental Science 1071 (2022) 012017 doi:10.1088/1755-1315/1071/1/012017

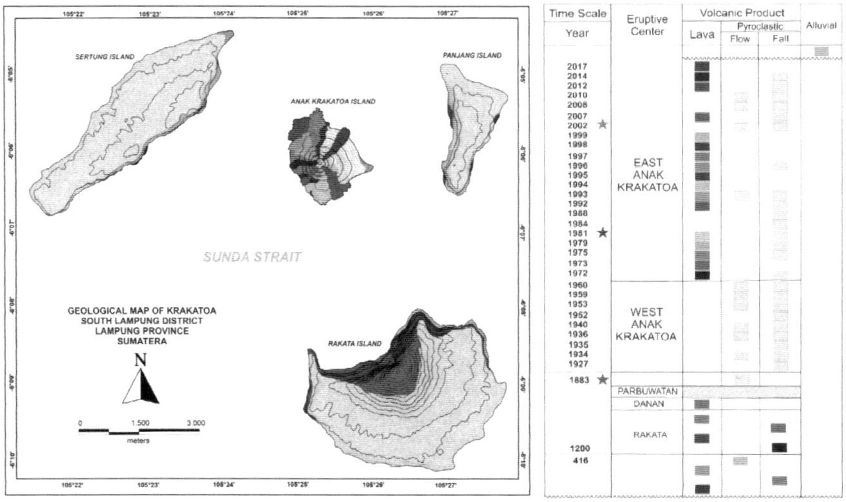

Fig 2. Geological map and stratigraphic column of Krakatau Volcanic Complex show the differences of volcanic products between different phases [9].

The first activity of Anak Krakatau started on December 29, 1927 through underwater eruption in the center of caldera Krakatau Complex. On January 20, 1927 Anak Krakatau activities were seen above the water surface. The activity was quite intensive until the first explosive eruption occurred above the water surface on February 15, 1929 and resulted in an eruption column reaching 800 meters [11]. According to [3], Anak Krakatau grows rapidly. Despite the significance growth, Anak Krakatau has a destructive phase which is mostly related to flank collapse such as 1950 [3] and 2018.

3. Samples and Analysis Method

Field observation and sample collection were conducted at seven stopsites in the northern flank of Anak Krakatau in August 2019 (Fig. 3a). Stratigraphic measurements were performed at stopsite 1, 4 and 6 and focused on the tephra deposits above the 1996 lava flow. The samples collected consist of lava, block ballistic, and tephra deposits.

The stratigraphic analyses were carried out on the scale 1:10 to determine layers thickness, grain size, components, and structure of the deposits.For petrographic analysis, 35 rock samples were selected (2 lavas, 4 blocks, 8 bulk of tephra deposits, and 21 scorias). The calculation of mineral abundance is carried out using point counting method according to [12] at 1000 points on each sample. Major and trace element were determined using ICP-AES/MS at the ALS Chemex Canada commercial laboratory to 15 rock samples (12 scorias, 2 blocks, and 1 lava). For the granulometry analysis, 24 primary deposits were selected to determine the deviation value (σ), the average value (μ) and the median value (Md ϕ) of the grain size to know the type of pyroclastic deposits using type of pyroclastics deposit diagrams [13].

4. Results

4.1. Stratigraphy

The stratigraphic sections examined in 3 sites (site 1, site 4 and site 6) approximately 960 meters far from the active crater. We identified primary and secondary tephra deposits based on the physical characteristics (depositional structures and tephra componentry). The secondary deposits resembles time gaps of active eruptive periods. From the tephra stratigraphic sequence, we classified the pyroclastic deposits into five type (A-E type) based on the characteristic of juvenile, grain size, structure, consolidation level, and sorting (Fig. 3).

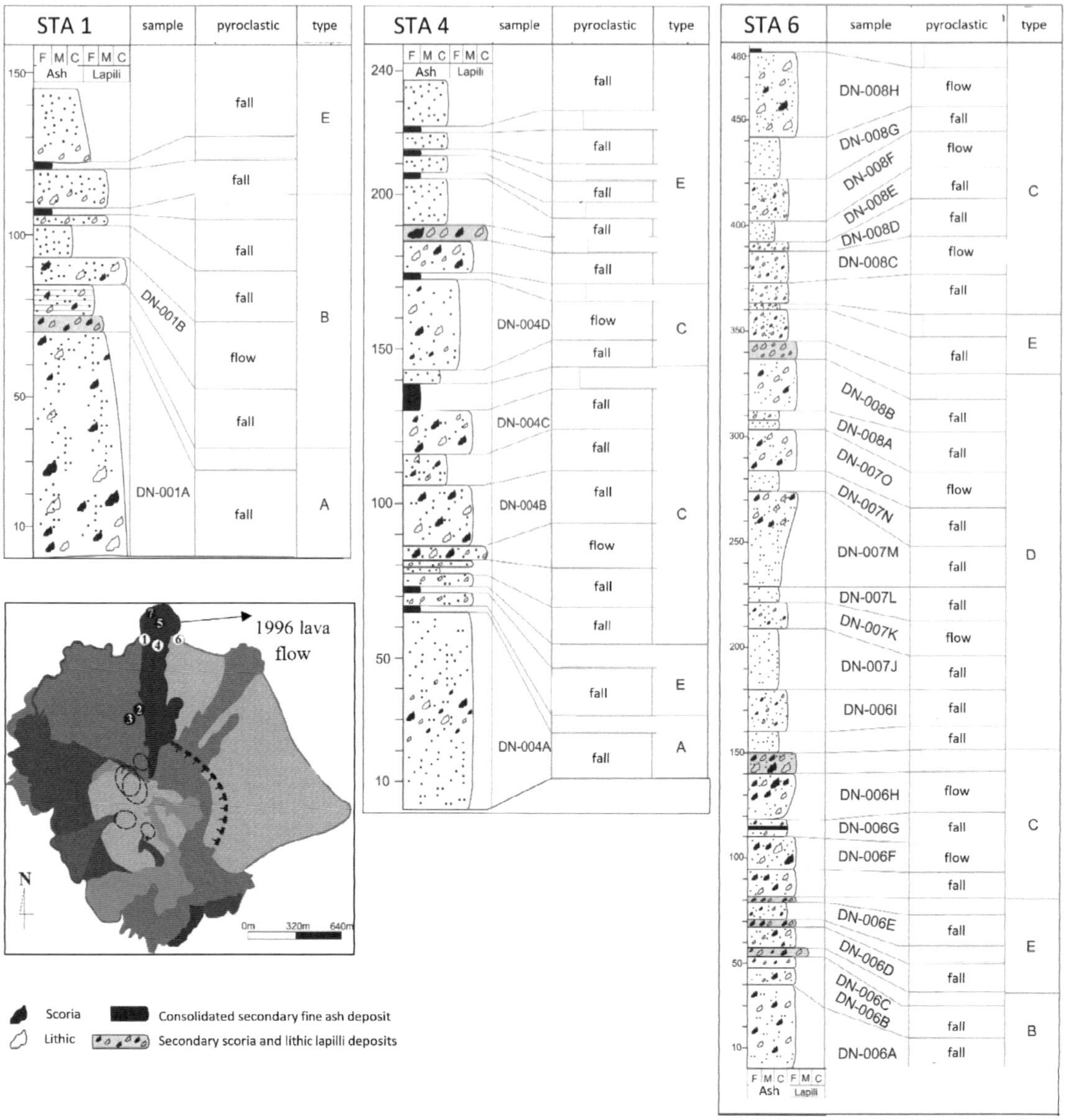

Fig 3. Tephra stratigraphic measurement location on site 1, 4 and 6. Observation points are plotted in the inset map. Tephra sequence are classified into five types of A, B, C, D, and E types based on the similarity of the thickness, grain size, and sortation. Size abbreviation F: Fine; M: Medium; C: Coarse.

Fig 4. (a) The field appearance of post-1996 products. (b) The components of post-1996 products resembling 3 different grain size of bomb/block, lapilli, and ash.

4.2. Granulometry

Table 1. Median, average and deviation value followed by the type of pyroclastic deposit.

Site	Sample	Median Value (Md ϕ)	Average Value (μ)	Deviation Value (σ)	Pyroclastic Type	Site	Sample	Median Value (Md ϕ)	Average Value (μ)	Deviation Value (σ)	Pyroclastic Type
1	001B	-0.9	-0.82	2.3	Flow	6	008B	-0.8	-0.93	1.9	Fall
1	001A	-3.4	-2.48	2.5	Fall	6	008A	-0.7	-0.43	1.59	Fall
4	004D	-0.54	-0.37	2.27	Flow	6	007O	-3	-1.81	2.35	Flow
4	004C	-2.8	-2.17	1.88	Fall	6	007N	0.3	0.28	2.14	Fall
4	004B	-2.4	- 2.06	2	Fall	6	007M	-2.2	-1.83	2.57	Fall
4	004A	-0.1	0.07	1.5	Fall	6	007L	-0.6	-0.52	2.21	Fall
6	008H	-0.4	0.32	1.95	Flow	6	007K	-1.3	-1.25	2.04	Flow
6	008G	0.7	0.83	1.76	Fall	6	007J	0.6	0.04	2.17	Fall
6	008F	-0.1	-0.02	2.18	Flow	6	006I	0.6	0.45	1.89	Fall
6	008E	-0.8	-0.9	1.9	Fall	6	006H	-1	-1.03	2.23	Flow
6	008D	-0.5	-0.6	2.25	Fall	6	006F	-0.4	-0.4	2	Flow
6	008C	-0.7	-0.7	2.25	Flow	6	006B	-3.1	-2.4	2.1	Fall
						6	006A	-0.5	-0.61	1.65	Fall

Physically, the pyroclastics deposits of post-1996 products composed of pyroclastic fall and pyroclastic flow. The differences between in the fall and flow deposits are mostly shown in the sorting and consolidation. The fall deposits have various sorting from well sorted to poorly sorted, besides the flow deposits have mostly poorly sorted. The pyroclastic fall deposits are more consolidated than those of flow deposits.

The components of pyroclastic deposits divided into three types, based on the grain size. The bomb/block components consist of scoria and non-vesicular lithic presented irregularly as a balistic. The lapilli components consist of scoria and non-vesicular lithic. Beside, the ash components consist of scoria, glass shard, free crystal, cognate lithic and accessories lithic. The field appearance and the components of the deposits are shown in **Fig 4.**

Granulometry analysis was used to find the grain size distribution by knowing from the histogram and chart of the grain size vs frequency and the grain size vs cumulative frequency. There were three grain size trend dominates the post-1996 products spesifically on phi -4 until -3, on phi 0 until 1 also on both of phi -4 until -3 and phi 0 and 1 at the same time. The result of median, average and deviation value of the grain size are shown in the **Table 1**. The post-1996 products of Anak Krakatau are relatively on a poorly sorting with deviation value (σ) less than 1. Refers to [14], the pyroclastic deposit generally have a poorly sorting, however the flow deposit have more poorly sorted than the fall one.

4.3. Petrography

The petrography analysis was performed on the lava, scoria, non-vesicular lithic, block and bulk tuff deposits as shown in the **Table 2**. The component of the post-1996 products consist of volcanic glass, vesicles and crystals. The highest vesicle content was found in the scoria sample reach up to ~30%, followed by block sample varying from 5% - 20% and the lowest vesicle content presented by non-vesicular lithic sample and bulk sample of tephra deposits which only less than 5%.

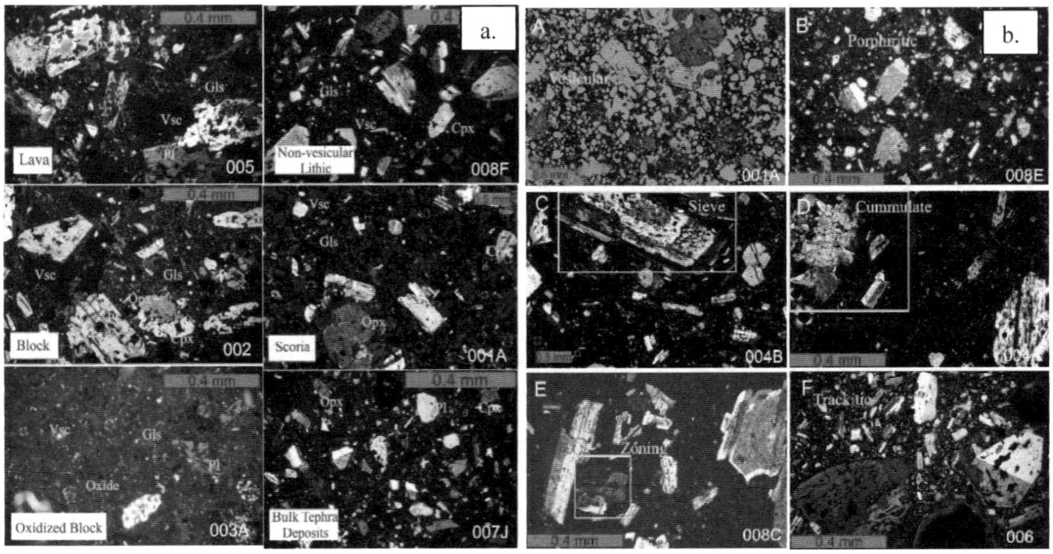

Fig 5. Thin section photographs of the post-1996 products of Anak Krakatau. (a) Lava, non-vesicular lithic and scoria fragments, block balistic (fresh and oxidized) and bulk tephra deposits. (b) Vesicular and porphyritic structure dominates all type of deposits, besides the mineral texture represented by sieve (major), cumulate, zoning and trachytic.

Thin section analysis on **Fig. 5a** shows that the lava flow, scoria, lithic and block samples consist of Basaltic Andesite with less than ~10% of phenocryst. The bulk tuff deposits consist of Vitric Crystal Tuff and Vitric Tuff. The post-1996 products generally consist of plagioclase, clinopyroxene and orthopyroxene. Plagioclase (10-30 vol%), clinopyroxene (1-12 vol%) and orthopyroxene (1-13 vol%). Opaque and oxide minerals were found only in block sample mostly as a minor constituent with percentage less than 2 vol% except in sample 3A block, the oxide constituent reaches up to 13 vol%.

From **Fig. 5b** shows the stucture dominates by vesicular and porphyritic structure and abundant in all sample types. The porphyritic structure consists of volcanic glass and plagioclase microlite as groundmass. Besides, the mineral texture is dominated by sieve texture and abundant in all sample.

4.4. Geochemistry

The geochemistry of the post-1996 products of Anak Krakatau were obtained from lava flow, block balistic and juvenile fragment (scoria) samples from the pyroclastic deposits (**Table 3**). Silica contents show narrow range of 52 – 55 wt.% inwhich scoria and ballistic block are higher thanon the lava product.

5. Discussion

5.1. The characteristics of the post-1996 deposits of Anak Krakatau

The post-1996 products on the northern side of Anak Krakatau Volcano were started with lava flow and followed by pyroclastic deposits. The pyroclastic deposits were present with different type and duration. Thick and massive pyroclastic fall deposit on top of the 1996 lava flow indicates that the initial explosive eruption was continupusly and relatively long in duration. As a result, there was massive thick deposit initiate the first pyroclastic deposits of the post-1996 products represented by Type A and Type B (**Fig 3**). After that, the stage was a series of short eruptions period with short pauses. The result is there were thin fall deposits (less than 10 cm) alternating with the secondary deposits from the secondary processes. The secondary processes generate erossion as a marker that there were no eruption activity. The product represent by Type E (**Fig 3**).

Table 2. Modal composition of the post-1996 products from thin section analysis. Pl = plagioclase; Cpx = clinopyroxene; Opx = orthopyroxene; Opq = opaque minerals; VG = volcanic glass; Vsc = vesicles; Oxd = oxide minerals. Name of non fragmental refer to Le Maitre IUGS (1989) and the fragmental refer to Cook (1965). Symbol "v" resembles relative abundance.

Site	Sample	Lithology	Phenocryst				VG	Vsc	Oxd	Mineral Texture	Name
			Pl	Cpx	Opx	Opq					
2	2	Block	vvvv	v	v	v	vvvv	vvvv		Sieve	Basaltic Andesite
3	3A	Block	v	v	v		vvvvv	vvvv	vvv		Basaltic Andesite
3	3B	Block	v	vv	v	v	vvvvv	vv	v		Basaltic Andesite
3	3C	Block	vvv	v	v		vvvvv	vvvv	v	Sieve	Basaltic Andesite
1	001B	Scoria	vvv	v	vv		vvvvv	vvvvv		Sieve	Basaltic Andesite
1	001A	Scoria	vvv	vv	v		vvvvv	vvvvv		Sieve, Cumulate	Basaltic Andesite
4	004D	Non-V Lithic	v	vv	vv		vvvvv	vv		Sieve	Basaltic Andesite
4	004C	Bulk tuff deposits	vvv	vv	v		vvvvv	vv		Sieve	Vitric Crystal Tuff
4	004B	Scoria	vvvvv	v	v		vvvvv	vvvvv		Sieve	Basaltic Andesite
4	004A	Scoria	vvv	vv	v		vvvvv	vvvvv		Sieve, Cumulate	Basaltic Andesite
6	008H	Scoria	vvv	v	v		vvvvv	vvvvv		Sieve	Basaltic Andesite
6	008G	Non-V Lithic	vvv	v	v		vvvvv	v		Sieve	Basaltic Andesite
6	008F	Non-V Lithic	vvvv	v	v		vvvvv	vv		Sieve, Cumulate	Basaltic Andesite
6	008E	Non-V Lithic	vvv	v	v		vvvvv	v		Sieve	Basaltic Andesite
6	008D	Scoria	vvv	v	v		vvvvv	vvvvv		Sieve	Basaltic Andesite
6	008C	Scoria	vvv	v	vv		vvvvv	vvvvv		Sieve, Cumulate, Zoning	Basaltic Andesite
6	008B	Scoria	vvv	v	v	v	vvvvv	vvvvv		Sieve	Basaltic Andesite
6	008A	Bulk tuff deposits	vvv	v	v		vvvvv	v		Sieve	Vitric Crystal Tuff
6	007O	Scoria	vvv	vvv	vv		vvvvv	vvvvv		Sieve	Basaltic Andesite
6	007N	Bulk tuff deposits	vvv	v	v		vvvvv	v		Sieve	Vitric Crystal Tuff
6	007M	Scoria	vvv	v	v		vvvv	vvvvv		Sieve	Basaltic Andesite
6	007L	Bulk tuff deposits	vvvv	vv	v		vvvvv	v		Sieve	Vitric Crystal Tuff
6	007K	Scoria	vvvv	vv	vvv		vvvvv	vvvvv		Sieve	Basaltic Andesite
6	007J	Bulk tuff deposits	vvvv	v	vv		vvvvv	vv		Sieve	Vitric Crystal Tuff
6	006I	Bulk tuff deposits	vvv	vvv	v		vvvvv	vv		Sieve, Cumulate	Vitric Crystal Tuff
6	006H	Scoria	vvv	vvv	v		vvvvv	vvvvv		Sieve	Basaltic Andesite
6	006G	Scoria	vv	vvvv	vv		vvvvv	vvvvv		Sieve	Basaltic Andesite
6	006F	Non-V Lithic	vvv	vvv	vvvv		vvvvv	vv		Sieve, Cumulate	Basaltic Andesite
6	006E	Scoria	vvv	v	vvv		vvvvv	vvvvv		Sieve	Basaltic Andesite
6	006D	Bulk tuff deposits	v	v	v		vvvvv	vvv		Sieve	Vitric Tuff
6	006C	Scoria	vvvvv	vv	v		vvvvv	vvvvv		Sieve, Cumulate	Basaltic Andesite
6	006B	Non-V Lithic	v	vvv	vvv		vvvvv	vv		Sieve, Cumulate	Basaltic Andesite
6	006A	Bulk tuff deposits	vvvvv	vv	vv		vvvvv	vvv		Sieve	Vitric Crystal Tuff
5	5	Lava	vvvv	v	v		vvvv	vvvvv		Sieve	Basaltic Andesite
7	6	Lava	vvvv	v	vv		vvvv	vvvvv		Sieve, Cumulate, Trachytic	Basaltic Andesite

The next stage was an eruption period with increasing intensity, represent by the thickening upward deposits shown by Type C (**Fig 3**). The grain size present in this period was quite coarse and become the highest intensity of all period. The following period occured on a long duration and only found on STA 6. The characteristics of this period were drastically increasing the intensity and then gradually followed by decreasing intensity. As a result, there were shown by the thickening deposit on the middle part and become thinning upward (Type D). This period present for a quite long duration reach up to 190 cm thickness. And the last period on the upper part of the stratigraphic measurement were shown with the Type E.

Table 3. Geochemical composition of the post-1996 products of Anak Krakatau. Major oxides (in wt%) were analyzed using ICP-AES; trace elements (in ppm) were analyzed using ICP-MS.

Sample	1	2	3	4	5	6	7	8	9	10	11	12	13	14	15
Sample Type	scoria	scoria	scoria	scoria	scoria	scoria	scoria	scoria	scoria	scoria	scoria	scoria	block	block	lava
Major elements (ICP-AES), wt%															
SiO_2	54.7	53.9	54.8	53.6	52.9	53.5	53.3	52.6	52.8	53.5	53.9	54.5	56.5	54.4	53.6
Al_2O_3	16.85	17.2	17.15	16.4	15.95	16.5	16.25	16.75	16.5	16.85	16.95	17.1	17.75	17.8	18.15
Fe_2O_3	9.83	10.75	9.81	10.1	9.7	9.31	9.6	9.44	10.15	9.82	10.45	10.25	9.65	10.55	10.75
CaO	7.82	8.34	7.95	7.68	7.51	7.72	7.48	08.09	7.67	7.99	8.14	8.21	7.15	08.05	8.37
MgO	3.96	4.33	3.93	4.15	3.87	3.92	3.8	4.27	3.81	4.16	3.79	04.02	03.09	04.06	4.32
Na_2O	3.56	3.45	3.65	4.15	04.09	4.22	4.28	4.23	3.59	3.75	3.52	3.61	04.04	3.72	3.67
K_2O	0.94	0.86	0.95	0.92	0.93	0.95	0.99	0.88	0.92	0.89	0.93	0.89	1.18	0.99	0.9
Cr_2O_3	6	6	6	7	6	6	5	7	5	6	5	8	3	7	7
TiO_2	01.04	01.01	01.05	0.99	0.97	1	01.01	0.99	01.01	01.02	01.04	01.06	01.02	01.05	01.02
MnO	0.17	0.18	0.17	0.17	0.17	0.17	0.17	0.17	0.17	0.17	0.17	0.18	0.18	0.19	0.19
P_2O_5	0.29	0.27	0.31	0.31	0.27	0.28	0.28	0.27	0.26	0.3	0.29	0.31	0.32	0.32	0.29
LOI	0.05	-0.18	0.1	1.91	1.76	1.98	1.92	2.21	1.18	1.14	0.91	0.33	0.32	-0.13	-0.19
Total	99.28	100.18	99.94	100.45	98.19	99.62	99.15	99.97	98.13	99.66	100.16	100.53	101.26	101.07	101.14
Trace elements (ICP-MS), ppm															
Ba	177	172	181.5	169.5	171	170.5	180.5	166	174.5	166	181.5	168.5	213	186	180
Ce	32.2	32.1	33.1	31.3	30.9	31.3	33	29.3	30.5	29.7	31.3	30.7	38.8	34.2	32.9
Cr	50	50	50	60	40	40	40	50	40	50	40	60	20	50	50
Cs	0.93	0.81	0.91	01.03	01.02	0.93	01.05	0.88	0.96	0.86	0.89	0.82	1.33	1.14	0.97
Dy	5.47	5.18	5.74	5.39	4.78	5.29	5.48	5.14	5.11	5.11	5.44	5.28	5.4	4.95	4.58
Er	3.46	3.36	3.52	3.15	3.34	3.35	3.72	2.94	3.38	3.26	3.35	3.51	3.66	3.29	03.04
Eu	1.59	1.38	1.47	1.42	1.44	1.42	1.53	1.47	1.51	1.44	1.5	1.43	1.54	1.41	1.44
Ga	17.7	18.1	18.9	17.3	17.5	17.8	18.4	18.1	18.3	18.5	18.1	18.8	18.9	18.8	18.8
Gd	5.88	5.35	5.84	5.78	5.46	5.83	5.83	5.38	5.53	5.32	5.69	5.7	06.04	5.49	5.41
Hf	3.2	3.1	3.4	3	3.1	3.4	3.4	3.1	2.9	2.9	3.4	3.4	4.1	3.4	3.2
Ho	1.13	01.03	1.13	1.1	01.03	01.04	1.11	01.01	1	01.07	01.08	01.08	1.28	1.11	01.02
La	13.9	14.1	14.5	13.3	13.1	13.5	14.2	12.5	13.1	13.1	13.5	12.8	16.8	14.9	14.2
Lu	0.48	0.43	0.43	0.47	0.44	0.44	0.49	0.44	0.47	0.42	0.45	0.47	0.53	0.48	0.5
Nb	3.6	3.4	3.5	3.6	3.1	3.3	3.5	3	3.4	3.2	3.4	3.3	4	3.4	3.2
Nd	19.5	19	20.8	19.5	18.2	19.4	20.1	18.7	19.2	18.5	19.5	19.6	22.9	20.8	20.4
Pr	4.28	4.29	4.46	04.07	4.27	4.25	4.39	04.06	4.14	04.06	4.3	4.18	05.08	4.74	4.39
Rb	24.5	21.9	24.6	31.7	24.4	27.5	25.7	24.8	23.7	22.7	22.8	23.2	30.8	24.3	22.2
Sm	05.07	05.01	5.17	4.86	4.65	5.39	4.99	4.73	4.84	4.91	4.91	4.88	5.81	5.46	05.09
Sn	3	4	2	5	11	3	9	3	8	3	6	3	5	3	4
Sr	345	356	360	343	325	339	334	348	347	348	357	352	372	394	414
Ta	0.3	0.3	0.2	0.5	0.3	0.3	0.3	0.3	0.2	0.3	0.3	0.3	0.2	0.2	0.2
Tb	0.83	0.83	0.87	0.77	0.79	0.77	0.84	0.76	0.81	0.82	0.8	0.81	0.95	0.85	0.82
Th	3.25	3.22	3.2	3.2	03.04	2.95	3.13	2.76	2.96	2.81	03.05	2.9	4.11	3.14	2.91
Tm	0.44	0.43	0.52	0.47	0.44	0.47	0.48	0.43	0.45	0.44	0.45	0.45	0.51	0.46	0.45
U	262	58.7	54.7	7.73	0.91	1.89	0.85	0.91	0.73	0.76	0.68	0.78	01.05	0.8	01.05
V	223	224	231	217	205	209	210	221	216	225	223	228	171	228	228
Y	28.2	26.9	29.2	27	26.8	27.5	29	26.5	27	26.9	27.7	27.4	33.7	29.6	28.3
Yb	03.09	3	3.17	3.11	2.91	2.94	2.94	2.78	2.98	2.76	03.07	3.13	3.64	2.96	2.97
Zr	123	112	125	119	117	122	128	113	119	116	121	118	150	127	117

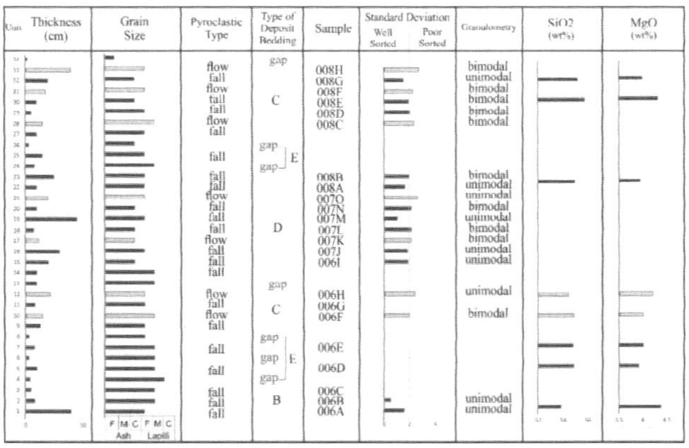

Fig 6. Summary of changes in layer unit thickness and sorting against the type of deposits bedding and chemical composition on STA 6 (the most representative site according to the thickness).

The fall deposits always initiates every eruption period and could be presented in one eruption period without any flow deposits as shown in **Fig 6**. This is indicate that the post-1996 eruptions tend to generate eruption column with high concentration of gases compared with the particles. This phase refers to [15] and [16] indicates the constructive phase of the Anak Krakatau. The sorting variation tend to follow the type of pyroclastic deposits. There is no special trend of the sorting or the relationship with the change of the layer unit thickness. The poor sorting dominates the flow deposits ($\sigma > 2$). The sorting is getting poorer followed by increasing intensity of coarse material, which is shown by bimodal type.

Generally, the differences between fall and flow deposits of the post-1996 products of Anak Krakatau Volcano were not significant. All deposits has deviation value > 1 and this is possible because the location research were around the eruption center. As a result of this, the materials has a vary varied size because there were no time to sorting. Therefore, other parameters were used to identify the characteristic of the deposits. The parameters are consolidation and deposit structure. The flow deposits has a middle to high consolidation due to high concentration of particles in high temperature, while the fall deposits has a low consolidation due to high concentration of gases. The structure present in flow deposits were massive and reverse gradation, while in fall deposits were massive and bedding.

To support the determination of pyroclastic deposits type, analysis on [13] diagram was carried (**Fig 7b**). The result shows that the post-1996 products were placed in the pyroclastic fall zone and the intersection zone. All of the products which fill the pyroclastic fall zone, all of them were pyroclastic fall deposits in accordance to the stratigraphic analysis. Those identified as pyroclastic fall deposits has better sortation than those of pyroclastic flow deposit (Fig. 7b). While the intersection one there were only 9 samples identified as a pyroclastic flow deposit (001B, 004D, 008H, 008F, 008C, 007O, 007K, 006H, 006F) and the remains are pyroclastic fall deposit. From **Fig 7a** we know that the eruption type of the post-1996 products were Vulcanian to Strombolian according to [17] with moderate grain size. Although there were a few samples which on a Plinian type, we indicated that those poorly sorted deposits were near the eruption centre.The pyroclastic deposits from post-1996 products are composed of pyroclastic fall and pyroclastic flow. The components consist of three sizes, block (10-15cm), lapilli (phi -3, phi 0 and phi 1) and ash. These components were considered to scoria and non-vesicular lithic. Whereas, the ash components consist of scoria, glass shard and cognate lithic as a juvenile fragment, free crystal, accessories lithic and altered materials.

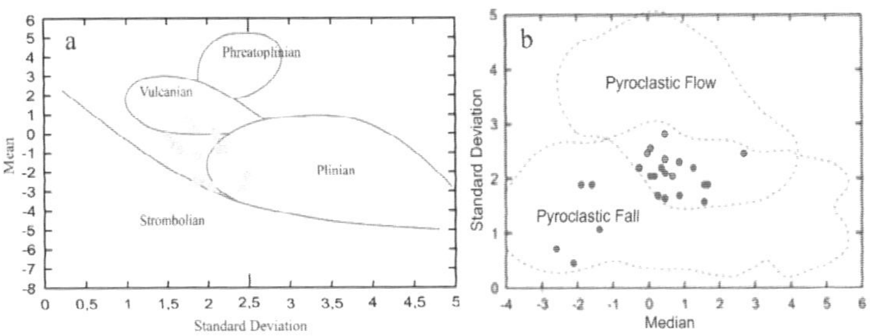

Fig 7. (a) The diagram by [17] showing eruption type of the post-1996 products range strombolian – vulcanian type. (b) The diagram by [13] showing the type of pyroclastic deposits, the solid grey circles represent those identified as a pyroclastic flow deposits in the field observation, whereas solid red circles are pyroclastic fall deposits.

5.2. The mineralogy and geochemistry characteristics of the post-1996 products of Anak Krakatau

Glass volcanic dominates the post-1996 products of Anak Krakatau as a result of rapid cooling. The porphyritic texture is clearly shown as indicated by bimodal distribution with one crystal larger than the others. The sieve texture on plagioclase of the post-1996 products have highly abundance and might indicate in-equilibrium environment. This texture always present in the early part of each period.

TAS diagram in **Fig 8a** shows that whole rocks compositions of post-1996 eruptive materials are basaltic andesite in composition. These results are overlaping with basaltic andesite of Anak Krakatau 1927 – 1980 by [10]. This means that the post-1996 products have the same characteristic with Anak Krakatau 1927 – 1980 or the first phase of Anak Krakatau. The alkalinity level diagram in **Fig 8b** shows Medium-K series. This series were also overlaping with Anak Krakatau by [10] and reference sample from the previous phase. This indicates that the post-1996 products derived from the same magma source with the older product of Anak Krakatau, which are 1883 product or Krakatau product represented by Rakata and Sertung. The alkalinity level diagram shows K_2O is positively correlated with SiO_2, meaning that the potassium (K) reacts as an incompatible element LILE (Rb, Ba, Sr) and increases along with the differentiation process of magma. From the harker diagram on **Fig 8c**, the post-1996 products of Anak Krakatau overlaps with the trend formed by the previous phases.

SiO_2 has relatively short range from 53% - 56% compared with Anak Krakatau by [10] but the post-1996 products were still overlapping with the trend from the previous phases. The relation between SiO_2 content and Al_2O_3, FeO, CaO and MgO contents indicate a the negative trend. While the relation between SiO_2 content and Na_2O and K_2O contents indicate a positive trend. According to [18] the compatible element such as FeO and MgO will be easier to become minerals so the concentration will decrease along with the differentiation processes. The short change of geochemical composition of the post-1996 products reflect an insignificant differentiation process. **Fig 6** shows a general increasing trend of SiO_2 composition, accompanied by decreasing MgO, from the oldest to the youngest deposit as an evidence of the differentiation process. However, decreasing SiO2 content, accompanied by increasing MgO, occurred occasionally. This event might resembles the new injection of new batch magma from deeper source.

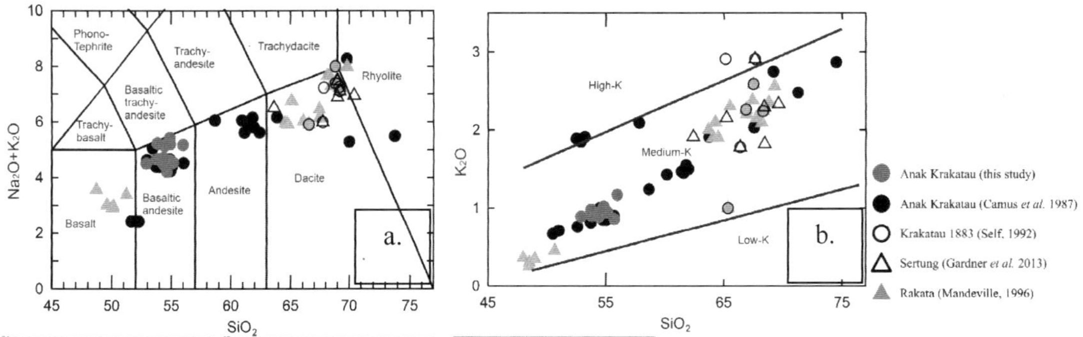

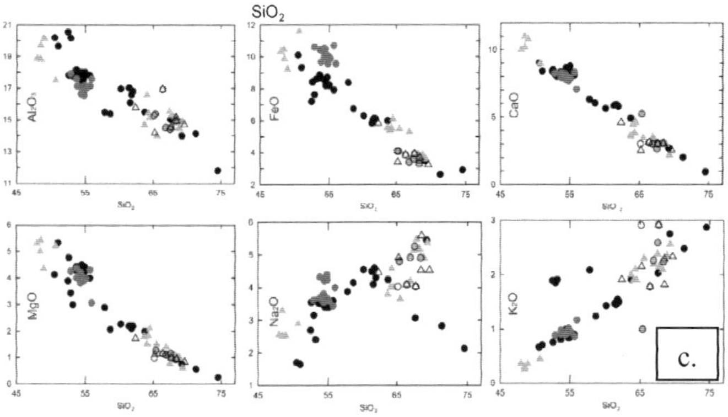

Fig 8. (a) TAS diagram shows Basaltic Andesite type. (b) Alkalinity level diagram shows Medium-K series. (c) Harker diagrams of major oxide to SiO_2 (The data plotted with major oxide of Krakatau, 1883 Krakatau and Anak Krakatau refer to basic geochemical Georoc data as a comparison) (http://georoc.mpch-mainz.de /georoc/).

6. Conclusions

a. The granulometry analysis from the post-1996 products shows the eruption type was strombolian to vulcanian. The pyroclastic deposits are divided into pyroclastic fall and pyroclastic flow which are composed of components scoria block and non-vesicular lithic (fresh and oxidized), scoria lapilli and non-vesicular lithic, scoria ash, glass shard, free crystal and accessories lithic.

b. The post-1996 volcanic products of Anak Krakatau show that the lava is basaltic andesite, while the fragmental products are vitric crystal tuff and vitric tuff containing basaltic andesite scoria. SiO_2 content ranges from 53 to 56 wt.% and MgO ranges from 3.82 to 4.37 wt.%. The short range of the geochemical content but still on the same trend with the previous phase shows that the post-1996 products of Anak Krakatau did not evolved significantly despite of the various eruption type.

References

[1] Tregoning P, Brunner F K, Bock Y, Puntodewo S S O, McCaffrey R, Genrich J F, Calais E, Rais J and Subarya C 1994 *Geophysical Research Letters **2** 19 2135-2138 p*

[2] Gardner M F, Troll V R, Gamble J A, Gertisser R, Hart G L, Ellam R B, Harris C, Wolff J A 2013 *Journal of Petrology **54** 1 149–182 p*

[3] Sutawidjaja I S 2006 *Jurnal Geologi Indonesia **1** 3*

[4] Achmad A R, Syifa M, Park S J and Lee C W 2019 *Journal of Coastal Research **90** 214-220*

[5] Zen M T 1969 *Growth and State of Anak Krakatau in September 1968* (Bandung: Bandung Institute of Technology)

[6] Walter T R, Haghighi M H, Schneider F M, Coppola D, Motagh M, Saul J, Babeyko A, Dahm T, Troll V R, Tilmann F *et al*. 2019 *Nature Communications **10** 4339*

[7] Susilohadi S, Gaedicke C and Djajadiharja Y 2008 *Tectonophysics* **467** 1–4 55-71

[8] Mulyana B 2006 *Bulletin of Scientific Contribution **4** 2*

[9] Abdurrachman M, Widiyantoro S, Priadi B and Ismail T 2018 *Geosciences **8(4)**, 111*

[10] Camus G, Gourgaud A and Vincent P M 1987 *Journal of Volcanology and Geothermal Research **33** 299-316*

[11] Sutawidjaja I S 1997 *Annuals of Disas. Prev. Res. Inst., Kyoto Univ., **40** B-1*

[12] Hutchison C S *1974 Laboratory Handbook of Petrographic Technique* (New York : John Wiley & Sons)

[13] Walker G P L 1971 *The Journal of Geology*

[14] Cas R A F and Wright J V 1988 *Volcanic Successons, Modern and Ancient* (London: Chapman & Hall)

[15] Darmawan H, Suratman W, Agung H, Haryo E W, Bachtiar W M, Nia H, Syarifuddin, Milya A U A, Raditya J and Wikanti A 2021 *(Preprint http://dx.doi.org/10.21203/rs.3.rs-230137/v1)*

[16] Darmawan H, Bachtiar W M, Wahyudi, Agung H, Haryo E W, Nia H, Mamay S, Syarifudin, Raditya J, Suratman W, Nia H, Syarifuddin and Wikanti A 2021 *Indonesian Journal of Geography **52** 3*

[17] Woods A W and Bursik M I 1991 *Buletin of Volcanology **53***

[18] Winter J D 2014 *Principles of Igneous and Metamorphic Petrology 2nd Edition* (USA: Pearson Education Limited)

[19] Ardian D N 2020 Characteristic of Deposits, Mineralogical and Geochemistry of the Post-1996 Anak Krakatau Product (Universitas Gadjah Mada undergraduate thesis:unpublised)

Acknowledgements

This study is funded by the Indonesian Ministry of Education and Research program of INSINAS number 14/INS-2/PPK/E4/2019. We gratefully thank Nature Conservation Agency of Bengkulu Province for the support during field campaign. Data of this manuscript has been used in [19; unpublished bachelor degree thesis].

IOP Conf. Series: Earth and Environmental Science 1071 (2022) 012018 doi:10.1088/1755-1315/1071/1/012018

A Preliminary Volcanological Study of North Eastern Kaba Volcano, Bengkulu Province, Indonesia

M E J Wijaya [1], **L D Setijadji** [1*]

[1] Department of Geological Engineering, Gadjah Mada University, Jl. Grafika No. 2 Yogyakarta 55281, Indonesia
*Corresponding author's e-mail : lucasdonny@ugm.ac.id

Abstract. Mount Kaba is one of the active volcanoes located in the eastern part of Bengkulu Province. This research was conducted in Gunung Kaba Complex because the distribution location of the last eruption product is close to some monogenetic volcano and residential area. This is interesting to study considering because can be a potential threat of disaster. The results of this study are expected to support providing an overview of the potential threat of volcanic disasters in the Mount Kaba Complex. The research methods used include geomorphological, stratigraphic, petrographic, and geochemical analyses. The geomorphological unit of Mount Kaba consists of 6 units. The types of volcanic rocks in the research area are dominated by lava and pyroclastic rocks. Based on the petrographic analysis of 10 samples, the names of the rocks in the research area are basalt scoria, basalt, and andesite. Based on the geochemical analysis of 8 samples, it is known that silica content in the rocks ranging from 44.40925 - 63.2993 %. Based on the AFM diagram, all samples are included in the tholeiitic series. Based on the TAS (Total Alkali Silica) diagram, it is known that the igneous rocks of the study area consist of andesite, basalt, tephrite, and trachyandesite.

1. Background

Previous research discussed the geology of Kaba Mountain in general, such as Gafoer (1992) and Kusnadi (2011). It is necessary to conduct research that examines further the stratigraphy and morphology of the Kaba volcanic complex. This research was conducted in the northeastern part of the Gunung Kaba Complex because based on the appearance of satellite imagery, the morphology of the volcano and the distribution of the product of the last eruption of Mount Kaba are often found in this area. This is interesting to study considering its location is close to a monogenetic volcano and is in a residential area. This can be a potential threat of disaster. The results of this study are expected to provide an overview of the formation of the Gunung Kaba Complex and the potential threat of volcanic disasters that can be utilized for disaster mitigation studies.

2. Regional Geology

The stratigraphy of the study area according to the Bengkulu geological map sheet (Gafoer et al., 1992) is shown in Figure 1. It consists of at least 2 units, namely the Volcanic Breccia Unit (Qhv) and the Andesite-Basalt Volcanic Rock Unit (Qv). The Volcanic Breccia Unit (Qhv) consists of volcanic breccia, lava and andesite-basalt tuff from Mount Kaba. Meanwhile, the Andesite-Basalt Volcanic Rock Unit (Qv) consists of andesite to basalt lava, tuff, and lava breccia. Kusnadi et al. (2011) carried out an integrated geological and geochemical investigations. They made a geological map of the Kepahiang Geothermal Area, a rock unit in the form of Mount Kaba eruption products consisting of alternating lava flows (dacitic, andesitic, and basaltic andesite), pyroclastic deposits (falls and flows), laharsandsurfacedeposition.

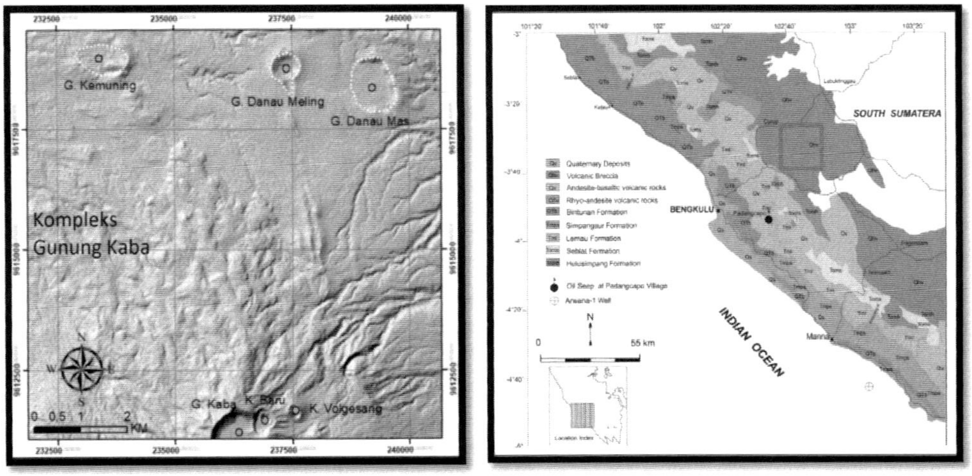

Figure 1. a. Map of the location of the research area (left), b.Geological map of Bengkulu (Gafoer et al., 1992) (right).

3. Research Method

This research includes field and laboratory researchs. Field research consists of geomorphological and stratigraphic analyses. It was conducted to understand the distribution, relationship, and position of volcanic rocks to other rocks by sampling and take pictures of outcrops found. Laboratory research includes petrographic and geochemical analyses, which includes major, trace, and rare-earth elements (REE). Petrographic analysis using polarized microscope and geochemical analysis were carried out at ALS Canada using the ICP-MS method (ME-ICP06) for the main elements and the ICP-MS method (ME-MS81) for trace and rare-earth elements (REE). Petrographic analysis using 10 samples and geochemical analysis using 15 samples. Select the samples by choose some samples from same description of rock samples. There is 15 locations analysed stratigraphically.

4. Research Results

Field research was conducted to obtain igneous rock samples for petrographic and geochemical analyses. Rock samples obtained include 30 locations spread from north to south of the research area. There are 10 rock samples for petrographic analysis and 8 samples for geochemical analysis.

4.1. Geomorphological Analysis

The geomorphological unit based on analysis of contour map (DEMNAS Indonesia) consists of 6 units, namely the Scoria Cone, the Danau Mas Lava Hill, the Volcanic Crater, the Lava-Pyroclastic Hill, the Pyroclastic Hill, and the Volcanic Plain Units. The lithology of the geomorphological unit consists of lava, pyroclastic, and scoria rocks. Morphographically, the geomorphological units have subparallel and subdendritic flow patterns, with relief forms on the lower and upper slopes. The valley shape is a V-U shape. These geomorphological units have an altitude ranging from 975 to 1825 m.a.s.l from the morphometric aspect. This geomorphological units have a slope included in the sloping class - a very steep class with a magnitude of 15 – 140 %. The dominant endogenous force forming this geomorphological units come from the volcanism process based on the morphogenetic aspect. While the exogenous forces that developed are in the form of weathering and erosional.

Table 1. Geomorphological units of research area

No	Geomorphological Unit	Geomorphological Characteristics								Lithology
		Morphography			Morphometry			Morphogenetics		
		Flowing Pattern	Relief Form	Valley Form	Elevation (m.a.s.l)	Slope		Endogen	Exogen	
						Class	%			
1	Danau Mas Lava Hill Unit	Subparallel	Down Slope	V-U	1075-1150	Little Steep	15%-30%	Volcanism	Weathering and Erosion	Lava
2	Scoria Cone Unit	Subparallel	Down Slope	V-U	1050-1100	Little Steep	15%-30%	Volcanism	Weathering and Erosion	Lava
3	Volcano Crater Unit	Subparallel	Upper Slope	V	1650-1825	Very Steep	70%-140%	Volcanism	Weathering and Erosion	Lava-Pyroclastic
4	Lava- Pyroclastic Hill Unit	Subparallel	Upper Slope	V	1250-1650	Steep	30%-70%	Volcanism	Weathering and Erosion	Lava
5	Pyroclastic Hill Unit	Subdendritic	Upper Slope	V	1075-1550	Little Steep	15%-30%	Volcanism	Weathering and Erosion	Lava-Pyroclastic
6	Volcano Plain Unit	Subdendritic	Down Slope	V-U	975-1050	Sloping	7%-15%	Volcanism	Weathering and Erosion	Lava-Pyroclastic

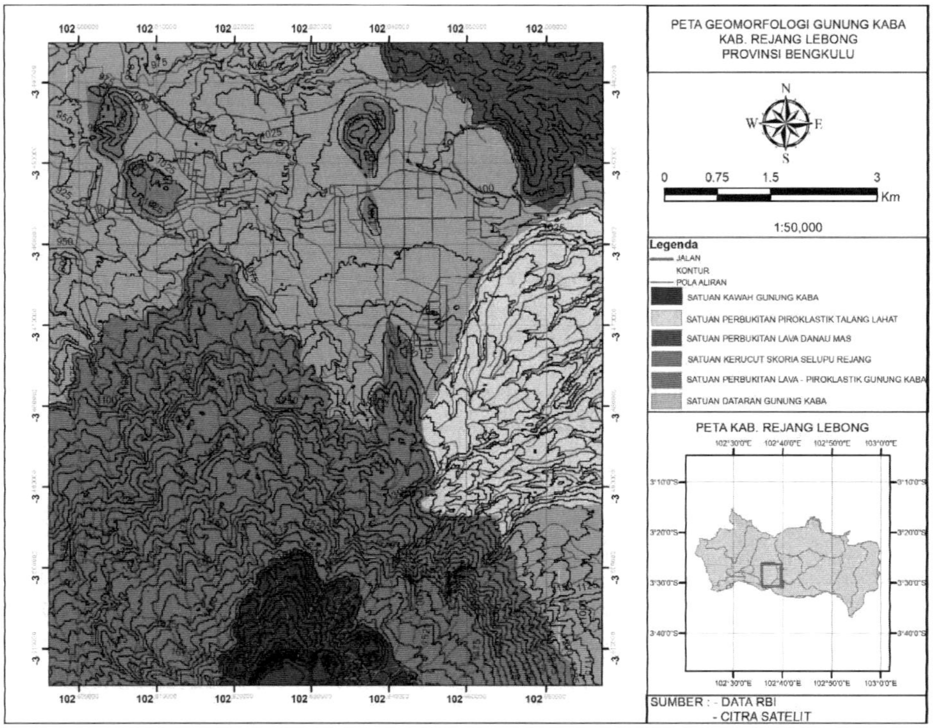

Figure 2. Geomorphological map of the research area

4.2 Stratigraphic Analysis

The type of volcanic rock in the study area is dominated by lava and pyroclastic rocks. Based on the the sampling location, megascopically the rocks at locations STA 1 to STA 6 are dominated by pyroclastic rocks and lava rocks around the Kaba volcano. At STA 7 - STA 9, there are lava rocks and pyroclastic deposit. While at STA 10 – STA 12 in the form of pyroclastic deposits. At STA 13 – STA 17 are dominated by lava rocks. The southern area consists of lava and pyroclastic rocks alternating to the top of the mount at STA 18 – STA 24. Then, at STA 25 – STA 30 are dominated by lava rock.

IOP Conf. Series: Earth and Environmental Science 1071 (2022) 012018 doi:10.1088/1755-1315/1071/1/012018

Table 2. List of outcrops in the research area

NO	OUTCROP	X (m)	Y (m)	NO	OUTCROP	X (m)	Y (m)	NO	OUTCROP	X (m)	Y (m)
1	Pyroclastic Rock and Lava (Scoria)	9618428	233727	11	Pyroclastic Deposits (Tuff)	9615738	239329	21	Pyroclastic Deposits (Tuff)	9612636	236750
2	Pyroclastic Rock and Lava (Scoria)	9619213	234092	12	Pyroclastic Deposits (Tuff)	9614877	238777	22	Pyroclastic Deposits	9612045	236617
3	Pyroclastic Rock and Lava (Scoria)	9618189	234386	13	Lava	9614701	238065	23	Pyroclastic Deposits and Lava	9611720	236458
4	Pyroclastic Rock and Lava (Scoria)	9618652	237403	14	Lava	9615358	238098	24	Pyroclastic Deposits	9611952	236889
5	Pyroclastic Rock and Lava (Scoria)	9618203	237515	15	Lava	9616155	237905	25	Lava	9616706	234911
6	Pyroclastic Deposits (Tuff)	9617614	237585	16	Lava	9615219	237494	26	Lava	9616022	234917
7	Lava	9618933	239198	17	Lava	9614190	238011	27	Lava	9615245	235336
8	Lava	9618638	239591	18	Pyroclastic Deposits (Tuff)	9613725	237460	28	Lava	9615790	233596
9	Pyroclastic Deposits (Tuff)	9617852	239297	19	Lava	9613333	237201	29	Lava	9614721	233915
10	Pyroclastic Deposits (Tuff)	9616556	239688	20	Pyroclastic Deposits (Tuff)	9613054	236936	30	Lava	9614004	233782

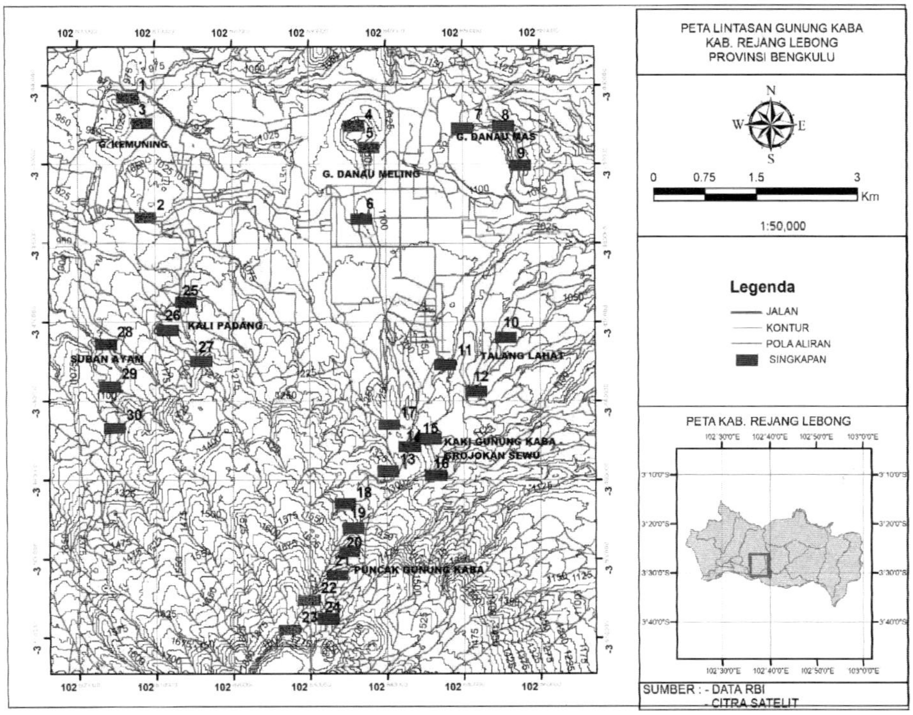

Figure 3. Outcrop map of the research area

Basalt scoria rocks has description of a grayish-black color, scoria texture, massive rock structure Basalt scoria rocks has composition of mafic minerals such as olivine, pyroxene, glass, and opaque minerals. Pyroclastic rocks have a description of ash-brown rocks with a round-grained texture. The grain size of sand (lapilli) is up to the size of ash. Fabric (grain packing) is open with bad sorting. Massive rock structure with a composition of mafic minerals such as olivine, pyroxene, glass, and opaque minerals. Lava rock has a description of rock blackish gray with an aphanitic texture. Massive rock structure with a composition of mafic minerals such as olivine, pyroxene, glass, and opaque minerals.

Based on the results of field and laboratory research in the form of geomorphological, stratigraphic, and petrographic and supported by geochemical analyses data, the research area, in general, can be divided into 7 geological units. The five units are andesite lava 1, andesite lava 2, basalt lava, scoria basalt lava, pyroclastic flow, pyroclastic fall, and lava-pyroclastic breccia. Andesite lava 2 and pyroclastic fall units occupy the largest area of research, while the scoria basalt lava unit has the smallest area.

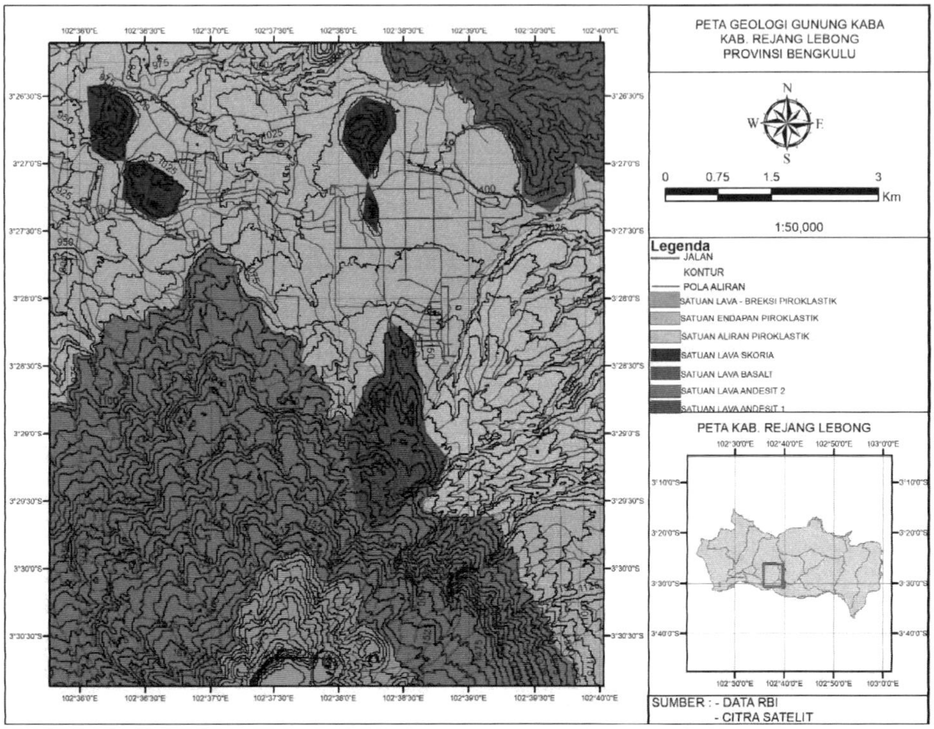

Figure 4. a. Geological map of the research area

4.3. Petrographic Analysis

Based on the petrographic analysis table, the thin sections are dark gray-black gray. Aphanitic-porphyritic rock texture with holohyaline (glass composition) and hypocrystalline of crystallinity degrees. Inequigranular grain uniformity with 0.1 – 0.5 mm crystal size and subhedral-anhedral crystal form. Specific textures range from vitrophyric, subophytic, flow, and intersertal. The structures are scoria and massive rocks. The rock composition for scoria consists of volcanic glass, vesicles, and crystal fragments evenly distributed in the thin section. While the rock composition for basalt and andesite consists of olivine, plagioclase evenly distributed in the thin section as groundmass and phenocryst, there is also amphibole, pyroxene, and glass. The names of the rocks in the research area are basalt scoria, basalt, and andesite (showed by figure 5).

Table 3. Petrographic analysis in the research area

Sample	Color	Texture	Degree of Cristallinity	Grain Uniform	Crystal Size (mm)	Form	Special Texture	Structure	ROCK NAME
1	Dark Gray	Aphanitic	Holohyaline	Inequigranular	0.1-0.5	Anhedral	Vitrophyric	Scoria	Basalt Scoria
5	Dark Gray	Aphanitic	Holohyaline	Inequigranular	0.1-0.5	Anhedral	Vitrophyric	Scoria	Basalt Scoria
7	Grayish Black	Porphyritic	Hypocrystalline	Inequigranular	0.1-0.5	Subhedral-Anhedral	Flow	Massive	Basalt
8	Grayish Black	Porphyritic	Hypocrystalline	Inequigranular	0.1-0.5	Subhedral-Anhedral	Flow	Massive	Basalt
13	Gray	Aphanitic	Hypocrystalline	Inequigranular	0.1-0.5	Subhedral-Anhedral	Subofitic	Massive	Andesite
17	Gray	Aphanitic	Hypocrystalline	Inequigranular	0.1-0.5	Subhedral-Anhedral	Porphyritic	Massive	Andesite
22	Gray	Aphanitic	Hypocrystalline	Inequigranular	0.1-0.5	Subhedral-Anhedral	Porphyritic	Massive	Andesite
23	Gray	Aphanitic	Hypocrystalline	Inequigranular	0.1-0.5	Subhedral-Anhedral	Porphyritic	Massive	Andesite
27	Gray	Aphanitic	Hypocrystalline	Inequigranular	0.1-0.5	Subhedral-Anhedral	Intersertal	Massive	Andesite
29	Gray	Aphanitic	Hypocrystalline	Inequigranular	0.1-0.5	Subhedral-Anhedral	Subofitic	Massive	Andesite

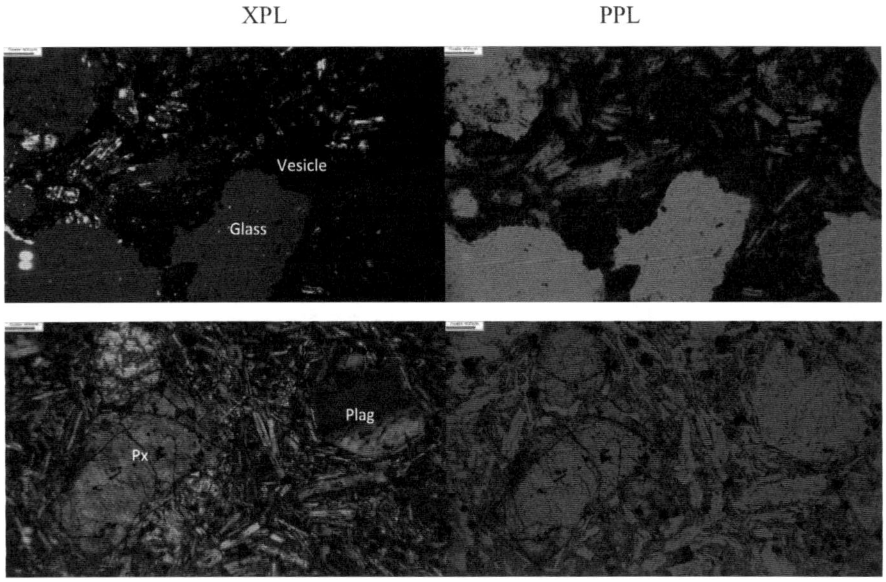

Figure 5. Petrography of the research area in XPL mode and PPL mode. a. Petrography of sample STA 4 (left), b. Petrography of sample STA 7 (right)

4.4. Geochemical Analysis

Based on the results of the main element analysis (Table 4), it is known that the rocks in the study area contain Silica ranging from 44.40925 - 63.2993 %. Based on SiO_2 content, the most acidic rock is andesite at STA 29 location, while the most alkaline rock is basalt at STA 4 location.

Magma affinity of igneous rocks is determined by plotting the main element data on several classification diagrams. Based on the AFM diagram (Figure 6) from Irvine and Baragar (1971), all samples of magma affinity rock are included in the tholeiitic series. Calc-alkaline magmas generally form in older island arcs and active continental margins. Tholeiitic magma generally forms in young island arcs or close to troughs in older island arcs (Wilson, 1991)

Table 4. Major, trace, and rare-earth elements in the research area

SAMPLE	SiO_2	Al_2O_3	Fe_2O_3	CaO	MgO	Na_2O	K_2O	Cr_2O_3	TiO_2	MnO	P_2O_5	SrO	BaO	Total
STA01	59.1967	15.50662	9.818975	4.949995	2.2123	3.869078	2.333522	0.00202	1.27283	0.181837	0.565702	0.040408	0.050509	100
STA03	58.45374	15.94708	9.61923	5.096913	2.286875	4.050867	2.42019	0.003077	1.30239	0.174341	0.553772	0.041021	0.051275	100
STA05	44.40925	21.55388	13.88416	5.54386	4.210357	2.446115	5.333119	0.00401	1.694168	0.220551	0.651603	0.02005	0.030074	100
STA08	47.41674	19.32951	11.22498	5.053861	5.587323	2.043693	0.946321	0.012081	1.09733	0.161079	0.231547	0.04027	0.020134	100
STA13	62.52466	16.54189	6.070355	4.552813	1.831224	4.239174	2.660839	0.001012	0.84985	0.121408	0.22258	0.040469	0.050586	100
STA20	61.6529	16.54135	6.014917	4.882206	2.095196	4.070175	2.53629	0.002005	0.822039	0.120301	0.200497	0.0401	0.040099	100
STA23	58.63879	16.64679	7.782062	6.171735	3.418939	3.74677	2.057327	0.005963	0.914367	0.139137	0.188837	0.039754	0.049694	100
STA29	63.2993	15.9242	6.111657	4.206767	2.014069	4.177002	2.817712	0.000992	0.853251	0.128981	0.19843	0.029765	0.049608	100

SAMPLE	Ba	Ce	Cr	Cs	Dy	Er	Eu	Ga	Gd	Hf	Ho	La	Lu	Nb	Nd	Pr	Rb	Sm	Sn	Sr	Ta	Tb	Th	Tm	U	V	W	Y	Yb	Zr
STA01	456	74.1	20	5.28	7.67	4.11	2.08	20	8.08	6.9	1.59	34.9	0.59	8.3	43.1	10	89.4	9.9	3	329	0.6	1.2	12	0.61	2.68	112	2	39.9	4.04	269
STA03	475	77.2	10	5.45	7.77	4.39	2.05	22.3	8.79	7.3	1.69	35.9	0.64	9.1	44.5	10.5	94.6	9.82	5	351	0.6	1.22	12.35	0.57	2.76	126	2	42.2	4.07	281
STA05	252	39	30	3.35	5.64	3.31	1.56	25.2	6.35	5.7	1.15	17	0.46	6.6	24.3	5.44	167.5	6.19	4	187.5	0.4	0.91	8.08	0.44	1.69	374	1	28.6	3.19	218
STA08	221	25.1	90	3.24	3.61	1.71	1.06	19.7	3.63	2.4	0.7	13	0.28	3.1	17.2	3.98	34.2	3.83	1	301	0.2	0.55	3.33	0.26	0.74	261	1	17.1	1.67	88
STA13	441	61.4	10	6.27	5.87	3.2	1.54	20.2	6.38	6.9	1.17	29.3	0.54	7.9	34.4	8.14	105	7.36	3	307	0.5	1.01	12.2	0.49	2.77	114	2	33.2	3.64	271
STA20	421	59.1	10	5.91	5.59	3.48	1.44	20	6.35	6.5	1.22	27.6	0.53	7.1	32.3	7.51	98.6	7.01	3	300	0.4	0.98	11.45	0.47	2.88	128	2	32.2	3.51	256
STA23	487	49.6	40	4.97	4.97	3.13	1.38	20.7	5.41	6	1.04	23.3	0.42	6.4	28.3	6.55	80.6	6.03	3	345	0.5	0.87	9.36	0.4	2.13	177	2	27.6	2.88	216
STA29	466	64.8	10	6.76	6.25	3.56	1.54	20.3	6.68	7.7	1.27	30.5	0.56	8.2	35.4	8.35	114	8.09	3	284	0.5	1.03	12.85	0.54	2.89	113	2	34.8	3.76	290

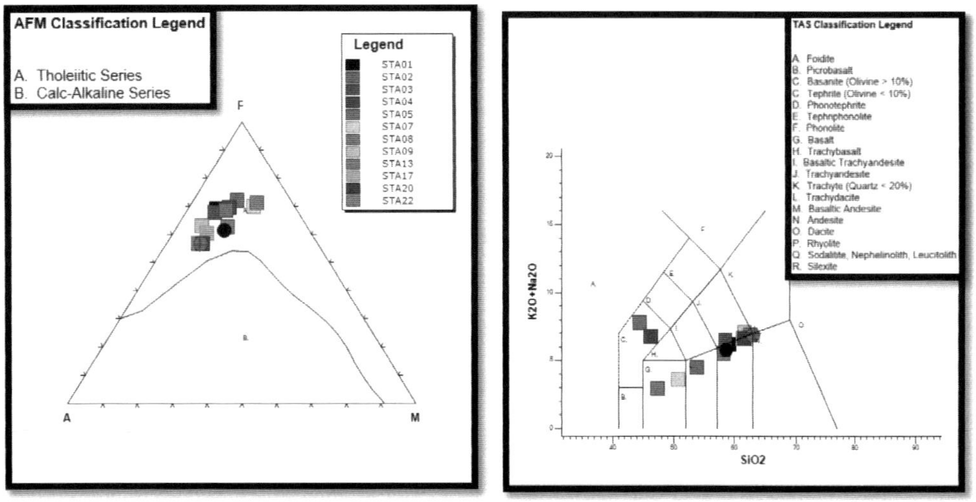

Figure 6. a. Plotting of rock geochemistry on the Irvine and Baragar (1971) Diagram at research area (left), b. Plotting of rock geochemistry on Le Bas et al., (1986) TAS diagram at research area (right)

Determination of rock types can also be done using the main elemental data SiO2, K2O, and Na2O. The classification diagram used is the Total Alkali Silica (TAS) diagram, according to (Le Bas et al., 1986). Based on the plotting results on the TAS diagram, it is known that the igneous rocks of the study area consist of andesite, basalt, tephrite, and trachyandesite.

The description of the main oxide content can also be seen in the variation diagram of Harker (1909) in Figure 7. The Harker diagram is used to determine the evolution of magma and fractional crystallization by comparing SiO2 and other major elements such as Al2O3, MgO, CaO, FeO, and others. Based on the Harker diagram (1909), it can be seen that there is a correlation between the location of the rock and the level of rock acidity (SiO2 content). The further south the rock is more acidic, the higher the Na2Oand K2O content. This is in contrast to the increasing content of MgO and Fe2O3 for rock positions to the north. The oxide content of Na2O, K2O, and SiO2 reflects the acidity of the rock, while the content of MgO and Fe2O3 reflects the alkalinity of the rock. The abundance of Fe2O3, MgO, CaO, and TiO2 compounds towards SiO2 shows a negative correlation trend. The abundance of K2O and Na2O compounds is positively correlated to the abundance of SiO2.

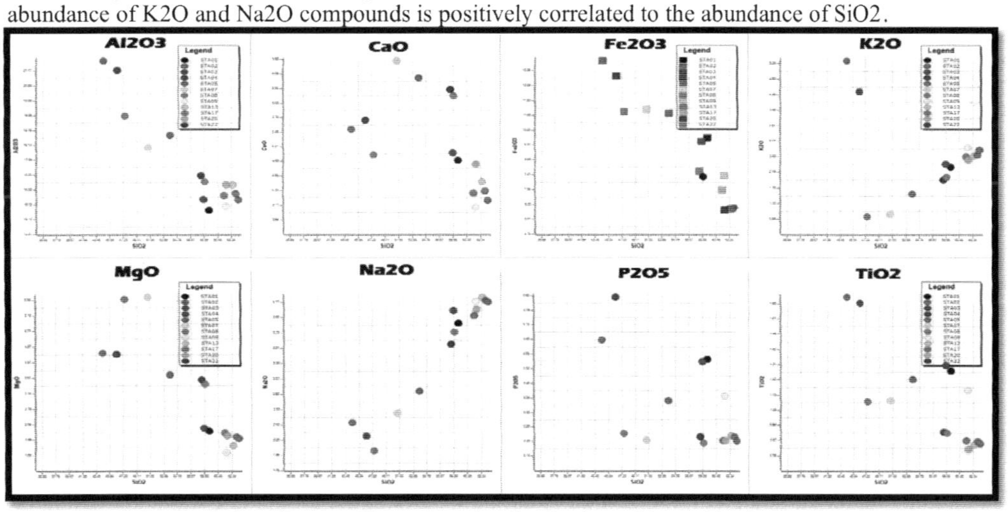

Figure 7. Harker variation diagram illustrating the distribution of the oxides in the research area a. Al2O3, b. CaO, c. Fe2O3, d. K2O, e. MgO, f. Na2O, g. P2O5, h. TiO2

The results of the analysis of trace elements of rock in the research area are presented in a spider diagram after normalizing to MORB (Pearce, 1982). The diagram shows a spike pattern on the elements Sr, K, Rb, Ba, and Th, which indicates enrichment of these elements in the rock compared to the element Ta until Yb. Elements Sr, K, Rb, and Ba are low ionic potential incompatible elements easily carried away by dilute fluids. Enrichment in the LILE group and depleted in the HFSE group indicate an island arc volcanic setting associated with magmatism in the subduction zone. The formation of magma in the subduction zone is influenced by the input of H2O, which causes a decrease in the boiling point so that the resulting magma is hydrous rather than magma in OIB and undergoes LILE enrichment (Pearce, 1982).

After normalization to chondrites (Nakamura, 1974), rare earth elements are presented in spider diagrams. Based on the spider diagram, it can be seen that there are two different patterns. Enrichment in the LREE and depleted in the HREE groups. The first pattern shows the enrichment of Light REE elements (La, Ce, Pr, Nd, Pm, Sm, and Eu). The pattern that shows LREE enrichment is that of calc-alkaline basalt. While the second pattern shows a lower LREE enrichment, this pattern is indicated as a pattern from island arc tholeiitic (Wilson, 1989).

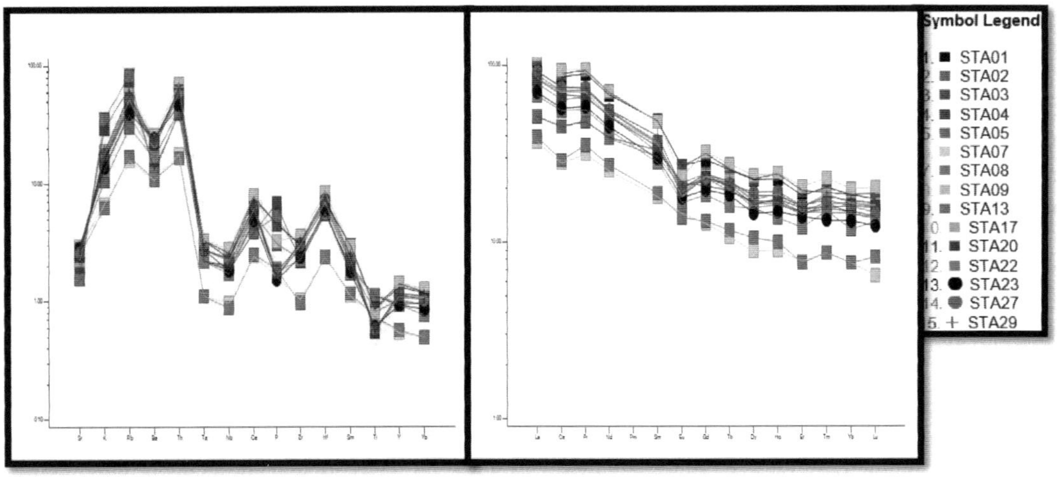

Figure 8. a. Normalized REE spider diagram against MORB (Pearce, 1982) (left), b. Normalized REE spider diagram against Chondrite REE Diagram (Nakamura, 1974) (right)

5. Summary

The geomorphological unit consists of 6 units, namely the Scoria Cone, the Danau Mas Lava Hill, the Volcanic Crater, the Lava Hill, the Pyroclastic Hill, and the Volcanic Plain Units. The types of volcanic rock in the research area are dominated by lava and pyroclastic rocks. Based on the petrographical analysis, the rock names are basalt scoria, basalt, and andesite. Geological units can be divided into 7 units, namely andesite lava 1, andesite lava 2, basalt lava, scoria basalt lava, pyroclastic flow, pyroclastic fall, and lava-pyroclastic breccia. Based on the results of geochemical analysis, the Silica contents of the rocks in the study area ranging from 44.40925 - 63.2993 %. Based on the AFM diagram (Irvine and Baragar, 1971), all samples are included in the tholeiitic series. Based on the TAS (Total Alkali Silica) diagram (Le Bas et al., 1986), it is known that the igneous rocks of the study area consist of andesite, basalt, tephrite, and trachyandesite. The analysis results of trace elements are presented in a spider diagram after normalizing to MORB (Pearce, 1982). Enrichment in the LILE group and depleted in the HFSE group indicate an island arc volcanic setting associated with magmatism in the subduction zone. Rare earth elements are presented in spider diagrams after normalization to chondrites (Nakamura,

1974). The pattern that shows LREE (La, Ce, Pr, Nd, Pm, Sm, and Eu) enrichment is calc-alkaline basalt (Wilson, 1989). The results of geological and geochemical data can show how big the explosive level of a volcano is. This can be used as supporting data in making disaster mitigation policies against potential volcanic disasters.

References

[1] BAS, M.L., Maitre, R.L., Streckeisen, A., Zanettin, B. and IUGS Subcommission on the Systematics of Igneous Rocks, 1986. A chemical classification of volcanic rocks based on the total alkali-silica diagram. *Journal of petrology*, *27*(3), pp.745-750.

[2] Brahmantyo, B. and Salim, B., 2006. Klasifikasi Bentuk Muka Bumi (Landform) untuk Pemetaan Geomorfologi pada Skala 1: 25.000 dan Aplikasinya untuk Penataan Ruang. Jurnal Geoaplika (2006) 1(2), pp.071 – 078.

[3] Gafoer, S., Amin, T.C. and R.. Pardede, 1992. Peta geologi lembar Bengkulu, Sumatra. Pusat Penelitian dan Pengembangan Geologi.

[4] Global Volcanism Program, 2013. Kaba (261220) in Volcanoes of the World, v. 4.8.8 (17 Apr 2020).Venzke,E(ed.).SmithsonianInstitution.Downloaded31May2020(https://volcano.si.edu/volcano.cf?vn=261220).

[5] Harker, A., 1909. *The natural history of igneous rocks*. Methuen & Company.

[6] Irvine, T.N. and Baragar, W.R.A., 1971. A guide to the chemical classification of the common volcanic rocks. *Canadian journal of earth sciences*, *8*(5), pp.523-548.

[7] Kusnadi, D. and Nurhadi, M., Suparman. 2011. Penyelidikan Terpadu Geologidan Geokimia Daerah Panas Bumi Kepahiang, Kabupaten Kepahiang, Bengkulu. *Prosiding Hasil Kegiatan Pusat Sumber Daya Geologi Tahun 2010*.

[8] Nakamura, N., 1974. Determination of REE, Ba, Fe, Mg, Na and K in carbonaceous and ordinary chondrites. *Geochimica et cosmochimica acta*, *38*(5), pp.757-775.

[9] Pearce, J.A., 1982. Trace element characteristics of lavas from destructive plate boundaries. *Orogenic andesites and related rocks*, pp.528-548.

[10] PVMBG. 2014. *Gunung Kaba*. Bandung: Pusat Vulkanologi dan MitigasiBencana Geologi. Data Dasar Gunungapi Indonesia

[11] Rollinson, H.R., 1993. Using geochemical data: evaluation, presentation, interpretation (No. 550.84 ROL).

[12] Suganda, O. K., Kusdinar, E., Irianto, A., Setiawan, T., 1990, Laporan Kegiatan Pemetaan Geologi Gunung Api Kaba

[13] Wilson, M., 1991; Igneous Petrogenesis, A Global Tectonic Approach, Harper Collins Academic, 2nd published, London, England, 466 pp

Acknowledgement

Thank you to my supervisors, Mr. Lucas Dony Setijadji and Mr. Haryo Edi Wibowo and the Department of Geological Engineering, Universitas Gadjah Mada for helping fund research.

IOP Conf. Series: Earth and Environmental Science 1071 (2022) 012019 doi:10.1088/1755-1315/1071/1/012019

Identification of Seismic Response to Aquifer System with A Synthetic Modelling Approach

Yana Udin[1] and Wiwit Suryanto[1*]

[1]Departement of Physics Gadjah Mada University, Yogyakarta
*Corresponding author's e-mail: ws@ugm.ac.id

Abstract. The microseismic method is relatively new in groundwater exploration, especially in Indonesia. This study will examine the response of microseismic to groundwater aquifer systems. This study aims to analyze the response of seismic waves and to interpret the seismic wave response characteristics to the aquifer system model. The modeling of the aquifer system was carried out using seismic stratigraphy data. This data was used as input data to model the exposed HVSR curve using the Microtrem program in MATLAB. The study of physical parameters and supporting parameters of the aquifer layer and the air contained in it was carried out to support this modeling. The aquifer system modeling was carried out with various parameters, including saturation, porosity, and thickness of the aquifer layer. Furthermore, an analysis of the model is carried out to determine the response characteristics of the aquifer system to microseismic waves. The results show that the frequency range as a marker of the presence of the aquifer ranges from 2-4 Hz, which is determined by parameters within a specific value.

1. Introduction

The demand for freshwater continues to increase with the development and increasing population for both consumption and industrial needs. With the high demand and the emergence of groundwater pollution, a cleaner and more stable water source is needed. Deep groundwater is rainwater that seeps into the ground through the process of absorption and filtration by rocks and minerals in the soil. Deep groundwater is usually found at 100-300 meters under the surface [1]. This groundwater source is a more stable water capacity. Deep groundwater can be used as a source of clean water for consumption and industry because it has a more abundant and stable amount of water.

The presence of groundwater causes an increase in the density of a rock or soil. In addition to providing benefits for living things, groundwater can also be one of the triggering factors for natural disasters, such as landslides and liquefaction. A groundwater survey is one of the efforts that can do to determine the presence of groundwater so that it can use to reduce disaster risk and fulfill consumption needs. One of the methods that can be used to determine groundwater's presence is the microseismic method.

Microseismic methods for groundwater exploration studies have not been widely used, except for hydrocarbon exploration, geotechnical engineering, landslides, and others. The microseismic method is a new method in groundwater exploration, especially in Indonesia. Research using this method needs to study first how the aquifer system responds to microseismic waves to determine the presence of groundwater in the aquifer. So, in this research, seismic forward modeling will be carried out to obtain the seismic wave response caused by the groundwater system. In addition, this study's conduction determines the parameters that respond to the existence of the aquifer system [2].

2. Methodology

2.1. Groundwater and Aquifer

Groundwater is water that lies under the surface. Groundwater is in the saturated zone and then moves to various layers of soil and rock until the water comes to the surface, such as lakes, ponds, rivers, and the sea [3]. The upper surface of the saturated zone is called the water table. Groundwater is divided into two types, shallow groundwater, and deep groundwater. Shallow groundwater is located not far from the surface and lies above an impermeable layer. Deep groundwater is usually found at 100-300 meters below the surface [1]. This groundwater source is more stable in water capacity than shallow groundwater.

Groundwater can be found in rock layers with unique hydrogeology characteristics, such as porosity, permeability, and several other hydraulic properties. Groundwater can be found in rock layers with unique hydrogeological characteristics, such as porosity, permeability, and several other hydraulic properties. Parameters that affect the output of the HVSR curve are saturation, porosity, and thickness of the aquifer layer. In addition, the speed of seismic waves (S waves) decreases in line with the addition of porosity and saturation values which causes the frequency value to decrease and the amplitude value to increase [2]. Rocks that have hydraulic properties can store and transmit water like an aquifer. The composition of the aquifer is measured from the groundwater-surface to the impermeable layer, which includes the aquiclude and aquifuge layers [4]. When viewed from the distribution composition and interpretation of rock layers, aquifers can be divided into three. These are unconfined, confined, and leaky aquifers [5].

Aquicludes is a layer of rock that can store water but cannot drain water at sufficient levels. Clay, shale, fine tuff, and silt are examples of aquicludes. The aquitard layer has almost the same characteristics as an aquifer. The aquitard layer can also store water like an aquifer. However, the difference is that this aquitard layer can only drain a limited amount of water. The aquitard is located between the aquifer and aquiclude layers—examples of aquitard layers such as shale, marl, and claystone. The aquifuge layer is a layer that is impervious to water, where this layer cannot drain and store water, such as granite and hard, dense, and compact rocks [4].

2.2. Microseismic

The passive seismic method is a geophysical method that utilizes microseismic waves. Passive seismic detects low-frequency ambient seismic noise to distinguish geological structures and find oil, gas, or other resources underground [6][7]. Microseismic is a passive seismic method. Microseismic is a natural ground vibration with an amplitude of 0.1-1 m and a period ranging from 0.05-0.1 seconds to 1-2 seconds by natural and artificial events [12]. Microseismic methods can determine the dynamic characteristics (dominant frequency and amplification) of an area [8][9]. Microseismic has also been used to analyze site effects and sediment characteristics. This method considers the dynamic characteristics of the surface layer from the frequency range of 0.5 – 20 Hz [13].

Microseismic methods can be used to detect and identify the presence of groundwater directly. The initial principle of this method is the nonlinear interaction of the fluid with the porous medium, which is measured passively and has a low-frequency value (2 to 4 Hz) [12]. The aquifer system modeling (Figure 1) was made using stratigraphic seismic data consisting of rock layer thickness, shear wave velocity value, and rock density. These parameters are then processed to produce a H/V curve. The obtained H/V curves were then analyzed to identify the presence of groundwater aquifers.

Table 1. Variation of *S* wave velocity values for input parameters in seismic stratigraphic data

Material	Unconfined aquifer (m/s)	Confined Aquifer (m/s)
Topsoil	300	300
Dry Sand	500	710
Wet Sand	200	200
Clay	930	930, 1600

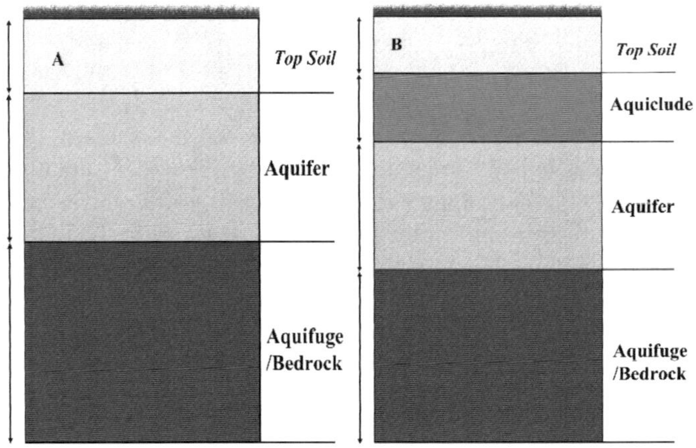

Figure 1. The aquifer system model used: A. Unconfined Aquifer and B. Confined Aquifer

Synthetic stratigraphic seismic data for seismic stratigraphic data for each model shown in **Figure 1** are as follows:

- Unconfined Aquifer
  For models with aquifers in the second layer: (300 m/s, 0 m), (200 m/s, 5 m), (930 m/s, 15 m). And models with layers without aquifers: (300 m/s, 0 m), (500 m/s, 5 m), (930 m/s, 15 m).
- Confined Aquifer
  For models with aquifers in the third layer: (300 m/s, 0 m), (930 m/s, 5 m), (200 m/s, 17 m), (1600 m/s, 27 m). And models with layers without aquifers: (300 m/s, 0 m), (930 m/s, 5 m), (710 m/s, 17 m), (1600 m/s, 27 m).

The data used only applies to the type of lithology and geological setting in table 1.

Figure 2 results from shear wave velocity modeling for an aquifer system model with three layers (free aquifer). Figure 2 on the left shows a lower shear wave velocity response. At the same time, Figure 2 on the right shows a higher shear wave velocity response. This shows that if there is no aquifer layer or a layer that is not filled with water (dry), then the shear wave velocity will be faster when compared to the shear wave velocity through the layer containing water or aquifer. This also applies to Figure 3.

2.3. HVSR Method (Horizontal to Vertical Spectral Ratio)

Horizontal to Vertical Spectral Ratio (HVSR) was first introduced by Nogoshi and Iragashi (1971). They stated that there was a relationship between the ratio of horizontal and vertical components of microtremor measurements which Nakamura later developed. The amplification factor or the amplitude value of horizontal and vertical movements on the sedimentary soil surface is based on seismic movements on the soil surface that are in direct contact with bedrock in the basin area [13].

Determine the structure of the subsurface layer, and it is done by correcting the ratio of the Fourier spectrum of the microtremor signal on the horizontal component to the vertical component [13]. According to Nakamura, this method aims to identify S-wave amplification in soft sediment by measuring microtremor. The spectral ratio of the H/V component of the recorded ambient noise is equivalent to the ratio of the S-waves of the earthquake recorded at the sediment surface to the sediment-bedrock interface at the bottom of the sediment layer.

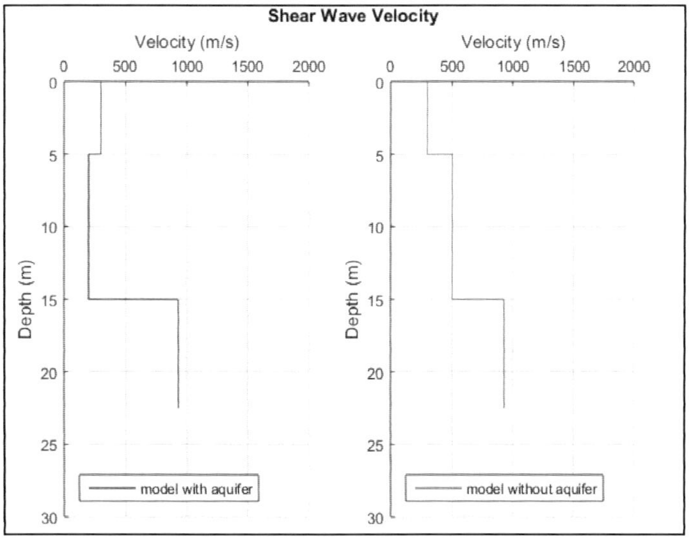

Figure 2. The Aquifer System Model with three layers as Unconfined Aquifer (V_S with Aquifer, left; and V_S without Aquifer, right).

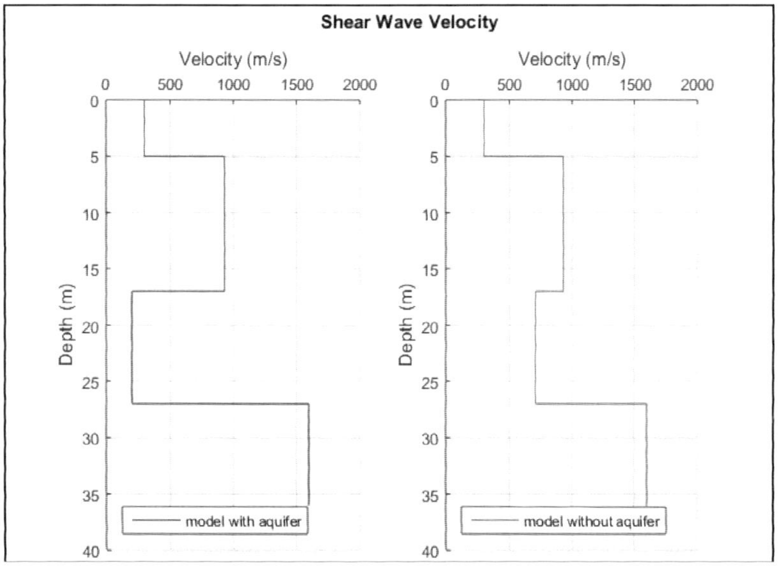

Figure 3. The Aquifer System Model with four layers as Confined Aquifer (V_S with Aquifer, left; and V_S without Aquifer, right).

The HVSR method is one method that can be used to understand the nature of the subsurface structure without causing damage to the structure. This method produces essential parameters, namely dominant frequency values and amplification related to subsurface physical parameters. In addition, the HVSR method can explain the phenomenon of the resonant frequency of the surface layer by taking measurements somewhere.

The calculation of the ratio of the horizontal component to the vertical component microtremor spectrum (HVSR) can be formulated as [11]:

$$HVSR = S_t = \frac{H_s}{V_s} = \frac{\sqrt{(H_{NS})^2 + (H_{EW})^2}}{V_s} \tag{1}$$

Based on equation (1), the H/V method depends on two main assumptions. That is: with S_t is the *site-effect* or *horizontal-to-vertical spectral ratio*, H_s is the spectrum of the horizontal component of the sediment layer, V_s. is the spectrum of the vertical component recorded in the sediment, H_{EW} is the spectrum of the horizontal component in the west-east direction, H_{NS} is the spectrum of the horizontal component in the north-south direction.

- Assumption 1: $\frac{H_b}{V_b} \approx 1$ i.e., microtremor measurements in the hard rock show the spectral ratio.
- Assumption 2: $V_s \approx V_b$ i.e., vertical movement propagates without disturbance through the sediment layer.

3. Result and Discussion

There are two models of aquifer systems, namely free aquifers and confined aquifers. Both models are made of synthetic seismic data and become input data to produce a shear wave velocity model and HVSR curve using MATLAB and Microtrem. Figure 4 is the HVSR curve comparing the model with an aquifer (water) and the model without an aquifer (dry). This three-layer aquifer system model shows an HVSR curve (blue) with an amplitude spectrum that shows anomalously high at a frequency of about 4 Hz, which identifies the presence of an aquifer. In contrast, the HVSR curve (red) has a low anomaly with a frequency value of about 8 Hz, which indicates the presence of a dry layer.

Figure 5 is the HVSR curve for an aquifer system model with four layers. The third layer is the aquifer layer. This curve identifies the presence of the aquifer in the third layer. HVSR curves (blue) are two amplitude spectra with different anomalies. The first anomaly has a frequency of 2 Hz, and the second anomaly has a frequency of 13 Hz. The appearance of two anomalies, low and high on the blue curve, indicates the presence of an aquifer located in the third layer. In comparison, the HVSR curve (red) has a low anomaly with a frequency value of around 16 Hz, which indicates the presence of a dry layer.

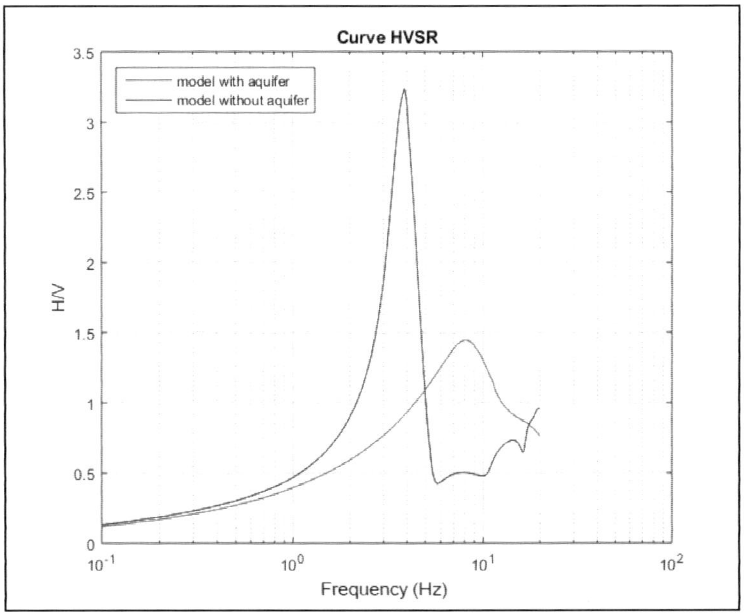

Figure 4. The Aquifer System Model for Unconfined Aquifer Model.

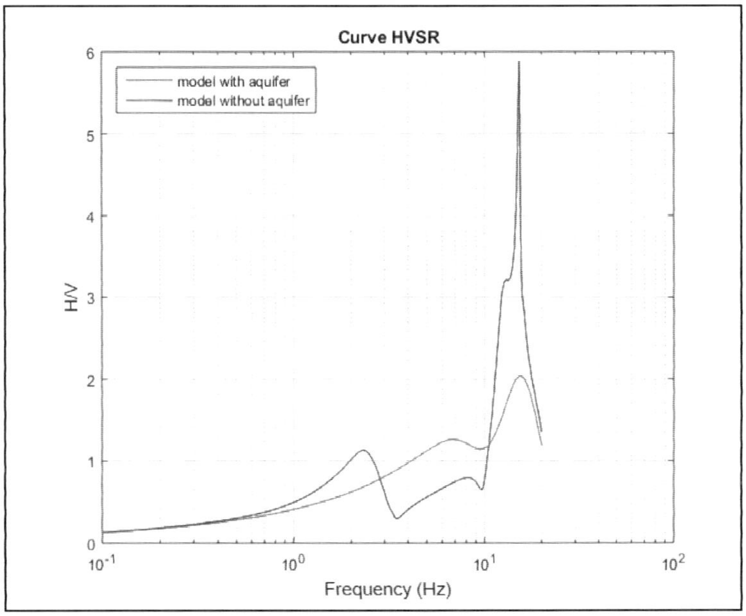

Figure 5. The Aquifer System Model for Confined aquifer.

4. Conclusion

Several parameters affect the HVSR curve: density, rock thickness (number of rock layers), and velocity. Of the three parameters, two parameters that affect the results of the H/V curve are rock thickness (number of rock layers) and velocity. Increasing the number of rock layers (thickness) will cause the emergence of an amplitude spectrum with different anomalies.

The existence of the aquifer based on the modeling results is indicated at a frequency of 2-4 Hz. This is reinforced by the graph of the speed of seismic waves (S waves) experiencing a decrease in speed due to a layer in which there is liquid. The decrease in the speed value causes the frequency value to decrease and the amplitude value to increase.

References

[1] Sutrisno, CT dan Suciastuti, E. 2010. Teknologi Penyediaan Air Bersih. Jakarta: Rineka Cipta.

[2] Udin, Y. (2022). Identifikasi Respon Seismik Terhadap Model Sederhana Sistem Akuifer 1D. Thesis. Universitas Gadjah Mada. Yogyakarta.

[3] Fetter, C.W. (1994). *Applied Hydrogeology. 4th Edition*. Macmillan College Publishing Company, New York.

[4] Salsabila, A., & Nugraheni, I. L. (2020). Pengantar Hidrologi. Bandar Lampung: AURA.

[5] Kruseman, G.P., De Ridder, N.A. & Verweij, J.M., 1970, Analysis and evaluation of pumping test data, International institute for land reclamation and improvement The Netherlands.

[6] Wahyudi, 2008, Aplikasi Mikroseismik Untuk Memindai Dan Mengidentifikasi Keberadaan Hidrokarbon, *BIMIPA*, **18**, 2, 114–122.

[7] Suryanto, W., Wahyudi, Adriansyah and Sardjito, 2009. A microseismic survey over Tambun oilfield, West Java, Indonesia: Mapping of low-frequency anomalies. *Proceedings, the 9th SEGJ Interrnational Symposium*.

[8] Kanai, K. and Tanaka, T. (1961) On Microtremors VIII. *Bulletin of Earthquakes Research Institute*, **39**(1), 97-114. https://doi.org/10.15083/0000033837.

[9] M. Farid, Sunarto, and W. Suryanto, —Mapping of potential areas tsunami-prone in Bengkulu city, *ARPN Journal of Engineering and Applied Sciences*, **vol. 11**, no. 7, pp. 4828–4832, 2016

[10] Motamed, R., Ghalandarzadeh, A., Tawhata, I., & Tabatabaei, S. H. (2007). Seismic microzonation and damage assessment of Bam city, southeastern Iran. *Journal of Earthquake Engineering*, **11**(1), 110–132. https://doi.org/10.1080/13632460601123164.

[11] Nakamura, Y. (2000). Clear identification of fundamental idea of Nakamura's technique and its applications. *Proceedings of the 12th World Conference on Earthquake Engineering*, **24**, 25–30. https://www.iitk.ac.in/nicee/wcee/article/2656.pdf.

[12] Ali, M.Y., Berteussen, K.A., Small, J. & Barkat, B., 2007, *A low frequency, passive seismic experiment over a carbonate reservoir in Abu Dhabi, first break*, 25, 11.

[13] Atakan K. (2009) The Need for Standardized Approach for Estimating the Local Site Effects Based on Ambient Noise Recordings. *In: Mucciarelli M., Herak M., Cassidy J. (eds) Increasing Seismic Safety by Combining Engineering Technologies and Seismological Data. NATO Science for Peace and Security Series C: Environmental Security.* Springer, Dordrecht. https://doi.org/10.1007/978-1-4020-9196-4_2.

Acknowledgments

The authors are grateful to the Department of Physics, Universitas Gadjah Mada, were provides facilities for conducting this research.

IOP Conf. Series: Earth and Environmental Science 1071 (2022) 012020 doi:10.1088/1755-1315/1071/1/012020

Parameters test on time migration with Kirchhoff method on 2D synthetic data with dipping event and its reflector structural changes

Aurum Datametriana[1] and Wiwit Suryanto[1*]

[1]Departement of Physics Gadjah Mada University, Yogyakarta
*Corresponding author's email: ws@ugm.ac.id

Abstract. This paper describes the effect of parameter variation with Kirchhoff migration in post-stack seismic data. Parameters tested covering velocity, aperture, and maximum dip. The data used in this work was generated from velocity and density models with gaussian syncline and anticline layer boundaries. This work is the effect of parameter changes on the model and conclude best applied parameters for Kirchhoff migration in certain model created.

1. Introduction

Migration is an important step in routine processing of seismic reflection data because it will return the reflector position to its true position below the surface. Events contained in the recorded seismic data are not always in their actual position. Seismic reflection recorded in gather data and stacked pre-migration occasionally not in actual position especially for inclined reflector. Seismic migration process returns reflectors to its actual position. One of the migration methods is the Kirchhoff migration which is the commonly used in seismic data processing. There are several parameters that can affect the results of the Kirchhoff migration process including aperture, maximum dip, and velocity that is used in the migration process. parameter selection error may cause the reflector to be repositioned incorrectly. In this study, we discuss in detailed the effect of various parameters on Kirchhoff migration using synthetic reflection seismic data. So that we can find out the effect of changes in each parameter on the migration results. Moreover, we can estimate the optimum parameters used in the migration process [2].

2. Methodology

2.1. Kirchhoff migration

Kirchhoff migration can be used for data with difficult topography, irregular recording geometry, post-stack migration and pre-stack migration [5]. Kirchoff migration corrects structural dips to 90°. However, lateral velocity variations cannot be corrected [4]. The principle of migration on seismic data using Huygens secondary source. Claerbout (1985) used a harbor barrier on the coast to describe the process of the Huygens effect.

The point aperture on the Huygens secondary source will cause wavefront amplitude to change depending on the angle that called obliquity factor. Apart from the obliquity factor, the change in amplitude is caused by spherical spreading, which depends on the distance from the source point to the wavefront (r) [9].

The scheme used in the Kirchhoff migration is diffraction summation. Before doing the diffraction summation, the amplitude and phase correction will be carried out first. The phase change that is caused by angle is called the obliquity factor. Obliquity factor correction is done by multiplying the amplitude by cosine θ. θ is the angle between the event recorded on the surface and the peak of the diffraction curve. Correction of spherical spreading done by multiplying amplitude 1/r. Furthermore, the input data that has been multiplied is filtered using the wavelet shaping factor. The result of the addition follows [9].

$$P_{out} = \frac{\Delta x}{2\pi} \sum_{x} \left[\frac{\cos\theta}{\sqrt{v_{RMSs} r}} \rho(t) * P_{in} \right] \quad (1)$$

where P_{out} is migrated data, x is the distance between the observation point and the diffraction peak, v_{RMS} is RMS velocity, ρ(t) is the rho filter, and P_{in} is input data. The filter ρ(t) can be calculated by time-derived of input data [margrave]. The summation of the diffraction hyperbolas must be in accordance with the equation [9]

2.2. *Kirchhoff migration parameters*

The diffraction Hyperbole curves summation is determined by the width of the migration aperture. The migration aperture represents number of traces that will be used in the summation. Width of the migration aperture is related to the horizontal displacement (dx).

$$d_x = \frac{v^2 t}{4} \frac{\Delta t}{\Delta x} \quad (2)$$

Δt is the height and Δx is the width of the unmigrated dipping reflector. the optimum aperture value is usually twice the migration halo. the migration halo is the extra area that must be added due to the lateral movement of the dip [3].

In addition, the parameter that affects the migration process is maximum dip. The maximum dip parameter is related to the aperture parameter. The smaller the maximum dip used the aperture will be narrower [3]. The curvature of the diffraction hyperbola curve that is used for amplitude summing in the migration process is influenced by the velocity function.

$$t^2 = \tau^2 + \frac{4x^2}{{v_{rms}}^2} \quad (3)$$

where x is the distance between the amplitude summation point on a particular trace to the apex of the hyperbole, v_{rms} is the rms velocity, t is the time before the migration process, and τ u is the time after the migration process [9].

2.3. *Seismic data synthetic*

The acquisition geometry in the forward model process is described in table 1. Each shooting consists of one source and 240 receivers with 120 receivers on the left side and 120 receivers on the right side of the source. There were 323 shots from the forward model. Shot gather of seismic synthetic will be processed into stack section as input for the post-stack time migration. The velocity that is used in the migration process is the RMS velocity from the velocity model. Synthetic data generation and Kirchhoff migration processing where functions provided by Crewes.

IOP Conf. Series: Earth and Environmental Science 1071 (2022) 012020 doi:10.1088/1755-1315/1071/1/012020

Table 1. Acquisition Parameter.

Spread Type	Split Spread
Receiver Number	240 channels
Receiver Interval	20 m
Source Interval	40 m
Near Offset	20 m
Far Offset	2400 m
Spread Length	4200 m
Fold Coverage	60
Source Depth	0 m
Receiver Depth	0 m

The classical wave equation is expressed in terms of

$$\nabla^2 \psi = v^{-2} \partial_t^2 \psi \tag{4}$$

where ψ is acoustic wave function, v is wave velocity propagation [5].

The method used in the forward model is finite difference in the acoustic wave equation according to the equation

$$\psi(x,z,t+\Delta t) = 2\psi(x,z,t) - \psi(x,z,t-\Delta t) + \frac{\Delta t^2 v^2}{\Delta x^2}[\delta_{xx}\psi(x,z,t) + \delta_{zz}\psi(x,z,t)] \tag{5}$$

with

$$\begin{gathered}\delta_{xx}\psi(x,z,t) = \psi(x+\Delta x,z,t) - 2\psi(x,z,t) + \psi(x-\Delta x,z,t) \\ \delta_{xx} = \delta_{zz}\end{gathered} \tag{6}$$

where Δt is discrete time interval and x and z are cartesian coordinate [6].

The model used in forward modelling is an acoustic model with variations in p-wave velocity and density. The model consists of 3 layers with each velocity about 2000, 2500, and 3500 m/s, respectively (Figure 1). The slope of the anticline is 20.61° while the syncline is 39.17° (Figure 1). The difference angle of anticline and syncline is useful for sensitivity testing of the maximum dip parameter. The density model is obtained by converting the velocity model using the Gardner equation according to

$$\rho = 0.21 {v_p}^{0.25} (g/cm^3) \tag{7}$$

where ρ is density and v_p is p-wave velocity [8]. The dimensions of the model are 12,890 kilometers with 8 kilometers deep (Figure 1 and 2).

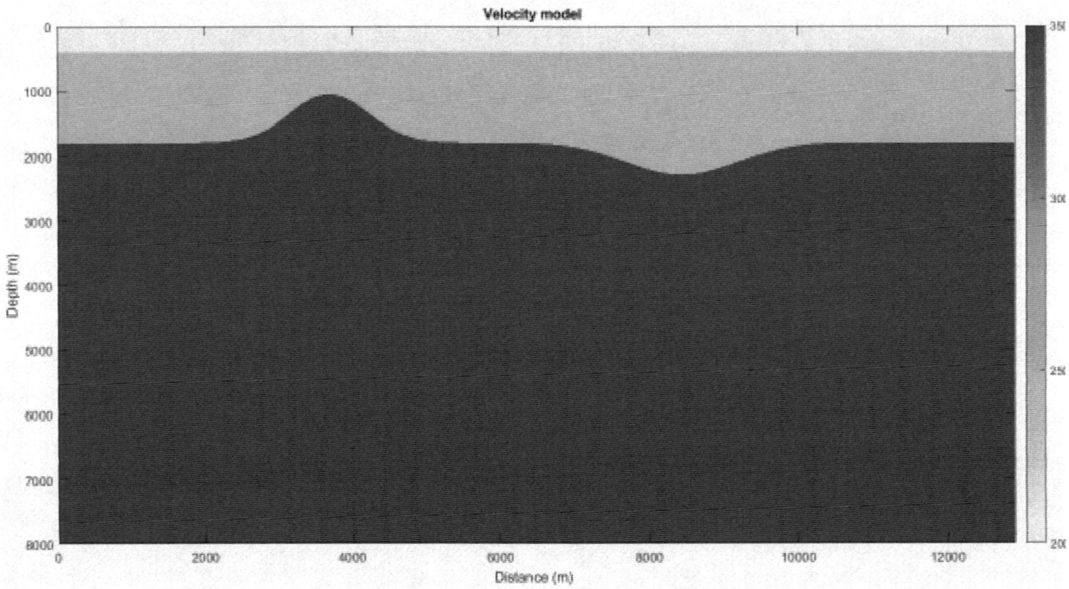

Figure 1. P wave Velocity Model. the yellow layer is 0-400m below the surface, green layer is 400-1800m below the surface, and blue layer is 1800-8000m below the surface.

Post stack time migration (PoSTM) is the migration process that is used in this work. The input data in this PoSTM is the stacked data. The shot gather data is converted to CMP gather. Then this CMP gather is processed by normal moveout (NMO) process. The NMO process is carried out to return the seismic reflector to the zero offset position according to

$$t_{NMO}(x) \approx \frac{o^2}{2t_0 v_{NMO}^2}$$

where t_{NMO} is the time for NMO correction, o is offset, t_0 is time at zero offset, and v_{NMO} is velocity NMO from velocity analysis process. The next process is the stack process. Stack is seismic trace summing are at the zero offset position at the same CMP position [7].

3. Result and Discussion

There are two reflectors in the stack section 400ms and 1550ms. The reflector at 2000ms is multiple of the reflector at 1550ms. This multiple will repeat every 400ms. This multiple is formed due to the reflection of waves between the surface and the reflector at a depth of 400ms. The "anticline reflector" at a depth of 2600-2200ms is the diffraction produced by the syncline structure at 1550ms. In this work, the stack data is only derived from shot gather results from the forward model. The multiple of the two reflectors is not eliminated because it will be discussed whether its effect on the migration process. From the stack raw results, there is diffraction caused by anticline and syncline structures. The diffraction of the anticline structure will be an anticline that looks wider than the actual structure. The diffraction of the syncline structure will be an "anticline reflector" at depth 2600-2200ms that is called bow tie. This diffraction effect will be corrected by using the Kirchhoff migration process. So that the reflector can return to its actual position.

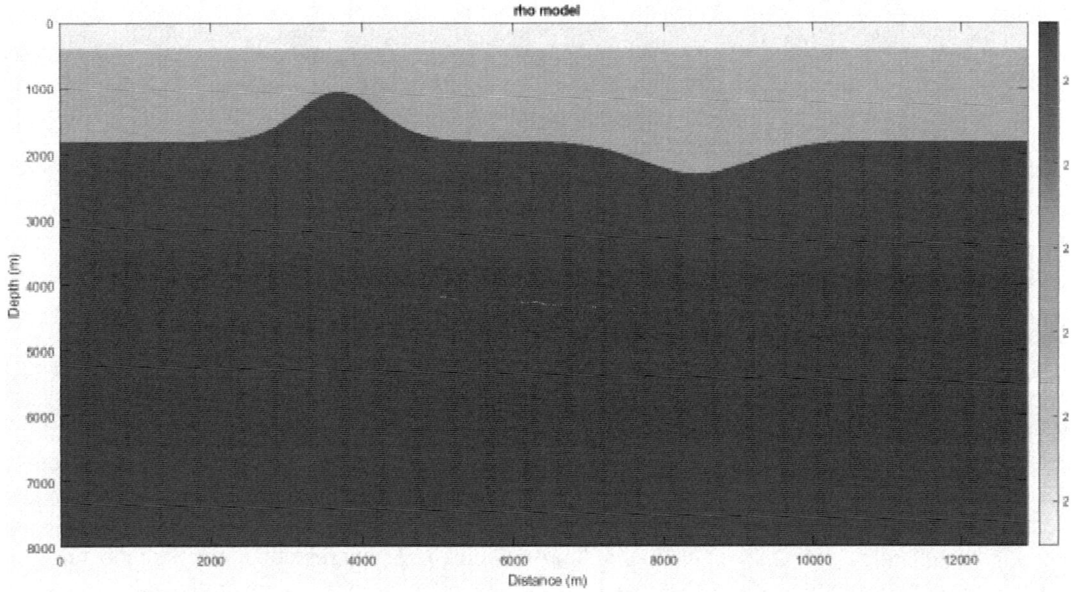

Figure 2. Density Model that generated from P-wave velocity model.

The migration process will not affect the straight reflector because there is no diffraction effect that appears on the reflector. The first parameter to be tested is aperture. There are 4 aperture values used in the migration process with a value range of 500 to 3500 with increment 1000. This aperture test value was chosen to observe the effect of the migration process on small to large apertures. A small aperture will use the less input data in the Kirchhoff migration process than a large aperture. The results of the Kirchhoff migration process with various aperture parameters can be seen in Figure 4. Small aperture will cause discontinuous reflection especially in syncline structure. The syncline becomes poorly reconstructed (yellow circle in the figure 4). The higher the aperture, the more the true shape of the structure will appear. The correct aperture used in the migration process is the aperture just before the migration result does not change when the aperture is increased. In this case, the best aperture value is 2500m. The accurate aperture can optimize running time of migration process.

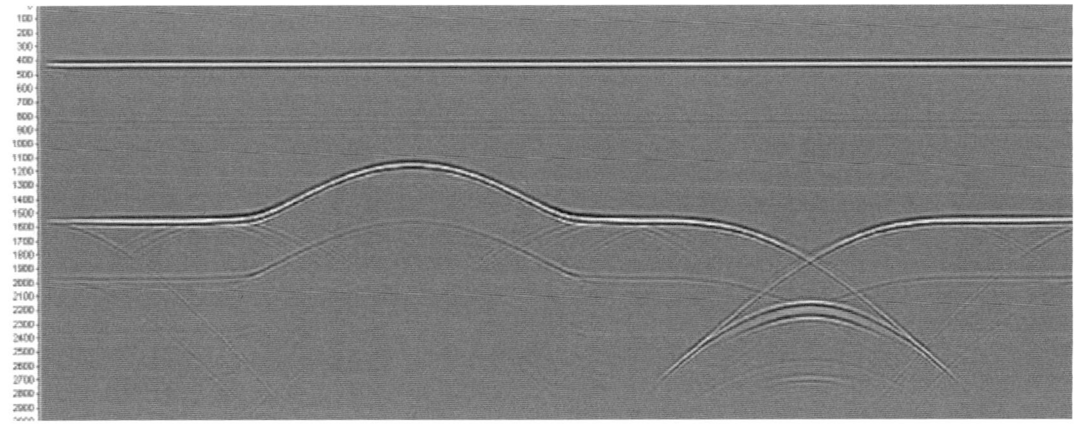

Figure 3. Stack raw from the shot gather that generated from the forward model process.

The second parameter is the maximum dip. There are 3 maximum dip value that is used in the Kirchhoff migration test. First maximum dip value is 10° (Figure 5a). At this aperture value a fake reflector will appear (yellow arrows in the Figure 5a). Second maximum dip value is 20° (Figure 5b). The fake reflector still appears but with a smaller amplitude and the shape is closer to the original reflection than the migration result with a maximum dip of 10 (red arrow in the Figure 5b). The 40° the maximum dip, the fake events will disappear (Figure 5c). If the maximum dip used is higher than the optimal dip, the migration result will not show any change. The optimal dip can be indicated by the final change from the maximum dip. The optimum dip used in this case is 40°. The optimal dip is close to the actual dip.

The third parameter is variation of RMS velocity. From Figure 6, 3 variations of the velocity used, namely 10% lower than the true velocity, the true velocity (100%) and 10% higher than the true velocity. There are 3 variations of RMS velocity that will be tested in the Kirchhoff migration process. The first is the true RMS velocity (100%) (Figure 6b). The results of the Kirchhoff migration process using this velocity show that the reflector is migrated to its actual position. The second speed variation is that the RMS velocity is 10% lower than the true RMS velocity (Figure 6a). Kirchhoff migration results using this velocity will appear undermigration artifacts (blue arrows in Figure 6). In addition, the use of a lower RMS velocity will cause structural changes. The anticline structure becomes wider, and the syncline structure becomes narrower than the true structure. The third velocity variation is that the RMS velocity is 10% higher than the true RMS velocity (Figure 6c). The results of the Kirchhoff migration process using this velocity will bring up overmigration artifacts (yellow arrow in Figure 6). In addition, the higher RMS velocity will cause anticline structure to be a narrower and the syncline structure to be wider than the true structure.

Multiple events at depths of 2600-2200 will be migrated. In testing the aperture parameter, the syncline structure is not properly corrected (Figure 4). At aperture 500 the syncline structure of the multiple is not visible (Fig. 4a). At higher apertures (Figures 4b, 4c, and 4d), the syncline structure will begin to appear but not like the true structure. This happens because the velocity at the multiple position is higher than the true velocity. In the anticline structure also migrated. However, the deeper the anticline, the anticline structure in the multiple will be the narrower and sharper (Figs. 4b, 4c, and 4d). This also occurs in the maximum dip parameter test.

In the velocity parameters test, variations in the RMS velocity that is used in the migration process will affect the shape of the multiple. When the RMS velocity is decreased by 10% (Fig. 6a), the anticline and syncline structure forms will appear and approach the true shape. When using the true RMS velocity (100%) and the RMS velocity is 10% higher than the actual speed, the syncline structure is not reconstructed properly. Meanwhile, the anticline structure at this both RMS velocities will become narrower and sharper.

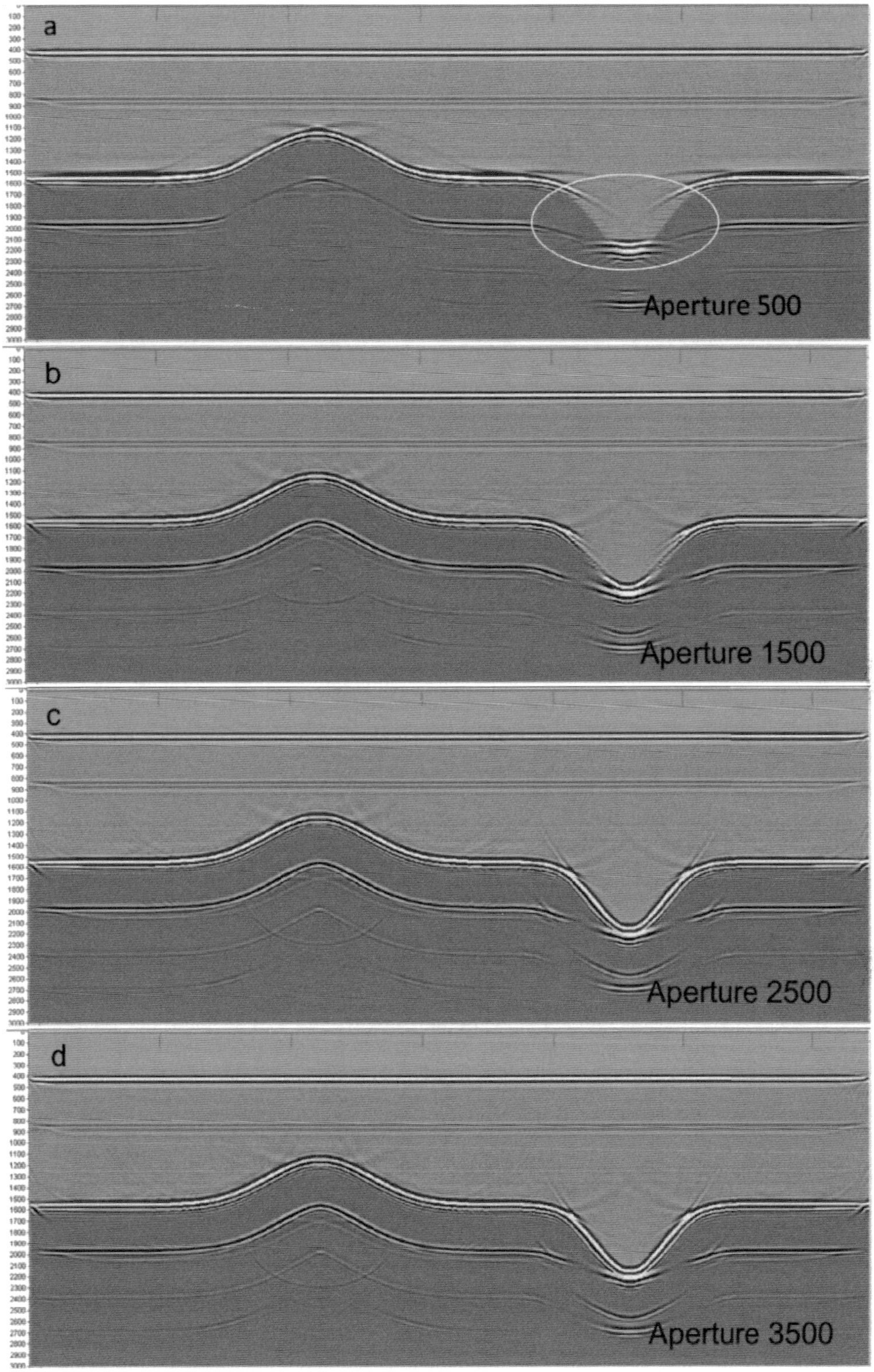

Figure 4. Post-stack time migration with aperture width a)500m, b)1500m, c)2500m, and d)3500m overlay with true model structure (colored).

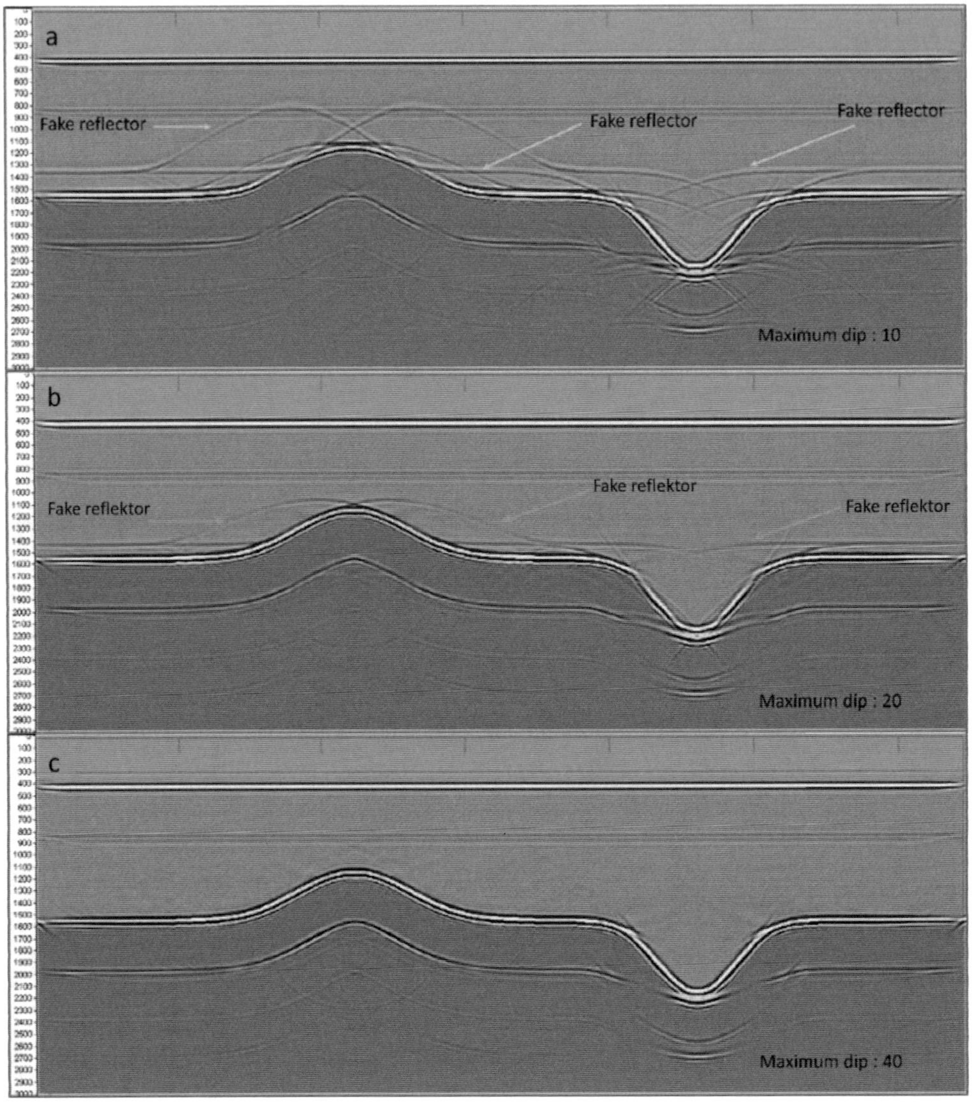

Figure 5. Post-stack time migration with maximum dip value a)10, b)20, and c)40 overlay with true model structure (colored).

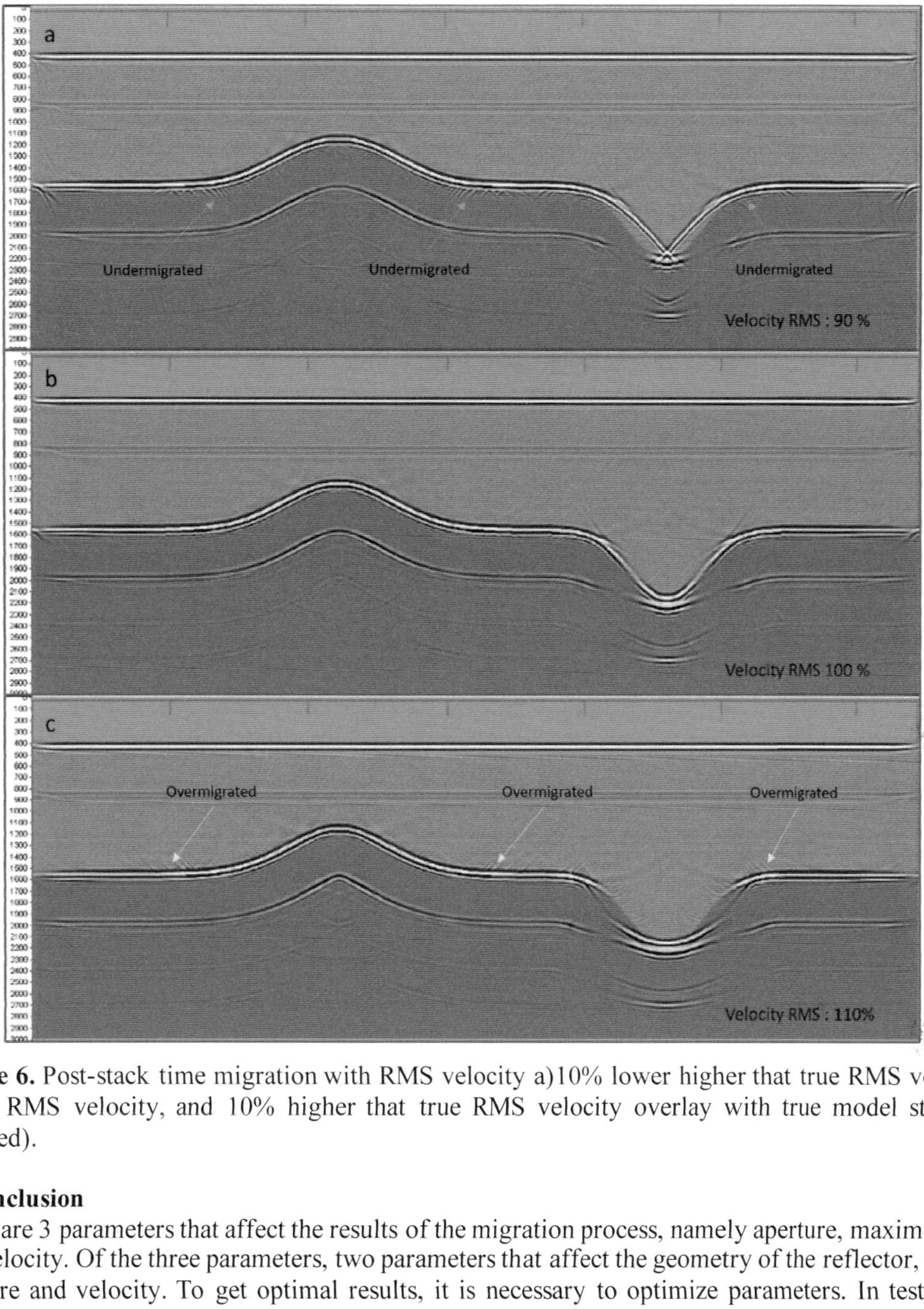

Figure 6. Post-stack time migration with RMS velocity a)10% lower higher that true RMS velocity, b)true RMS velocity, and 10% higher that true RMS velocity overlay with true model structure (colored).

4. Conclusion

There are 3 parameters that affect the results of the migration process, namely aperture, maximum dip, and velocity. Of the three parameters, two parameters that affect the geometry of the reflector, namely aperture and velocity. To get optimal results, it is necessary to optimize parameters. In testing the aperture from low to high values, the optimum aperture should be used in the Kirchhoff migration process when there is no change precisely in the result of the migration process. The maximum dip parameter will be optimum when the estimated true structure dip is known. Velocity parameter will be optimum when there is no migration artifact. You have to be careful when picking velocity in the velocity analysis process. When doing the migration process, multiple will also be migrated but not return to the desired form. so that it must be ensured before the migration process, the multiple has been corrected so that it does not affect the interpretation.

References

[1] Claerbout, J. F. (1985). *Imaging the Earth's Interior*. Blackwell Scientific Publications. https://doi.org/10.1111/j.1365-246X.1986.tb01086.x

[2] Datametriana, A. (2022). Pengujian Parameter Proses Migrasi Kawasan Waktu Menggunakan Metode Kirchhoff pada Data Sintetik 2D dengan Dipping Events dan Conflicting Dips. Thesis. Universitas Gadjah Mada. Yogyakarta

[3] Dondurur, D. (2018). Acquisition and Processing of Marine Seismic Data. In *Acquisition and Processing of Marine Seismic Data*. https://doi.org/10.1016/C2016-0-01591-7

[4] Fang, K. (1995). Prestack Kirchhoff migration. CREWES Research Report, 7. https://www.crewes.org/

[5] Margrave, G. F. (2003). *Numerical Methods of Exploration Seismology with algoritms in MATLAB*. Calgary University. https://www.crewes.org/

[6] Mitchell, A., & Griffiths, D. (1980). The Finite Difference Method in Partial Differential Equations. In *A Wiley-Interscience Publication*.

[7] Theodosius Marwan Irnaka, Wahyudi Wahyudi, Eddy Hartantyo, Adien Akhmad Mufaqih, Ade Anggraini, Wiwit Suryanto, "SEISGAMA: A Free C# Based Seismic Data Processing Software Platform", *International Journal of Geophysics*, **vol. 2018**, Article ID 2913591, 8 pages, 2018. https://doi.org/10.1155/2018/2913591

[8] Quijada, M. F., & Stewart, R. R. (2007). Density estimations using density-velocity relations and seismic inversion. *CREWES Research Report*, **19**. https://www.crewes.org/

[9] Yilmaz, Ö. (2001). Seismic Data Analysis: Processing, Inversion, and Interpretation of Seismic Data. *Society of Exploration Geophysicists*. https://doi.org/10.1190/1.9781560801580.

Acknowledgments

The autors is grateful to Department of Physics, Universitas Gadjah Mada, were provides facilities in conducting this research.

Geochemistry and geochronology of intrusive units in the Suyoc epithermal deposit: Constraints on magma fertility in the Mankayan Mineral District, Philippines

J A S Gabo-Ratio[1*], K D Jabagat[2], O B Soberano[3], K Yonezu[4], and Y H Lee[2]

[1]National Institute of Geological Sciences, College of Science, University of the Philippines, Diliman, Quezon City, Philippines
[2]Department of Earth and Environmental Sciences, National Chung Cheng University, Chia-Yi, Taiwan, R.O.C.
[3]Graduate School of International Resource Sciences, Akita University, Japan
[4]Department of Earth Resources Engineering, Kyushu University, Fukuoka, Japan
*Corresponding author's e-mail: jgratio@nigs.upd.edu.ph

Abstract. The Mankayan Mineral District in northern Luzon, Philippines is host to world-class Cu-Au hydrothermal deposits such as the Far Southeast porphyry copper deposit and the Lepanto, Victoria, and Teresa epithermal deposits. It also hosts the Suyoc epithermal prospect in the southern portion of the district. This study focuses on the petrography, whole rock geochemistry, mineral chemistry and U-Pb dating of the three massive batholitic intrusive units (gabbro-diabase basement, tonalite-granodiorite, and hornblende quartz diorite) in Suyoc. U-Pb dating results reveal that the tonalite is Late Eocene in age (37.20 ± 4.70 Ma), while the hornblende quartz diorite was dated 3.18 ± 0.77 Ma (Middle Pliocene). The gabbro-diabase exhibits tholeiitic signature while the tonalite-granodiorite and hornblende quartz diorite are calc-alkaline. Trace element plots for the three host rocks indicate formation in a subduction setting. Furthermore, discrimination diagrams point to an adakitic character for the younger hornblende quartz diorite and typical arc rock signatures for the older gabbro and tonalite-granodiorite. The results indicate that hydrothermal mineralization is attributed to the hornblende quartz diorite, which coincides with the Pliocene mineralization recognized in the Mankayan Mineral District.

1. Introduction

Recent research emphasized the importance of understanding the geochemistry of igneous rocks in determining fertility of magmatic systems that form porphyry copper deposits [1, 2]. Barren and fertile intrusive rocks would have formed under varied geochemical compositions and magmatic conditions, which would provide constraints on ore deposit formation [3]. An ideal natural laboratory to study these magmatic systems are in highly mineralized areas like the Mankayan Mineral District.

The Mankayan Mineral District (MMD) in northern Luzon, Philippines is considered as one of the most productive mineral districts in the southwest Pacific [4]. It hosts a world-class fossil hydrothermal system consisting of the Far Southeast porphyry copper deposit, Lepanto high sulfidation epithermal deposit and the Victoria intermediate sulfidation epithermal deposit. In addition, the MMD also consists of other hydrothermal prospects that have yet to be investigated in detail, and this includes the Suyoc epithermal vein prospect located ~4 kms south of the Far Southeast-Lepanto-Victoria system (Figure 1a). Intrusive rocks to the west of the Suyoc epithermal veins were investigated in this study to understand magma fertility in relation to the various stages of igneous activity in the area.

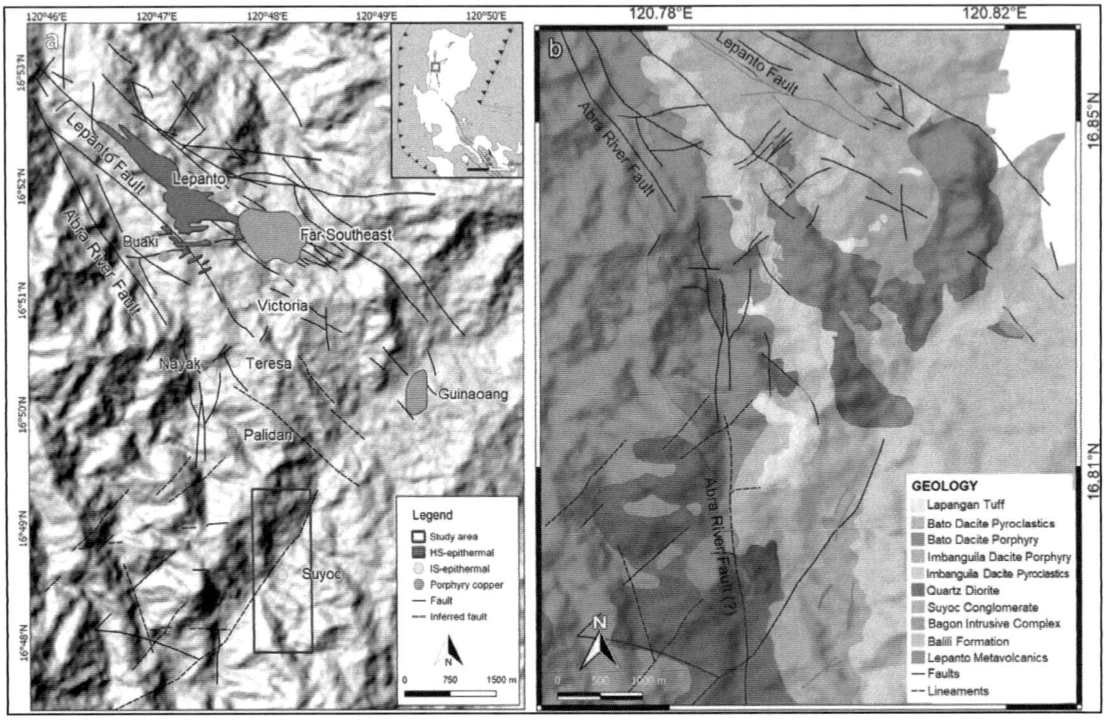

Figure 1. a) Distribution of porphyry copper and epithermal gold deposits in the Mankayan Mineral District. Suyoc is in the southern portion of the study area. b) Geologic map of the district adopted from [5].

2. Geologic Outline

The Mankayan Mineral District (MMD) is part of the Northern Luzon segment of the Taiwan-Luzon volcanic arc. It was formed from subduction of the South China Sea Basin along the Manila Trench [6]. In the Pliocene, subduction of the Scarborough Seamount Chain, an extinct mid-oceanic ridge, resulted in tectonic uplift and magmatic activity [6]. This is believed to result in the formation of Pliocene-Pleistocene hydrothermal mineralization in the MMD [7]. The geology and mineralization of the MMD shares similarities with the well-known Baguio Mineral District, which is 55 km south of Mankayan [7].

The spatial distribution of mineralization in the the MMD is believed to be controlled by the Abra River Fault, which is a splay of the Philippine Fault Zone [8]. The NS- to NNW-trending Abra River Fault is a left-lateral fault with thrust component. One of its NW-trending splays is the Lepanto Fault, which hosts the Lepanto epithermal deposit.

The district is underlain by Late Cretaceous to Early Eocene Lepanto Metavolcanics basement (Figure 1b) composed of basaltic to andesitic volcanic rocks that have undergone low grade greenschist metamorphism [10]. Middle Eocene to Early Oligocene volcanic and shallow sedimentary rock sequences of the Balili Formation overly this basement rock. Major magmatism, which formed intermediate composition batholithic intrusions of the Bagon Intrusive Complex, occurred during Late Oligocene [11]. Associated with the magmatism is the deposition of terrestrial to shallow marine volcano-sedimentary sequences of the Suyoc Conglomerate comprised of conglomerate, sandstone, shale, limestone and interbeds of volcanic flows and tuffs [5]. Magmatism in the Pliocene resulted in the intrusion of Quartz Diorite and the deposition of the Imbanguila Dacite. Continued magmatism during the Pleistocene formed the Bato Dacite. The youngest unit in the area is the Late Pleistocene Lapangan Tuff [5].

3. Methodology

Least altered igneous rocks were collected from surface exposures in the study area. Sample preparation of double-polished thin sections was done at the University of the Philippines – National Institute of Geological Sciences (UP-NIGS) using a Pelcon Automatic Thin Section machine and a Topper Diamond Polishing unit. Petrographic analysis was conducted using the Olympus BX53-P polarizing microscope.

Samples for whole rock major composition determination were powdered and made into pressed pellets. These were analyzed using a Rigaku RIX3100 X-Ray Fluorescence (XRF) spectrometer at the Department of Earth Resource Engineering, Kyushu University, Japan. Prior to analysis, loss on ignition (LOI) values were measured. For trace element compositions, powdered samples were subjected to four-acid digestion technique then analyzed using an Inductively Coupled Plasma Optical Emission Spectrometer (Agilent ICP-OES 5100) and Inductively Coupled Plasma Mass Spectrometer (Agilent 7700x ICPMS) at the Department of Earth and Environmental Sciences, National Chung Cheng University, Taiwan.

Zircons separated from the igneous rocks were subjected to U-Pb isotopic dating using a laser ablation inductively coupled plasma mass spectrometer (LA-ICPMS) at the National Chung Cheng University in Taiwan. The laser ablation was set at 70 seconds with less than 1000 counts per second. Calibration was made prior to analysis using a GJ-1 standard. Isotopic ratios were calculated using the GLITTER 4.4.2 (GEMOC) software. Probability curves, density plots, and weighted mean ages were prepared using Isoplot v. 3.0.

4. Results and Discussion

Three major igneous units were identified in Suyoc: the gabbro-diabase, tonalite-granodiorite, and hornblende quartz diorite. The gabbro-diabase is part of the Cretaceous Lepanto Metavolcanics, the basement unit in the area. It exhibits low-grade deformation and alteration. The gabbro-diabase is composed of allotriomorphic-granular plagioclase and actinolite with minor magnetite and pyrite grains. The tonalite-granodiorite is considered part of the Bagon Intrusive Complex. It consists of varying amounts of plagioclase, quartz, potassium feldspar, biotite, and hornblende with minor magnetite. Zircon grains from this unit yielded an age of 37.20 ± 4.70 Ma (Late Eocene). Meanwhile, the hornblende quartz diorite is dated as Middle Pliocene (3.18 ± 0.77 Ma). This is the first time that a Middle Pliocene intrusive is recognized in Mankayan. This age can be correlated to the Black Mountain Diorite in the Baguio Mineral District. The hornblende quartz diorite dominantly consists of plagioclase, hornblende, and quartz ± magnetite.

The gabbro-diabase plot in the basalt to basaltic andesite fields in the total alkali vs silica diagram (Figure 2A) consisting of 48-55 wt% SiO_2. The granodiorite-tonalite are classified as dacites and rhyolites with 72-76 wt% SiO_2 while the hornblende quartz diorite plot in the andesite field with 57-63 wt% SiO_2. The AFM diagram [12] indicates that the gabbro-diabase exhibit tholeiitic signatures (Figure 2b). In contrast, both the granodiorite-tonalite and hornblende quartz diorite show a calc-alkaline fractionation trend. Furthermore, the two younger intrusive units reveal negative correlation of FeO and TiO_2 in Harker diagrams. Trace element compositions normalized to primitive mantle values indicate that all the intrusive rocks indicate strong subduction signatures with enriched large ion lithophile elements and strong negative Nb and Zr anomalies. When plotted in the chondrite-normalized REE diagram, the tonalite-granodiorite and hornblende quartz diorite rocks have elevated LREE with the hornblende quartz diorite exhibiting a negative Eu anomaly. In contrast, the gabbro-diabase exhibits positive Ti anomaly and depleted LREE values.

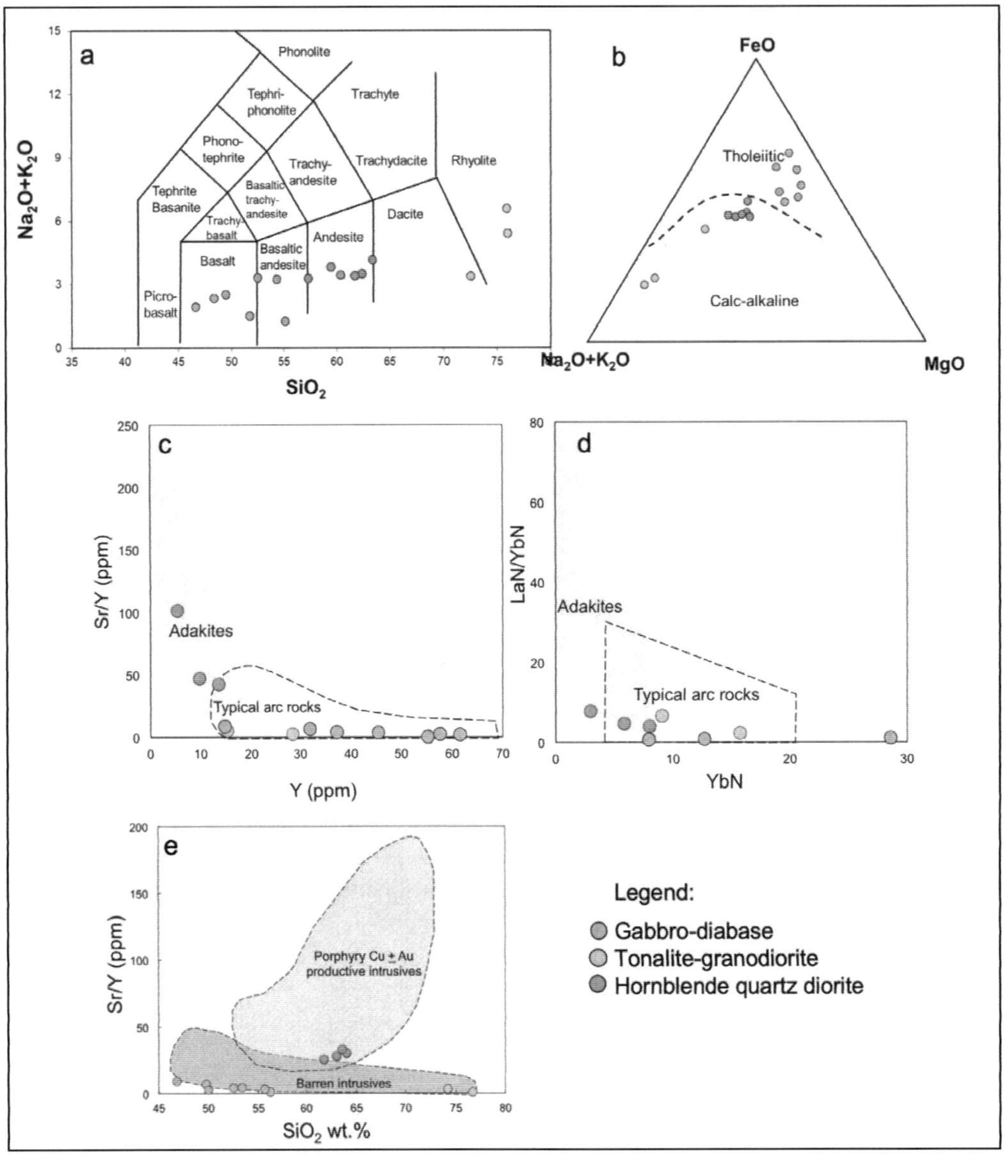

Figure 2. a) Total alkali vs. silica discrimination diagram showing the compositions of the gabbro diabase, tonalite-granodiorite, and hornblende quartz diorite. b) AFM diagram [12] showing a tholeiitic character for the gabbro-diabase and calc-alkaline signatures for the tonalite-granodiorite and hornblende quartz diorite. c-d) Adakite discrimination diagrams [13] revealing typical arc rocks signature for the gabbro-diabase and tonalite-granodiorite but an adakitic character for the hornblende quartz diorite. e) Diagram [14] classifying the hornblende quartz diorite as a fertile or productive intrusive while the two older units are considered as barren.

When plotted in discrimination diagrams for adakites [13], the older gabbro-diabase and tonalite-granodiorite units plot in the field for typical arc rocks while the hornblende quartz diorite are classified as adakites (Figure 2c-d). In addition, the elevated Sr/Y values of the hornblende quartz diorite when plotted against SiO_2 [14] classify it as a fertile intrusive in contrast to the two older intrusive rocks that plot in the barren intrusive field.

The gabbro-diabase unit of the Lepanto Metavolcanics represents the basement rocks in the Mankayan Mineral District. Its tholeiitic character points to a mid-oceanic ridge basalt (MORB) or back-arc setting for Mankayan in the Cretaceous. Meanwhile, the calc-alkaline tonalite-granodiorite of the Bagon Intrusive Complex represents a subduction-related magmatic event in the Late Eocene. The recognition of a Late Eocene intrusive in Mankayan supports tectonic models indicating Eocene subduction along the proto-East Luzon Trench beneath Luzon, which later experienced arc reversal that resulted in subduction at the Manila Trench [15]. The calc-alkaline hornblende quartz diorite represents an adakitic pulse of magmatism in the Pliocene. This magmatic event is associated with subduction of the Scarborough ridge along the Manila Trench. It represents the fertile magmatism associated with the ubiquitous hydrothermal mineralization in the Mankayan Mineral District.

5. Summary and Conclusions

The results of the study are summarized in Table 1. The tholeiitic Cretaceous Lepanto Metavolcanics gabbro-diabase unit was possibly generated in a MORB or back-arc setting. On the other hand, the calc-alkaline signatures of the Late Eocene Bagon Intrusive Complex tonalite-granodiorite could represent subduction along the proto-East Luzon Trough. Meanwhile, the Pliocene hornblende quartz diorite correlate with the Black Mountain Diorite could represent subduction in the Manila Trench that is affected by the Scarborough Seamount Chain. Furthermore, the results indicate that hydrothermal mineralization is only attributed to the hornblende quartz diorite, which coincides with the Pliocene mineralization recognized in the Mankayan Mineral District.

Table 1. Summary of the characteristics of the intrusive units in Suyoc.

	Gabbro-diabase	**Tonalite-granodiorite**	**Hornblende quartz diorite**
Age	Cretaceous	Late Eocene	Pliocene
Geochemical affinity	Tholeiitic	Calc-alkaline	Calc-alkaline; adakitic
Tectonic Setting	MORB/ back-arc?	Island arc	Island arc
Potential for mineralization	✗	✗	✓

References

[1] Richards J 2003 Tectono-magmatic precursors for porphyry Cu-(Mo-Au) deposit formation *Econ Geol* **98** 1515–1533.

[2] Wilkinson J 2013 Triggers for the formation of porphyry ore deposits in magmatic arcs *Nat Geosci* **6** 917–925.

[3] Loucks R 2014 Distinctive composition of copper-ore-forming arc magmas *Aust J Earth Sci* **61** 5–16.

[4] Chang Z, Hedenquist J, White N, Cooke D, Roach M, Deyell C, Garcia J, Gemmell J, McKnight S, Cuison A 2011. Exploration tools for linked porphyry and epithermal deposits: Example from the Mankayan intrusion-centered Cu-Au district, Luzon, Philippines *Econ Geol* **106** 1365−1398.

[5] Soberano O, Gabo-Ratio J, Queano K, Dimalanta C, Yumul G, Andal E, Yonezu K, Boyce A 2021 Mineral chemistry, fluid inclusion and stable isotope studies of the Suyoc epithermal veins: Insights to Au-Cu mineralization in southern Mankayan Mineral District, Philippines *Ore Geol Rev* **131** 104035.

[6] Balce G, Encina Y, Momongan A, Lara E 1980 Geology of the Baguio District and its implication on the tectonic development of the Luzon Central Cordillera, Philippines *Geol Paleontol SE Asia* **22** 265–287.
[7] Stephan J, Blanchet R, Rangin C, Pelletier B, Letouzey J, Muller C 1986 Geodynamic evolution of the Taiwan-Luzon Mindoro belt since the late Eocene *Tectonophysics* **125** 245–268.
[8] Manalo P, Imai A, Subang L, de los Santos M, Yanagi K, Takahashi R, Blamey N 2018 Mineralization of the Northwest quartz-pyrite-gold veins: Implications for multiple mineralization events at Lepanto, Mankayan Mineral District, Northern Luzon, Philippines *Econ Geol* **113** 1609–1626.
[9] Jabagat K, Gabo-Ratio J, Queano K, Yonezu K, Dimalanta C, Lee Y, Yumul G 2020 Petrogenetic constraints on magma fertility in the Baguio Mineral District, Philippines: Probing the mineralization potential of the igneous host rocks in the Sangilo epithermal deposit *Ore Geol Rev* **125** 103703.
[10] Wolfe J 1981 Philippine geochronology *J Geol Soc Phil* **35** 1-30.
[11] Peña R, Reyes M 1970 Sedimentological study of a section of the "Upper Zigzag" formation along Bued River, Tuba, Benguet *J Geol Soc Phil* **24** 1-19.
[12] Irvine T, Baragar W 1971 A guide to the chemical classification of the common volcanic rocks. *Can J Earth Sci* **8** 523–548.
[13] Defant M, Drummond M 1990 Derivation of some modern arc magmas by melting of young subducted lithosphere *Nature* **347** 662–665.
[14] Loucks R 2014 Distinctive composition of copper-ore-forming arc magmas Aust. *J Earth Sci* **61** 5–16.
[15] Yumul G, Dimalanta C, Tam T, Ramos E 2008 Baguio Mineral District: An oceanic arc witness to the geological evolution of northern Luzon, Philippines *Isl Arc* **17** 432–442.

Acknowledgements
The authors would like to acknowledge the Itogon Suyoc Resources Inc. for the logistical support during the fieldwork component of the study. Members of the Rushurgent Working Group from UP NIGS are also thanked for their assistance and guidance.

Band processing of Landsat 8-OLI multi-spectral images as a tool for delineating alteration zones associated with porphyry prospects: A case from Suyoc, Benguet, Philippines

N.E. Parcutela[1], C.B. Dimalanta[1*], L.T. Armada[1], R.S. Austria[1], J.A. Gabo-Ratio[1], and G.P. Yumul Jr. [2]

1 Rushurgent Working Group, National Institute of Geological Sciences, University of the Philippines, Diliman, Quezon City, Philippines
2 School of Environmental Science and Management, University of the Philippines, Los Baños, Laguna, Philippines
*Corresponding author's e-mail: cbdimalanta@up.edu.ph

Abstract. The Mankayan Mineral District in Luzon is marked by several porphyry and epithermal mineralizations. The possibility of blind porphyry mineralization in Mangga, Basig, and Palidan was forwarded based on the results of recent ground magnetic surveys in Suyoc, Mankayan, Benguet. These areas exhibited the diagnostic magnetic anomaly patterns expected of porphyry copper deposits. This signature, a magnetic core surrounded by a non-magnetic rim, is due to the alteration haloes common to porphyry systems. The magnetic core corresponds to the potassic zone dominated by iron-oxide-bearing minerals (e.g., magnetite). Meanwhile, the abundance of non-magnetic clay minerals at the periphery of the mineralized region corresponds to propylitic-argillic alteration zones. Delineation of alteration zones using remote sensing was done to further support the suggested mineralization prospects. Mixture tuned matched filtering (MTMF) was carried out to reduce the effects of vegetation cover from Landsat 8 data. Principal Components Analysis (PCA) was applied to generate bands exploiting the greatest difference between the clay and magnetite alterations with background earth materials. A classified composite map depicts the altered regions. The three sites previously identified from magnetic anomalies also display the expected alterations. Another prospect in the Lapangan area is being proposed based on magnetic anomalies and mapped alterations.

1. Introduction

The use of remote sensing techniques for mineral exploration is focused on detecting rock alterations that can be used as markers for mineralization prospects [1, 2]. Spectral curves of alteration minerals are distinct in comparison with unaltered rocks [3, 4]. Iron oxide-bearing minerals and hydroxyl-bearing minerals such as clay minerals and micas have diagnostic absorption features in the visible, near-infrared, and short wavelength infrared bands of Landsat 8 multi-spectral images [5]. This enables remote sensing techniques to aid in the identification of mineralization sites.

Through examination of the spatial distributions and genetic relationships between remotely sensed alteration minerals, favorable metallogenic regions may be delineated. Because of their very distinct alteration haloes, alteration mapping through remote sensing technologies has been widely applied for porphyry mineralization [6, 7]. The potassic zone, which almost immediately lies on top of the mineralization, is dominated by magnetic iron-oxide-bearing minerals (e.g., magnetite). On the other hand, the periphery of the mineralized region affected by argillic alteration is mainly composed of less magnetic clay alteration minerals [8].

2. Geology and Mineralization

The Mankayan Mineral District (MMD) in Luzon Cordillera is known for several epithermal and porphyry copper mineralization (Figure 1). Porphyry copper and epithermal mineralization in thedistrict are hosted in the Lepanto Metavolcanics, Balili Formation, and the Imbanguila Dacite. Mineralization is mainly influenced by the NW- and NE-trending structures in the district [9, 10]. Suyoc is located in the southern part of the MMD. This is underlain by the Lepanto Metavolcanics, Balili Formation, Suyoc Conglomerate, Imbanguila Dacite and Lapangan Tuff (Figure 1). A recentground magnetic survey in Suyoc, Mankayan, Benguet revealed three prospects for blind porphyry mineralization. These are located in Mangga, Basig, and Palidan. These prospects were identified based on magnetic anomaly patterns consistent with porphyry deposits.

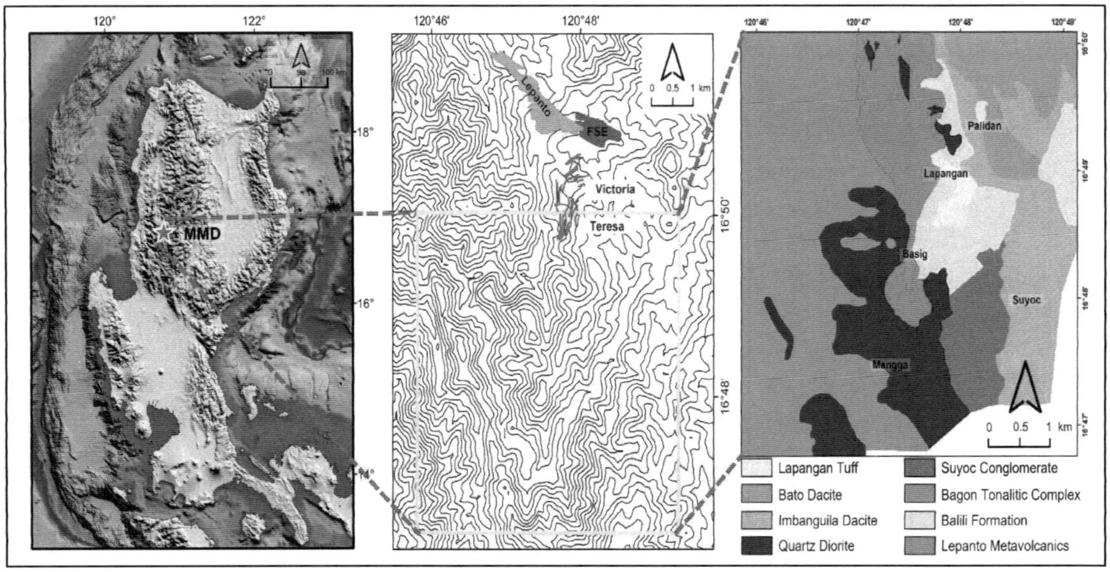

Figure 1. (Left) Location of the Mankayan Mineral District (MMD) (red star) in Luzon Island. **(Center)** Index map showing the location of the study area (yellow box) in the MMD relative to the well-known Lepanto and Far Southeast (FSE) deposits and the Victoria and Teresa veins. **(Right)** Geologic map of Suyoc adopted from [11].

3. Data and Methods

3.1. Data

Spectral signatures related to mineral alterations were identified using multi-spectral images captured by Landsat 8-OLI. Daytime scenes with an acquisition date of 30 August 2021 with minimum cloud cover were selected. Individual bands were downloaded as GeoTIFF files using the earth explorer website of the United States Geological Survey (USGS).

3.2. Methodology

Common pre-processing algorithms were applied to the dataset. The dynamic numbers of the original raster file downloaded were converted to reflectance values using the FLAASH algorithm hosted in ENVI 5.3 software. After which, orthographic correction was applied to correct any geometric distortion of the downloaded scenes. Reduction of topographic effects was also carried out using a digital elevation model (DEM) of the study area derived from the shuttle radar topography mission (SRTM) [12]. The application of remote sensing techniques in tropical regions is limited by dense vegetation cover. We identified spectral signatures associated with vegetations through mixture tuned matched filtering (MTMF) technique. These were then subtracted from the total pixel spectrum of the downloaded scenes, resulting in images with reduced contributions from vegetation signatures. We selected several bands that can effectively be used to differentiate the spectral signatures of iron oxide and clay alterations. Landsat 8-OLI's bands 2, 4, 5, 6 can help delineate iron oxide alterations whereas bands 2, 5, 6, and 7 are useful in detecting clay alterations [7, 13]. Principal Components Analysis (PCA) was applied to further highlight the spectral signatures of the alteration minerals. The PCA bands were stacked to generate a classified image from the false color composite to map the altered areas in the region. Figure 2 shows the summary of the pre-processing and processing techniques that were employed in this study.

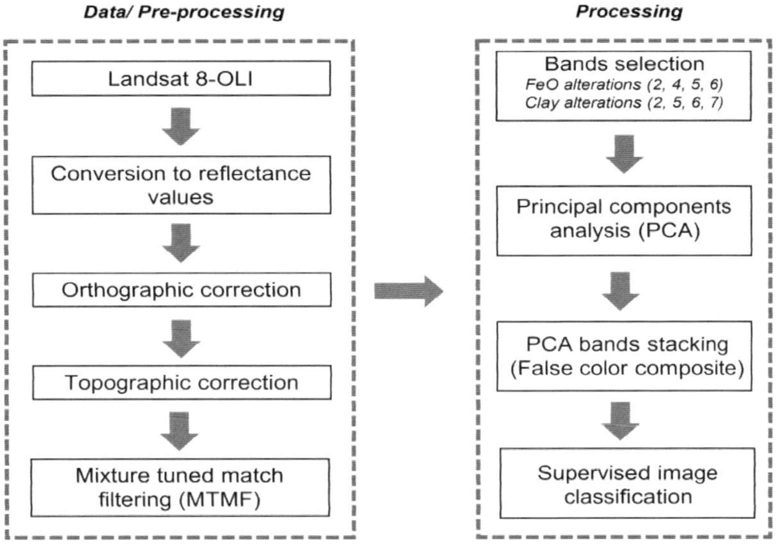

Figure 2. Summary of the pre-processing and processing techniques applied for delineation of alteration zones associated with porphyry prospects using Landsat 8-OLI data. An additional pre-processing layer, the MTMF, was applied to suppress the influence of the spectral signatures coming from vegetation cover in the region.

4. Results and Discussion

The selected PCA bands for each of the band combinations highlighting iron oxide alterations (Figure 3A) and clay mineral alterations (Figure 3B) are presented. For both grayscale images, areas with the respective alterations are shown as brighter white pixels. Gray pixels correspond to vegetation cover in the area. Darker regions correspond to bare ground regions not displaying the specific alteration. Iron oxide alterations are more dominant compared to clay mineral alterations. There are also regions where both iron oxide and clay mineral alterations are observed. The association of both mineral alterations in one area is a possible indication of the occurrence of alteration mineral haloes which could be used as markers for prospecting porphyry targets.

A classified image of the stacked PCA bands more clearly shows the configuration of these alterations. Areas with both iron oxide and clay alterations are depicted by the yellow pixels (Figure 4A). These were compared to the reduced-to-equator (RTE) magnetic anomaly map of the area (Figure 4B). In the RTE anomaly map, magnetic areas are depicted by RTE lows (blue-green) whereas non-magnetic regions are depicted by RTE highs (red-pink). Porphyry copper systems have distinctive magnetic anomalies due to the associated alteration haloes - a magnetic core (potassic zone) and non-magnetic rim (phyllic, argillic or propylitic zones). From the observed anomalies, [14] identified three possible prospects for porphyry mineralization (white-dashed circles). These areas also show the occurrence of iron oxide and clay mineral alterations. An additional prospect isproposed in the Lapangan area (black-dashed circle). This region is characterized by magnetic anomalies associated with porphyry systems while displaying the signatures of iron oxide and clay alterations delineated from the remote sensing technique. Furthermore, the Lapangan area has been identified to be composed of copper disseminations hosted in a heavily fractured and argillized pebble breccia body [14]. For the previously identified regions and the newly identified Lapangan area, we also considered the occurrence of field manifestations of porphyry mineralization. There are also other regions that roughly show the distinctive magnetic anomaly patterns akin to porphyry copper deposits such as those shown in Esperanza, northeast of Pongo, and southeast of Cop-copit (Figure 4B). However, these regions were not considered as prospect regions due to several reasons. In the case of Esperanza, there is no magnetic data east of the region (Figure 4B). Because of this, there is a possibility that the configuration of this anomaly might change once the data gap is filled as magnetic anomaly values are interpolated values from known station readings. Furthermore, mineralization within the Esperanza area is manifested as a vein-type mineralization in quartz-calcite veins with prominent clay alterations proximal to the mineralized veins. In the case of the regions northeast of Pongo and southeast of Cop-copit, the observed magnetic anomalies roughly conform with the magnetic signature of porphyry copper deposits. However, the abundance of vegetation cover (green pixels) and bare ground regions (brown pixels), and the minor distribution of alteration minerals (yellow pixels) (Figure 4A) were deemed not sufficient to consider these regions as prospect areas. Furthermore, field data in these regions did not suggest the occurrence of porphyry mineralization.

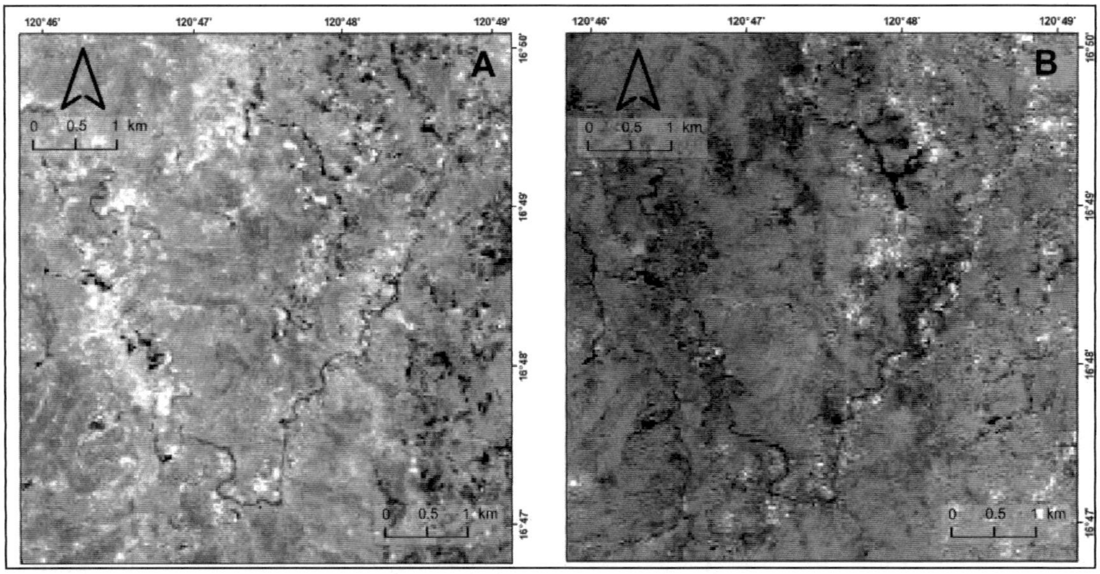

Figure 3. (A) PCA band selected for the band combination for iron oxide alterations. **(B)** PCA band selected for band combination for clay mineral alterations. For both figures, white or lighter pixels correspond to alteration minerals (iron oxide alterations for 3A and clay mineral alterations for 3B); gray pixels correspond to vegetation cover; and black or dark pixels correspond to bare ground in the region.

The selected PCA bands for each of the band combinations highlighting iron oxide alterations (Figure 3A) and clay mineral alterations (Figure 3B) are presented. For both grayscale images, areas with the respective alterations are shown as brighter white pixels. Gray pixels correspond to vegetation cover in the area. Darker regions correspond to bare ground regions not displaying the specific alteration. Iron oxide alterations are more dominant compared to clay mineral alterations. There are also regions where both iron oxide and clay mineral alterations are observed. The association of both mineral alterations in one area is a possible indication of the occurrence of alteration mineral haloes which could be used as markers for prospecting porphyry targets.

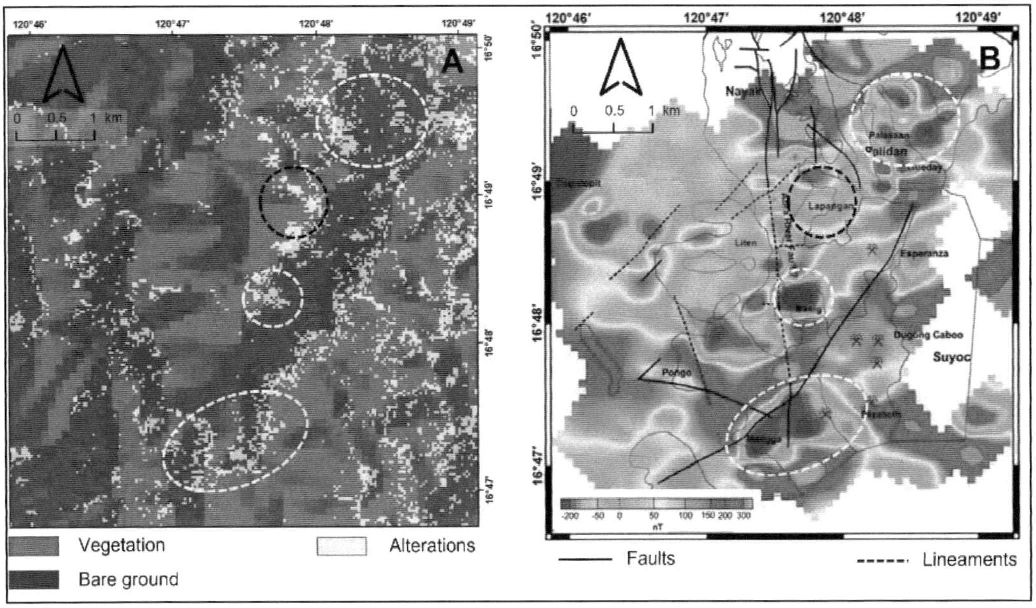

Figure 4. (A) Classified image depicting the mapped alterations in yellow pixels. **(B)** Reduced-to-equator (RTE) magnetic anomaly map adopted from [14]. White-dashed circles show the locations of porphyry prospects previously delineated using magnetic anomalies. Black-dashed circle marks the location of an additional porphyry prospect being proposed based on magnetic anomalies and mapped alterations.

5. Conclusions

Application of remote sensing techniques for geologic applications in tropical regions is heavily hindered by dominant vegetation cover. This limitation is amplified in mineral alteration mapping since this involves generally smaller areas and exploiting more minute differences in spectral signature from background earth materials. Relatively newer pre-processing techniques such as the MTMF can effectively be used to suppress signals coming from vegetations that can dilute other spectral signals in the region. Through the results of our study, we showed that the combined results from alteration mapping using Landsat 8 data and the available magnetic anomaly data provide better constraints on the identification of possible porphyry copper targets in a mineralized region.

References

[1] Sabins F 1999 Remote sensing for mineral exploration *Ore Geol Rev* **14** 157-183.

[2] Di Tomasso I, Rubinstein N 2007 Hydrothermal alteration mapping using ASTER data in the Infiernillo porphyry deposit, *Argentina Ore Geol Rev* **32** 275-290.

[3] Hubbard B, Crowley J 2005 Mineral mapping on the Chilean-Bolivian Altiplano using co-orbital ALI, ASTER, and Hyperion imagery: Data dimensionality issues and solutions *Remote Sens Environ* **99** 173-186.

[4] Ducart D, Crosta A, Filio C 2006 Alteration mineralogy at the Cerro La Mina epithermal prospect, Patagonia, Argentina: field mapping, short-wave infrared spectroscopy, and ASTER images *Econ Geo* **101** 981-996.

[5] Zahra M, Mohammad B, Sasan B, 2015 Digital image processing and analysis techniques for detection of hydrothermal alteration zones: A case study in Siah-Jangal Area, north of Taftan Volcano, Southeastern Iran *J Indian Soc Remote Sens* **43 (2)** 363-377.

[6] Safari M, Maghsoudi A, Pour A 2017 Application of Landsat-8 and ASTER satellite remote sensing satellite data for porphyry copper exploration: a case study from Shahr-e-Babakm Kerman, south of Iran *Geocarto Int* **33 (11)** 1186-1201.

[7] Pak U, Pak C 2021 Detection of hydrothermal zones using Landsat 8 OLI image: A case study of gold prospecting in Nyongwon Area, DPR Korea *J Indian Soc Remote Sens* **49 (9)** 2249-2259.

[8] Sillitoe R 2010 Porphyry Copper Systems *Econ Geol* **105** 3-41.

[9] Sajona F, Izawa E, Motomura Y, Imai A, Sakakibara H, Watanabe K 2002 Victoria carbonate-base metal gold deposit and its significance in the Mankayan mineral district, Luzon, Philippines *Resour Geol* **52 (4)** 315-328.

[10] Manalo P, Imai A, Subang L, de los Santos M, Yanagi K, Takahashi R, Blamey N 2018 Mineralization of the northwest quartz-pyrite-gold veins: implication for multiple mineralization events at Lepanto Mankayan, mineral district, northern Luzon, Philippines *Econ Geo* **113 (7)** 1609-1626.

[11] Soberano O, Gabo-Ratio J, Queaño K, Dimalanta C, Yumul Jr G, Andal E, Yonezu K, Boyce A 2021 Mineral chemistry, fluid inclusion and stable isotope studies of the Suyoc epithermal veins: insights to Au-Cu mineralization in southern Mankayan Mineral District, Philippines *Ore Geol Rev* **131**.

[12] Farr T, Kobrick M 2000 Shuttle Radar Topography Mission (SRTM) produces a wealth of data *Eos Trans AGU* **81** 583-593.

[13] Aydal E, Ardal E, Dumanlilar O 2007 Application of the Crosta technique for alterationmapping of granitoidic rocks using ETM+ data: Case study from eastern Tauride belt (SE Turkey) *Int J Remote Sens* **28 (17)** 3895-3913.

[14] Arellano C, Armada L, Dimalanta C, Queaño K, Andal E, Yumul Jr G 2021 Interpretation of ground magnetic data in Suyoc, Mankayan Mineral District, Philippines *Resour Geol* **71 (4)** 363-376.

Screening criteria of underground coal gasification (UCG): a case study from Mangunjaya Area, South Sumatra Basin, Indonesia

F Anggara[1,2*], M Riasetiawan[3], F R Assamarqandi[4], B F Sartika[4], I J Rizaldi[4], V Syahra[5], M A Ibrahim[6], and S B Rahmat[6]

[1]Department of Geological Engineering, Universitas Gadjah Mada, Yogyakarta 55281, Indonesia
[2]Unconventional Geo-resources Research Group (UGRG), Faculty of Engineering, Universitas Gadjah Mada, Yogyakarta 55281, Indonesia
[3]Department of Computer Science and Electronics, Faculty of Mathematics and Natural Sciences, Universitas Gadjah Mada, Yogyakarta 55281, Indonesia
[4]Undergraduate Program, Department of Geological Engineering, Universitas Gadjah Mada, Yogyakarta 55281, Indonesia
[5]Cloud & Grid Technology Working Group, Universitas Gadjah Mada, Yogyakarta 55281, Indonesia
[6]Center for Mineral, Coal, and Geothermal Resources, Geological Agency; Bandung 40254, Indonesia
*Corresponding author's email: ferian@ugm.ac.id

Abstract. Underground coal gasification (UCG) is a technology that can increase the value of coal. This technology turns coal into gas below the surface through the gasification process. UCG offers many applications for power generation, heating, and the chemical industry. The Mangunjaya area is one prospecting area, located in the South Sumatra Basin, which has been actively producing coal for a long time. This study focused on determining the potential location of UCG in Mangunjaya. Screening criteria were extracted from previous studies including coal thickness, depth of the coal seam, depth and thickness ratio, coal rank, the dip of a coal seam, roof and floor rock type and thickness, and geological structure. Thirteen boreholes were examined to determine their potential, namely MJ1 to MJ13. The evaluation results show that the Mangus coal seam at a depth of 260,4 meters in the MJ2 borehole and the surrounding area in 15 km^2, matches the criteria and has the most potential coal for UCG development.

1. Introduction

Coal is one of the most abundant geological resources in Indonesia. According to the coal resource balance in 2020, there are 143,73090 billion tons of total coal resources and 38,80548 billion tons of total surface coal reserves [1]. Meanwhile, data shows 43,533 billion tons of resources and 173,51 million tons of total reserves are coal below the surface [1]. Coal resources are classified as high and low ranks based on their calorific value. Low-rank coal resource is not well developed and has an enormous potential to use [2]. Subsurface low-rank coal that cannot be mined by open-pit mining can be made to increase the value of coal in the form of Underground Coal Gasification (UCG).

UCG is a method of converting subsurface coal to produce gas through the gasification process for the energy source. Two wells must be drilled to accommodate UCG, the first well is used to pump water or air, and the other is used to drain gas [3].The higher pressure of gas is an advantage, UCG enhance the storage capacity of the coal seam to store CO_2 [4][5]. This technology offers many other advantages, including increased worker safety, no surface disposal of ash and coal tailings, low dust and noise pollution, low water use, greater coal exploitation, savings in gas transportation and storage costs, and low methane gas emissions into the atmosphere [4][5]. The gas produced through the UCG process is called syngas, composed of H_2, CO, CO_2, CH_4, and H_2S. These gases will be further processed depending on their use for the chemical industry, heating in industrial activities, power generation, hydrogen production, synthetic natural gas, or other fuels [5]. Numbers of countries are developing this technology, such as the USA, Australia, United Kingdom, Spain, New Zealand, South Africa, Turkey, China, Pakistan, Vietnam, and Indonesia [3]. Indonesia has several advantages for UCG, including its significant coal resources which are mostly low-rank coal (lignite to sub-bituminous) [6]. Also, abundant deep-seated coal (depth 100-500 m below surface) resources up to 40 billion tones spread over South Sumatra Basin, Kutai Basin, Tarakan Basin, Barito Basin, and Pasir Basin [7]. One of the areas targeted by the UCG project is the Mangunjaya area in South Sumatra Province.

Not all coal can be utilized as UCG. This technology effectively utilizes coal without mining and requires careful site selection to ensure the quality and quantity of produced gas during the life of a project [8]. Some criteria suggested by Durucan et al. [8] include geological conditions for site selection and subsurface engineering, panel design, reagent use, and process engineering aspects. This paper evaluates various coal criteria based on the geological conditions for UCG site selection in the Mangunjaya area, using surface and sub-surface data conducted by the Center of Mineral, Coal, and Geothermal Resources Geological Agency of Indonesia (PSDMBP Badan Geologi).

2. Methodology

The method used in this paper is a screening criteria for UCG referring to existing publications. The data used in this paper are surface and subsurface data obtained from 13 boreholes consisting of MJ1 - MJ13 (Figure 1). The data is processed based on the criteria made to obtain the potential location and coal seam for UCG.

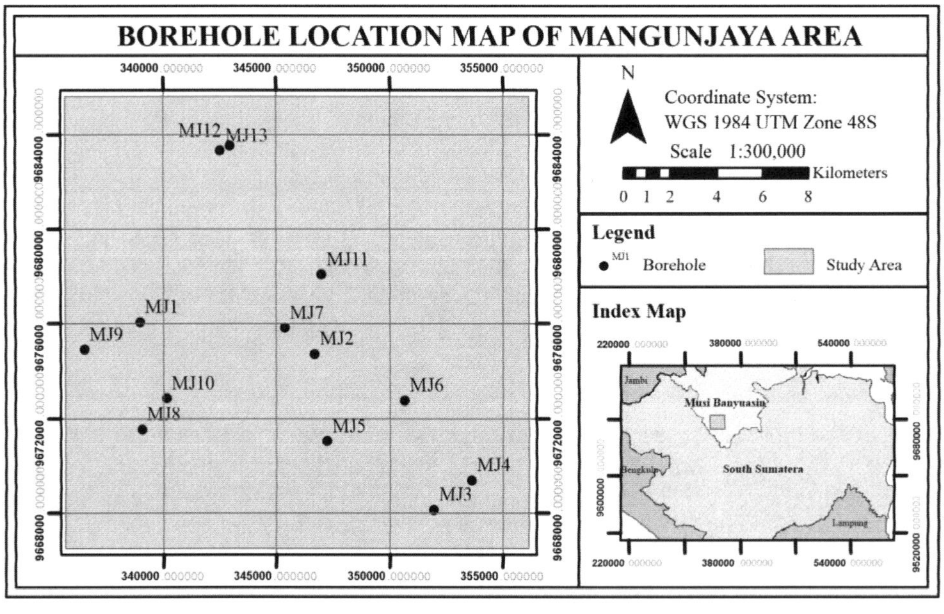

Figure 1. Borehole location map of Mangunjaya Area (base map referred to [9]).

3. Regional geology

Mangunjaya Area is located in the southern Musi Banyuasin Regency, South Sumatra Province (Figure 1), and is geologically situated in the South Sumatra Basin. This basin is a back-arc basin with the boundaries of Tigapuluh Mountains in the north, Lampung High in the south, the Bukit Barisan Mountains in the west, and the Sunda Shelf in the east. Pulunggono et al. [10] stated that the South Sumatra Basin was formed by tectonic activity that occurred between the Sunda Shelf and the Indian Ocean Plate during the Tertiary Period. The regional geological structure of the South Sumatra Basin consists of grabens, normal faults trending north-south and northwest-southeast, anticlines, and synclines trending northwest-southeast [11].

According to Darman and Sidi [12], the stratigraphic order of South Sumatera Basin are the metamorphic and igneous rock of pre-tertiary age as bedrock, Lahat Formation (Late Paleocene - Late Oligocene), Talang Akar Formation (Late Oligocene – Early Miocene), Baturaja Formation and (Early Miocene), Gumai Formation (Early Miocene – Middle Miocene), Air Benakat Formation and Muara Enim Formation (Middle Miocene – Late Miocene). Muara Enim Formation is a coal-bearing formation in the study area [13]. This formation consists of claystone, siltstone, sandstone, and coal with a shallow marine – non-marine depositional environment. There are several coal seam group in the Muara Enim Formation namely: Kladi, Merapi, Petai, Suban, Mangus, Burung, Benuang, Kebon, Babat or Benakat, Lematang, and Niru [14]. Benakat, Kebon, Benuang, Mangus, Burung, and Suban seams can be found in the study area with coal rank ranging from lignite B – A and sub-bituminous.

4. Result and discussion

4.1. Criteria for UCG

This study evaluates subsurface data from several boreholes from exploration drilling conducted by PSDMBP. Screening criteria for UCG based on published papers summarized in Table 1 and some detail descriptions are as follow:

4.1.1. Coal thickness. There is no fixed maximum and minimum thickness of a coal seam that can be utilized as UCG since the thickness of the coal varies in each region. As in Indiana, coal seams suitable for UCG are more than 2 meters in thickness. Those coal seams in this area tend to be thin, less than 1.5 m [15]. On the other hand, in Powder River Basin, the coal seams can reach 10 - 30 m in thickness. According to GasTech (2007) in [16], the ideal coal seam thickness utilized for UCG is greater than 10 m, whereas the ideal thickness of lignite coal is in the range of 2 - 4 meters [3].

4.1.2. Depth of coal seam. The minimum depth of the coal seam varies depending on the condition of the basin of the coal seam. In the USA, the average depth of coal used for UCG ranges from 50 - 300 meters, while in Europe, UCG production can reach a depth of 1300 meters [17]. Referring to Burton et al. [17], the minimum depth for UCG is 12 meters, but it is more recommended if the depth is >150 m, especially for lignite rank coals [3].

4.1.3. Coal rank. UCG will be more effective on low-rank coal because it tends to shrink when heated and increases the permeability and connectivity between the two boreholes [17]. Thus, the coal that is effective for UCG is lignite to sub-bituminous (calorific value <7000 Kal/gr) [18]. Coal with a higher rank requires further coking tests. Coking coal have tendency to block gas flow through the coal bed, thus hindering the gasification process [5].

4.1.4. Depth and thickness ratio. A depth thickness ratio of coal is needed to avoid subsidence due to the loss of part of the coal seam due to the gasification process [19]. The minimum ratio value is inversely proportional to the depth. Dwitama et al. [19] suggested that the depth and thickness ratios are 22, 18, and 15. The minimum ratio of 22 for coal seam with a minimum thickness of 5 m and the depth range of 120 – 200 m. The minimum ratio of 18 for coal seam with minimum thickness of 11 m and the depth range of 300 – 400 m. There are two conditions for a minimum ratio of 15; firstly, for the depth range of 300 – 400 m, the coal seam must have minimum thickness of 20 m, and if the depth exceeds 400 m, coal seam must have 27 m minimum in thickness.

4.1.5. Dip of the coal seam. Gas tech (2007) in [16] recommends coal seams with a slope of 0-20°. Many sites implement UCG on gentle coal slopes [20]. The slope will facilitate the drainage, maintain the hydrostatic stability during the gasification process, and minimize the damage to the down dip production well from material that is moved in association with the UCG process [16]. However, this slope value is not absolute. The UCG project in Russia has succeeded in producing the UCG with a 65° dip slope. There are also several successful projects with steeping dip [19][17].

4.1.6. Roof and floor rock type and thickness. Roof and floor rock prevent contamination from the gasification process to the groundwater. The impermeable rock strongly influences the hydrology and gasification process in all parts of the layer above and below it. The subsurface temperature and the pressure can make the rock around the coal seam crack and cause gas leaks. Roof and floor rocks are impermeable rocks with a suggested thickness of 10-20 m [3]. Rocks with low permeability are claystone, siltstone, and tuff, alteration of impermeable rocks with sandstone may be included. Another suggestion given by [19] considered thickness of the roof rock is twice the thickness of the coal and for the floor is equal to coal thickness.

4.1.7. Geological structure. The recommended location for UCG exploration should have a simple geological structure where there are no major faults that could affect the coal seam directly [3]. It would be better if the structure of the coal were in the form of syncline folds [19].

Table 1. Screening criteria for UCG.

No	Criteria	Parameters
1	Coal rank	Lignite – sub-bituminous, calorific value <7000 Kal/gr, if calorific value >7000 Kal/gr coking analysis must be carried out [18]
2	Coal thickness	2 – 4 meters [3]
3	Coal depth	>150 meters [3]
4	Depth/thickness ratio	depth 120-200 meters ratio 22, depth 200-300 meters ratio 18, depth 300-400 meters ratio 15, depth >400 ratio 15 [19]
5	Dip	<20° preferred, maximum dip no exceed than 65° [16] [19]
6	Roof and floor rock type	Impermeable rock (siltstone, claystone, and tuff) [3] [19]
7	Roof and floor thickness	Roof twice as coal thickness and floor equivalent to coal thickness [19]
8	Geological structure	Simple geological structure feature, no major fault effects coal seam. Syncline is preferred [3] [19]

4.2. Site selection for UCG

After the screening criteria are determined, the data is examined and compared with the parameters summarized in Table 1. The screening results for each borehole are shown in Table 2. Since the data are based on exploration drilling, geological structure and dip data are limited. Based on regional geology, the study area is considered to have a simple geological structure. The evaluation results for each borehole are as follows:

4.2.1. Borehole MJ1. This borehole has a total depth of 416,38 m with 13 coal layers. The coal thicknesses vary from 0,82 to 13,5 meters. The roof and floor rocks are siltstone, claystone, sandstone, and alteration of sandstone and claystone. The coals came from different seams, namely Benakat, Kebon, Benuang, and Mangus. Based on ASTM, this coal is ranked as lignite A and B.

4.2.2. Borehole MJ2. With a total depth of 483,4 m, eight coal seams with thicknesses ranging from 0,8 m to 10,7 m were found from 44,5 m to 282,7 m depth. The coals belong to Suban, Benuang, Burung, and Mangus seams. The roof and floor rocks are siltstone, claystone, and sandstone with various grain sizes. According to ASTM classification, coals are mainly classified as lignite B, and only coal seam five is classified as lignite A.

Table 2. Evaluation of coal layers from each borehole.

Borehole	Coal layer	Criteria										
		Coal rank (ASTM)	Calorific value (Kal/gr)	Coal thickness (m)	Coal depth (m)	Depth/thickness ratio	Roof rock type	Roof thickness (m)	Roof and coal ratio	Floor rock type	Floor thickness (m)	Floor and coal ratio
MJ1	1	Lignite B	3284,70	6,42	73,53	11,45	2	1,91	0,30	1	0,23	0,04
	2	Lignite B	3344,34	2,82	80,18	28,43	1	0,23	0,08	1	3,92	1,39
	3	Lignite B	3344,34	3,78	86,92	22,99	1	3,92	1,04	1	11,75	3,11
	4	Lignite B	3293,88	2,69	141,31	52,53	1	16,06	5,97	1	0,24	0,09
	5	Lignite B	3285,25	13,50	158,00	11,70	1	4,80	0,36	1	14,50	1,07
	6	Lignite B	3310,13	1,12	186,00	166,07	1	14,50	12,95	1	1,38	1,23
	7	Lignite B	3419,53	0,83	197,10	237,47	2	2,60	3,13	1	6,59	7,94
	8	Lignite B	3348,47	1,68	204,52	121,74	1	6,05	3,60		2,07	1,23
	9	Lignite B	3333,90	1,64	212,69	129,69	1	1,86	1,13	1	37,22	22,70
	10	Lignite B	3449,61	6,45	251,55	39,00		2,05	0,32	2	5,64	0,87
	11	Lignite A	3520,72	0,82	315,68	384,98	1	16,18	19,73	2	15,00	18,29
	12	Lignite B	3495,10	1,93	389,00	201,55	1	2,29	1,19	2	0,54	0,28
	13	Lignite A	3508,92	8,93	391,47	43,84	2	0,54	0,06	1	7,15	0,80
MJ2	1	Lignite B	3380,30	0,95	44,50	46,84	1	15,17	15,97	2	33,85	35,63
	2	Lignite B	3456,78	1,88	79,30	42,18	2	33,85	18,01	2	4,42	2,35
	3	Lignite B	3424,11	6,25	93,00	14,88	1	5,80	0,93	1	4,00	0,64
	4	Lignite B	3431,07	5,68	117,65	20,71	1	18,40	3,24	2	7,30	1,29
	5	Lignite A	3584,02	1,38	152,52	110,52	2	0,36	0,26	1	28,55	20,69
	6	Lignite B	3491,35	0,80	228,16	285,20	1	2,01	2,51	1	2,27	2,84
	7	Lignite B	3492,10	10,70	260,40	24,34	1	20,04	1,87	1	10,20	0,95
	8	Lignite A	3518,41	1,40	281,30	200,93	1	8,15	5,82	1	15,25	10,89
MJ3	1	Lignite B	3388,38	1,50	11,60	7,73		1,45	0,97	2	0,15	0,10
	2	Lignite B	3349,34	1,35	13,25	9,81	2	0,15	0,11	2	0,25	0,19
	3	Lignite B	3349,34	2,65	14,85	5,60	2	0,25	0,09		1,10	0,42
	4	Lignite B	3349,34	1,33	18,60	13,98		1,10	0,83		1,07	0,80
MJ4	1	Lignite B	3240,55	2,28	3,44	1,51	2	0,01	0,00	1	0,08	0,04
	2	Lignite B	3359,15	2,30	5,80	2,52	1	0,08	0,03	2	2,70	1,17
	3	Lignite B	3359,15	2,70	8,10	3,00	2	2,30	0,85	2	2,25	0,83
	4	Lignite B	3359,15	2,25	10,80	4,80	2	2,70	1,20	1	4,85	2,16
	5			1,10	17,90	16,27	1	4,85	4,41	1	1,50	1,36
	6			0,85	49,45	58,18	2	20,18	23,74	1	4,00	4,71
	7	Subbituminous B	5625,62	1,00	54,30	54,30	1	4,00	4,00	1	1,20	1,20
MJ5	1			2,00	9,00	4,50	1	5,15	2,58	1	27,80	13,90
	2	Lignite B	3395,56	0,81	71,90	88,77	1	3,30	4,07	2	2,04	2,52
	3	Lignite B	3395,56	2,04	72,71	35,64	2	0,81	0,40	2	5,60	2,75

Table 2. Evaluation of coal layers from each borehole (cont.).

MJ6	1			0,55	14,80	26,91	2	0,40	0,73	2	2,22	4,04
	2	Lignite B	3254,47	11,33	17,57	1,55	2	2,22	0,20	2	1,25	0,11
	3			1,25	28,90	23,12	2	11,33	9,06		4,85	3,88
	4			1,85	35,00	18,92	1	4,85	2,62		12,35	6,68
MJ7	1			1,14	58,46	51,28	1	0,36	0,32	1	0,40	0,35
MJ8	1			1,35	23,20	17,19	1	10,20	7,56	2	1,55	1,15
	2			3,90	26,10	6,69		1,55	0,40	2	2,00	0,51
	3			1,00	32,00	32,00		2,00	2,00		1,00	1,00
	4			1,90	34,00	17,89		1,00	0,53	1	2,60	1,37
MJ9	1			1,46	9,50	6,51	2	0,60	0,41	1	0,04	0,03
	2			1,85	11,00	5,95	1	0,04	0,02	1	0,09	0,05
	3			1,56	12,94	8,29	1	0,09	0,06	2	0,78	0,50
	4			0,78	14,50	18,59	2	1,56	2,00	1	0,69	0,88
	5			4,53	15,97	3,53	1	0,69	0,15	1	12,00	2,65
	6			3,20	49,00	15,31	1	0,50	0,16	2	0,20	0,06
	7			3,60	54,90	15,25	1	2,10	0,58	1	16,00	4,44
MJ10	1	Lignite B	3239,33	1,45	8,35	5,76	1	1,95	1,34	2	4,70	3,24
	2	Lignite B	3263,27	1,65	17,40	10,55	1	2,90	1,76	1	0,17	0,10
	3	Lignite B	3263,27	0,18	19,22	106,78	1	0,17	0,94	1	13,00	72,22
	4			0,27	40,95	151,67	1	7,25	26,85	1	0,08	0,30
	5			0,70	41,30	59,00	1	0,08	0,11	2	0,15	0,21
	6			1,30	42,15	32,42	2	0,15	0,12	2	2,45	1,88
	7			2,45	43,45	17,73	2	1,30	0,53	2	2,60	1,06
	8			2,60	45,90	17,65	2	2,45	0,94	2	5,80	2,23
MJ11	1	Lignite B	3371,92	1,25	31,45	25,16	2	0,20	0,16	1	0,08	0,06
	2	Lignite B	3371,92	1,37	32,78	23,93	1	0,08	0,06	1	2,18	1,59
MJ12	1	Lignite B	3325,73	3,10	41,25	13,31	1	2,75	0,89	1	2,90	0,94
	2	Lignite B	3426,63	5,95	47,25	7,94	1	2,90	0,49	1	0,30	0,05
MJ13	1			1,20	45,60	38,00	1	6,75	5,63	2	0,20	0,17
	2			2,25	47,00	20,89	2	0,20	0,09	2	0,40	0,18
	3			0,20	77,15	385,75	1	2,30	11,50	1	0,03	0,15
	4			1,44	77,38	53,74	1	0,03	0,02	1	0,03	0,02
	5			3,75	78,85	21,03	1	0,03	0,01	1	0,40	0,11

Match
Not match
No data

*rock type = 1: impermeable rock 2: other rock type

4.2.3. Borehole MJ3. This borehole has a total depth of only 69,5 m with four coal seams. Coal with thicknesses ranging from 1,33 m to 2,65 m, was found at depth of 11,6 m to 19,93 m. All coal seams belong to the Mangus seam, with roof and floor rocks being coaly shale. Core loss and coal core loss are found in this borehole. Coal seams are classified as lignite B rank based on the ASTM classification.

4.2.4. Borehole MJ4. This borehole has a total depth of 107,5 m. There are seven coal seams with thicknesses ranging from 0,85 to 2,7 m found from a depth of 2,44 m to 55,3 m. Coal is included in the Benakat and Kebon seam. Based on ASTM classification, the four coal seams found in this borehole mostly are lignite B in rank, and the deepest seam is sub-bituminous B. Roof and floor are classified as siltstone, claystone, sandstone, and coaly shale.

4.2.5. Borehole MJ5. This borehole has a total depth of 80,75 m. There are three coal seams with a thickness of 2; 0,81; 2,04 m found from 9 m to 74,75 m depth. Based on the ASTM classification, the coal seam is lignite B in rank. Roof and floor rocks are claystone and sandstone

4.2.6. Borehole MJ6. There are four coal seams with thicknesses ranging from 0,55 m to 11,33 m found at depth of 14,8 to 36,85 m. The coal belong to Benakat seam and according to the ASTM classification has lignite B in rank. At this borehole, roof and floor rocks are claystone, coaly shale, and alteration of sandstone and mudstone.

4.2.7. Borehole MJ7. This borehole has a total depth of 109 m. There is only one layer of coal with thickness of 1,114 m at a depth of 58,46 – 59,6 m. This coal seam belongs to Benakat seam with roof and floor are claystone and coaly shale.

4.2.8. Borehole MJ8. The borehole MJ8 has a total depth of 76 m. There are four coal seams found in this borehole with a thickness vary from 1 m to 3,9 m. The coal at this location is Mangus seam. Most roof and floor rocks cannot be determined due to core loss.

4.2.9. Borehole MJ9. This borehole has a total depth of 77,5 m. In this layer, seven coal seams originate from Benakat and Kebon seams. The thickness of the coal in this seam varies from 0,78 m to 4,53 m. The majority of the coal seam's impermeable roof and floor rock is claystone.

4.2.10. Borehole MJ10. This borehole contains eight coal seams with varying thicknesses from 0,18 m to 2,6 m. The total depth of this borehole is 59,5 m. Based on ASTM, the top 3 coal seams are coal with lignite B in rank The coal seam of this borehole originates from the Burung seams with roof and floor rocks which are predominantly claystone and sandstone.

4.2.11. Borehole MJ11. Two coal layers were found in this borehole at 31,45 and 32,78 meters in depths. The seams' thicknesses are 1,25 and 1,37 meters with a coal rank of lignite B. These coal seams belong to the Mangus seam. Roof and floor rocks are siltstone, claystone, and coaly shale.

4.2.12. Borehole MJ12. The MJ12 borehole has a total depth of 53,5 m. Two coal seams were found in this borehole at 41,25 - 44,35 m and 46,25 – 53,2 depth with a thicknesses of 3,1 m and 5,95 m. The coal in this location is part of the Benuang seam with roof, and floor rocks are claystone and tuff. Based on ASTM classification, the coal in MJ12 is lignite B.

4.2.13. Borehole MJ13. This borehole has a total depth of 83 m, contains five coal seams with thicknesses ranging from 0,2 m to 3,75 m. The coal at this location is part of the Benakat and Kebon seam. The roof and floor rocks of coal at this location are claystone, sandstone, and mudstone, some of which are alterations.

After boreholes evaluation have been carried out, one coal seam was found to have the potential for UCG. It is coal seam from borehole MJ2 at a depth of 260,4 meters which has 10,7 meters in thickness; 24,34 depth/thickness ratio and impermeable roof and floor rocks.

4.3. A potential area for UCG

Since the potential site for UCG has been found based on drilling data, the area of potential UCG needs to be calculated. The potential area for UCG in this study is obtained by delineating the coal seam based screening criteria mentioned previously in sub-chapter 4.1&4.2. Delineation is based on the value of the coal's strike and dip. Since the borehole does not provide these data for subsurface coal, data of coal outcrop around borehole MJ2 are used. The delineation follows the dip direction that implies the coal seam's continuity. This delineation is carried out to determine the extending area of the subsurface coal seam area and still matches the criteria for UCG based on the depth/thickness ratio of coal. The result of this delineation is around 15 km^2 as shown in Figure 2.

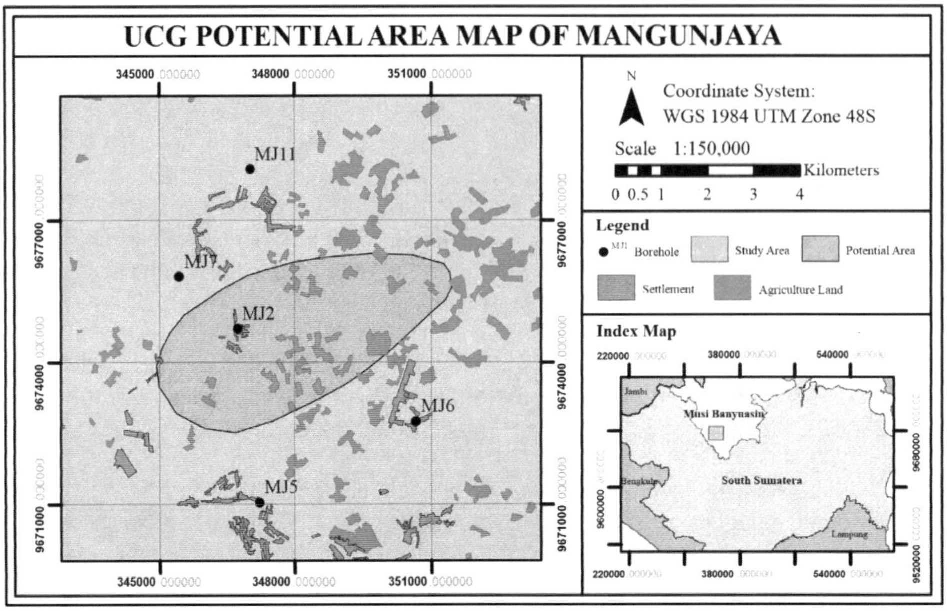

Figure 2. UCG potential area map of Mangunjaya (base map referred to [9][21]).

5. Conclusion

Screening criteria for coal in the Mangunjaya area have been conducted at 13 drilling well and outcrop data. The criteria used are coal thickness, depth of the coal seam, depth and thickness ratio, coal rank, the dip of a coal seam, roof and floor rock type and thickness, and geological structure. Based on the evaluation results, there is only one coal layer in the MJ2 borehole at a depth of 260,4 meters that matches most of the criteria for UCG. This coal layer belongs to the Mangus seam. The potential area is 15 km^2 around the MJ2 borehole.

References

[1] Sunuhadi D W, Heditama D M, Muksin I, Eddy H R, Ibrahim M A, Juarsa A, Oktaviani P, Hidayat R, Hermawan D, Mustofa S A, Permana L A and Rustina T S 2020 *Neraca Sumber Daya dan Cadangan Mineral, Batubara, dan Panas Bumi Indonesia Tahun 2020*, (Bandung: PSDMBP) 120 p

[2] Otoshi A, Sasaki K, and Anggara F 2021 Screening of UCG chemical reactions and numerical simulation up-scaling of coal seam from laboratory models. *J. Combustion Theory and Modelling* **2** p. 1–25. https://doi.org/10.1080/13647830.2021.198458

[3] Bielowicz B and Kasiński J R 2014 The possibility of underground gasification of lignite from Polish deposits *Int J Coal Geol* **131** p. 304–18. https://doi.org/10.1016/j.coal.2014.06.025

[4] Khadse A, Qayyumi M, Mahajani S and Aghalayam P 2006 Underground coal gasification: A new clean coal utilization technique for India. *J. Energy* **32** p. 2061–71. https://doi.org/10.1016/j.energy.2007.04.012

[5] Bhutto A W, Bazmi A A and Zahedi G 2013 Underground coal gasification: From fundamentals to applications. *J. Progress Energy Combustion Science* **39** p. 189–214. https://doi.org/10.1016/j.pecs.2012.09.004

[6] Anggara F and Amijaya 2015 D H CBM and UCG Energy Potential to Increase The National Energy Supply *Department of Geological Engineering, Universitas Gadjah Mada* p. 1–7

[7] Fatimah, Suryana A, Wibisono S A 2014 Potensi Deep Seated Coal di Indonesia, *Majalah Mineral dan Energi* **12** 28 p.

[8] Durucan S, Korre A, Shi J Q, Idiens M, Stanczyk K, Kapusta K, Dabrowska A R, Kempka T, Wolf K H, Younger P, Zavsek S, Poulsen N E, Bojda D, Franzsen S, Muresan M, Gao J,

Beath A and Mastalerz M 2014 TOPS: Technology options for coupled underground coal gasification and CO2 capture and storage. *J. Energy Procedia* **63** p. 5827–35. https://doi.org/10.1016/j.egypro.2014.11.616

[9] Anonymous 2021 *Database of Global Administrative Areas* https://gadm.org/download_country_v3.html (accessed December 2, 2021)

[10] Pulunggono A, Haryo A, Kosuma C G 1992 Pre-Tertiary and Tertiary Fault Systems as A Framework of The South Sumatra Basin; A Study of SAR-Maps *Proc. IPA. Annual Convention* **21** p. 339–60

[11] Bishop M G 2001 South Sumatra Basin Province, Indonesia: The Lahat/Talang Akar-Cenozoic Total Petroleum System *USGS Open-File Report 2001* 13 p

[12] Darman H and Sidi F H 2000 *An Outline of the Geology of Indonesia* (Jakarta: Lereng Nusantara) 200 p

[13] Yesa R, Nurdrajat, Firmasyah Y, Gani RMG, Ibrahim MA and Wibisono SA 2020 Estimation of Coal Resources in Mangunjaya Area, Musi Banyuasin Regency, South Sumatra Province. *Padjajaran Geosci J* **4** p. 524–41

[14] Susilawati R, Ward CR. Metamorphism of mineral matter in coal from the Bukit Asam deposit, south Sumatra, Indonesia. *Int J Coal Geol* **68** p. 171–95. https://doi.org/10.1016/j.coal.2006.02.003

[15] Shafirovich E, Varma A, Mastalerz M, Drobniak A and Rupp J 2008 *The Potential for Underground Coal Gasification in Indiana* Indiana Center for Coal Technology Research (CCTR) Final Report 46 p

[16] Shafirovich E and Varma A 2009 Underground coal gasification: A brief review of current status *J. Ind Eng Chem Res* **48** p. 7865–75. https://doi.org/10.1021/ie801569r

[17] Burton E, Friedmann J and Upadhye R 2019 Practices in Underground Coal Gasification (Livermore) *Larence Livermore National Laboratory Report* 119 p

[18] Standar Nasional Indonesia1998 *Klasifikasi Sumberdaya dan Cadangan Batubara* 15 p

[19] Dwitama E P, Ramdhani M R and Maria R 2021 A Preliminary Evaluation of Underground Coal Gasification Potential in South Sumatra Basin: A Case Study of The Muara Enim Coal Formation *Buletin Sumberdaya Geologi* **16** p. 83–97

[20] Sury M, White M, Kirton J, Carr P, Woodbridge R, Mostade M, Chappell R, Hartwell D, Hunt D and Rendell N 2004 *Review of environmental issues of underground coal gasification* (Birmingham: WS Atkins Consult Ltd, Univ Liège Belgium, FWS Consult Ltd) 126 p

[21] Badan Informasi Geospasial 2017 *Peta RBI Format shp* https://tanahair.indonesia.go.id/portal-web/download/perwilayah# (accessed December 4, 2021)

Gold recovery from the residue of jewelry waste recycling: pre-treatment using shaking table separation and flotation

A P Jiga, J Pasithbhattarabhorn, O Juntarasakul, J Pimolrat, N Soonthornwiphat and T Phengsaart*

Department of Mining and Petroleum Engineering, Faculty of Engineering, Chulalongkorn University, Bangkok, 10330, Thailand
*Corresponding author's e-mail: theerayut.p@chula.ac.th

Abstract. Gold is one of the precious metals used all over the world. It is used in various ways, such as in electrical and electronic equipment, as well as in jewelry. Gold is constantly depleted, so recycling is a very important process for recovering valuable materials as secondary raw materials. In this study, the residue from a jewelry waste recycling process was used. Shaking table separation and flotation as the pre-treatment before the hydrometallurgical processes were investigated. For shaking table experiments, the effects of table angle and water flow rate were investigated and a batch-type mechanical flotation was used for the flotation experiment. The results showed that for the shaking table separation, gold recoveries were high, while the purity of gold was higher, with lower recovery.

1. Introduction

Gold is one of the precious metals that is widely used for monetary standards in coins, bars, and decoration due to its rarity, color, and luster [1]. It is also used in electrical and electronic equipment as well as aviation industries due to its characteristics such as resistance of corrosion, malleability, ductility, and durable properties [2,3].

Nowadays, many researchers study various methods for recovery and purification of gold from secondary resources like e-waste and jewelry waste since the concentration is very high compared with the primary resources like gold ores from the mine. In the last two decades, researchers have developed various low-pollution technologies like physical separation as pre-treatment were to maximize the gold recovery and minimize environmental impact [4,5]. In this study, the gold recovery from the residue of jewelry waste recycling using shaking table separation and flotation was investigated.

2. Material and methods

2.1. Sample characterization

The sample used in this study is from a jewelry company in Thailand. It is a residue from the gold recycling process of jewelry waste using physical separation and pyrometallurgy. After these recovery processes, the recovered gold alloys will be purified by other processes while two residues (i.e., slag and dust) remain. In this study, dust was used. The size distribution of the raw dust sample was analyzed by the laser diffraction particle size analyzer (Mastersizer 3000, Malvern Panalytical Ltd., UK) in water. For chemical composition, all elements contained in raw dust sample were analyzed by the x-ray fluorescence (XRF) spectrometer (Epsion 4, Malvern Panalytical Ltd., UK), respectively.

2.2. Shaking table separation experiments
The shaking table separation experiments were conducted under the following conditions: sample of 500 g of, table frequency of 300 rpm, stroke displacement of 10 mm, table slope at 8, 10, and 12 degrees, and water flow rate at 4, 5, and 6 L/min. After the experiments, the chemical composition of the products (heavy, medium, and light fractions) was analyzed by XRF spectrometer.

2.3. Flotation experiment
For flotation experiments, batch-type mechanical flotation was used with the following conditions: a sample of 50 g, the water of 1 L, the impeller speed of 1000 rpm, methyl isobutyl carbinol (MIBC) as a frother of 40.5 ppm, and a flotation time of 10 min without any collector (the reagents used to modify or improve the hydrophobicity and floatability of materials e.g., kerosene). After the experiments, the chemical composition of froth and tailing products were analyzed by XRF spectrometer.

3. Results and discussion

3.1. Sample characterization
Figure 1 shows the particle size distribution of the sample characterized by the laser diffraction particle size analyzer. The particle size at 50% of cumulative passing (D_{50}) is 21.7 µm. Table 1 shows the chemical composition of the sample characterized by the XRF spectrometer. The results showed that the major elements are silicon (Si), aluminum (Al), calcium (Ca), and iron (Fe), which may be sourced the composition of the furnace used for the metallurgical recovery process. In addition, minor elements of titanium (Ti), potassium (K), and sulfur (S) are also present. The precious elements include silver (Ag) and gold (Au), with concentration 0.39% and 0.04%, respectively.

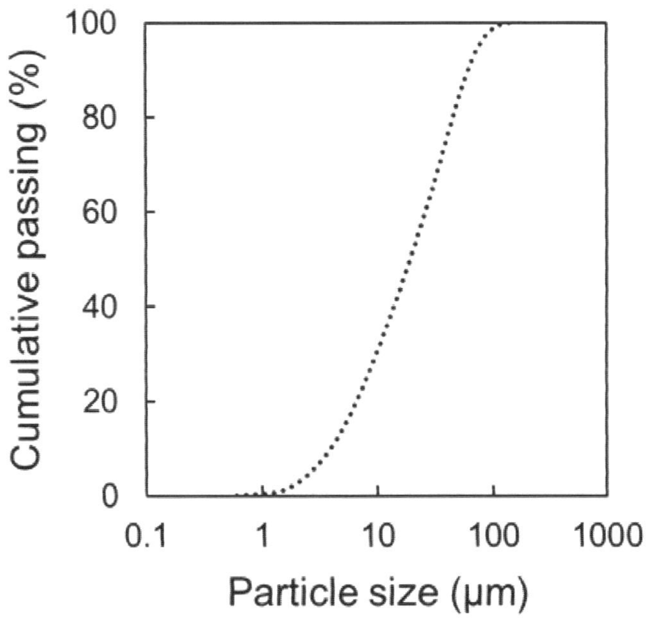

Figure 1. Size distribution of sample.

Table 1. Chemical composition of sample.

Elements	Amount (%)
Si	37.42
Al	29.33
Ca	11.75
Fe	10.05
Ti	3.21
K	2.17
S	2.01
Ag	0.39
Au	0.04
Others	3.63

3.2. Shaking table separation experiments

Table 2 shows the yield, Au purity, and Au recovery of heavy products after shaking table separation experiments at table slopes of 8, 10, and 12 degrees and varying water flow rates (at 4, 5, and 6 (L/min)). Due to the high specific gravity of Au, it should be mainly recovered in heavy products. The results showed that Au recoveries in heavy products were high with a high yield. However, Au purities in heavy products at all conditions were similar to the feed, indicating that the shaking table separation could not concentrate the gold. These results could be attribute to the particle size of the sample that was too fine. This further suggests that the table slope and water flow rate do not give significant effect in separation process.

Table 2. Yield, Au purity, and Au recovery of heavy products after shaking table separation experiments at table slope of 8, 10, and 12 degree and water flow rate of 4, 5, and 6 (L/min).

		Yield (%)			Au purity (%)			Au recovery (%)		
Water flow rate (L/min)		4	5	6	4	5	6	4	5	6
Table slope (°)	8	57.20	57.67	50.00	0.05	0.04	0.04	68.98	76.69	66.36
	10	51.60	50.78	43.58	0.04	0.04	0.03	68.02	60.48	90.00
	12	34.17	40.70	30.22	0.05	0.06	0.03	51.84	66.76	32.44

3.3. Flotation experiment

The flotation experiment, the Au purity increased from 0.04% in feed into 0.37% in froth product even no collector was used indicating that Au in the sample is inherently hydrophobic. However, the Au recovery was only 42.50%, lower than most of Au recovery obtained from shaking table experiments (Table 2). This result suggests that flotation is more suitable for the pre-treatment before the hydrometallurgy mainly for fine particle size. It could be recovered by flotation even the collector was not used. However, to increase the Au recovery, the use of collectors may give better results.

4. Conclusions

For shaking table separation experiments, Au recoveries in heavy products were high but the Au purities could not increase. For flotation experiments, Au purity was highly increased however, the Au recovery was low since the collector was not used.

5. References

[1] Syed S 2012 Recovery of gold from secondary sources-A review *Hydrometallurgy* **115–116** 30–51
[2] Wang J, Lu Y and Xu Z 2019 Identifying extraction technology of gold from solid waste in terms of environmental friendliness *ACS Sustain. Chem. Eng.* **7** 7260–7267
[3] Hainfeld J F, Dilmanian F A, Slatkin D N, Smilowitz H M 2008 Radiotherapy enhancement with gold nanoparticles *USA Asian J. Pharm. Pharmacol.* **60(8)** 977–985.
[4] Ding Y, Zhang S, Liu B, Zheng H, Chang C and Ekberg C 2019 Recovery of precious metals from electronic waste and spent catalysts: A review *America Resour. Conserv. Recycl.* **141** 284–298.
[5] Aydın B and Gül A 2021 Environmentally friendly process instead of cyanide leaching on recycling of gold and silver from jewellery scraps and wastes *Turkey Waste Manag. Res.* **39** 233–241

6. Acknowledgements

The authors gratefully acknowledge the editors and anonymous reviewers for their valuable inputs to this paper. The authors also wish to thank the members of Department of Mining and Petroleum Engineering, Faculty of Engineering, Chulalongkorn University for their support of this research.

Heat flow and gravity anomalies in some presumed hidden geothermal prospects in Java

P D Maghfira[1], P Utami[1,2*], and S W Niasari[2,3]

[1]Geological Engineering Department, Faculty of Engineering, Universitas Gadjah Mada
[2]Geothermal Research Centre, Faculty of Engineering, Universitas Gadjah Mada
[3]Physics Department, Faculty of Mathematics and Natural Sciences, Universitas Gadjah Mada
*Corresponding author's e-mail: p.utami@ugm.ac.id

Abstract. The currently developed geothermal fields in Java are hosted by Quaternary volcanic complexes in the central zone of the island. Those fields have many vigorous thermal manifestations. However, seismic tomography shows the presence of an east-west-trending, low P-wave velocity (V_p) zones along the island. This indicates the presence of some hot, molten mass at depth, and provides the possibility of the existence of subsurface heat energy resource that can be utilized despite the rarity of surface manifestations. We investigate the heat flow and gravity anomalies in the areas lacking thermal manifestations, namely Paguyangan, Candi Umbul-Telomoyo, Pablengan, and Parang Wedang. The first three are associated with Slamet, Telomoyo, and Lawu volcanoes, respectively. Parang Wedang is far from the active Merapi volcano, but close to the Opak Fault – one of the major strike-slip faults of Java. This research is a preliminary effort in exploring those presumed hidden geothermal prospects. The results shows that all the investigated areas have high heat flow anomaly (145 – 160 mW/m^2). Their gravity anomaly varies. The highest (>70 mGal) and medium values (12 - 24 mGal) are found in Pablengan, and Candi Umbul-Telomoyo, respectively. The lowest values (<5 mGal) occur in Parang Wedang.

1. Introduction

Java has geothermal resource potential of 8,557 MWe [1]. The total installed capacity in the island is 1,312 MWe, generated from the Kamojang (235 MWe), the Darajat (270 MWe), the Wayang Windu (287 MWe), the Karaha (30 MWe), the Patuha (55 MWe), the Awibengkok field (375 MWe), and the Dieng fields (MWe) [2]. These fields are associated with volcanogenic geothermal systems and characterized by abundant surface manifestations.

Tomographic map of Java shows low P wave velocity (V_p) values trending east-west in the middle parts of island. This low V_p value indicates the occurence of subsurface hot rocks or media [3] which are thought to correlate with volcanoes. Low V_p values were not found in the eastern part of Java due to lack of earthquake observation stations. In geothermal context, low V_p value is an indication of the presence of a high temperature rock masses in the subsurface that may serve as heat source.

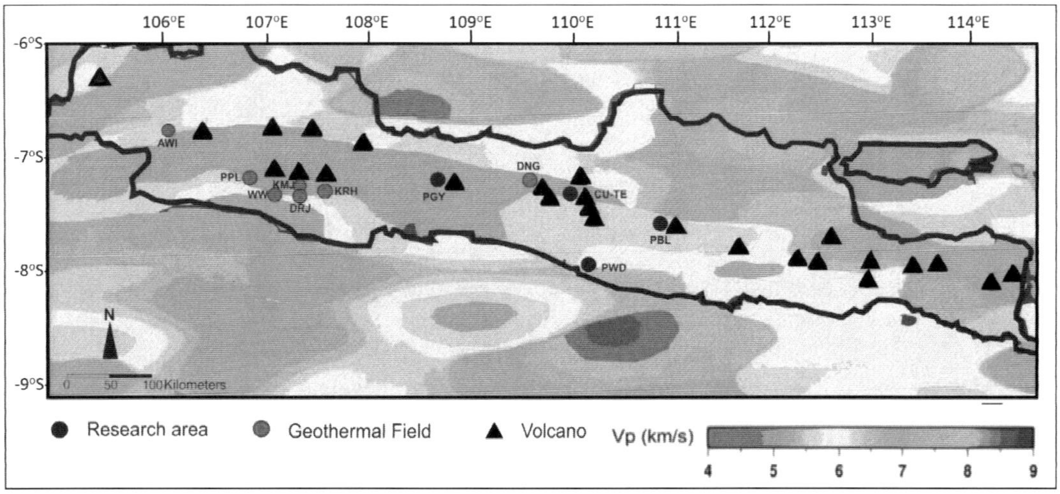

Figure 1. Tomography map of Java [3]. Research areas: PGY= Paguyangan, CU-TE=Candi Umbul-Telomoyo, PWD=Parang Wedang, PBL=Pablengan. Producing geothermal fields: AWI=Awibengkok, PPL=Patuha, WW=Wayang Windu, KMJ=Kamojang, DRJ=Darajat, KRH=Karaha, DNG=Dieng. Base map from DEMNAS [4].

As well as being associated with low V_p values, heat source of a geothermal system is also characterized by high heat flow anomaly and high gravity anomaly values [5]. In this research, we calculate heat flow and gravity anomalies in four areas lacking of manifestations, i.e. Paguyangan, Candi Umbul-Telomoyo, Parang Wedang dan Pablengan (respectively marked as CU-TE, PWD, and PBL in Figure 1). The purpose is to determine whether in the research area there are high heat flow and high gravity anomalies, which may indicate the presence of hot rock masses in the subsurface.

2. Geology

2.1 Regional geology

The subduction of the Indo-Australian under the Eurasian plates which started in the Late Miocene and continue to the present, has resulted in magmatic activities and formed the Sumatra-Java and the Banda volcanic arcs [6]. The volcanoes in the arcs are the potential hosts of geothermal systems [7].

The physiography of Java is divided into 7 zones, namely the Quaternary volcanic zone, the alluvial plains of the northern Java, the Rembang-Madura anticlinorium, Bogor-North Serayu and Kendeng anticlinorium, domes and ridges in the central depression, southern mountains, and central depression of Java and Randublatung zones [8]. Geothermal fields that have been developed are located in the Quaternary volcanic physiographic zone.

Figure 2 shows the physiography and major structural features of Java. Java is intersected by three major structures namely the strike-slip faults of Cimandiri (SW-NE) and Citandui (SW-NE) in West Java, and the Central Java strike-slip fault (SW-NE). Other important faults are Baribis, Kendeng, Semarang, Pati-Rembang, Pasuruan, Probolinggo, and Opak faults [9].

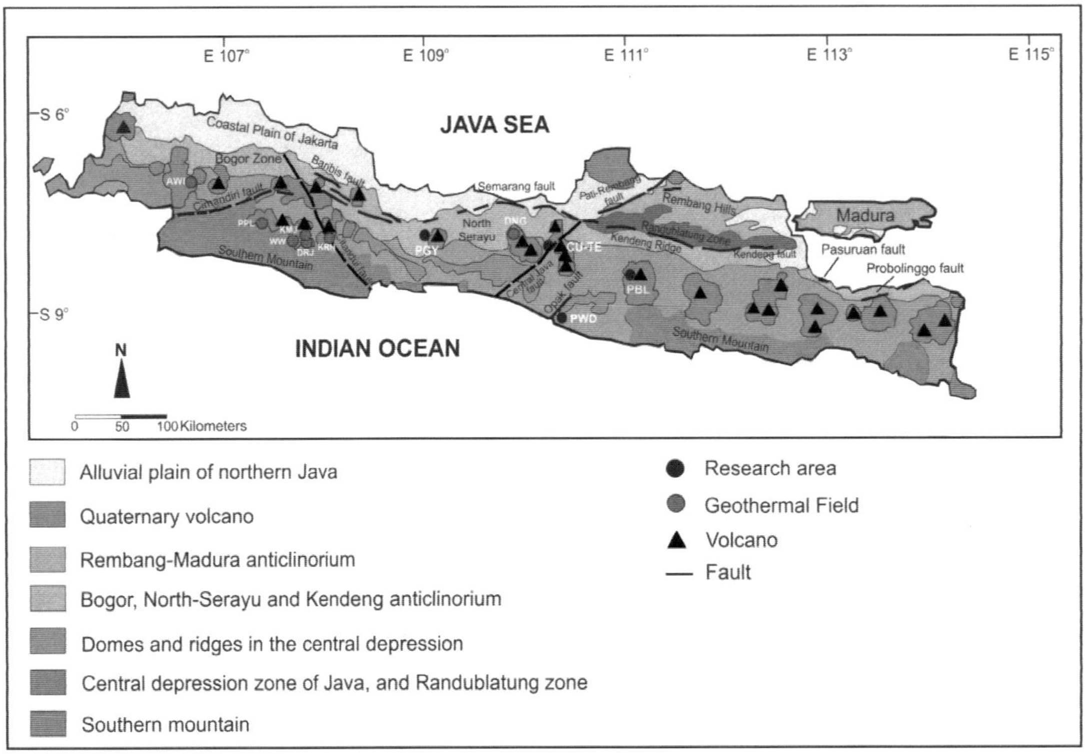

Figure 2. Compilation maps of physiographic [8] and major structural features in Java [9]. Research areas: PGY= Paguyangan, CU-TE=Candi Umbul-Telomoyo, PWD=Parang Wedang, PBL=Pablengan. Producing geothermal fields: AWI=Awibengkok, PPL=Patuha, WW=Wayang Windu, KMJ=Kamojang, DRJ=Darajat, KRH=Karaha, DNG=Dieng. Base map from DEMNAS [4].

2.2 Paguyangan

Paguyangan thermal manifestation area is located in Brebes Regency, Central Java. The closest volcano is Slamet, which is 22 km to the west. Among the thermal manifestations associated with the Slamet volcano, Paguyangan hot springs have the highest temperatures (70°C) [10]. Based on the geological map of Purwokerto and Tegal sheets [11], Paguyangan is associated with the Rambatan Formation which is composed of Tertiary sedimentary rocks. Paguyangan hot springs are categorized as neutral-pH chloride waters [12].

2.3 Candi Umbul-Telomoyo

Candi Umbul-Telomoyo is located in Magelang and Temanggung Regency, Central Java. Candi Umbul warm spring is located 15 km to the west of the inactive Telomoyo volcano. Other thermal manifestations are Pakis Dadu and Candi Dukuh warm springs. Candi Umbul warm spring is associated with the Kaligetas Formation which is composed of volcanic breccias [13]. The occurrence of the spring coincides with a N-S trending fault [14] The temperature of the warm spring is 30 to 35°C. This manifestation is categorized as neutral pH bicarbonate water [15].

2.4 Parang Wedang

Parang Wedang warm springs are located in Bantul Regency, Special Region of Yogyakarta. Parang Wedang warm springs are associated with a major structure, namely the Opak fault which is about 2 km to the west. The closest volcano from Parang Wedang is Mount Merapi, 50 km to the north. Parang Wedang warm spring is associated with the Wonosari Formation which is composed by carbonate sedimentary rocks [16]. The manifestation of Parang Wedang is warm spring that have temperature of 40°C and pH range of 6-7. Parang Wedang warm spring is categorized as chloride water. The high Cl concentration (7,121 ppm) was due to seawater input into hydrology of the springs [17].

2.5 Pablengan
The Pablengan area is one of the thermal manifestations associated with Lawu volcano. Pablengan is located in Karanganyar Regency, Central Java. Based on the geological map of Lawu volcano [18]. The springs are situated in a plain, and associated with the Pre-Lawu volcanic rocks. The temperatures of the Pablengan springs range from 30-40°C. Their pH values are about 6. Pablengan warm springs are categorized as near neutral-pH chloride waters [19].

3. Methodology

3.1 Heat flow
In general, the temperature increases with depth. The heat flow q_s (mW/m^2) is defined as results of two measurable quantities (equation 1), namely the temperature gradient $\frac{dT_{(z)}}{dZ}$ and the thermal conductivity k (Wm^{-1}K^{-1}). The negative sign indicates that heat is flowing upwards. The average value of heat flow in the earth crust is 80 mW/m^2 [20].

$$q_s = -k\frac{dT_{(z)}}{dZ} \quad (1)$$

3.2 Gravity
The gravity method is one of geophysical methods that can determine variation of the density of the subsurface rocks [21]. To cover a wide area, satellite gravity is more effective compared to land survey measurement. Satellite gravity uses GeoSAT and ERS-1 which provide altimeter data. Satellite gravity shares the same principle with airborne gravity. We use TOPEX satellite gravity data. TOPEX satellite works by sending radar signals to the earth's surface and reflect them back to the satellite [22].

4. Data and analyses

4.1 Heat flow
We combine heat flow data from oil and gas wells [23] and the results of heat flow calculations (Table 1) based on geothermal wells temperature data from Kamojang [24], Patuha [25], Dieng [26], and Ijen [27]. The result is visualized in the form of heat flow map by using Oasis Montaj™ software (Figure 3).

4.2 Gravity
TOPEX satellite gravity data is of public domain [28]. The data consists of free air gravity anomaly, elevation, and geographic coordinates. The resolution of the TOPEX gravity data is 1.8 km [22].

Gravity data processing is carried out to get complete Bouguer anomaly. The data processing starts with Bouguer correction to remove the gravity effect caused by the elevation difference between the station and sea level. The terrain correction is then carried out to get complete Bouguer anomaly [29]. To show gravity anomaly at shallow depth, residual anomaly calculated by subtracting the complete Bouguer anomaly with the regional gravity anomaly [30].

Table 1. Temperature at the sea level in several geothermal wells in Java and the results of heat flow calculations.

Geothermal fields	Temperature (°K) at 0 masl	Heat flow (mW/m^2)
Kamojang [24]	463	178
Patuha [25]	430	162
Dieng [26]	438	166
Ijen [27]	397	151

5. Results

High heat flow anomalies (>120 mW/m^2) are located beneath the volcanic zone of Java (Figure 3). High heat flow anomalies are also present in all research areas, namely Paguyangan, Candi Umbul-Telomoyo, Parangwedang and Pablengan. The low heat flow anomaly (25-80 mW/m^2) is located in the northern part of Java which is associated with the northern Java basin. Low-medium anomalies were found in the volcanic ranges in East Java except Lawu and Ijen volcano. The absence of high heat flow anomaly in Kelud, Kawi, Arjuno, Bromo, Semeru, Lamongan, and Argopuro volcanoes is due to lacking of measurement.

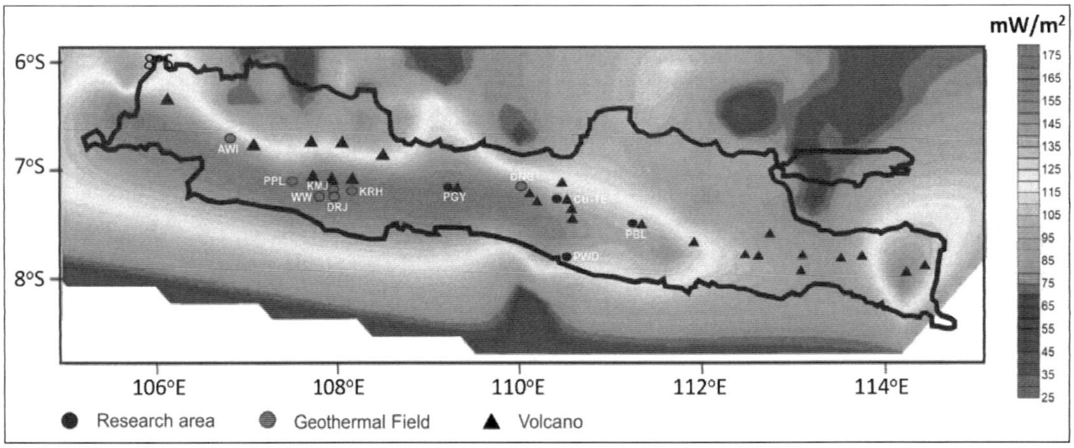

Figure 3. Heat flow anomaly map of Java. Research areas: PGY= Paguyangan, CU-TE=Candi Umbul-Telomoyo, PWD=Parang Wedang, PBL=Pablengan. Producing geothermal fields: AWI=Awibengkok, PPL=Patuha, WW=Wayang Windu, KMJ=Kamojang, DRJ=Darajat, KRH=Karaha, DNG=Dieng. The absence of high heat flow anomaly in Kelud, Kawi, Arjuno, Bromo, Semeru, Lamongan, and Argopuro volcanoes is due to lacking of measurement. Base map from DEMNAS [4].

The complete Bouguer anomalies range from -852 to 1044 mGal. High anomalies (>230 mGal) are located in the volcanic areas. The medium anomalies (4-30 mGal) are found in the northern part of the island. Low anomalies (-500 to 3 mGal) sit in the Indian Ocean and part of East Java (Figure 4).

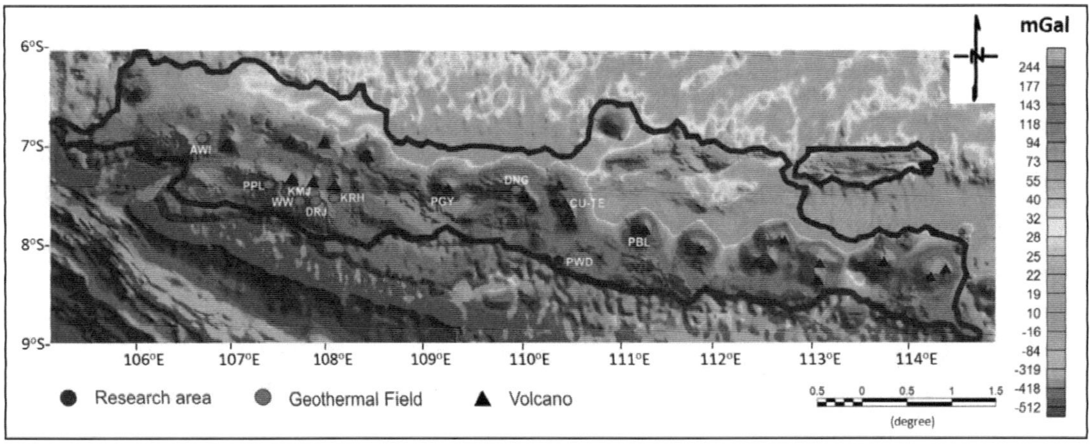

Figure 4. Complete Bouguer anomaly map of Java. Research areas: PGY= Paguyangan, CU-TE=Candi Umbul-Telomoyo, PWD=Parang Wedang, PBL=Pablengan. Producing geothermal fields: AWI=Awibengkok, PPL=Patuha, WW=Wayang Windu, KMJ=Kamojang, DRJ=Darajat, KRH=Karaha, DNG=Dieng. Base map from DEMNAS [4].

The residual gravity anomalies of the Paguyangan area range from -33 to 95 mGal (Figure 5a). High anomalies (25-95 mGal) are located in the body of Slamet volcano, Guci, and Baturraden manifestations, which composed by igneous rocks. Paguyangan manifestation has gravity anomalies range from 12 to 24 mGal. We interpret the medium anomalies are controlled by the Rambatan sedimentary rock formation.

The residual gravity anomalies of the Candi Umbul-Telomoyo area range from -26 to 68 mGal. The high anomalies are located in the body of the inactive Telomoyo volcano (Figure 5b). In addition, medium anomalies (-5 to 24 mGal) are found in Candi Umbul and Pakis Dadu which are located in the western part of Telomoyo volcano. Low anomalies (<-5 mGal) are found in the southwest and northeast of Telomoyo.

The residual gravity anomalies of the Parang Wedang area range from -12 to 11 mGal (Figure 5c). The high anomalies (>6 mGal) are found in the northwest of Parang Wedang. Low gravity anomalies (<4 mGal) are located along the Opak fault.

The residual gravity anomalies of the Pablengan area range from -62 to 78 mGal (Figure 5d). High anomalies (>50 mGal) are located in the body of Lawu volcano and the associated geothermal manifestations. Pablengan area has the highest anomaly (60-65 mGal) compared to the other research area.

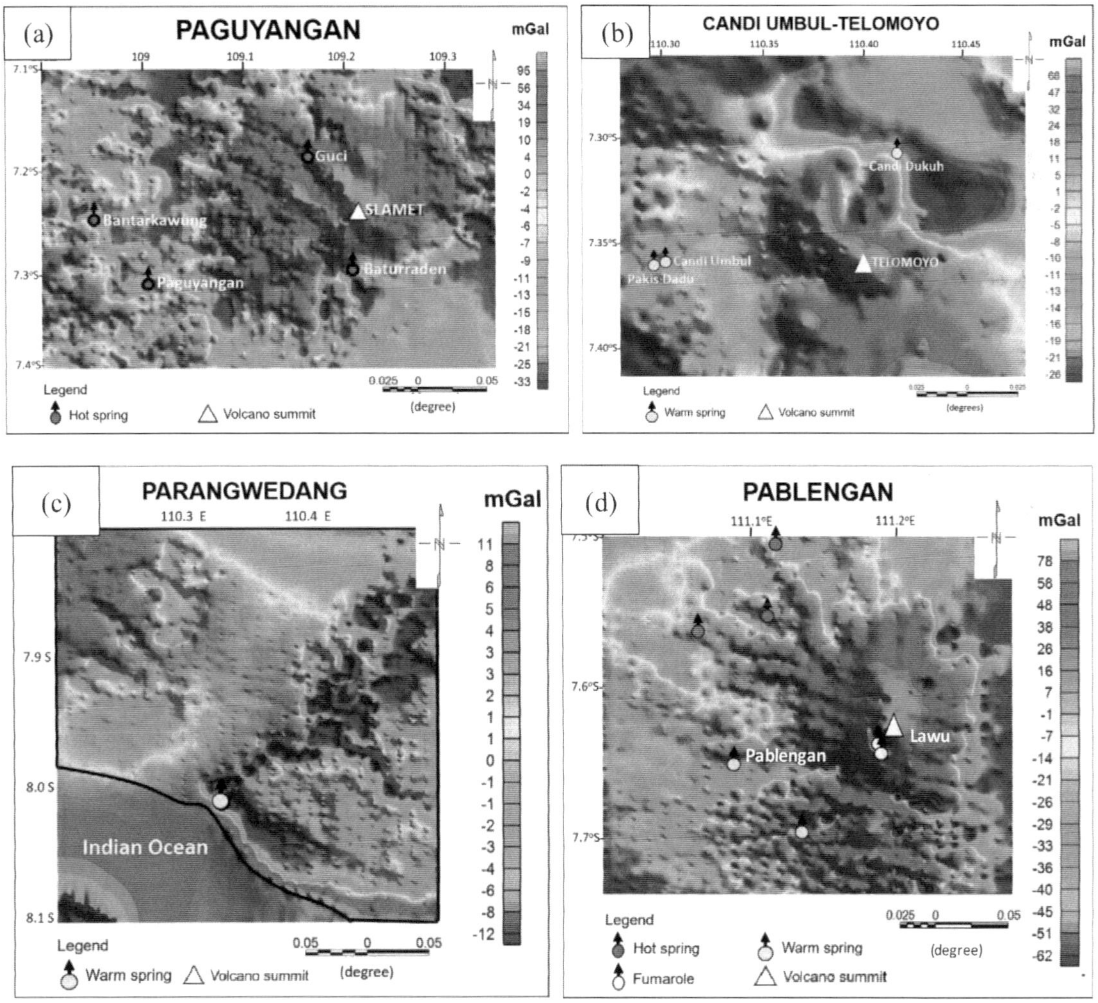

Figure 5. Residual gravity anomaly map of Paguyangan (a), Candi Umbul-Telomoyo (b), Parang Wedang (c), and Pablengan (d).

6. Conclusions

There is a correlation between the high residual gravity anomaly and the volcanic arc in Java. The research areas which have high heat flow anomalies, namely Pablengan, Paguyangan, and Candi Umbul are located in the Quaternary volcanic arc. However, the gravity anomaly values vary. High gravity anomalies are found in the body of Slamet, Telomoyo, Lawu volcanoes, and Pablengan hot spring area which forms part of the Lawu volcano. Meanwhile, the medium gravity anomalies are located in Paguyangan and Candi Umbul. Parang Wedang area, which is not associated with volcanoes, has the lowest gravity anomaly value.

Besides heat flow and gravity anomalies, volcanogenic geothermal prospects can be identified by their volumes (>50 km^3) and their ages (0.05-0.2 Ma) [10]. The volumes of the volcanoes in the research area are: Slamet (500 km^3) [11], Lawu (300 km^3) [17], and Telomoyo (35 km^3) [15]. Slamet and Lawu volcanoes are therefore highly recommend for further exploration. Meanwhile, the Telomoyo volcano, whose volume is less than 50 km^3, is considered to have low geothermal potential. The volume of a volcano reflects the size of the magma chamber [10].

The order of priority for further exploration is shown in Table 2. The first priority is the Pablengan prospect, as it has the highest values of heat flow and gravity anomaly. The second priority is the Paguyangan prospect. Although Paguyangan has a high heat flow anomaly but its gravity anomaly is low. The third priority is the Candi Umbul prospect, which has low values of both heat flow and gravity anomaly. The Pablengan, Paguyangan, and Candi Umbul prospects are located in the Quaternary volcanic terrains, meaning that they likely have magmatic heat sources. The last priority is the Parang Wedang prospect which also has low values of heat flow and gravity anomaly. The Parang Wedang prospect is not associated with Quaternary volcanism.

In order to determine the thermal and the hydrological structure of the presumed hidden geothermal prospect in the research areas, we recommend that detailed geological mapping on the scale of 1:25,000 or larger to be conducted. The results will be useful to guide the selection of the more detailed geophysical surveys, and to identify the geological elements that may control the heat and mass transfer in the geothermal prospects.

Table 2. The order of priority for further exploration.

Priority	Research area	Heat flow anomaly (mW/m^2)	Gravity anomaly (mGal)
1	Pablengan (PBL)	136	60-65
2	Paguyangan (PGY)	168	15-24
3	Candi Umbul-Telomoyo (CU-TE)	110	12-16
4	Parang Wedang (PWD)	92	3-6

References

[1] Ministry of Energy and Mineral Resources 2019 PLTP Lumut Balai Unit 1 Ditargetkan COD Agustus 2019 Jakarta. *http://ebtke.esdm.go.id/post/2019/07/04/2276/pltp.lumut.balai.unit.1.ditargetkan.cod.agustus.2019*

[2] Ministry of Energy and Mineral Resources 2021 Pembangkit Listrik Tenaga Panas Bumi di Jawa Barat. *http://esdm.jabarprov.go.id/daftar-pltpb-jawa-barat/*

[3] Chasanah U and Santosa B J 2017 3D Tomographic Imaging of P Wave Velocity Structure Beneath Java Island using Fast Marching Tomography Method (*Journal of Science*) **2** p 5

[4] Geospatial Information Agency 2018 DEMNAS (Digital Elevation Model Nasional) Jakarta

[5] Garg S K Pritchett J W and Combs J 2010 Exploring for Hidden Geothermal Systems, *Proceedings World Geothermal Congress Bali Indonesia* p 7

[6] Katili J A 1975 Volcanism and plate tectonics in the Indonesian island arcs (*Tectonophysics*) 26 pp 165- 188

[7] Hamilton W B 1979 Tectonics of the Indonesian Region U.S. Geology Survey Washington, DC
[8] van Bemmelen R W 1949 The Geology of Indonesia Vol. 1A: General Geology of Indonesia and Adjacent Archipelagos, Netherlands
[9] Marliyani, G.I., Helmi, H., Arrowsmith, J.R., Clarke, A., 2020, Volcano morphology as an indicator of stress orientation in the Java Volcanic Arc, Indonesia, *Journal of Volcanology and Geothermal Research*.
[10] Wohletz K and Heiken G 1992 *Volcanology and Geothermal Energy* (University of California Press: United States of America)
[11] Triastuti H Mulyana I Sumaryadi M Alfianti H Ipmawan V Rusdi M Kriswati E Sulton F 2020 Studi perbandingan aktivitas Gunung Slamet: periode krisis 2019 dengan erupsi 2014 *Prosiding Kolokium Kajian Kegunungapian 2019* (Pusat Vulkanologi and Mitigasi Bencana Geologi: Bandung)
[12] Djuri M Samodra H Amin T C and Gafoer S 1996 Peta Geologi Lembar Purwokerto dan Tegal Skala 1:100.000 (Pusat Penelitian and Pengembangan Geologi: Bandung)
[13] Juhri S and Harijoko A 2016 Karakteristik Geokimia Air Panas Bumi Di Sekitar Gunung Slamet, *Proceeding Seminar Nasional Kebumian Ke-9: Peran Penelitian Ilmu Kebumian Dalam Pemberdayaan Masyarakat.*
[14] Thanden R E Sumadirdja H Richards P W Sutisna K Amin T C 1996 Peta Geologi Lembar Magelang dan Semarang, Jawa. Skala 1:100.000. (Pusat Penelitian dan Pengembangan Geologi: Bandung)
[15] Hermawan D and Rezky Y 2011 Delineasi Daerah Prospek Panasbumi Berdasarkan Analisis Kelurusan Citra Landsat di Candi Umbul-Telomoyo, Provinsi Jawa Tengah (*Buletin Sumber Daya Geologi*: Bandung) 6 pp 1-10
[16] Rahardjo W Sukandarrumidi Rosidi H R D 1995 Peta Geologi Lembar Yogyakarta, Jawa. Skala 1:100.000. (Pusat Penelitian dan Pengembangan Geologi: Bandung)
[17] Hermawan D Anna Y Kusnadi D 2010 Penyelidikan Terpadu Geologi and Geokimia Daerah Panas Bumi Gunung Lawu, *Prosiding Hasil Kegiatan Pusat Sumber Daya Geologi*, p 509-525.
[18] Abdurrachman K Suganda O K Hendrastro M Irianto 1996 Peta Geologi Gunung Api Lawu 1:50.000 (Pusat Penelitian and Pengembangan Geologi: Bandung)
[19] Hermawan D Permana L A 2018 Keprospekan Panas Bumi Gunung Lawu Berdasarkan Kajian Vulkanostratigrafi (*Buletin Sumber Daya Geologi*: Bandung) **13** pp 199-213
[20] Ranalli G and Rybach L 2005 Heat transfer and lithosphere rheology in geothermal areas: Features and examples (*Journal of Volcanology and Geothermal Research*) **148** pp 3-19
[21] Blakely RJ 1996 *Potential Theory in Gravity and Magnetic Applications* (Cambridge University Press: Edinburgh)
[22] Sandwell D T R D Müller W H F Smith E Garcia R Francis New global marine gravity model from CryoSat-2 and Jason-1 reveals buried tectonic structure (*Science*) **346** pp. 65-67
[23] Faculty Engineering of University of North Dakota *Global Heat Flow Database http://www.heatflow.und.edu*
[24] Utami P 2000 Characteristics of the Kamojang Geothermal Reservoar (West Java) as Revealed by its Hydrothermal Alteration Mineralogy (*Proceedings of World Geothermal Congress 2000*: Japan)
[25] Utama P P 2020 Geologi Bawah Permukaan and Alterasi Hidrotermal Lapangan Panas Bumi Patuha. Unpublished thesis (Faculty of Engineering Universitas Gadjah Mada: Yogyakarta)
[26] Shalihin M G Utami P and Nurpratama M 2020 The Subsurface Geology and Hydrothermal Alteration of the Dieng Geothermal Field, Central Java: A Progress Report (*IOP Conf. Series: Earth and Environmental Science*) **417** p 12
[27] Daud Y Nuqramadha W A Fahmi F Pratama S A Rahman K R Subroto W 2017 Discovering "Hidden" Geothermal Reservoir in Blawan-Ijen Geothermal Area (Indonesia) Using 3-D Inversion of MT Data, (*Proceeding of The 1st Geo Electromagnetic Workshop (Geo-EM)*: Bandung)
[28] University of California San Diego *Extract XYZ-Grid Topography or Gravity https://topex.ucsd.edu/cgi-bin/get_data.cgi.*

[29] Grant F S and West G F 1965 *Interpretation Theory in Applied Geophysics* (McGraw-Hill Book Company: New York)
[30] Telford W M Geldart L P and Sheriff R E 1990 *Applied Geophysics: 2nd Edition* (Cambridge University Press: New York)

Acknowledgements
The authors thank Geophysics Study Program, Department of Physics, Faculty of Mathematics and Natural Sciences Universitas Gadjah Mada for providing the Oasis MontajTM software. University of San Diego and University of North Dakota provide free access to gravity satellite and heat flow data, respectively.

Mineralogy and geochemistry of Kliripan manganese ore deposit, Hargorejo Village, Kokap District, Kulonprogo Regency

B Islamadina and I W Warmada*

Geological Engineering Department, Engineering Faculty, Gadjah Mada University, Jl. Grafika No. 2 – Bulaksumur, Yogyakarta 55281 Indonesia
*Corresponding author's e-mail: warmada@ugm.ac.id

Abstract. This study discusses the mineralogy and geochemistry of manganese deposits in the Kliripan area, where manganese had been mined since the Dutch occupancy until the 1970s. Although mining has been discontinued since then, the demand for manganese has increased in various industries, prompting exploration to continue to meet this demand. We performed petrographic, XRD, and XRF analysis of the ore and wall rock samples. Jasper and grainstone from the Sentolo Formation are the two lithologies where manganese mineralization occurs. Manganese-bearing minerals include pyrolusite, manganite, rhodochrosite, and lithiophorite, while the gangue minerals consist of quartz, calcite, hematite, and clay minerals. We interpret that Jasper is a product of alteration and has MnO content as high as 5.02 wt.%. Meanwhile, the economic grade manganese ore in grainstone has an MnO content up to 48.51 wt.%. The elements Ce, Zr, Co+Ni+Cu, are important discriminants to determine the type of manganese deposit. The results showed that the diagenetic process in reductive water formed the manganese ores. Oxidation of divalent manganese ion formed pyrolusite and manganite. Rhodochrosite crystallized when dilute manganese solution reacts with CO_2 from limestone diagenesis. We concluded that the Kliripan manganese ore deposit has both hydrothermal and diagenetic deposits.

1. Introduction

Manganese is the 10th most abundant element in earth's crust [1]. The most common use of manganese is in the steel industry to reduce the iron's brittleness and increase its strength. According to Cannon [2], the following products need manganese to produce: battery cathode, electronic device, water treatment chemicals, car base paint, ceramic and glass colorant, as well as cosmetic products. With its wide utilization within modern industries, manganese resources are needed to secure the demand.

In Kliripan, Hargorejo Village, Kokap District, Kulonprogo Regency, D.I. Yogyakarta Province, there was once manganese mining activity starting from the Dutch occupation period until the 1970s. According to the D.I. Yogyakarta Culture Conservation Centre [3], mining in Kliripan went up and down during the colonial occupation, then continued after Indonesia's independence before fading and disappearing in 1972-1976. Nevertheless, exploration for manganese resource continued to this day and the need to research its mineralogy and geochemistry of the known deposit is relevant. Such study is helpful to understand which deposit has the most manganese ores of economic grade and what geological processes involved in its accumulation.

2. Geological setting

Kliripan, Hargorejo Village, is a part of Kulonprogo Mountains. This morphology has an elliptical shape that extends northeast to southwest. The oldest sedimentary rock in Kulonprogo is the Eocene–Early Oligocene Nanggulan Formation which consists of sandy marl, claystone, and limestone [4]. Vulcanic rocks of Early Miocene, called Old Andesite Formation (OAF) is the major component of Kulonprogo Mountains and thus giving it hilly to undulating morphology. This unit consists of andesitic volcanic breccias, tuff and lapilli tuff, agglomerate, andesitic lava, and sandstone. On top of the OAF lies Jonggrangan Formation in the centre and northeast which comprises sandstone, lignite, and limestone. This study focuses on the Sentolo Formation, which overlays the OAF in the south-eastern part of Kulonprogo Mountains.

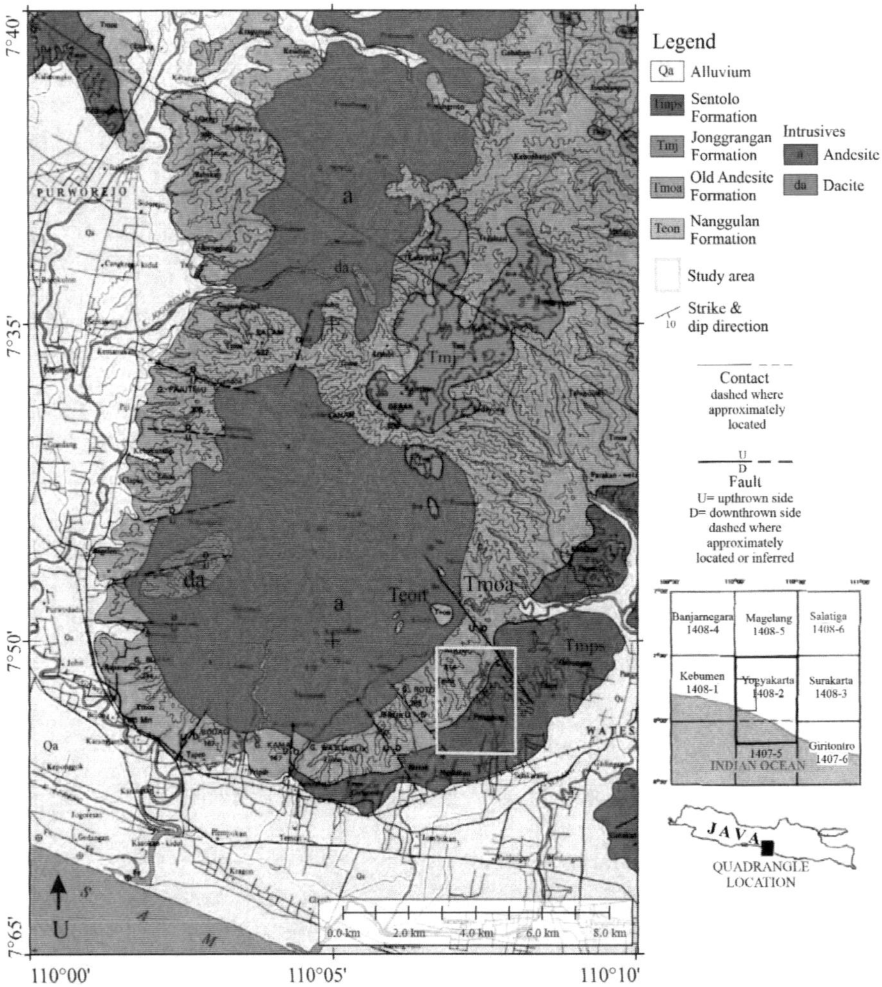

Figure 1. Regional geology map of the study area in Kliripan and its surrounding, Hargorejo Village, about 11 km to the west of Wates, Kulonprogo Regency, modified from Regional Geology Map of Yogyakarta [4].

Surface geological mapping in the study area reveals that the Sentolo Formation consists of polymictic breccia, which superimposed epiclastic breccia of OAF. This bed is followed by a gradual succession of gravelly sandstone, siltstone and allochemic mudrock. On top of the mixed siliciclastic–carbonate rocks, lies massive grainstone. Manganese ores mainly found at the lower part of the grainstone in Kliripan and Anjir Hamlet. Below the manganese-bearing layer, we found jasper that lies unconformably above siltstone. Jasper is only found in Kliripan and Anjir and covers a very limited area. Figure 2 shows the measured section of a crop in Kliripan where the most complete succession from gravelly sand to grainstone is found.

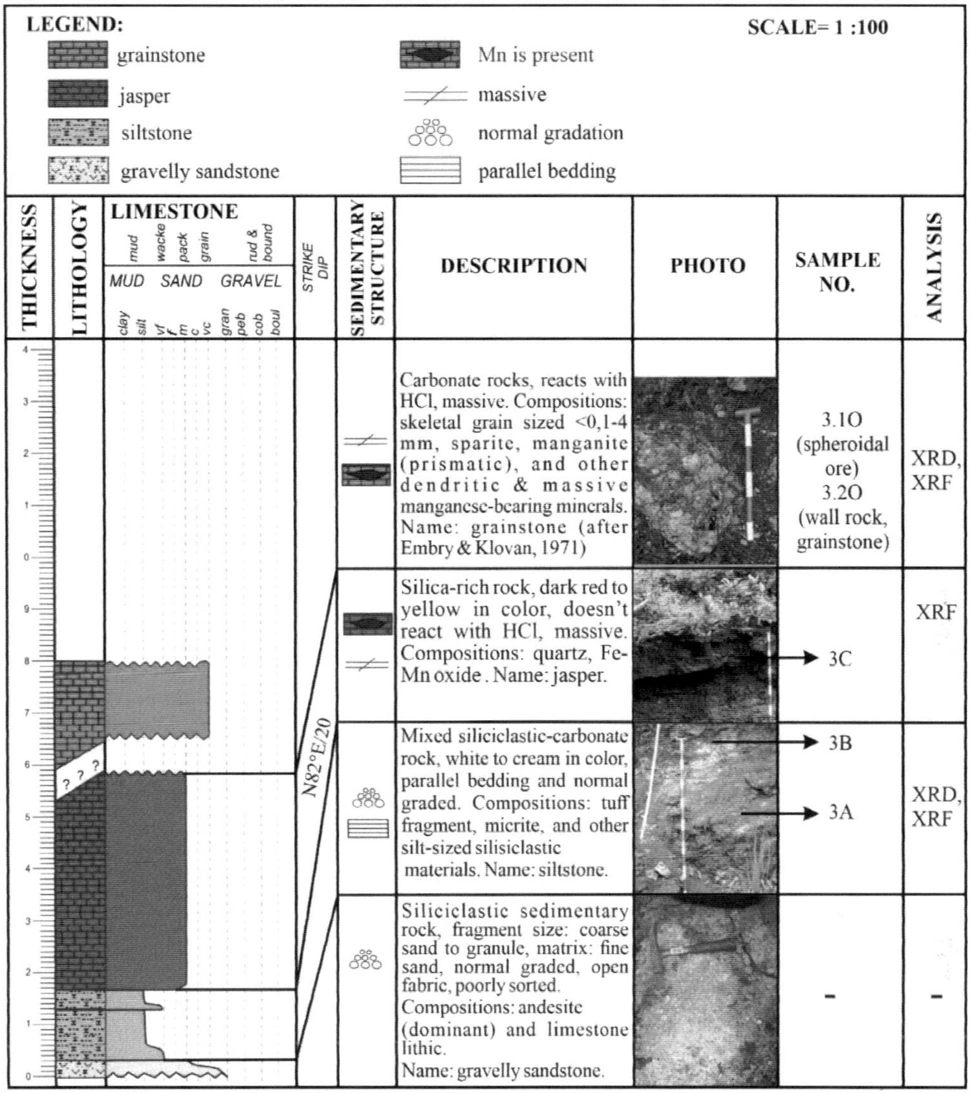

Figure 2. Measured section of a crop in Kliripan abandoned mine.

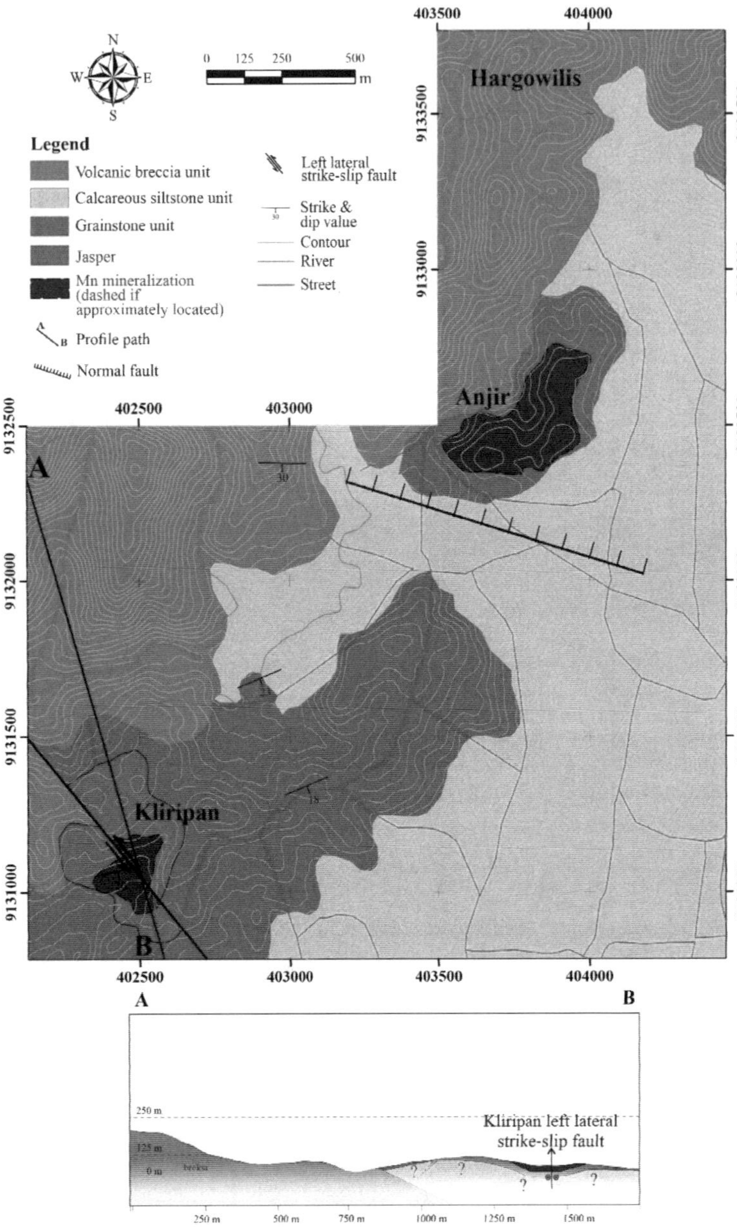

Figure 3. Geology, alteration, and manganese mineralization map of the study area, base map from [5].

3. Methods

We collected rock samples from the Kliripan abandoned mines, Hargorejo Village, Kokap District, Kulonprogo Regency. Then we observed 8 thin sections of manganese ores, wall rocks, and barren rocks to determine textural changes between them, ore textures, and the gangue minerals present. We also performed X-Ray Diffraction (XRD) analysis to identify any manganese bearing minerals. We used [6] as a guide in identifying mineral phase. To obtain the geochemical data of both ores and wall rocks, we performed Loss in Ignition (LOI) and X-Ray Fluorescence (XRF) analysis. To determine the deposit type(s), we compared trace elements concentration in our sample to other known deposits in several discriminant diagrams. Table 1 lists XRF samples of this study and sediment reference materials [7]. The reference material is from a river valley 10 km east of Bogor.

Table 1. XRF samples and its source.

Sample no.	Rock type	Source
1A	Jasper	
2A	Jasper	
3A	Siltstone	
3B	Siltstone	
3C	Jasper	
3.2O	Grainstone with Mn	
3.1O	Spheroidal Mn ore	This study
4A	Grainstone	
18O	Massive Mn ore	
18A	Grainstone	
9A	Siltstone	
9B	Polymetallic nodule	
SED-A	Marl	
SED-B	Volcaniclastic sandstone	[7]
SED-C	Mudstone	

4. Results

4.1. Alteration

Jasper thin section found below the manganese ores layer shows several indications of hydrothermal alteration. Among these are the microcrystalline quartz that exhibits moss texture and saccharoidal texture, shown in Figure 4. Moss texture, in which an aggregate of quartz shows a grape-like spheroidal grains reflects amorphous silica gel transformation to quartz [8]. Meanwhile, saccharoidal texture is an aggregate of fine-grained quartz whose appearance resembling sugar grains upon inspection in hand specimens. This texture forms when massive granular carbonate reacts with silica-bearing fluid and gets replaced by quartz crystallites through disordered crystallographic deformities [9]. Apart from those textures, there are iron oxide mineral and opaque massive mineral, possibly manganese-bearing mineral.

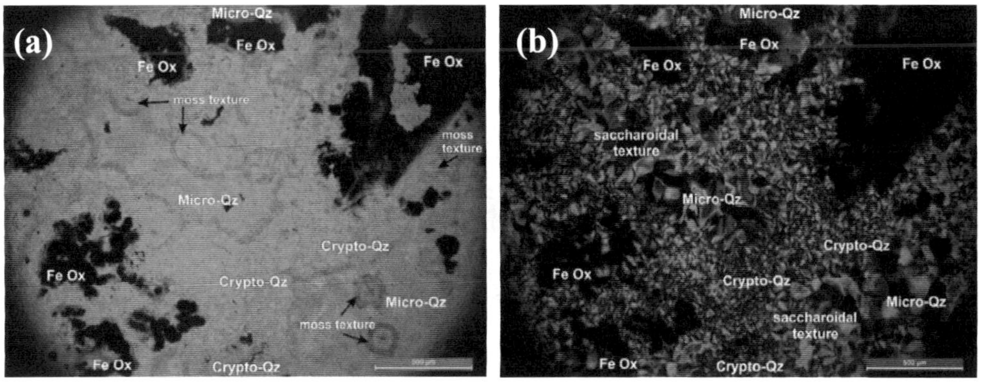

Figure 4. Photomicrograph of jasper: (a) plane polarization (b) crossed polarization; Fe Ox= iron oxide; Crypto-Qz = cryptocrystalline quartz; Micro-Qz = microcrystalline quartz.

IOP Conf. Series: Earth and Environmental Science 1071 (2022) 012026 doi:10.1088/1755-1315/1071/1/012026

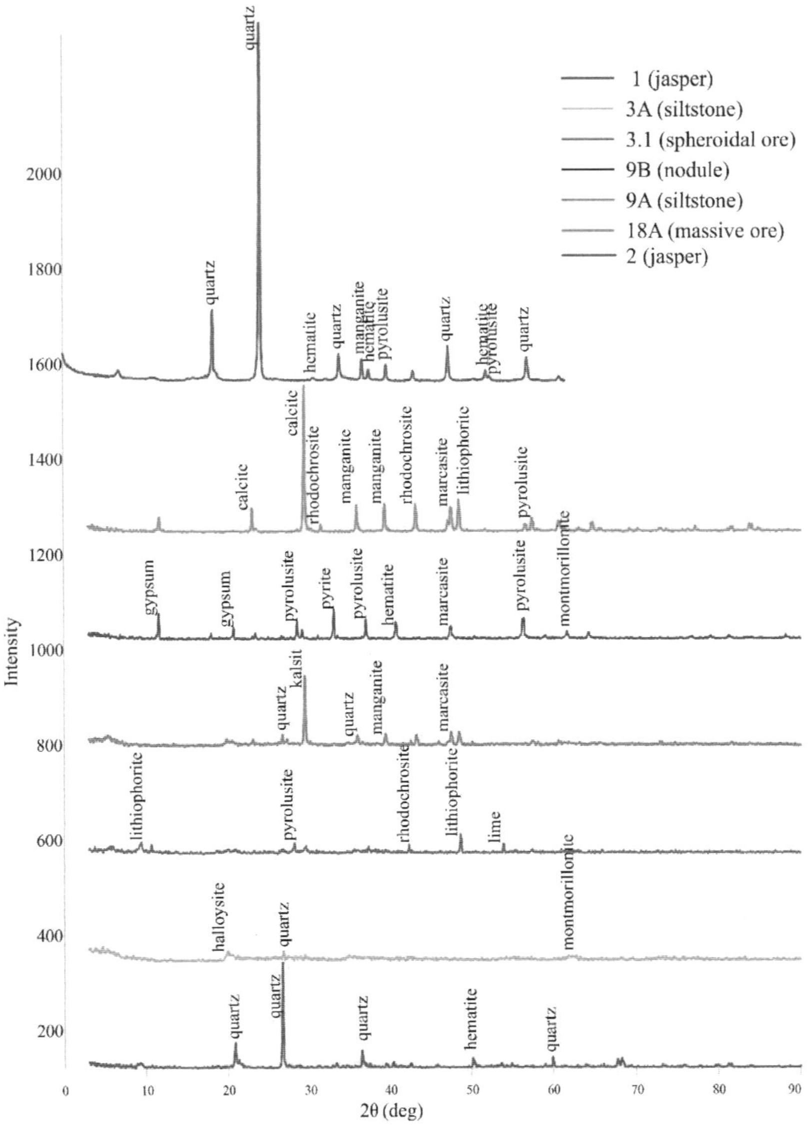

Figure 5. XRD diffractograms of 7 samples.

We also looked for hydrothermal alteration in the siltstone, which forms a sharp contact with the jasper above it. We identified two layers of siltstone below the jasper. In the hand specimen, the siltstone easily crumbles and appears to have no sign of alteration, which may change its texture. There is no difference in texture among the siltstone right below the jasper and the lower siltstone layer. XRD analysis of the siltstone sample (labeled STA 3) detected halloysite and montmorillonite (Figure 5). Halloysite is a low-temperature clay mineral [10]; meanwhile, montmorillonite is thought to be of detrital origin, due to the lack of hydrothermal alteration texture.

In the Al vs. Si diagram [11], [12], see Figure 6 and Table 2 for major oxides concentrations, we plotted grainstone, jasper and Mn ore. All jasper samples fall into the hydrothermal category. Meanwhile, other samples fall into the detrital-diagenetic region (3.1O and 18O samples, respectively). Sample 9B is a polymetallic nodule and falls into the hydrogenous-detrital category. Furthermore, sample 9A has proximity to the Si-Al concentration of marine sediments. The other three samples, i.e., 3.2O, 4A, and 18A, are challenging to put into one category because of the low Si concentration. This is reasonable because 4A and 18A are limestone, while 3.2O is manganese ore found in grainstone. Thus, the Al vs. Si diagram is not meant to distinguish limestone wall rock origin.

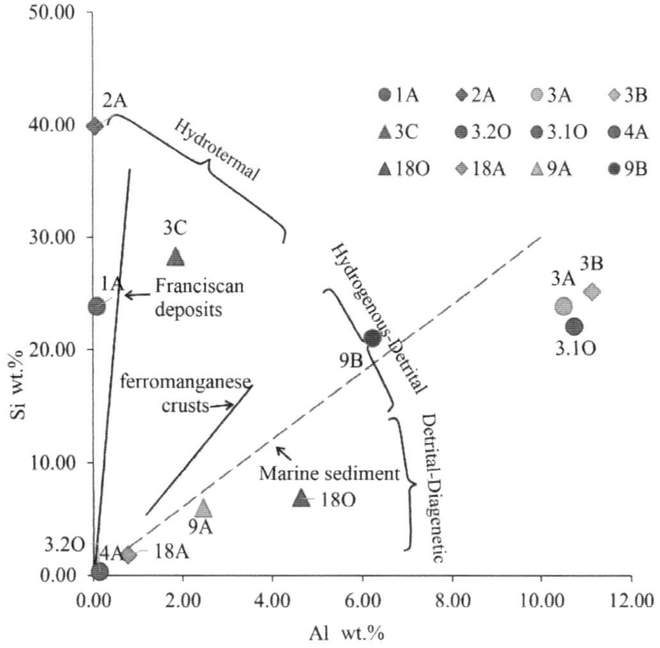

Figure 6. Al-Si discriminant diagram plot of grainstone, jasper and Mn ore samples of this study along with lines showing the trend of marine sediment, ferromanganese crust, and Franciscan ophiolithic manganiferous chert deposit [12].

4.2. Mineralization
Manganese mineralization occurred in both jasper and grainstone. Manganese-bearing minerals detected in jasper are pyrolusite and manganite. Above jasper, there is a manganese ore layer in which the ores form an aggregate of spheroidal grains bounded by calcite. Above the spheroidal ores, manganese occurs in grainstone primarily as prismatic and dendritic crystals. XRD analysis of this bed identified the presence of manganese-bearing minerals, i.e., rhodochrosite, pyrolusite, and lithiophorite. Figure 7a is a photomicrograph of the spheroidal ores, which shows Mn ore, calcite, glauconite and neither sparite nor micrite. Grainstone from the Kliripan mine as a wall rock is shown in Figure 7b, which displays several parts of *Litophyllum sp.* and *Lepidocyclina sp.* have been replaced by manganese ores instead of undergoing micritization. In Figure 7c, there is less sparite than Mn ore filling the intragranular and intergranular space in the grainstone from Anjir mine. XRD analysis of the Anjir grainstone identified rhodochrosite, manganite, pyrolusite, and lithiophorite as the manganese-bearing minerals. For comparison, Figure 7d is grainstone without manganese ore, exhibiting some voids that serve as intergranular and intragranular porosity.

Table 2 lists major oxides of the Kliripan manganese deposit while Table 3 lists sedimen reference material [7]. The highest MnO content in jasper is at 5.02 weight %. Meanwhile, the ores from Anjir mine has MnO content as high as 48.51 weight %. Figure 8 is Post Archean Australian Shale (PAAS)-normalized the trace element distribution of jasper and grainstone (a) as well as manganese ores (b) from Kliripan along with sediment references. In general, the ores are depleted in Cr, Y, Nb, Nd, Pb, Th, and U. Meanwhile there seems to be high Sc concentration in the grainstone wall rock.

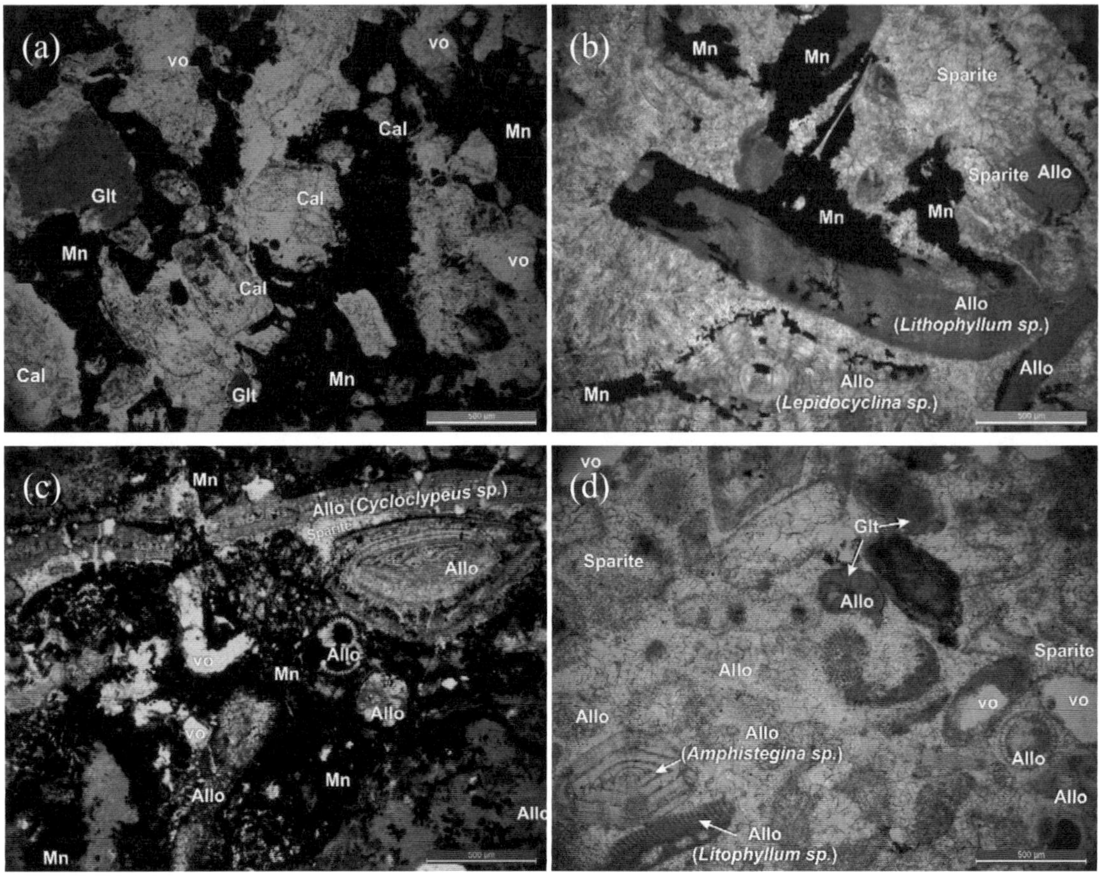

Figure 7. Plane polarized photomicrograph of (a) spheroidal ore; (b) wall rock grainstone from Kliripan; (c) wall rock grainstone from Anjir mine; and (d) grainstone without Mn ore; Allo = allochem; Cal = calcite; Glt = glauconite; Mn = manganese-bearing mineral; vo = void.

Table 2. Major oxides content of the Kliripan manganese deposit.

	1A	2A	3C	3A	3B	9B	4A	18A	3.2O	3.1O	18O	9A	**PAAS[a]**
Major oxides (wt.%)													
SiO_2	40.9	86.3	46.42	42.91	38.74	33.64	0.31	2.01	0.42	44.97	9.64	8.15	62.8
Al_2O_3	0.14	0.09	2.66	16.62	15.08	8.78	0.14	0.77	0.14	19.28	5.72	2.96	18.9
TiO_2	0.01	0.01	0.07	0.89	1.1	0.59	-	0.01	-	0.32	0.06	0.12	1
Fe_2O_3	42.61	8.91	28.62	19.3	12.82	6.67	0.12	0.45	0.01	3.72	0	51.15	-
MnO	0.25	5.02	0.66	0.15	0.34	0.05	0.18	0.47	3.04	16.52	48.51	0.01	0.1
CaO	0.04	0.37	0.09	1.65	1.93	23.09	50.4	48.7	49.49	6.11	0.31	4.3	1.3
MgO	b.d.l.	0.66	0.23	2.79	2.47	1.03	0.26	0.15	0.24	2.96	0.28	0.53	2.2
Na_2O	b.d.l.	b.d.l.	b.d.l.	b.d.l.	b.d.l.	0.07	b.d.l.	b.d.l.	b.d.l.	0.46	b.d.l.	1.22	1.2
K_2O	0.11	0.28	0.39	1.3	0.43	1.03	b.d.l.	0	0.06	0.8	0.07	0.26	3.7
P_2O_5	0.05	0.06	0.09	0.04	0.04	0.2	0.05	0.08	0.05	0.16	0.53	0.03	0.2
S	0.01	0.01	0.01	0	b.d.l.	0.19	0.02	0	0.03	0.01	0.01	35.8	
Cl	0	0.01	0	0	0	0	0	0	0	0.01	0	0.01	
H_2O^-	1.78	2.21	2.25	8.04	14.3	5.85	1.67	1.63	1.77	6.71	2.36	8	-
H_2O^+	3.97	4.17	4.08	18.3	12.9	25.5	38.3	37.0	39.2	16.7	10.65	32.2	
Total	89.8	108	85.6	112	100	106.7	91.5	91.3	94.5	118.7	78.1	144	97.9
FeO	38.34	8.02	17.37	11.54	25.75	0.01	3.35	0.11	0	0.41	46.02	6	6.5[b]
Mn	0.24	3.86	0.67	0.14	0.36	0.05	0.27	0.69	4.41	13.49	57.74	0.01	
Fe	32.77	6.85	22.01	14.84	9.9	5.13	0.09	0.35	0	2.86	0	39.34	
Al	0.09	0.05	1.84	10.51	11.14	6.24	0.14	0.78	0.13	10.75	4.65	2.46	
Si	23.9	39.9	28.3	23.9	25.2	21.0	0.3	1.8	0.4	22.1	6.9	6.0	

b.d.l. = below detection limit

[a]PAAS [13]

[b]FeOT, total Fe as FeO

Table 3. Major oxides content of sediment reference material [7].

Sample	**SiO_2**	**Al_2O_3**	**Fe_2O_3**	MgO	CaO	**Na_2O**	**K_2O**	**TiO_2**	MnO	**P_2O_5**	LOI	Total
SED-A (marl)	15	5.61	4.94	8.63	28.7	0.55	0.34	0.21	1.64	0.14	33	98.7
SED-B (volc.-clastic sandstone)	51.4	20.7	6.61	2.1	6.73	2.93	0.65	0.78	0.11	0.17	6.47	98.6
SED-C (mudstone)	55.7	22.3	6.53	2.13	0.72	0.36	2.01	0.87	0.05	0.05	9	99.8

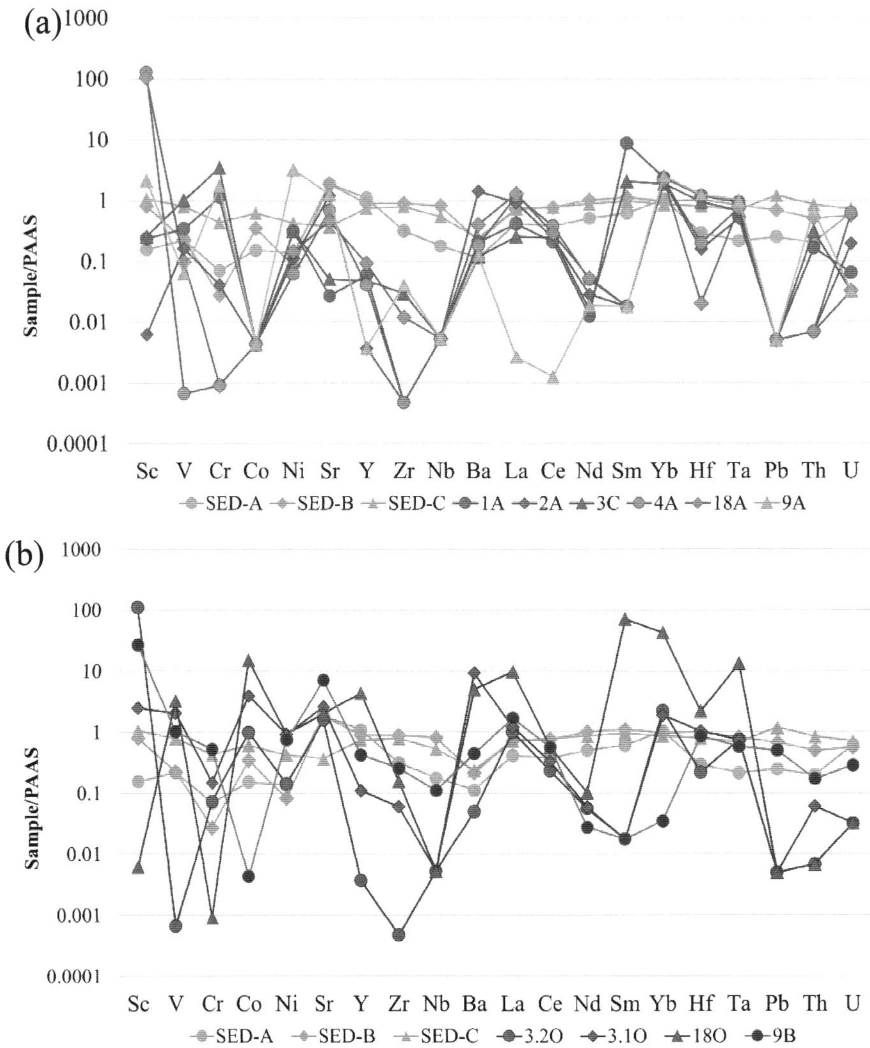

Figure 8. PAAS-normalized trace element distribution of (a) wall rock samples and (b) Mn ores. PAAS value from [13].

In Mn-Fe-(Ni+Co+Cu)×10 [11], [12] ternary diagram (Figure 9a), three Mn ore samples fall into either hydrothermal or diagenetic. The remaining grainstone and jasper samples fall into hydrothermal category. Figure 9b is Zn-Ni-Co [14] ternary diagram and shows ore samples are neither hydrothermal nor hydrogenetic, while the rest of the samples are plotted as hydrothermal. We inferred that, except for jasper, it is unwise to plot the grainstone as the wall rock to these diagrams. This is because in general limestone is depleted in those metals.

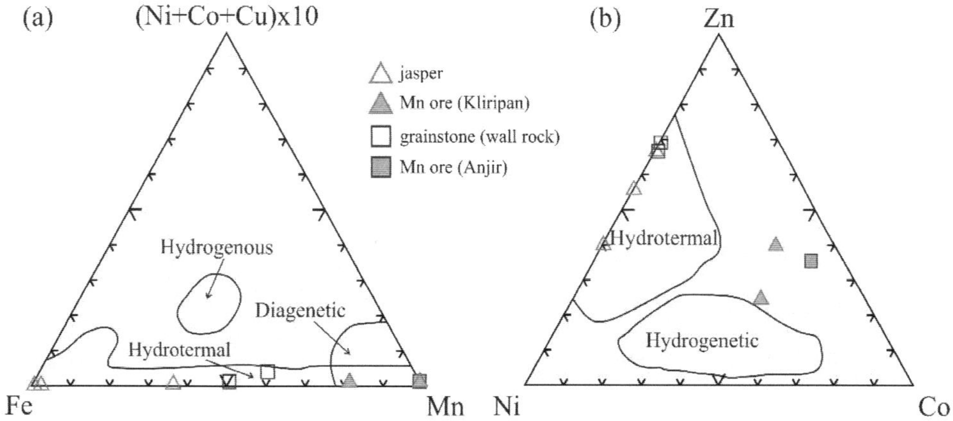

Figure 9. (a) Plot of Mn-Fe-(Ni + Co + Cu)x10 diagram [11], [12]; and (b) Zn-Ni-Co [14].

In the Ce vs Zr and Ce vs (Co+Ni+Cu)/1000 diagrams (Figure 10), we also plotted several diagenetic [15], [16], [17], [18]; hydrothermal [19], [20]; and hydrogenetic [21] Fe-Mn deposits. Our ore samples pave a similar area with the Nikopol manganese oxides [15], Kara Sea [16] and the Gulf of Finland [18] ferromanganese nodules.

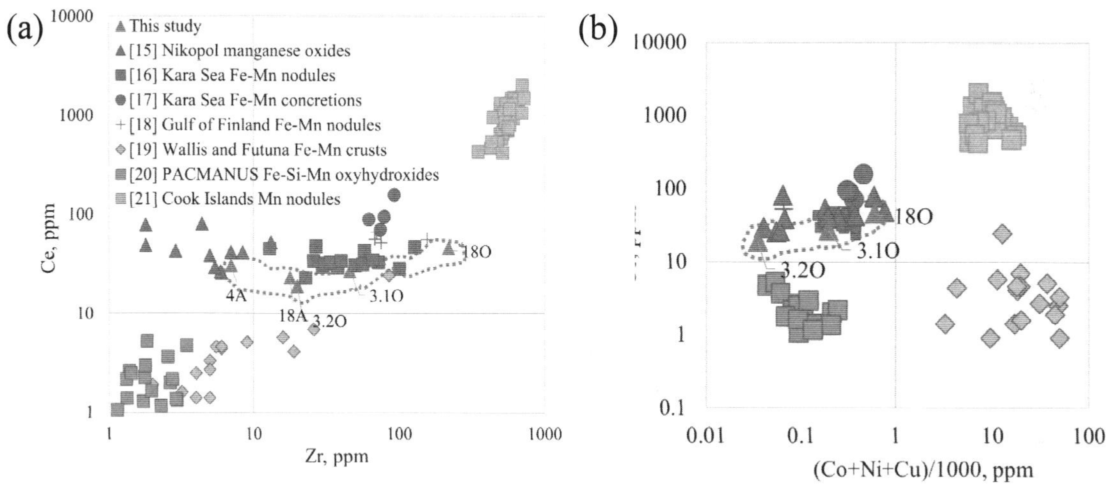

Figure 10. (a) Plot of Zr vs Ce and (b) Co + Ni + Cu)/1000 vs Ce diagram.

5. Discussion

The earliest presence of manganese mineralization in the study area is found in jasper, in the form of pyrolusite and manganese. The highest MnO content in the jasper sample is only 5.02 wt.%. However, the MnO content in this jasper sample is higher than the three reference samples for sedimentary rocks of the Sunda arc. Meanwhile, sample 2A and 3C both had higher concentrations of MnO than PAAS, although they are lower than the reference sample SED-A (marl). This means that there is MnO enrichment in jasper. The first evidence that corroborates jasper is of hydrothermal origin is the discovery of moss and saccharoidal texture. Moss texture preserves the spheroidal shape of amorphous silica gel, and each grain is defined by impurities [8]. The impurities may contain early Fe-Mn phase. In general, saccharoidal texture has been regarded as the product of calcite replacement [9]. Silica gel texture preservation and the indication of calcite replacement suggested that this jasper might originally was a carbonate rock. Further hint of the latter proposition is that the jasper has sharp contacts with the underlying siltstone and exhibits fractures and vugs due to silica gel shrinkage as dense quartz begin to crystallize [22]. We recommend the term jasperoid to refer this jasper.

The lower MnO content in other jasper samples could be due to the diagenetic remobilization. Metal remobilization occurs during the diagenesis process on the ocean floor [23]. Reductive condition is required to dissolve metal cations from the rock that incorporates them. To deposit manganese, cations need to be remobilized and deposited in relatively shallower areas which support oxidation. Manganese formation in the sea water requires disproportionation reaction, a reaction in which a species is both reduced and oxidized [24]. Based on the Pourbaix diagram (figure 11), Mn^{2+} or Mn(II) is the dominant phase of Mn in seawater [25], [26]. This explains the formation of manganite, pyrolusite, and rhodochrosite in the Mn ore. Pyrolusite or Mn(IV)–MnO_2 precipitates due to the oxidation of Mn(II) to Mn(IV). Meanwhile, manganite or Mn(III)–MnO(OH) crystallizes because Mn(II) oxidizes to Mn(III). Rhodochrosite is formed when Mn(II) reacts with CO_2 possibly from grainstone diagenesis. We have yet to discover the formation of lithiophorite–$(Al,Li)MnO_2(OH)_2$, because the Al_2O_3 content in grainstone itself is relatively low compared to PAAS. Moreover, the concentration of Li was not determined in the XRF analysis, but lithiophorite might be an indication of weathering [27].

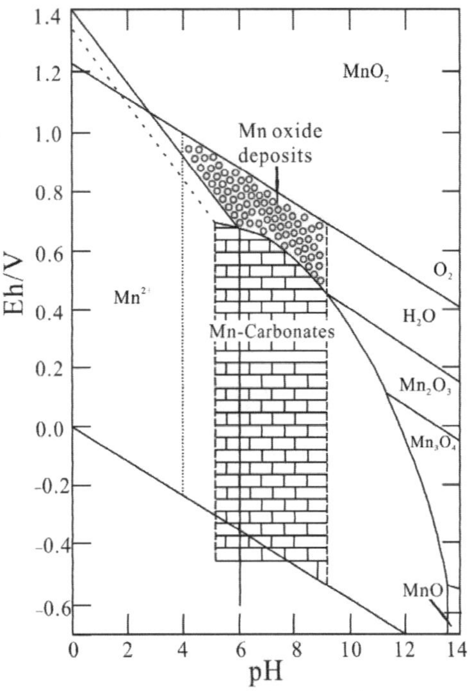

Figure 11. The Pourbaix diagram shows Mn oxide and Mn carbonate deposits stability in natural water [25], [26].

Diagenesis process as a mechanism for the formation of manganese ore in Kliripan is also supported by manganese-bearing minerals that partially replaced and filled allochems' cavity. Grainstone in study area has intergranular and intragranular porosity and lack of micrite. Both porosities accommodate the accumulation of manganese in limestone. In addition, the three ore samples belong to the diagenetic category in the Mn-Fe-(Ni+ Co+Cu)x10 diagram. Furthermore, in the Zr vs Ce and (Co+Ni+Cu)/1000 vs Ce plots, the three ore samples cover the same area as other diagenetic manganese deposits.

6. Conclusions

Manganese ore deposits in Kliripan Hamlet, Hargorejo Village are divided into two types: hydrothermal deposit which comprises jasper and the diagenetic deposit with grainstone as the host rock. These two deposits are collectively referred to as hydrothermal-diagenetic manganese deposits. Kliripan deposit contains pyrolusite, manganite, rhodochrosite and lithiophorite as the manganese-bearing mineral, while the gangue minerals are calcite, quartz, hematite, and clay minerals. The hydrothermal deposit has higher Si concentration and low concentration of Zr, Ce, Ni+Co+Cu than the diagenetic deposit.

References

[1] Kuleshov V 2017 *Isotope Geochemistry: The origin and formation of manganese rocks and ores* (Oxford: Elsevier) p 437
[2] Cannon W F 2014 *Manganese—It Turns Iron into Steel (and Does So Much More)* (U.S. Geological Survey Fact Sheet 2014–3087) p 2
[3] Balai Pelestarian Cagar Budaya Provinsi D I Yogyakarta 2019 Rumah Arsip Eks Tambang Mangaan Kliripan (cited February 2021). Available from: http://kebudayaan.kemdikbud.go.id/bpcbyogyakarta/rumah-arsip-eks-tambang-mangaan-kliripan/
[4] Rahardjo W, Sukandarrumidi, Rosidi H M D 1977 *Peta Geologi Lembar Yogyakarta: Pusat Penelitian dan Pengembangan Geologi, skala 1:100.000*
[5] Badan Informasi Geospasial (BIG) 2017 Rupa Bumi Indonesia (cited 20th October 2020) Available from: https://portal.ina-sdi.or.id/downloadaoi/
[6] Chen P Y 1977 *Table of Key Lines in X-ray Powder Diffraction Patterns of Minerals in Clays and Associated Rocks* (Bloomington: Authority of the State of Indiana) p 67
[7] Handley H K, Blichert-Toft J, Gertisser R, Macpherson C G, Turner S P, Zaennudin A, and Abdurrachman M 2014 *Geo. et Cos. Act.* **139** 205–26
[8] Morrison G W, Jaireth S, and Guoyi D 1995 *Textural zoning in epithermal quartz veins* (Townsville, Klondike Exploration Services)
[9] Dong G, Morrison G, and Jaireth S 1995 *Econ. Geol.* **90** 1841–56
[10] Bauluz B 2015 Rev.de la Real Acad. de Cie. **70** 1-33
[11] Bonatti E, Kraemer T, Rydell H 1972 *Classifcation and genesis of submarine iron manganese deposits. In: Horn, D. (Ed.), Ferromanganese Deposits on the Ocean Floor: International Decade of Ocean Exploration* (Washington, DC: National Science Foundation) pp 149–66
[12] Crerar D A, Namson J, Chyi M S, Williams L, and Feigenson M D 1982 *Econ. Geol.* **77** 519–40
[13] Taylor S R and McLennan S M 1985 *The Continental Crust: Its Composition and Evolution* (Oxford: Blackwell)
[14] Choi J H and Hariya Y 1992 *Econ. Geol.* **87** 1265–74
[15] Sasmaz A, Zagnitko V M and Sasmaz B 2020 *Ore Geol. Rev.* **126** 103772
[16] Vereshchagin O S, Perova E N, Brusnitsyn A I, Ershova V B, Khudoley A K, Shilovskikh V V and Molchanova E V 2019 *Ore Geol. Rev.* **106** 192–204
[17] Baturin G N 2011 *Oceanology* **51** 148–56
[18] Baturin G N 2009 *Lith. and Min. Res.* **44** 411–26
[19] Josso P, Pelleter E, Pourret O, Fouquet Y, Etoubleau J, Cheron S and Bollinger C 2017 *Ore Geol. Rev.* **87** 3–15
[20] Zeng Z, Ouyang H, Yin X, Chen S, Wang X and Wu L 2012 *J. Asian. Earth. Sci.* **60** 130–46
[21] Hein J R, Spinardi F, Okamoto N, Mizell K, Thorburn D and Tawake A 2015 *Ore Geol. Rev.* **68** 97–116
[22] Lovering T G 1962 *Econ. Geol.* **57** 861–889
[23] Bonatti E 1975 *Ann. Rev. E. Plan. Sci.* **3** 401–31
[24] Gramm-Osipov L M 1997 *in* Nicholson K, Hein J R, and Buhn B *Manganese Mineralization: Geochemistry and Mineralogy of Terrestrial and Marine Deposits* (London: The Geological Society) p 301–8
[25] Roy S 2006 *Earth Sci. Rev.* **77** 273–305
[26] Duan J, Fu Y, Zhang Z, Xiao J, and Wu C 2020 *J. Geochemical Explor.* **214** 106541
[27] Varentsov I M 1996 *Manganese Ores of Supergene Zone: Geochemistry of Formation* (Dordrecht: Springer Science+Business Media) p 354

Acknowledgments
The authors would like to thank the Department of Geological Engineering, Faculty of Engineering, Gadjah Mada University for providing auspices to aid this research as well as the staffs of GetIn-CICERO Laboratory, in which the petrographic and XRF analysis were performed.

IOP Conf. Series: Earth and Environmental Science 1071 (2022) 012027 doi:10.1088/1755-1315/1071/1/012027

Application of Hydraulic Flow Unit (HFU) and Windland R35 Methods in Sand-5 Reservoir, Field 'S', Talang Akar Formation, Asri Basin, Indonesia

S S A S Dondo[1] and S S Surjono[1*]

[1]Department of Geological Engineering, Faculty of Engineering, Universitas Gadjah Mada, Yogyakarta, Indonesia
*Corresponding author's email: sugengssurjono@ugm.ac.id

Abstract. The Sand-5 Reservoir in the Talang Akar Formation is the main oil and gas producer in the Field 'S', Asri Basin, Indonesia. Reservoir rock type (RRT) can be applied to optimize the performance of this reservoir. This study aims to compare the RRT methods between Hydraulic Flow Unit (HFU) and Windland R35 to find the best RRT method. Both methods utilized the porosity and permeability to classify the number of RRT, subsequently validated by the regression value (R2) of core permeability with predicted permeability. Four RRT values have been identified for both methods that are RRT-1 represent the very good reservoir quality, RRT-2 represents the good reservoir quality, RRT-3 represents the moderate reservoir quality and also RRT-4 represents the poor reservoir quality. The numbers of porosity and permeability for HFU method is 32-40% and 19.3-29.4 Darcy (RRT-1); 31-39% and 5.3-18.3 Darcy (RRT-2); 20-36% and 0.4-6.8 Darcy (RRT-3); 15-34% and 0.01-0.4 Darcy (RRT-4), while for Windland R35 method is 31-40% and 7.7-29.4 Darcy (RRT-1); 30-39% and 4-10 Darcy (RRT-2); 24-38% and 0.2-6.8 Darcy (RRT-3); 15-33% and 0.01-0.2 Darcy (RRT-4). Based on the relationship between core permeability and predicted permeability from each RRT, the result of the Winland R35 method is better than the HFU method.

1. Introduction

This conducted research in the Field 'S' in Asri Basin, Indonesia (Figure 1) is located in an offshore area about 90 km northwest of Jakarta with a research area of 25 km2. Field 'S' is the largest oil-producing field in the Asri Basin. Oil production in this area was started in December 1990 after drilling a well to a depth of 3735 feet (SSTVD) [1]. The series of reservoir targets has an excellent property with an average porosity value of 29% and permeability up to 30 Darcy. Sand-5 Reservoir is a lower part of the series. To determine the reservoir quality in Sand-5 Reservoir, the reservoir rock type (RRT) method is applied. RRT values can be approached by the flow character similarity using Hydraulic Flow Unit (HFU) and Windland R35.

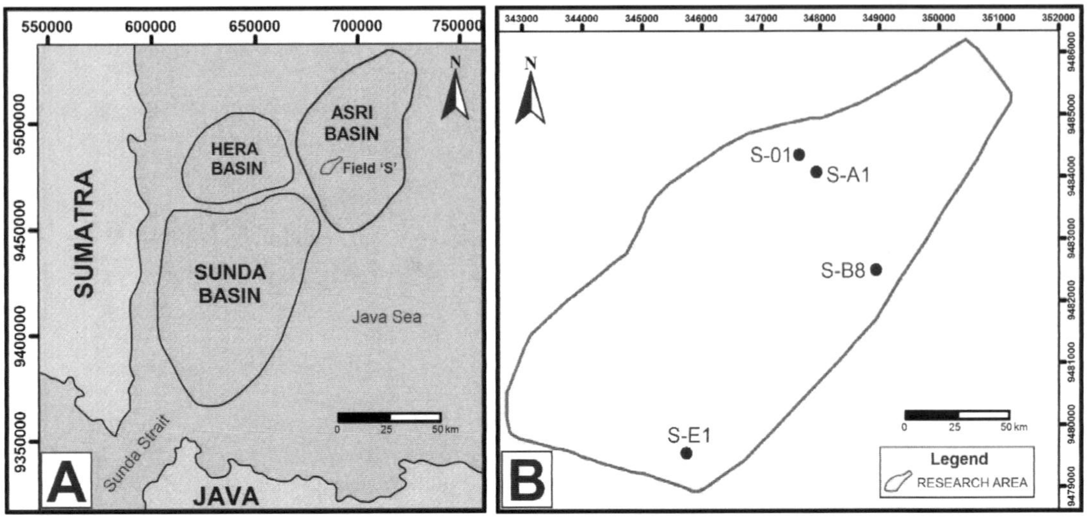

Figure 1. (a) Location of Field 'S' in Asri Basin, and (b) Location of research is limited by four wells (modified from [2]).

2. Regional Geology of The Research Area

The Asri Basin is a complex extensional basin formed by the intracratonic extensional force, [3,4]. During the Oligocene, due to the collision between the Indian and Eurasian plates, some expansions within the Eurasian plate were created. One of these expansions formed the Asri Basin which consists of a series of half-grabens with an eastward slope of the layer. As a result of this expansion, the half-grabens formed developed asymmetrically until the expansion process is complete.

As a part of The Gita Member of the Talang Akar Formation, Sand-5 Reservoir was formed in the syn-rift phase (Figure 2), begins with the removal and erosion of previously deposited sediments in the Early Miocene [3,5]. The Gita members are interpreted as fluvial deposits formed during the transgression phase with relatively weak energy and having sandstone, claystone, siltstone, and coal as their constituent lithology. The sandstone that was deposited in the fluvial environment generally has coarse to very coarse grain sizes with medium sorting, which characterizes the presence of clastic deposits deposited on river bodies. The lithology of the sandstones in the Gita Reservoir varies from very coarse to fine-grained with a mineral composition of about 75% of the sandstones are quartz arenite with the remaining 25% between sublitharenites and subarkose. In this case, the mineral composition of the rock can affect the porosity and permeability of the sandstone as the main lithology that is composed of the Gita Reservoir. The presence of fine-sized materials such as clay minerals and coal inserts in the Gita reservoir can reduce the porosity and permeability values. Overall, the reservoir has a good quality, with an average porosity value of 29% and a permeability value in the range of 10-20 Darcy. Towards the end of the Gita Member Deposit, thick coal began to be deposited due to the effect of transgression.

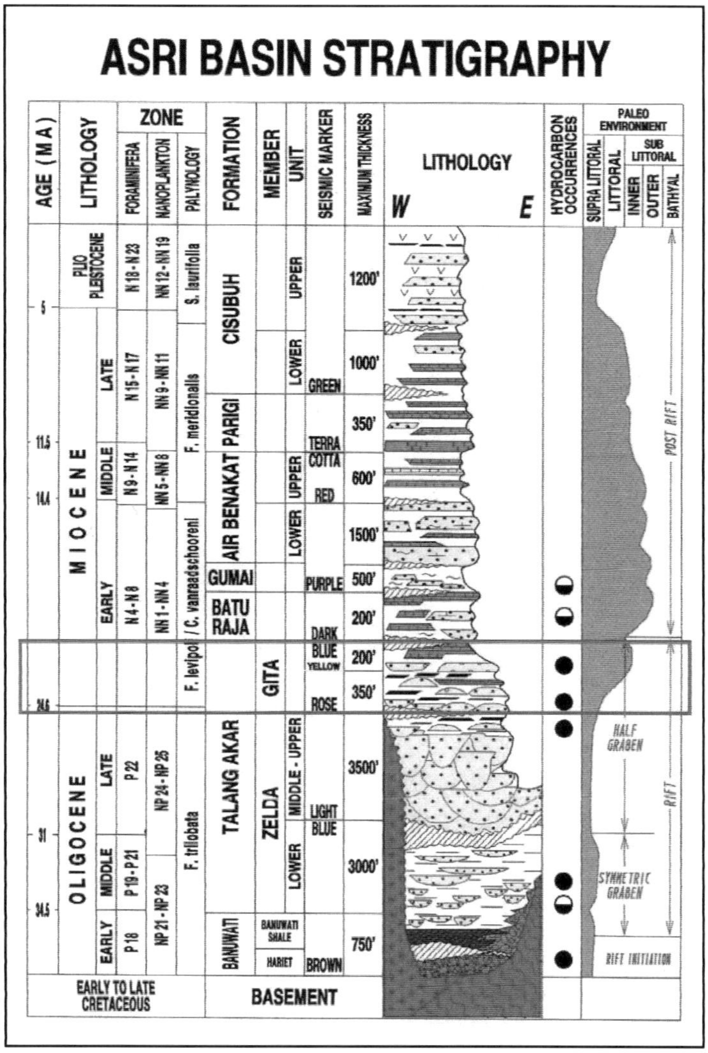

Figure 2. Asri Basin Regional Stratigraphy [3]. The Sand-5 Reservoir in the field 'S' is composed of Gita Members (red box) of the Talang Akar Formation, which was formed during the Syn Rift phase.

3. Research Methods

Identification of reservoir rock type (RRT) in this research using two RRT methods which are Hydraulic Flow Unit (HFU) and Windland R35 by petrophysical analysis through pore geometry determination (type and size) which has a relationship between porosity and permeability with quantitative calculation. This approach is also related to the determination of lithofacies that has been carried out by previous research [2]. By doing this, the best method of RRT can be validated by the regression value (R^2) of core permeability with predicted permeability for each RRT. The data used in this research consist of four wells in the Sand-5 Reservoir which have logs and core data, namely S-01, S-A1, S-B8, and S-E1 well (Figure 3).

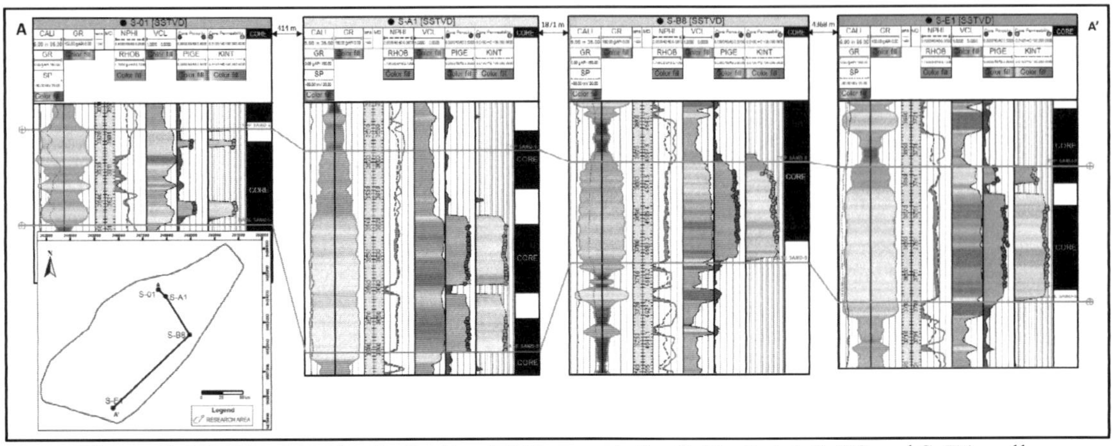

Figure 3. Well logs section in the Sand-5 Reservoir for S-01, S-A1, S-B8 and S-E1 well.

3.1. Hydraulic Flow Unit (HFU)

The Hydraulic Flow Unit (HFU) method is used to classify reservoir rocks into more detailed units where each unit is deposited in a similar geological condition and has undergone the same diagenetic alteration. [6,7] developed the concept of HFU to divide rock facies as a result of the considerable variation of permeability. This method is based on mineralogy and texture which defines similar fluid flow characteristics that are independent of lithofacies. Later, [6] suggested a methodology of using core and log data for the identification of Flow Zone Indicator (FZI) for the purpose of permeability prediction and can be expressed as follow:

$$\text{RQI} = 0.0314 \times \sqrt{\frac{\text{k}}{\text{Øe}}} \tag{1}$$

$$\text{Øz} = \left(\frac{\text{Øe}}{1-\text{Øe}}\right) \tag{2}$$

$$\text{FZI} = \frac{\text{RQI}}{\text{Øz}} \tag{3}$$

where RQI: Reservoir Quality Index, FZI: Flow Zone Indicator, k: permeability, Øe: effective porosity, and Øz: normalize porosity.

Normal probability analysis using FZI can be used to classify reservoir rock type (RRT) by a graph of normal probability on the FZI log, where the X-axis is the FZI log and the Y-axis is the FZI probability. With this analysis, reservoir rock type can be determined by looking at the slope changes of the regression line.

3.2. Windland R35

The Windland R35 method was used to classify reservoir rocks using calculation of the pore throat size in 35% mercury saturation while testing mercury injection capillary pressure. This method can be used to optimize of flow unit performance capacity and discriminate the flow units interval with similar pore throat radius with different inflow performance [8]. Windland's equation is powerful methods used to utilize the reservoir properties as porosity and permeability to classify number of RRT [9]. It proves that the method can use as a net pay cut-off to exclude very low porosity and permeability. Pore throat size

(R35) determining depending on the core permeability and porosity in reservoir units provides the best basis for defining reservoir flow units and is utilized to define petrophysical units as follows:

$$Log\ (R35) = 0.732 + 0.588\log(k) - 0.864\log\ (\emptyset) \quad (4)$$

where R35: port radius (micron) at 35% mercury saturation, k: permeability (mD), and $\emptyset$: core porosity (%).

Normal probability analysis using R35 can be used to classify RRT by using a semi-log plot of R35, where the X-axis is the pore throat, and the Y-axis is the Sample Number. Windland R35 value is sorted from smallest to largest, while the serial number is a sequence of numbers. With this analysis, RRT can be determined by looking at slope changes of the regression line.

4. Results and Discussion

Reservoir rock type (RRT) identification using Hydraulic Flow Unit (HFU) and Windland R35 method can be carried out based on the unique relationship between the porosity and permeability values. Therefore, this relationship cannot be relied upon for the accurate estimation of permeability values from porosity data. As a lower part of the reservoir series in the research area, Sand-5 Reservoir shows the cores data from four wells with porosity values ranging from 30-40% and permeability values up to 29.5 Darcy.

In the first assessment, the HFU method is used to classify RRT with several validations by using equation (1) and (2), where the function for Reservoir Quality Index (RQI) versus the logarithm of the Normalized Porosity ($\emptyset z$) consequently, for various values of the Flow Zone Indicator (FZI). All points with similar FZI lines have the same pore throat (Figure 4). Based on the determination of the FZI value on equation (3), four RRT have been identified in the Sand-5 Reservoir based on the relationship between the porosity and permeability value. Figure 5 shows the RRT distribution with the RRT-1 represents the very good reservoir quality with porosity values ranging from 32-40% and permeability values varying between 19.3-29.4 Darcy. The RRT-2 represents the good reservoir quality with porosity values ranging from 31-39% and permeability values varying between 5.3-18.3 Darcy. The RRT-3 represents the moderate reservoir quality with porosity values ranging from 20-36% and permeability values varying between 0.4-6.8 Darcy. And the RRT-4 represents the poor reservoir quality with porosity values ranging from 15-34% and permeability values varying between 0.01-0.4 Darcy.

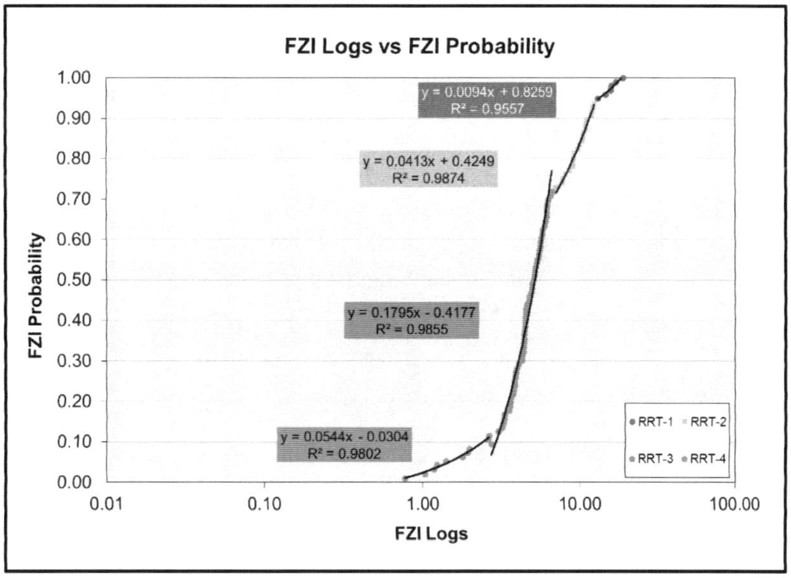

Figure 4. The results of reservoir rock type (RRT) classification based on the relationship between FZI logs and FZI probability (modified from [2]).

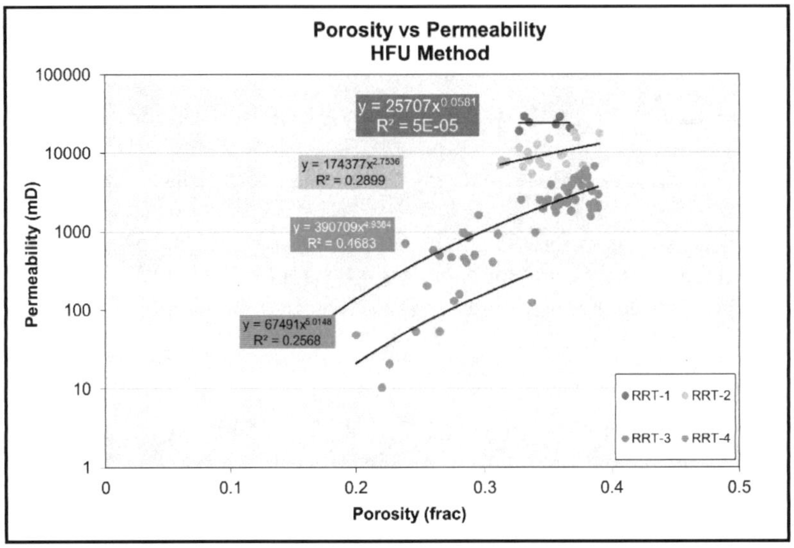

Figure 5. The results of reservoir rock type (RRT) classification based on the relationship between porosity and permeability (modified from [2]).

In the Second assessment, the Windland R35 method is used to classify reservoir rock type (RRT) using equation (4) as a validation. In this equation, RRT can be classified by using a semi-log of R35 between pore throat versus sample number and shows slope change of the regression line. Each change in the slope indicates the difference in RRT (Figure 6). There are four RRT have been identified in the Sand-5 Reservoir based on the relationship between the porosity and permeability value (Figure 7). The RRT-1 represents a very good reservoir quality with porosity values ranging from 31-40% and permeability values varying between 7.7-29.4 Darcy. The RRT-2 represents good reservoir quality with porosity values ranging from 30-39% and permeability values varying between 4-10 Darcy. The RRT-3 represents the moderate reservoir quality with porosity values ranging from 24-38% and permeability values varying between 0.2-6.8 Darcy. And the RRT-4 represents the poor reservoir quality with porosity values ranging from 15-34% and permeability values varying between 0.01-0.2 Darcy.

Based on two assessments of reservoir rock type identification (RRT) in this study, RRT-4 from both methods represents the reservoir with poor quality even though it has a high porosity value ranging from 15-34% and a low permeability value ranging from 0.01-0.2 Darcy. This is due to the porosity of the sandstone resulting from geological processes that affect the sedimentation process. These processes occur at the time processes during deposition and processes after deposition, which controls at the time of deposition are influenced by the texture of the sandstone. Processes after deposition that affect porosity are caused by physical and chemical influences, which are a function of temperature, effective pressure, and time.

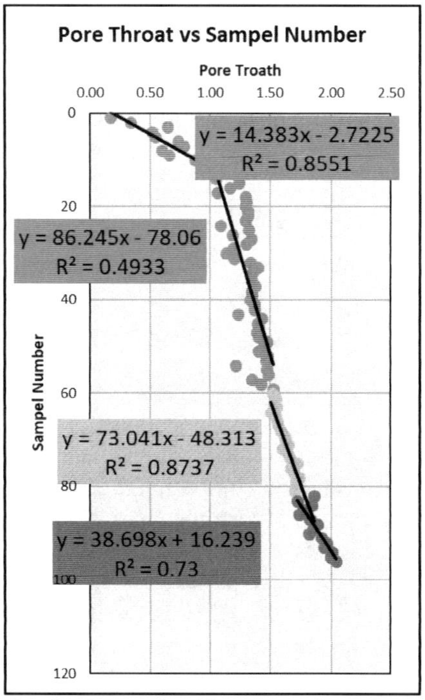

Figure 6. Semi-log plot of Windland R35 to classify reservoir rock type (RRT) between pore throat and sample number.

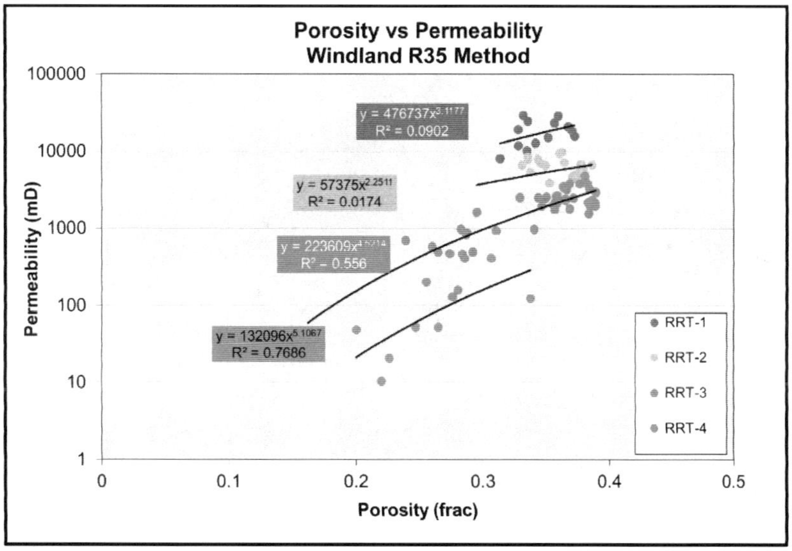

Figure 7. The results of reservoir rock type classification based on Windland R35 method in the correlation between porosity and permeability.

The result of the reservoir rock type (RRT) identification using the Hydraulic Flow Unit (HFU) and Windland R35 method based on the relationship between porosity and permeability have different regression values (R2) and the coefficient correlation values of each RRT. The other values indicate unfavorable results due to the scattered data, which resulted in varying permeability for the same porosity value. The existing trendline only represents the mean value of the permeability. Table 1 shows the coefficient correlation between porosity and permeability of each RRT method by using power trendline regression.

Table 1. The coefficient correlation of permeability and porosity of each reservoir rock type method.

RRT	Reservoir Rock Type Methods	
	HFU	Windland R35
RRT-1	$y = 25707x^{0.0581}$	$y = 476737x^{3.1177}$
RRT-2	$y = 174377x^{2.7536}$	$y = 57375x^{2.2511}$
RRT-3	$y = 390709x^{4.9364}$	$y = 223609x^{4.5214}$
RRT-4	$y = 67491x^{5.0148}$	$y = 132096x^{5.1067}$

In this research, the coefficient correlation of porosity and permeability (Table 1) can use to determine the predicted permeability value by comparing the core permeability as validation by using the regression value (R2) for each reservoir rock type (RRnT) from Hydraulic Flow Unit (HFU) and Windland R35 method. Based on the regression value (R2) of each RRT from both methods by comparing the core permeability and predicted permeability shows, the Windland R35 method has a good gradient value for each RRT rather than the HFU method (Figure 8 and Figure 9). The higher value of regression for each RRT from both methods indicates that the trendline match has been close to the actual condition of core data. Table 2 shows the regression value (R2) and coefficient correlation between HFU and Windland R35 method for each RRT using power trendline regression.

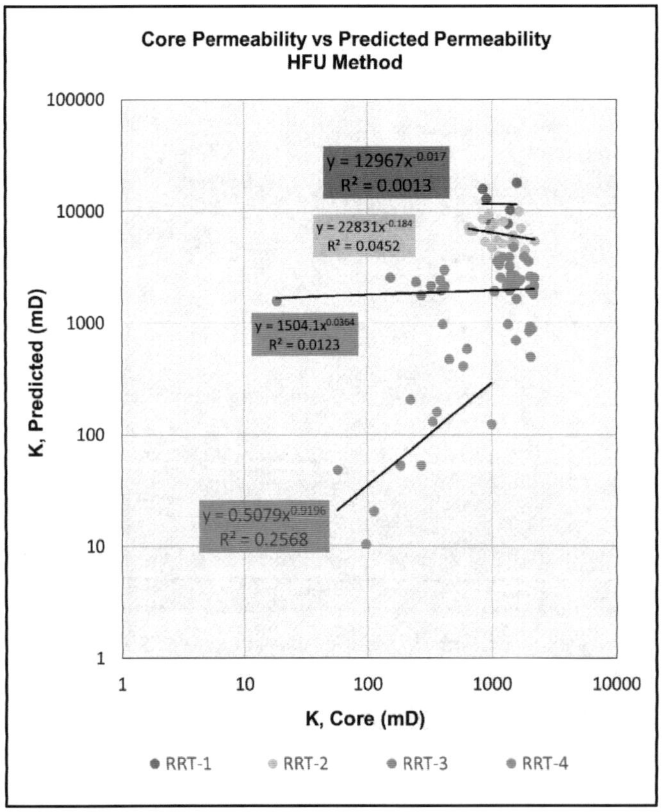

Figure 8. The cross-plot between core permeability and predicted permeability based on the Hydraulic Flow Unit (HFU) method.

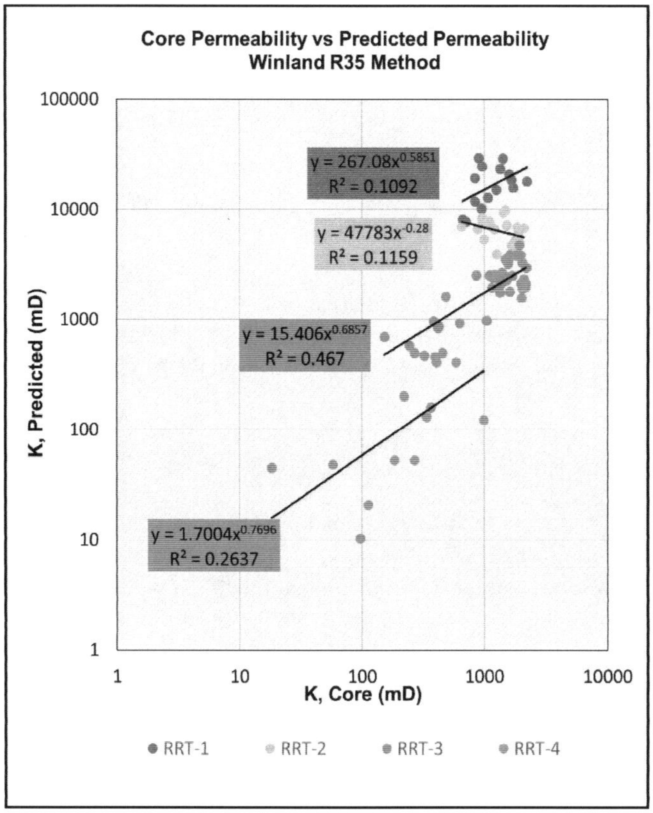

Figure 9. The cross-plot between core permeability and predicted permeability based on Windland R35 method.

Table 2. The coefficient correlation and regression value (R^2) of predicted permeability and core permeability for each reservoir rock type from HFU and Winland R35 method.

RRT	HFU		Windland R35	
	Coefficient Correlation	Regression Value (R^2)	Coefficient Correlation	Regression Value (R^2)
RRT-1	$y = 12967x^{0.017}$	0.0013	$y = 267.08x^{0.5851}$	0.1092
RRT-2	$y = 22831x^{0.184}$	0.0452	$y = 47783x^{0.28}$	0.1159
RRT-3	$y = 1504.1x^{0.0364}$	0.0123	$y = 15.406x^{0.6857}$	0.467
RRT-4	$y = 0.5079x^{0.9196}$	0.2568	$y = 1.7004x^{0.7696}$	0.2637

5. Conclusion

Four reservoir rock types (RRT) have been identified in the Sand-5 Reservoir depending on the Hydraulic Flow Unit (HFU) and Windland R35 method by regression value of porosity and permeability relationship for each RRT. Based on the relationship between core permeability and predicted permeability for each RRT, the Windland R35 method is the best reservoir rock type method with a regression value (R2) of 0.1092 for RRT-1, 0.1159 for RRT-2, 0.467 for RRT-3, and 0.2637 for RRT-4. These traits indicate that the Sand-5 Reservoir has the potential to become a good quality reservoir proved, with porosity values ranging from 15-40% and permeability values up to 29.5 Darcy. The application of RRT using the HFU or Windland R35 method in this research can be maximized by adding the number of wells and samples of the cores used to describe the reservoir condition in more detail and representative.

Acknowledgments
The authors would like to show gratitude to the PT. Pertamina Hulu Energi OSES for their permission and support in the making of this research in the Sand-5 Reservoir, Field 'S', Asri Basin.

Reference

[1] Young R 1991 Widuri field, offshore Southeast Sumatra: sandbody geometries and the reservoir model *Proceeding Indonesian Petroleum Assocociation, 20th Annual Convention* Twentieth Annual Convention (Indonesian Petroleum Association (IPA))

[2] Dondo S S A S 2021 *Distribusi Properti Reservoir dengan Metode Petrophysical Rock Type pada Lapangan 'S', Formasi Talang Akar, Cekungan Asri* (Yogyakarta: Universitas Gadjah Mada)

[3] Sukanto J, Nunuk F, Aldrich J B, Rinehart G P and Mitchell J 1998 Petroleum systems of the Asri basin, Java Sea, Indonesia *Proceeding Indonesian Petroleum Association, 26th Annual Convention* Twenty-Sixth Annual Convention (Indonesian Petroleum Association (IPA)) pp 291–312

[4] Aldrich J B, Rinehart G P, Ridwan S and Schuepbach M A 1995 Paleogene Basin Architecture of the Sunda and Asin Basins and Associated Non-Marine Sequence Stratigraphy *Proceedings of the International Symposium on Sequence Stratigraphy in S.E. Asia* (Indonesian Petroleum Association (IPA)) pp 261–87

[5] Primadani G S, Watkinson I M, Gunawan H and Ralanarko D 2018 Tectonostratigraphy of The Asri Basin, Se Sumatera, Indonesia: Unlocking the Hidden Potential of Oligo-Miocene Reservoirs and Implications for Hydrocarbon Prospectivity *Proceeding Indonesian Petroleum Association, 42nd Annual Convention* (Indonesian Petroleum Association (IPA))

[6] Amaefule J O, Altunbay M and Keelan D K 1993 Enhanced Reservoir Description: Using Core and Log Data to Identify Hydraulic (Flow) Units and Predict Permeability in Uncored Intervals/Wells *SPE annual technical conference and exhibition* the 68th Annual Technical Conference and Exhibition of the Society of Petroleum Engineers p 16

[7] Guo G, Diaz M A, Paz F, Smalley J and Waninger E A 2005 Rock Typing as an Effective Tool for Permeability and Water-Saturation Modeling: A Case Study in a Clastic Reservoir in the Oriente Basin *SPE Annual Technical Conference and Exhibition* p 15

[8] Martin A J, Solomon S T and Hartmann D J 1997 Characterization of Petrophysical Flow Units in Carbonate Reservoirs *AAPG Bulletin* **81** 734–59

[9] Al-Qenae K J and Al-Thaqafi S H 2015 New Approach for the Classification of Rock Typing Using a New Technique for Iso-Pore Throat Lines in Winland's Plot *All Days* SPE Annual Caspian Technical Conference & Exhibition (Baku, Azerbaijan: SPE) pp 1–15

Geology and the effect of compaction on base material disposal for the stability of coal disposal slopes in South Kalimantan, Indonesia

B A Abdullah* and I G B Indrawan

Department of Geological Engineering, Faculty of Engineering, Gadjah Mada University, Yogyakarta, Indonesia
*Corresponding author's e-mail: barlianaditama@mail.ugm.ac.id

Abstract. A mining activity generally removes overburden to extract minerals. Therefore, a certain area is needed to place the overburden material. One of the parameters that can affect the stability of the disposal is the base or the foundation of the disposal area which will later be stockpiled. This study is observe influence of base disposal to the stability of slope. Based on the lithological characteristics and observations of the outcrop, the writer arranges the stratigraphy of the study area in order to determine the stratigraphic arrangement and the distribution of the rock units laterally. In general, the study area is dominated by quartz sandstone, claystone, conglomerate and loose material from older rocks. The stratigraphic sequence of the study area from old to young includes Warukin Sandstone Unit (Tmw), Dahor Layang Member Conglomerate (TQdt) and Dahor Sandstone Unit (TQd). In general, the research area is included in the structural landform unit with geomorphological control that influences the tectonic process (endogenous). Based on the results of the study, there was a significant increase related to the stability of the coal disposal slope with compaction efforts carried out on the base material.

1. Introduction

The stability for the construction of the waste dump area (disposal) is critical for the long term of a mining activity, both in terms of the operational support aspects of active and post-mining activities. Geotechnical investigations and slope stability analysis must be carried out prior to stockpiling to ensure adequate and safe disposal [1]. The research location is in Sungai Raya Selatan Village, Sungai Raya District, Hulu Sungai Selatan Regency, South Kalimantan Indonesia [2]. The research location is a PKP2B consensus area owned by PT Antang Gunung Meratus in block 4 area where the area is planned to be used as a waste dump area.

As for the research area to be carried out, the disposal area gets considerable attention. This is because the availability of land to be used as a landfill for waste material is not available in large quantities related to regulatory, social and technical problems. One of the parameters that can affect the stability of the disposal is the base or the foundation of the disposal area which will later be stockpiled. In this research, we will try to find out the effect of soil compaction at the bottom of the disposal area on the volume of disposal that can be stockpiled in that area. By increasing the capacity of the soil, it is expected to increase the volume of disposal. In addition, compaction of the base disposal will make the subgrade more impermeable so that groundwater infiltration can be reduced and strengthen the embankment.

Research on soil bearing capacity for mine disposal previously researched by Andriani [4] in Kutai Kartanegara Regency. This study resulted in the maximum capacity of the soil bearing capacity in the available disposal area. However, this research has not discussed the potential for increasing the ability of the soil to withstand disposal loads.
Another researcher, Suherman [4] conducted a study related to efforts to increase the bearing capacity of the subgrade by using dynamic compaction techniques applied to the construction of toll roads. This research resulted in a conclusion that dynamic compaction can increase the bearing capacity of the soil from gravel sand to plastic clay with varying degrees of increase depending on the thickness and energy applied.

Based on some of the considerations above, this research is considered considered important as a solution in handling disposal in coal mining.

2. Geological Setting

Physiographically, the Barito Basin is located in the southeastern part of the island of Borneo. This basin is an asymmetrical basin (Figure 1). The Barito Basin is bounded to the west by the Schwaner Mountains as a representation of the bedrock of the Sunda Shelf and to the east by the Meratus Mountains, which separates the Barito Basin and the Asam-Asam Basin [5].

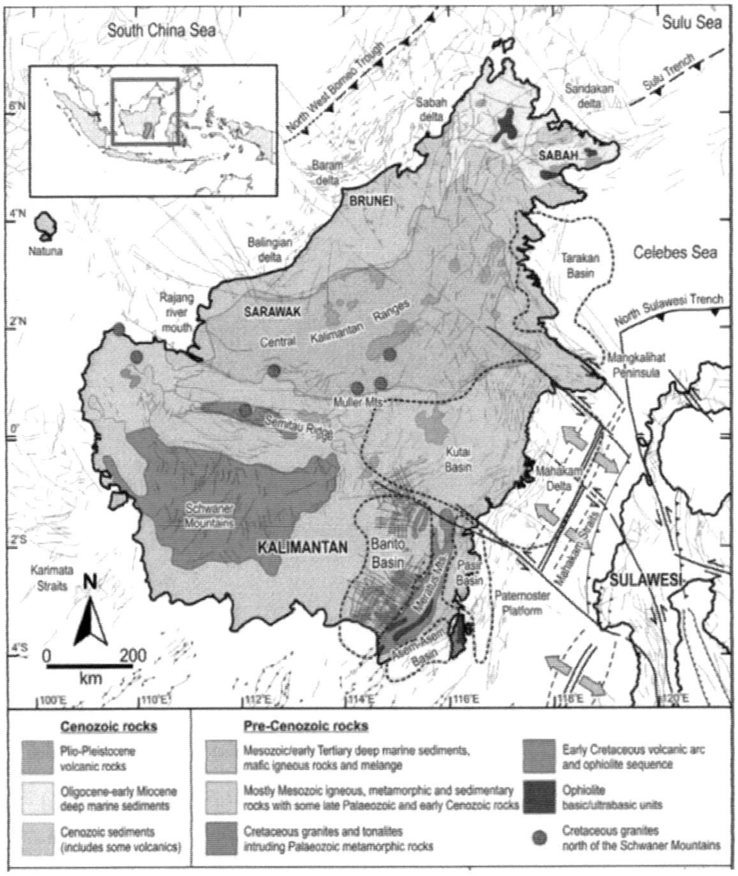

Figure 1. Simplified geological and physiographic map of Borneo (Sapiie and Rifiyanto. 2017).

To the north, the Adang Flexure separates the Barito Basin from the Kutai Basin. Meanwhile, in the southern part, the boundary becomes a little more vague or appears to narrow towards the south to the Java Sea due to the absence of any observable appearance of structures or changes in lithofacies [6].

The sequence of sedimentation events in the Barito Basin is inconsistently overlapping the Paleocene bedrock and beyond [7]. Sedimentation event sequences consist of five formations that record a transgressive to fully regressive cycle. The oldest sedimentary rocks were introduced into the Tanjung Formation and deposited on the fluviotidal coastal plain to the marine environment boundary, this causes the formation to be increasingly influenced by the marine environment.

3. Methodology

Geological mapping is carried out by taking into geological conditions, lithology, geomorphology and geological structures. This mapping is used to determine the lithology variation, distribution and thickness of each rock unit. The map of the observation location shows the coordinates or the location of the observation of geological data in the form of lithology distribution, formation boundary forecasts and soil sampling for later analysis in the laboratory (see appendix A and B).

Furthermore, in a spesific study on the effect of compaction on disposal stability, the Properties Index test, Size Distribution Analysis, and Atterberg Limit Test were first carried out which were distributed at 5 observation points or sampling locations in the planned disposal area. The selection of sampling locations is based on interest of area planned disposal to be built. This is done to find out how the variability of soil properties as a whole is.

To evaluate the strength of soil and rock, soil compaction test was carried out using the laboratory CBR test with 3 variations of loading, based on SNI 1744:2012 (10 blows, 25 blows and 56 blows) [8]. The California Bearing Ratio (CBR) test is a penetration test used to evaluate the strength of a subgrade. The CBR value is expressed in percent [9]. The results from the CBR lab are then converted using equation to get the cohesion value (c) for the calculation of disposal stability. Based on the three loading variations in the CBR test above, a slope design modeling comparison between the compacted and uncompacted base material was carried out.

The CBR value will then be converted to a soil cohesion value (Cu) by assuming the material as a cohesive soil to be further modeled in slide 6.0 software. Gregory and Cross [10] conducted a study on the correlation between CBR values and soil shear strength parameters for cohesive and non-cohesive soils. This research is based on the force acting with the CBR value at the penetration of 2.54 mm (0.1 inch) which is considered as the bearing capacity of the soil (q ultimate). The general soil bearing capacity equation for vertical loading, proposed by Meyerhof is as follows:

$$q_{ult} = c\, N_c s_c d_c + q'\, N_q s_q d_q + 0.5\, \gamma\, B\, N_y s_y d_y \quad (1)$$

For cohesive soils ($\varphi = 0$), the third term of the equation decreases to zero, and the second term decreases to q'. The value of s_c in the first term is reduced to 1.2. For cohesive soils where is zero, the terms q' and d_c can be neglected with little error. The bearing capacity equation is then reduced to 1.2 N_c. For the case of $\varphi = 0$, N_c=2+π, or 5.14, and the bearing capacity equation becomes $q_{ult} = 6.2c$. Since the q_{ult} is equal to the stress in the CBR test at 2.54-mm (0.1 inch) penetration, the expression for the correlation of the shear strength of the undrained soil with the CBR becomes

$$CBR = (q_{ult} \times 100) / 6895 = (6.2\, c \times 100) / 6895 = 0.09\, c \quad (2)$$

From the above calculation, it is obtained the value that will be used for slope modeling on the subsoil material in the planned disposal area.

Disposal slope modeling with and without compaction is distinguished based on the variation of loading performed when testing CBR on subgrade. Cohesion and shear strength value variations of the soil obtained after the analysis can be used as a comparison between the subgrade with and without compaction.

There are two types of data, namely primary and secondary which are distinguished based on the way the data is obtained and organized. Primary data is data obtained directly through observation or laboratory tests. The primary data obtained include the lithology of the study area, as well as several attributes of bedrock for disposal such as index properties, size distribution, atterberg limits and the value of variations in subgrade CBR. While secondary data obtained from companies such as basemap contour map of the research area and soil properties of disposal slope modeling.

Analysis of slope stability using the limit equilibrium method. This method was chosen because it is considered more suitable to be used to determine the factor of safety value directly for simple slope designs and the type of material that tends to be uniform. The software used in this research is Slide 6.0 (Rocscience).

4. Result and Discuccion

Based on lithological characteristics and observations of stratigraphic outcrops, the research area was arranged following the classification and nomenclature of Heryanto and Sanyoto (1994) [11], in order to determine the stratigraphic composition and distribution of the rock units laterally.

Rock units are characterized using informal units that refer to the nomenclature groupings in accordance with the rules of Sandi Statigrafi Indonesia [12]. In general, the research area is dominated by quartz sandstone, claystone, conglomerate and loose material from older rocks (appendix A and B). The stratigraphic order of the study area from old to young:

1. Warukin Sandstone Unit (Tmw)
2. Dahor Layang Member Conglomerate (TQdt)
3. Dahor Sandstone Unit (TQd)

The Warukin Sandstone Unit is generally dominated by fine-grained clastic sedimentary rock lithology in the form of silicaan sandstone and clay. Outcrops are found in valley areas with high weathering rates. Furthermore, the Dahor Layang Member Conglomerate unit refers to previous researchers, also based on direct observations made in the field with basic physical characteristics, namely lithology in the form of loose material and conglomerates of various older rock fragments such as igneous rocks and sandstones with a fairly large size. i.e. pebbles. While the name Dahor comes from the name of the Dahor formation which is based on previous researchers in the 1994 Amuntai sheet by Heryanto and Sanyoto [11], this unit is called the Dahor Layang Member Conglomerate.

And the last rock unit is the Dahor sandstone unit. This unit is distinguished based on direct observations made in the field with lithological physical characteristics in the form of sandstone with coarser grain sizes than the Warukin formation, conglomerates and loose material found locally with fairly large sizes such as gravel (Pebble – Cobble) (Figure 2).

Subsoil in the planned disposal area becomes an important matter to be studied in depth, considering its function as the main material in sustaining the stockpile material. If the subsoil material has good soil strength quality, it is expected to increase the volume of the disposal pile [13].

Figure 2. Some of the rock outcrops in the study area (a) The appearance of the silicate sandstone in LP 8 (Azz photo N326°E) in the Warukin sandstone unit (b) The appearance of a polymic conglomerate outcrop on LP 39 (Azz photo N5°E) in the Layang Dahor conglomerate unit (c) The appearance of the silicaan sandstone outcrop at LP 16 (Azz photo N61°E) Dahor Sandstone unit.

The value of the index properties of the soil sample (subsoil) in this study can be observed in the following table (Table 1), while the comparison graph between the liquid limit and the plasticity index can be observed in Figure 3 [14].

Table 1 Index of subsoil material properties at the location of the planned disposal.

No	Name	Observation Location		Specific Gravity (Gs)	Atterberg limits (%)			Soil type
		x	y		LL	PL	PI	
1	Dsp_1	306717	9683557	2,62	29,50	20,13	9,37	CL
2	Dsp_2	306948	9683486	2,60	34,25	23,06	11,19	CL
3	Dsp_3	306725	9683213	2,61	44,20	30,75	13,45	ML
4	Dsp_4	306556	9683391	2,61	56,15	28,04	28,11	CH
5	Dsp_5	306728	9683428	2,62	39,85	23,01	16,84	CL

Soil classification using the unified classification system (ASTM 1982)

In the table and figure above, we can observe that the soil samples in the study area are classified as CL, CH and ML soil types which indicate that the soil is clay or inorganic silt with low to medium plasticity (lean clays) [13]. The laboratory CBR value test was carried out on the Dsp_5 sample because the soil with CL type was the most dominant and the location of the sampling point was in the middle of the disposal plan.

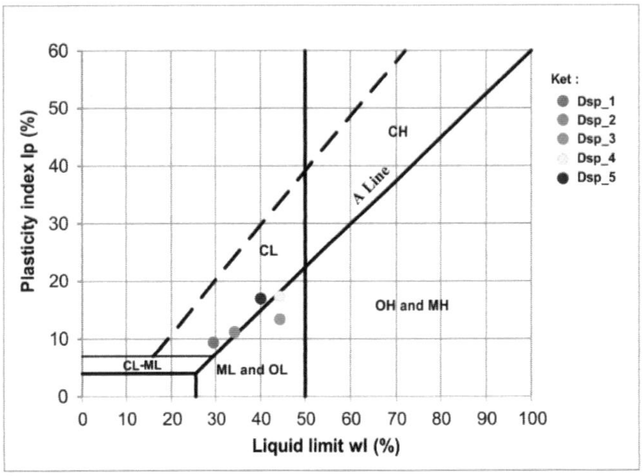

Figure 3. USCS soil type of research area.

The design of the disposal plan with subsoil without any compaction effort uses the cohesion value parameter obtained from the CBR test with 10 blows (CBR=4.91%). By considering the material as a cohesive soil based on previous research, the value of undrained soil cohesion (Cu) is 54.56°.

The next step is to make a slope design model with material properties for the compacted base material. This material property is obtained from the conversion of Cu values with the number of blows 25 times in the laboratory CBR test. Based on this, the value of undrained soil cohesion (cu) is 94.44°.

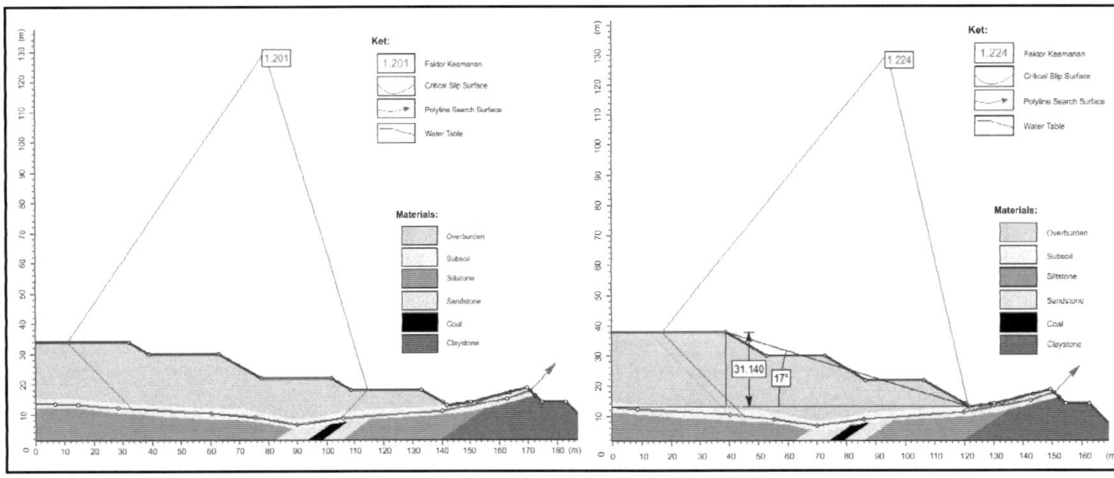

Figure 4. Disposal modeling with subsoil before and after compaction in the M section.

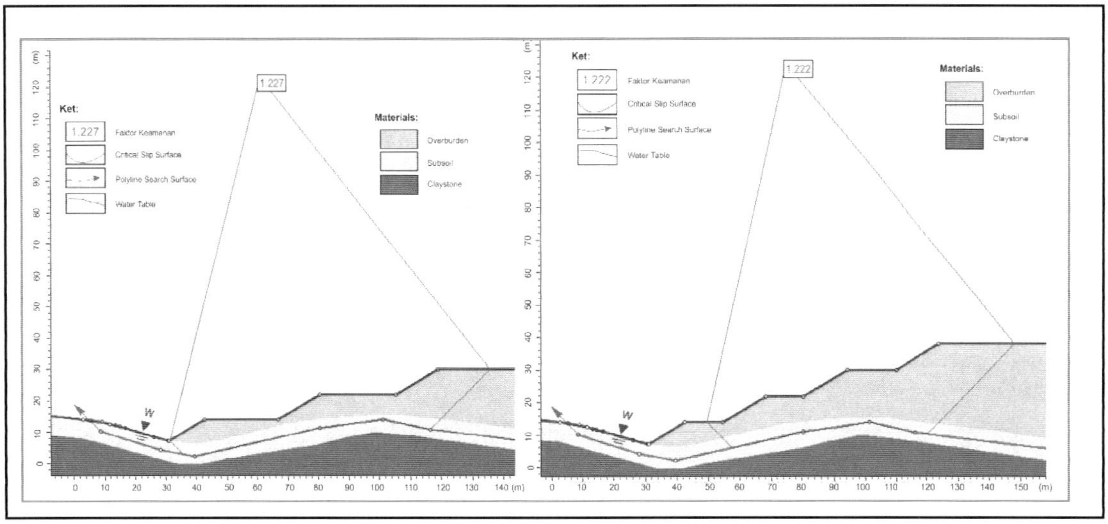

Figure 5. Disposal modeling with subsoil before and after compaction in the N section.

If we observe the comparison between the slope design with and without compaction (Figures 4 and 5), it can be seen that after compaction the slope can be optimized for height and slope. In the M section from 11° slope to 17° while the N section from 13° slope to 18°, the height of the M incision from 26 m to 31 m while the N incision from 33 m to 38 m. The increase in the value of slope stability at the incision before and after compaction is believed to be due to the effect of compaction carried out. By compacting the soil, the soil grains become denser, thereby increasing the bearing capacity of the soil.

5. Conclusion

In general, this study aims to determine how the influence of the base of disposal on its stability. Stability of slope disposal more or less affected by the strength of base material, therefore base of disposal is supposed to be getting more considered in designing the slope model.

Based on the results of the research that has been done and analyzed in the previous chapter, it can be concluded several things:

1. The stratigraphy of the research area from old to young consists of the Warukin Sandstone Unit (Tmw), the Dahor Layang Member Conglomerate (TQdt), and the Dahor Sandstone Unit (TQd) which are all Middle Miocene to Early Pleistocene (16.2-4 m.ya) .
2. The results of the property index analysis of 5 soil samples at the research site showed a specific gravity (Gs) value of 2.60-2.62 and soil types CL and ML which indicated that the soil in the study area was classified as clay to silt with low to moderate plasticity. .
3. The CBR soaked test value in the study area was divided into 3 variations of loading resulting in a CBR value of 4.91%, 8.50% and 12.57% respectively, while the optimum value for the maximum dry density was CBR 12.25%.
4. After compacting the subsoil, the height and slope of the slope can be optimized. By compacting the soil, the soil grains become denser, thereby increasing the bearing capacity of the soil.

IOP Conf. Series: Earth and Environmental Science 1071 (2022) 012028 doi:10.1088/1755-1315/1071/1/012028

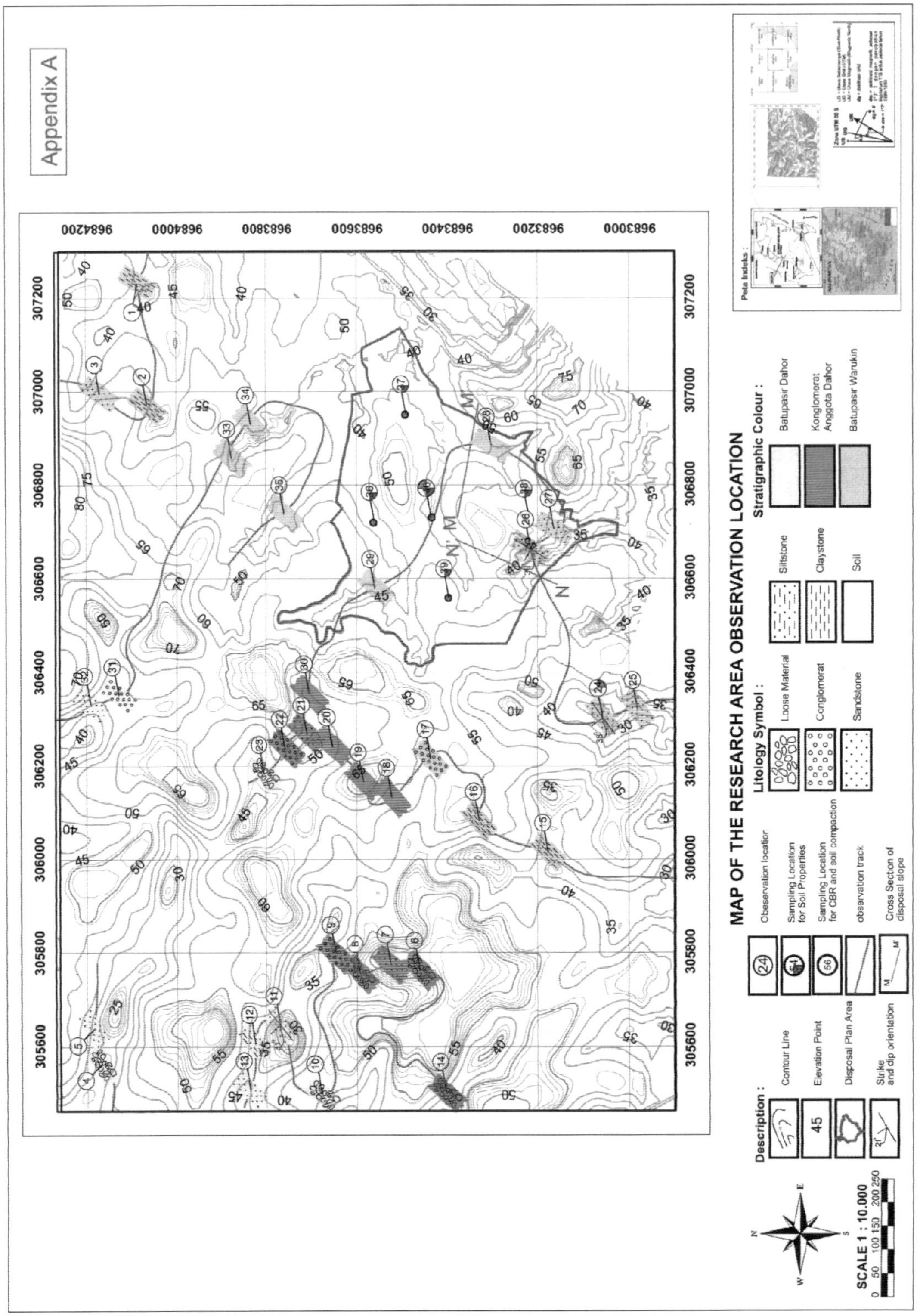

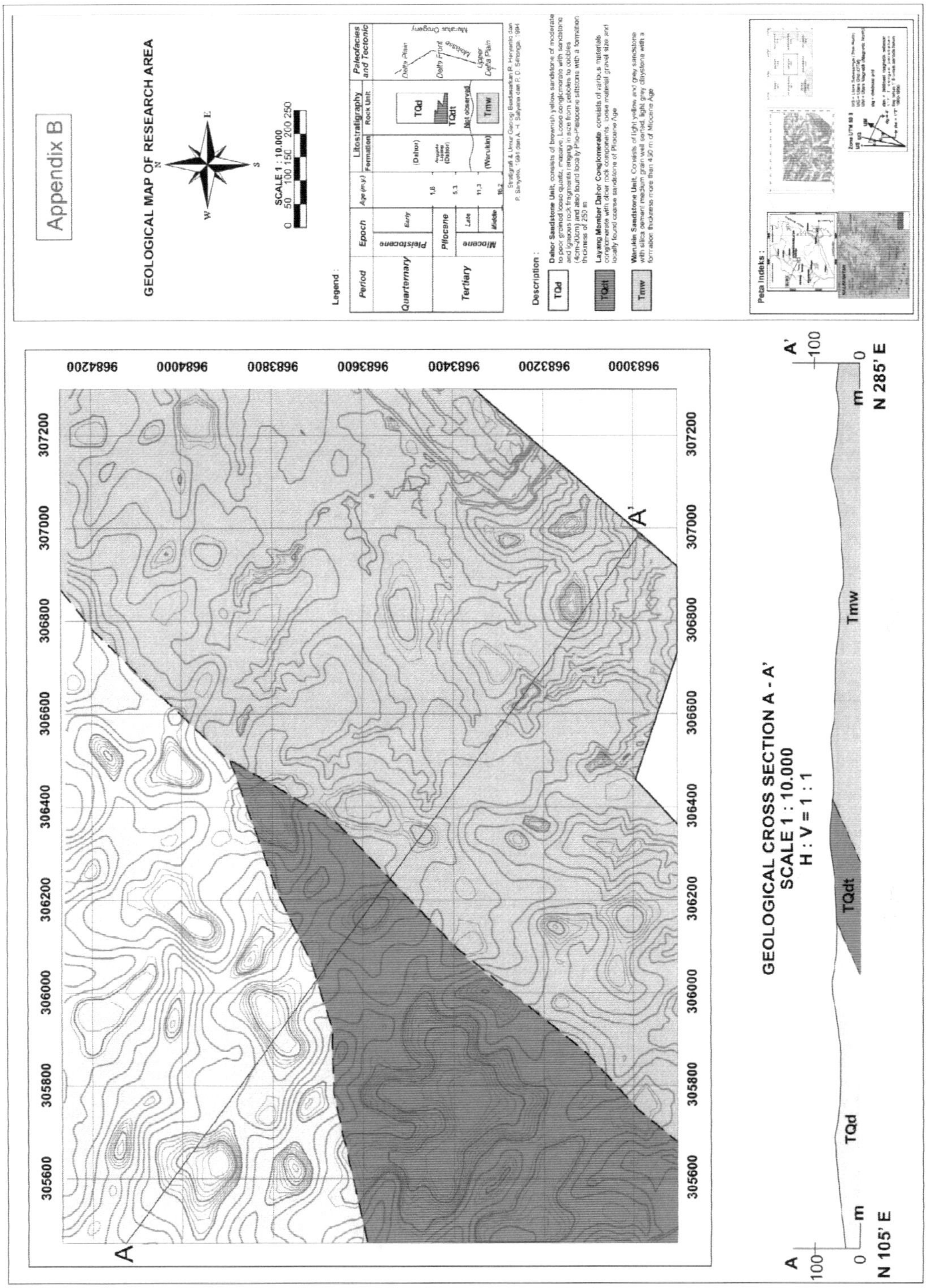
Appendix B
GEOLOGICAL MAP OF RESEARCH AREA
SCALE 1 : 10.000
0 50 100 150 200 250
Legend :
Period
Epoch
Litostratigraphy
Paleofacies and Tectonic
Quarternary
Pleistocene
Pliocene
Miocene
Tertiary
TQd
TQdt
Tmw
Not observed
Description :
Peta Indeks :
GEOLOGICAL CROSS SECTION A - A'
SCALE 1 : 10.000
H : V = 1 : 1
N 105' E
N 285' E
A
A'

6. References

[1] Muzaidi I 2016 Pengaruh Penggunaan Bahan Ecocure21 Terhadap Kestabilan Pada Area Disposal dan Area Hauling Road PT. Adaro Indonesia, Kalimantan Selatan Tesis RC 142501 Institut Teknologi Sepuluh Nopember Surabaya

[2] BSSR 2017 *Paparan Publik Tahunan PT Baramulti Suksessarana Tbk* 15 Maret 2017

[3] Andriani T, Zakaria Z, Muslim D, and Oscar AW 2018 *Daya Dukung Tanah Untuk Disposal di Tambang Batuabara Daerah Purwajaya, Kecamatan Loa Janan Kabupaten Kutai Kartanegara*. Padjajaran Geoscience Journal i-ISSN: 2597-4033 Vol.2, No.2,

[4] Suherman M 1999 *Peningkatan Daya Dukung Tanah Dasar dengan Teknik Pemadatan Dinamis* Jurnal Pusat Litbang Jalan Vol 16 pp 61-70.

[5] Darman, Herman and Sidi F H 2000 *An Outline of The Geology of Indonesia* (Jakarta: Ikatan Ahli Geologi Indonesia)

[6] Sapiie B and Rifiyanto A 2017 *Tectonics and Geological Factors Controlling Cleat Development in the Barito Basin, Indonesia* Journal Engenering Technology and Science Vol 49 No 3 2017 p 322-339

[7] Sikumbang N 1986 *Geology and Tectonics of Pre-Tertiary Rocks in The Meratus Mountains, South-East Kalimantan, Indonesia* PhD Thesis University of London.

[8] BSNI, 2012, *Metode Uji CBR (California Bearing Ratio) Laboratorium* SNI 1744:2012 (Badan Sertifikasi Nasional, Jakarta)

[9] Barnas E and Karopeboka B 2015 *Penelitian Kekuatan Tanah Metode CBR (California Bearing Ratio) Di SPBG Bogor 1 Bubulak Jl Kh R Abdullah Bin Nuh*. Universitas Borobudur. Jakarta

[10] Gregory G H and Cross S A 2007 *Correlation of CBR with Shear-Strength Parameters*. 9th International Conference on Low-Volume Roads June 24-27 2007 Austin Texas

[11] Heryanto R and Sanyoto P 1994 *Geological Map of The Amuntai Quadrangle Kalimantan* Geological Research and Development Centre Jl. Diponegoro 57, Bandung

[12] IAGI 1996 *Sandi Stratigrafi Indonesia* Ikatan Ahli Geologi Indonesia, ed Putrohari R D

[13] Das B M 1995 *Mekanika Tanah (Prinsip-prinsip Rekayasa Geoteknis)* Jilid 1 (Penerbit Erlangga Jl. H. Baping Raya No. 100 Ciracas, Jakarta 13740) ed Endah N and Mochtar I B

[14] ASTM 1982 *American Society for Testing and Materials Standards* Part 19 (Philadelphia Pa)

Acknowledgments

This paper and the research behind it would not have been possible without the exceptional support of my supervisor in Geological Engineering Department of Gadjah Mada University for their patient guidance, enthusiastic encouragement and useful critiques of this research work. We would also like to show our gratitude to the Geothecnical Department in Antang Gunung Meratus for all help that has been given until the end of this research.

IOP Conf. Series: Earth and Environmental Science 1071 (2022) 012029 doi:10.1088/1755-1315/1071/1/012029

Seasonal comparison of groundwater level and quality in the catchment of the Tonle Sap Lake, Cambodia

Kong Chhuon[1*], Khy Eam Eang[1], and Vuthy Chork[1]

[1]Faculty of Hydrology and Water Resources Engineering, Institute of Technology of Cambodia, Russian Federation Blvd., P.O. Box 86, Phnom Penh, Cambodia.
*Corresponding author's e-mail: chhuon.k@itc.edu.kh

Abstract. Groundwater in the floodplain of Tonle Sap Lake has been used for domestic, agriculture, and small industries. The variation of surface water availability and increasing demand, cause the increasing of groundwater extraction especially during dry season. So far, the study on groundwater resources is extremely limited comparing to surface water in the Tonle Sap catchments. So that, this study focusses on groundwater quality and level from 27 wells in Kompong Thom, Siem Reap, and Preh Vihear province. Seasonal variation of groundwater level shows that groundwater level in Siem Reap is about 1.5m and 4.3m below the ground surface in rainy season and dry season respectively while in Kampong Thom province it is about 0.7m and 2m below the ground surface in rainy season and dry season, respectively. from in situ measurement, some parameters such as pH ranged from 4.1 to 8.1, total dissolved solid from 21 to 1200 mg/L, turbidity from 0 to 88 NTU, temperature from 28 to 31 °C, ORP from -2.2 to 424mV, and EC from 35 to 2000 μSm/cm. The concentrations of major ions (Ca^{2+}, Mg^{2+}, K^{+}, Na^{+}, Cl^{-}, SO_4^{2-} and HCO_3^{-}) were approximate or under the permissible value and it was suitable for irrigation.

1. Introduction

Groundwater is one of the reliable and extensively used natural resources, it constitutes about twenty percent of the world's fresh water supply, which is about 0.61% of the entire world's water [1]. Tonle Sap Lake (TSL) has been known as the seasonally inundated freshwater via the unique hydrological system of inflow and out flow between TSL and Mekong River which attracts the multi-studies that have been carried out in various themes such as flooding, fishing, water balance analysis, ecosystem, socioeconomic, agricultural use, and contaminated water. The water quality of TSL has been recorded with surface water so far, but well water quality rarely researches in floodplain and upstream areas.

In Cambodia, people prefer to use groundwater for their agriculture, especially in dry season. Groundwater is also the major source for drinking water supply in Cambodia, 53% of Cambodian households drink from groundwater sources in the dry season. About 270,000 tube-wells with hand pump are functioning for drinking water purpose [2].

The objective of this research is to observe quality of groundwater and its water level in different seasons and to analysis the differences of the results between rainy season and dry season. The article also highlighted the range of physicochemical parameters value with comparison to the drinking water standards of Cambodia and WHO. It also indicated the relation among the parameters and to assess the water quality for irrigation purpose.

2. Methodology

2.1. Sampling sites

Twenty-seven water samples from different well in the floodplain and upstream area of Tonle Sap lake catchment were collected during rainy and dry seasons in 2019 and 2020. Groundwater level was collection by measuring the depth from ground surface to groundwater table in each well along the boundary of the Tonle Sap lake floodplain located in Kampong Thom province (KTP), Siem Reap province (SR) and Preah Vihear province (PV) of Cambodia (figure 1). The water level and wells depth were measured by using wells water level meter from HT HYDROTECHNIK. In-situ measurement and ex-situ measurement or laboratory analysis were applied for groundwater quality data collection in this study. YSI EXO2 Multi-Parameter Water Quality Sonde was used for in-situ water quality measurement. Groundwater from each well was sampled and stored properly then brought to analyse for some parameters in laboratory at Institute of Technology of Cambodia. The wells which were measured in Kampong Thom and Preah Vihear have depth ranging from 2.5m to 8.5m while the wells in Siem Reap have depth ranging from 3.7m to 70m. The depth of the wells which were measured this study are varied by different well types including dug wells and tube wells. The groundwater from the 27 wells is used mainly for domestic purposes and some for small scale irrigation in the village.

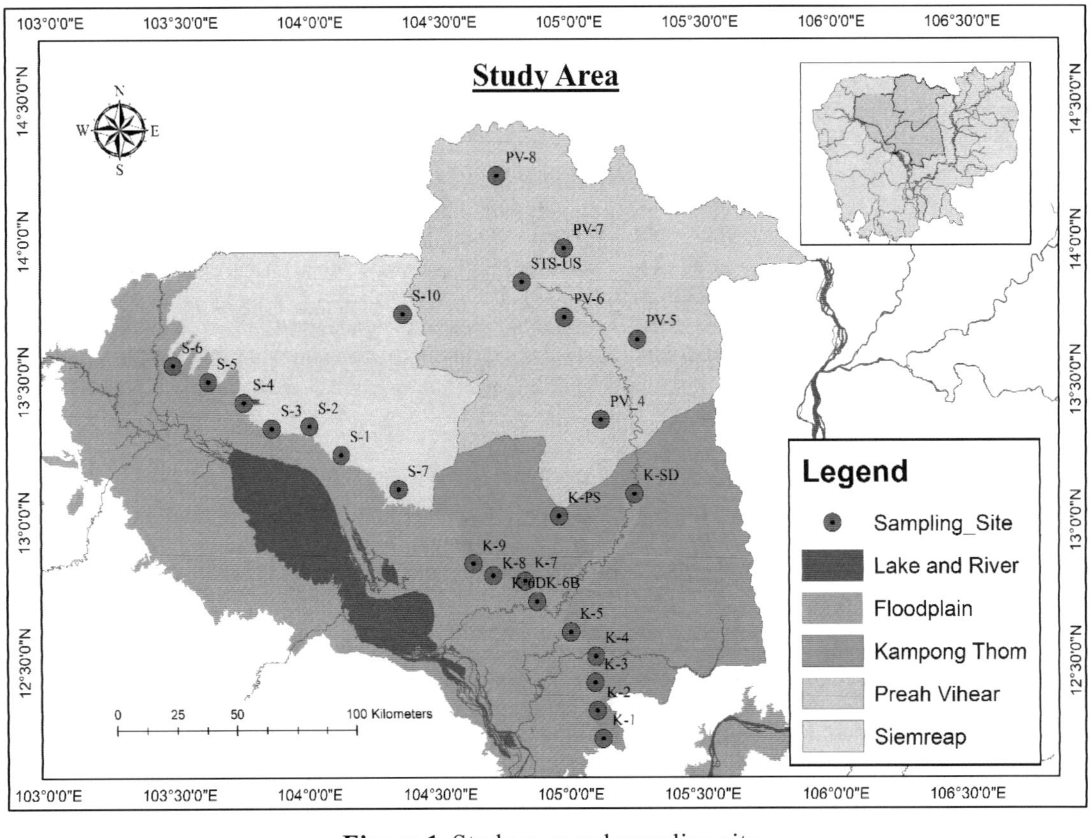

Figure 1. Study area and sampling site.

2.2. Laboratory testing

The physical and chemical parameter such as pH, TDS, temperature, EC were analyzed in-situ with YSI EXO2 Multi-Parameter Water Quality Sonde. Six heavy metals such as Mn, Fe, Cu, Pb, Cd and Cr were measured with atomic adsorption spectrophotometry (AAS), while major ions like Ca^{2+}, Mg^{2+}, K^{+}, Na^{+}, NO_3^{-}, Cl^{-}, and SO_4^{2-} were analyzed with ion chromatography (IC). IC and AAS measurement were analyzed at SATRAPS laboratory and chemical laboratory of Institute of Technology of Cambodia. Particularly, HCO_3^- was calculated from the alkalinity and this alkalinity was measured by titration with 0.01M sulfuric acid in the laboratory immediately after receiving the samples from the site.

2.3. Analytical Methods

To study the current status of water quality, all parameters have used to compare with Cambodia Drinking Water Standard (DWS) established by Ministry of Industry, Mine and Energy [3] and with drinking water standard by World Health Organization (WHO) [4]. For irrigation purpose, sodium adsorption ratio (SAR), and sodium percentage (%Na) were calculated based on US salinity laboratory (USSL) [5], Food Agriculture Organization (FAO) [6], and Wilcox (1948) [7].

3. Result and Discussion

3.1. Groundwater level

Groundwater level variation was compared only in Kampong Thom province (KTP) and Siem Reap province (SR). The groundwater level is measured the depth from land surface to the surface of water in the wells. The average groundwater level which measured from all wells in KTP were 2m and 0.7m below ground surface for dry season and rainy season respectively and the standard deviation of 0.8m and 0.3m in dry season and rainy season, respectively. The average groundwater level which measured from all wells in SR were 4.3m and 1.5m below ground surface for dry season and rainy season respectively and the standard deviation of 3.7m and 1.2m in dry season and rainy season, respectively. The groundwater level variation by season as indicated in the figure 2 were the average groundwater level from all wells which were separated from each province. In KTP, groundwater tables increase about 1.3m in average from dry season to rainy season while groundwater tables in SR increase about 3m in average from dry season to rainy season. From all wells observed in in SR showed that the maximum absolute increasing groundwater table between dry season and rainy season was 8m while the minimum absolute increasing was observed to be 0.5m. From all wells observed in KTP showed that the maximum absolute increasing groundwater table between dry season and rainy season was 2.5m while the minimum absolute increasing was observed to be 0.6m. In conclusion, groundwater tables in SR have more significant fluctuation between rainy season and dry season than in KTP. The significant change of groundwater table in SR between rainy season and dry season indicates more groundwater extraction during dry season.

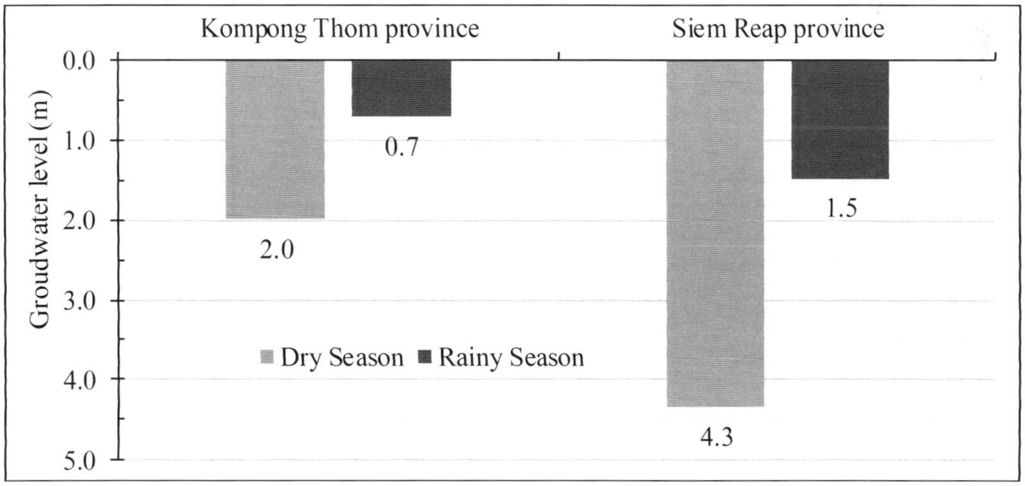

Figure 2. Seasonal groundwater level variation.

3.2. Physical parameters

The pH is one of the most important parameters indicating drinking water suitability. pH ranged from 4.1 to 8.1 in rainy season and range from 3.8 to 8.1 (Table 1). It was about 48% and 70% of samples exceeding the DWS in rainy and dry seasons. Temperature of the well water ranged from 28.4 to 31.6 °C in rainy season and 26.3 to 31 °C in dry season. TDS varied from 21 to 1251 mg/L in rainy season and 23 to 1390 in dry season. Only three samples from rainy season such as K-2, PV-4 and PV-US had higher TDS of 1048.4 mg/L, 1277.1 mg/L, 840 mg/L respectively and exceeded the permissible limit (<800 mg/L). Two samples from dry season such as K-3 (1390 mg/L) and S-7 (861 mg/L) had greater TDS values than the DWS and WHO. The value of turbidity in Siem Reap are higher than value of turbidity in Kompong Thom and Preh Vihea. About 62% of sample from Siem Reap exceeded the permissible limit (≤5 NTU). For ORP, 29% of sample were over the permissible limit (±400mV). In addition, electrical conductivity (EC) ranged from 35.3 to 2079.5 μS/cm in dry season and from 37.6 to 2294 μS/cm in rainy season.

Table 1. Ranges of physicochemical parameters and their comparison with drinking water standards of MIME (2004) and WHO (2017).

Parameter	Unit	Range		Permissible Limit (PL)	Percentage of sample exceeded PL	
		Rainy	Dry	MIME, WHO	Rainy (n=27)	Dry (n=27)
pH	-	4.12 - 8.17	3.89 - 8.13	6.5-8.5	55%	48%
Temp	°C	28.4 - 31.6	26.3 - 31	NA	NA	NA
EC	μS/cm	35.3 - 2079.5	37.6 - 2294	1500	4%	4%
TDS	mg/L	21 - 1251	23 - 1390	800	7%	13%
Turb	NTU	0 - 88.74	0.2 - 70.65	<=5	44%	55%
ORP	mV	-2.20 - 424.7	17.6 - 417.5	±400	7%	4%
Mn	mg/L	0 - 2.22	0 - 2.7	0.1	22%	22%
Fe	mg/L	0 - 23.91	0 - 0.83	0.3	48%	11%
K^+	mg/L	0 - 702.8	0 - 706.5	12	33%	25%
HCO_3^-	mg/L	0 - 702.89	0 - 634.4	600	0%	7%
NO_3^-	mg/L	0 - 34.91	0 - 34.4	50	0%	5%

3.3. Major ions

The concentration of major ion (Ca^{2+}, Mg^{2+}, K^+, Na^+, Cl^-, SO_4^{2-}, and HCO_3^-) were summarized in Table 1. Heavy metals such as Cd, Cu, Cr and Pb concentrations were under permissible limit except that the concentrations of Fe and Mn were greater than permissible limit and ranged from 0 to 23 mg/L and 0 to 2.7 mg/L, respectively. The concentration of Fe is more different between rainy and dry season. It ranged from 0 to 23 mg/L in rainy season and 0 to 0.6 mg/L in dry season. The major concentrations of cations and anions of the samples in both seasons were in respective order: Ca > Na+K > Mg and HCO_3 > SO_4 > Cl, suggesting that the hadrochemical facies of the water sample are dominated by Ca-HCO_3, Ca-Cl and Na-Cl.

3.4. Correlation matrix

In Table 2, the arrangement of correlation matrix in rainy season shows that Mg^{2+} had good correlation (where r is correlation coefficient) with EC (r = 0.66) and TDS (r = 0.67) respectively. HCO_3^- also had good correlation with pH (r = 0.82). For Table 3, the correlation of SO_4^{2-} with HCO_3^-, pH, TDS, and EC in dry season were strong with r = 0.67, 0.7, 0.77, and 0.81, respectively. The EC was also positively correlated with pH (r = 0.56) and TDS (r = 0.96).

Table 2. Correlation Matrix in rainy season.

Param	Turb	Temp	EC	TDS	pH	ORP	Mn	Fe	Ca^{2+}	Mg^{2+}	K^{+}	Na+	HCO_3^-	Cl^-	NO_3^-	SO_4^{2-}
Turb	1.00															
Temp	0.03	1.00														
EC	0.11	-0.30	1.00													
TDS	0.12	-0.31	**0.99**	1.00												
pH	0.02	-0.33	0.42	0.41	1.00											
ORP	-0.17	0.08	-0.10	-0.11	-0.39	1.00										
Mn	0.08	-0.02	0.14	0.15	-0.22	0.02	1.00									
Fe	0.28	0.04	-0.37	-0.35	-0.38	-0.08	0.32	1.00								
Ca^{2+}	0.15	-0.12	0.40	0.42	0.28	-0.16	0.06	-0.19	1.00							
Mg^{2+}	-0.02	-0.38	**0.66**	**0.67**	0.30	0.03	0.20	-0.31	0.44	1.00						
K^{+}	0.16	-0.16	0.40	0.39	0.06	0.11	0.25	-0.10	-0.04	0.21	1.00					
Na^{+}	-0.19	-0.21	-0.27	-0.27	-0.27	0.24	-0.19	0.14	-0.26	-0.10	-0.11	1.00				
HCO_3^-	0.06	-0.27	0.49	0.47	**0.82**	-0.36	-0.13	-0.44	0.21	0.30	0.13	-0.22	1.00			
Cl^-	-0.10	0.13	-0.15	-0.15	0.00	0.12	-0.23	0.10	-0.10	-0.06	-0.24	0.24	-0.01	1.00		
NO_3^-	-0.16	-0.05	0.16	0.15	0.06	0.13	0.04	-0.24	0.08	0.29	0.24	-0.11	0.09	-0.18	1.00	
SO_4^{2-}	-0.13	-0.17	0.46	0.45	0.42	0.02	-0.05	-0.43	0.23	0.35	0.20	-0.32	0.37	0.01	0.35	1.00

Note: Param refers to parameter

Table 3. Correlation Matrix in dry season.

Param	Turb	Temp	EC	TDS	pH	ORP	Mn	Fe	Ca^{2+}	Mg^{2+}	K^{+}	Na+	HCO_3^-	Cl^-	NO_3^-	SO_4^{2-}
Turb	1.00															
Temp	0.31	1.00														
EC	-0.16	-0.49	1.00													
TDS	-0.13	-0.53	**0.96**	1.00												
pH	-0.09	-0.27	**0.56**	0.53	1.00											
ORP	-0.49	-0.38	0.09	0.05	0.24	1.00										
Mn	-0.18	-0.07	-0.15	-0.11	-0.51	-0.11	1.00									
Fe	0.22	0.26	-0.71	-0.67	-0.71	-0.22	0.28	1.00								
Ca^{2+}	-0.02	-0.05	0.27	0.31	0.27	-0.05	-0.07	-0.26	1.00							
Mg^{2+}	-0.24	-0.56	0.49	0.53	0.35	0.31	0.15	-0.41	0.27	1.00						
K^{+}	0.00	-0.26	0.18	0.15	-0.07	0.26	-0.06	-0.04	-0.48	0.00	1.00					
Na^{+}	-0.43	-0.26	-0.34	-0.34	-0.43	0.43	0.43	0.44	-0.26	0.17	0.09	1.00				
HCO_3^-	-0.06	-0.24	0.57	0.54	0.76	0.17	-0.37	-0.80	0.17	0.28	-0.04	-0.39	1.00			
Cl^-	-0.05	0.05	-0.32	-0.32	0.05	0.23	-0.26	0.19	0.00	0.05	-0.35	0.43	-0.02	1.00		
NO_3^-	-0.45	-0.27	0.27	0.24	0.05	0.45	0.07	-0.30	-0.31	0.35	0.37	0.43	0.31	0.05	1.00	
SO_4^{2-}	-0.22	-0.51	**0.81**	**0.77**	**0.70**	0.11	-0.28	-0.66	0.22	0.37	0.13	-0.26	**0.67**	-0.07	0.22	1.00

3.5. Assessment for irrigation purpose

The percentage Na (Na%) chart is used check the suitability of water quality based on electrical conductivity value. The samples were plotted into different five classifications ranging from excellent to unsuitable. Most of the samples were acceptable for irrigation purpose. There are only a few samples that could not use for irrigation and there is only one sample in Siem Reap province classified as unsuitable for irrigation (figure 3). On the other hand, figure 4 illustrates the Sodium Absorption Ratio during rainy season in Kompong Thom varies from 0 to 1.25 meq/L, Siem Reap range from 0 to 6.37 meq/L. In Preh Vihea, the data of SAR is equal to zero for every location due to Na concentration is not detected. The water quality for irrigation in dry season was also acceptable as shown in figure 5, while the SAR range from 0 to 1.55 meq/L, 0 to 0.33 meq/L in Kompong Thom and Siem Reap respectively. Most samples are located in C1S2, C2S1, and C3S1, which were suitable for crop and irrigation in both seasons.

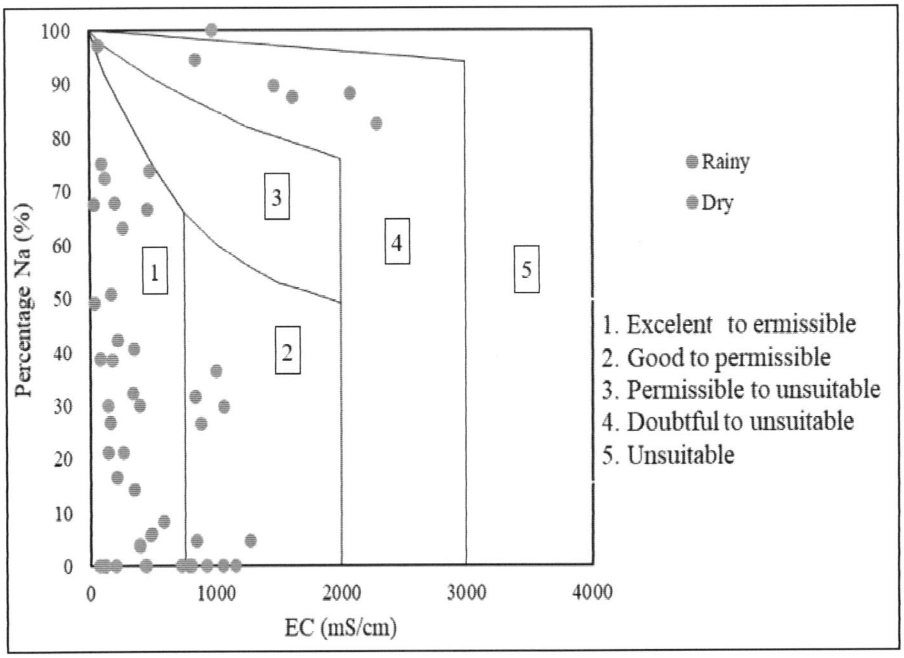

Figure 3. Rating of well water samples on the basis of electrical conductivity and percent sodium.

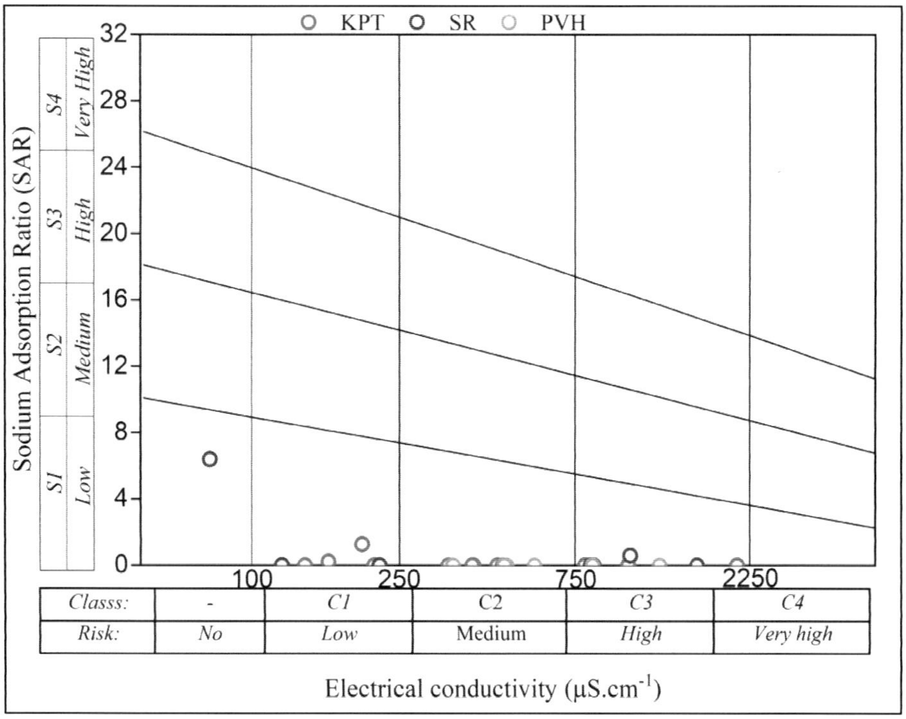

Figure 4. Diagram rating of well water samples in relation to salinity hazard and sodium hazard in rainy season.

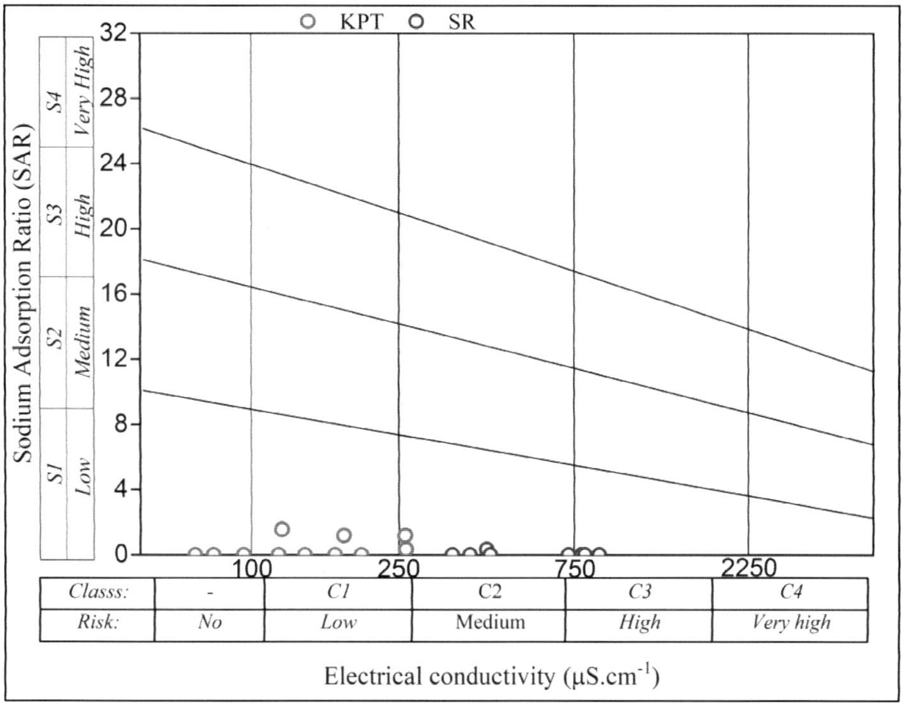

Figure 5. Diagram rating of well water samples in relation to salinity hazard and sodium hazard in dry season.

4. Conclusion

In Kompong Thom province, groundwater level decrease in dry season ranging from 0.6m to 3m in different wells. In average, groundwater level in Kampong Thom province drops about 1.3m in dry season. Groundwater level in Siem Reap province drops in dry season ranging from 0.5m to 8m in different wells. In average, groundwater level in Siem Reap drops about 3m in dry season. In Siem Reap province, the dropping of groundwater level is more significant which is the indication of more groundwater extraction than in Kompong Thom province. Regarding to the quality, the higher values of EC, TDS, Mn, and Fe were observed in rainy season while higher HCO_3^- was recorded in dry season. Generally, this study shows that groundwater quality in rainy season is better than in dry season. The groundwater quality of the study area had a principal issue with low pH. The water should be treated before consumption especially for the parameters of concerns such HCO_3^-, Fe, Mn and K, while the salt water should be checked before irrigated on soil or plants. For a better interpretation and clarification on geochemistry controlling the well water quality, saturation index (SI) may be needed and geological formation of the well profile is one of the important information to be suggested for further study.

References

[1] Alsalme, A., Al-zaqri, N., Ullah, R., & Yaqub, S. (2021). Saudi Journal of Biological Sciences Approximation of ground water quality for microbial and chemical contamination. *Saudi Journal of Biological Sciences*, **28** (3), 1757–1762. https://doi.org/10.1016/j.sjbs.2020.12.017

[2] Sok, S. (2011). *Groundwater Research in Cambodia*. https://archive.iges.or.jp/en/natural-resource/groundwater/PDF/activity20110602/S1-3_Mr.Sok-Sophally_GW_Cambodia.pdf

[3] Ministry of Industry Mines and Energy (2004). Kingdom of Cambodia Drinking Water Quality Standards. Drinking Water Quality Standards, 1–20.

[4] Herschy, R. W. (2012). Water quality for drinking: WHO guidelines. *Encyclopedia of Earth Sciences Series*, 876–883. https://doi.org/10.1007/978-1-4020-4410-6_184.

[5] Elbashier, M. M. A., Ebrahim, M. H. A., Musa, A. A., Ali, A. A. S., & Mohammed, M. A. A. (2016). Efficiency of Two Models for Prediction of Exchangeable Sodium Percentage from Sodium Adsorption Ratio on Saline and Non-Saline Soil. *Universal Journal of Agricultural Research*, **4** (1), 32–36. https://doi.org/10.13189/ujar.2016.040106.

[6] Simmons, H. B. (2000). *Salinity Problems*. In *Coastal Engineering Proceedings* (**Vol. 1**, Issue 2). https://doi.org/10.9753/icce.v2.7.

[7] Wilcox, L. V. (1948). *The quality of water for irrigation use*. In Technical Bulletin 962, U.S. Department of Agriculture (Issue Tech. Bull. 962, U.S. Dep. Agric.). ageconsearch.umn.edu.

Acknowledgments

This work is financially supported by Science and Technology Research Partnership for Sustainable Development (SATREPS) project through Japan International Cooperation Agency (JICA).

IOP Conf. Series: Earth and Environmental Science 1071 (2022) 012030 doi:10.1088/1755-1315/1071/1/012030

The Correlation between Rainfall, Temperature, Relative Humidity, and Rice Field Productivity in Aceh Besar

S Chairani[1]

[1]Department of Agricultural Engineering, Faculty of Agriculture, Syiah Kuala University, Banda Aceh, Indonesia, 23111.
*Corresponding author's e-mail : susi.chairani@gmail.com

Abstract. Various factors could affect rice field productivity, such as climate, management practices, and soil properties. Aceh Besar had experienced long drought, higher temperature, shifted seasons, and the decrease yield of rice productivity due to climate change. This research aimed to analyze the correlation between climate variables and rice field productivity, such as rainfall and mean, minimum, and maximum temperatures, relative humidity in Aceh Besar District. The monthly climate data and the rice field productivity data were employed for 10 (2011-2020) and 8 (2011-2018) consecutive years, respectively. The correlation between the climate variables were calculated using Pearson coefficient correlation. The results showed that rainfall and maximum temperature were positively correlated, as well as rainfall and relative humidity. In contrary, rainfall and mean temperature, rainfall and minimum temperature, rainfall and rice field productivity were negatively correlated. The latest indicating that rainfall did not impact the rice field productivity in Aceh Besar. It was quite contradictive to the reality in the field that significantly experiencing the long drought, higher temperature, shifted seasons and the decrease yield of rice field productivity. This was due to the lack of climate data employed that required longer period preferably 30 to 50 years which was not available.

1. Introduction

Climate is conditon of weather in a vast region and occurs in a long period of time (30 years or more). Climate is influenced by many variables, such as rainfall, temperature, relative humidity, wind speed, and solar radiation. The pattern of these climate variables varies in accordance with geography and time. The changes of climate variables can be analyzed statistically. Climate change occurs naturally due to volcano eruption and solar radiation. It also can occur due to human activities directly or indirectly (UNFCC, 2011), such as green house effect, global warming, ozone layer depletion, land use change, use of fossil fuels, deforestation, uncontrolled use of chlorofluorocarbons, and industrial exhaust gases. Climate change resulted the change in rainfall pattern and intensity, long dry season, reduced water sources, the occurrance of hurricane, and rise the sea level (Directorate General of Climate Change Control of Indonesia, 2019). Climate change has become a global issue that threatens the human well-being in different parts of the world (Kaletapwa *et.al*, 2012). Aceh Besar District is one of the districts in Aceh Province, Indonesia (Figure 1). The district covers 2,903.50 km^2 or 290,350 ha with 31,627 ha of agricultural land. The rice field supplied by irrigation system was 20,445 ha and rainfed rice field was 11,088 ha (Central Bureau of Statistics for Aceh Besar District, 2018). Its main commodity is rice. Due to climate change, Aceh Besar experienced long drought. The water supply for rice field decreased and dried out. It also increased the temperature in this region. The pattern and schedule planting for rice field were also shifted. The yield and farmers' income gradually decreased. This research aimed to analyze the correlation between climate variables, such as rainfall, temperature, relative humidity, and rice field productivity.

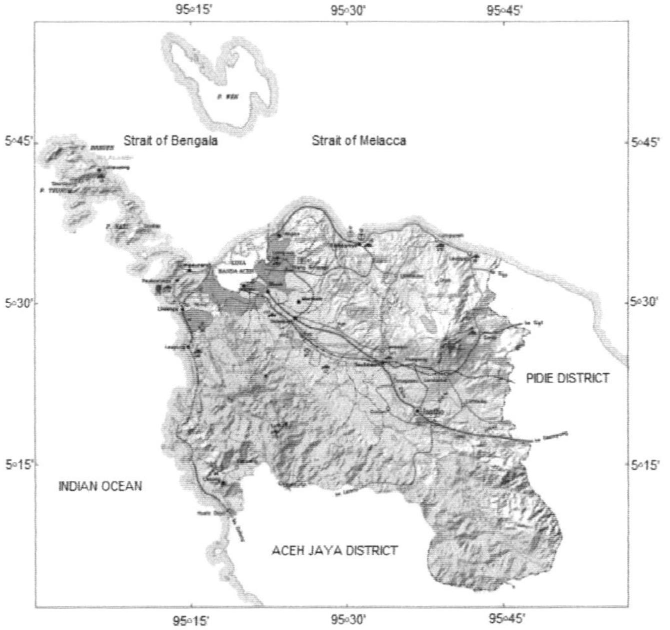

Figure 1. Map of Aceh Besar (Government of Aceh Besar District, 2016).

Numerous researchs on factors which affected rice field productivity have been carried out. These factors could be divided into two factors, environmental and management practices factors. The environmental factors were rainfall, relative humidity, and temperature. Hossain *et.al* (2013) investigated environmental factors that affected rice field productivity in Bangladesh. They found that humidity had positive impact on all type of rice production (AUS, AMAN, and BORO). Rainfall only had positive impact on BORO production. In contrary, temperature had negative impact on rice production. For management practices factors, Chowdury (2014) mentioned that mechanisation, irrigation, fertilizer, experience and farm size had positive and significant impact on rice field productivity. Meanwhile, human labour and education had negative influence. Kasmin *et.al.* (2020) also agreed that land area significantly had an impact on rice field productivity. Yet, quantity of seeds, the amount of urea and NPK fertilizers, the amount of pesticides, and the number of workers had no impact on rice field productivity. In order to increase the low rice field productivity in Malaysia, Mailena *et.al* (2014) suggested three factors that significantly affect the rice field efficiency which were household size, land ownership, and farmers' education.

2. Research Method

Monthly rainfall, mean temperature, minimum temperature, maximum temperature, and relative humidity data were obtained from Central Bureau of Statistics for Aceh Besar District for 10 consecutive years, from 2011 to 2020. The rice field productivity data were also derived from Central Bureau of Statistics for Aceh Besar District and only available for 8 consecutive years, from 2011 to 2018. The correlation between rainfall and mean temperature and the correlation between rainfall and other variables were calculated using Pearson correlation coefficient (Table 1), r_{xy} (equations 1 and 2). All calculations for statistical analysis utilized Microsoft Excel.

$$r_{xy} = \frac{Cov\,(x,y)}{\sigma_x\,\sigma_y} \tag{1}$$

where:
Cov(x, y) = Covariance between x and y
σ_x = Standard deviation of x
σ_y = Standard deviation of y

$$r_{xy} = \frac{n.\sum_{i=1}^{n} x_i y_i - (\sum_{i=1}^{n} x_i)(\sum_{i=1}^{n} y_i)}{\sqrt{n.\sum_{i=1}^{n} x_i^2 - (\sum_{i=1}^{n} x_i)^2 .\ n.\sum_{i=1}^{n} y_i^2 - (\sum_{i=1}^{n} y_i)^2}} \tag{2}$$

where:
n = number of sample

Table 1. Pearson correlation coefficient.

Pearson Correlation Coefficient	
$r_{xy} = 1$	The dependent variable y is perfectly correlated positively with the independent variable x
$0.8 < r_{xy} < 1$	The dependent variable y is strongly correlated positively with the independent variable x
$0.3 < r_{xy} < 0.6$	The dependent variable y is moderately correlated positively with the independent variable x
$0 < r_{xy} < 0.3$	The dependent variable y is weakly correlated positively with the independent variable x
$r_{xy} = 0$	The dependent variable y does not have any kind of linear correlation with the independent variable x
$0 < r_{xy} < -0.3$	The dependent variable y is weakly correlated negatively with the independent variable x
$-0.3 < r_{xy} < -0.6$	The dependent variable y is moderately correlated negatively with independent variable x
$-0.8 < r_{xy} < -1$	The dependent variable y is strongly correlated negatively with the independent variable x
$r_{xy} = -1$	The dependent variable y is perfectly correlated negatively with the independent variable x

Source: Profillidis *et. al.* (2019).

Instead of r_{xy}, the Pearson correlation coefficient in this paper would be denoted as r.

3. Results and Discussion

3.1 Rainfall

In 10 years of observation, from 2011 to 2020, the lowest rainfall occurred in June 167.20 mm and the highest rainfall occurred in November 578.40 mm in Aceh Besar District. The probability distributions of Q_1, Q_2, and Q_3 and the descriptive statistics of rainfall in mm were presented in Figure 2 and Table 2, respectively.

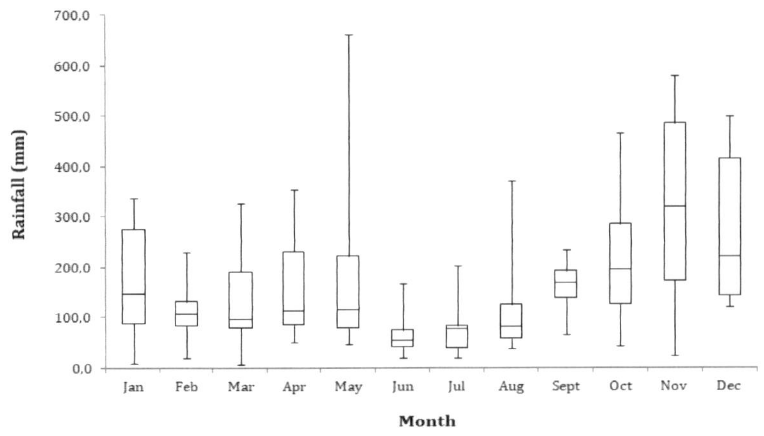

Figure 2. Probability Distributions of Rainfall in Aceh Besar District from 2011 to 2020.

Table 2. Descriptive Statistics of Rainfall in Aceh Besar District from 2011 to 2020.

	Jan	Feb	Mar	Apr	May	Jun
Max	337.90	229.60	327.50	354.90	660.90	167.20
Min	8.00	20.20	7.00	50.60	47.90	19.80
SD	115.29	55.63	94.08	113.58	195.24	42.73
Mean	176.37	112.88	133.16	160.59	196.61	65.17
CV	0.65	0.49	0.71	0.71	0.99	0.66

	Jul	Agt	Sept	Oct	Nov	Dec
Max	202.20	370.00	233.70	466.50	578.40	498.80
Min	20.30	38.00	66.10	41.80	24.60	120.00
SD	51.38	125.35	52.86	133.90	193.24	156.34
Mean	74.55	132.61	156.78	212.73	323.31	267.02
CV	0.69	0.95	0.34	0.63	0.60	0.59

3.2 Temperature

3.2.1 Mean Temperature. The lowest mean temperature was 25.7 °C which occurred in the months of January and November. Meanwhile, the highest mean temperature was 30 °C, occurred in the months of May and August. The probability distributions of Q_1, Q_2, and Q_3 and the descriptive statistics of mean temperature in Celcius degree (°C) were presented in Figure 3 and Table 3, respectively.

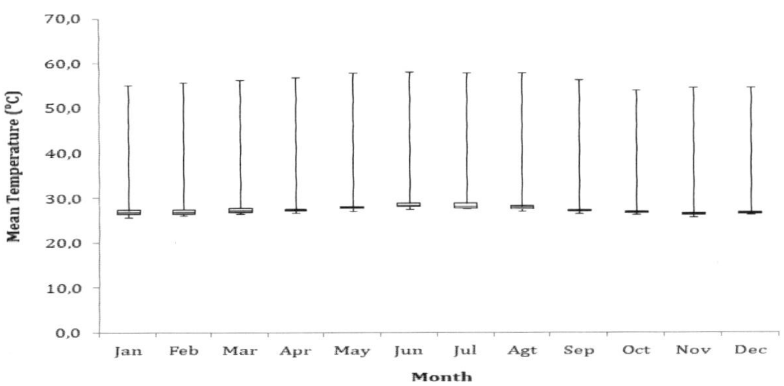

Figure 3. Probability Distributions of Mean Temperature in Aceh Besar District from 2011 to 2020.

Table 3. Descriptive Statistics of Mean Temperature in Aceh Besar District from 2011 to 2020.

	Jan	Feb	Mar	Apr	May	Jun
Max	28.50	28.90	29.10	29.60	30.00	29.80
Min	25.70	26.00	26.50	26.60	27.00	27.40
SD	0.80	0.90	0.80	0.90	0.80	0.70
Mean	26.80	27.00	27.30	27.50	27.90	28.30
CV	0.030	0.033	0.029	0.032	0.028	0.026

	Jul	Agt	Sept	Oct	Nov	Dec
Max	29.70	30.00	29.20	27.20	28.20	28.00
Min	27.60	27.00	26.50	26.20	25.70	26.30
SD	0.70	0.80	0.70	0.30	0.70	0.50
Mean	28.30	28.00	27.30	26.70	26.50	26.70
CV	0.025	0.029	0.027	0.012	0.026	0.018

3.2.2 Minimum Temperature. The lowest minimum temperature was 19.2 °C occurred in February and the highest minimum temperature was 24.5 °C occurred in the months of April and June. The probability distributions of Q_1, Q_2, and Q_3 and the descriptive statistics of minimum temperature in Celcius degree (°C) were presented in Figure 4 and Table 4, respectively.

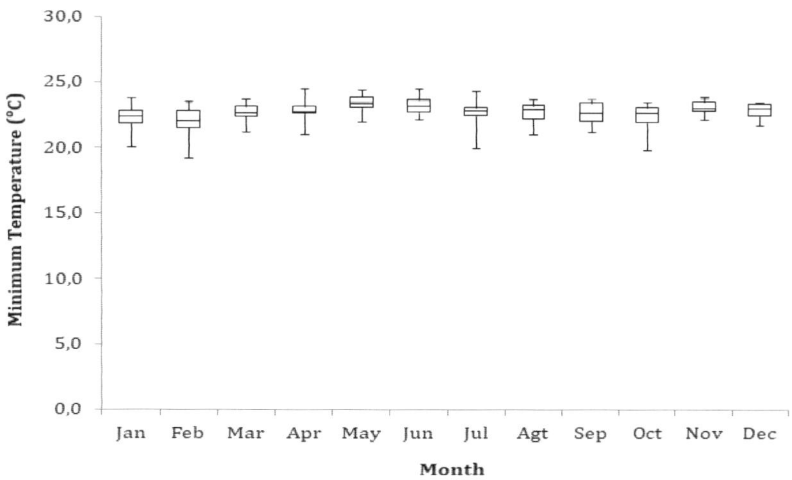

Figure 4. Probability Distributions of Minimum Temperature in Aceh Besar District from 2011 to 2020.

Table 4. Descriptive Statistics of Minimum Temperature in Aceh Besar District from 2011 to 2020.

	Jan	Feb	Mar	Apr	May	Jun
Max	23.80	23.60	23.70	24.50	24.40	24.50
Min	20.10	19.20	21.20	21.00	22.00	22.20
SD	1.20	1.20	0.70	0.90	0.80	0.70
Mean	22.20	22.20	22.60	22.80	23.30	23.30
CV	0.053	0.057	0.033	0.039	0.033	0.030

	Jul	Agt	Sept	Oct	Nov	Dec
Max	24.30	23.70	23.70	23.50	23.90	23.50
Min	20.00	21.00	21.20	19.80	22.20	21.70
SD	1.20	0.80	0.90	1.10	0.50	0.60
Mean	22.70	22.70	22.60	22.40	23.10	22.90
CV	0.052	0.035	0.038	0.050	0.022	0.026

3.2.3 Maximum Temperature. The lowest maximum temperature was 21 °C occurred in February and the highest maximum temperature was 37.2 °C occurred in July. The probability distributions of Q_1, Q_2, and Q_3 and the descriptive statistics of minimum temperature in Celcius degree (°C) were presented in Figure 5 and Table 5, respectively.

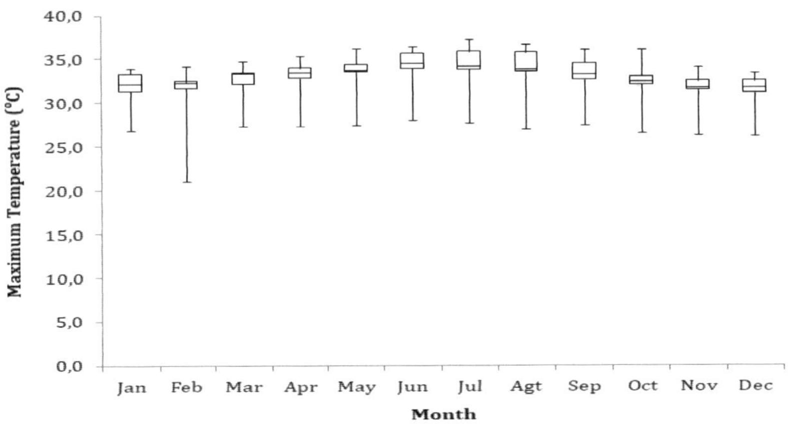

Figure 5. Probability Distributions of Maximum Temperature in Aceh Besar District from 2011 to 2020.

Table 5. Descriptive Statistics of Maximum Temperature in Aceh Besar District from 2011 to 2020.

	Jan	Feb	Mar	Apr	May	Jun
Max	34.00	34.20	34.80	35.40	36.20	36.40
Min	26.90	21.00	27.30	27.30	27.40	28.00
SD	2.10	4.10	2.10	2.20	2.30	2.40
Mean	32.00	30.80	32.70	33.00	33.50	34.20
CV	0.065	0.132	0.064	0.067	0.069	0.069

	Jul	Agt	Sept	Oct	Nov	Dec
Max	37.20	36.60	36.00	36.00	34.10	33.40
Min	27.60	27.00	27.40	26.50	26.30	26.10
SD	2.70	2.80	2.40	2.50	2.10	2.00
Mean	34.30	33.80	33.10	32.40	31.60	31.40
CV	0.079	0.082	0.071	0.076	0.066	0.064

3.3 Relative Humidity

The lowest relative humidity was 64 % occurred in August and the highest relative humidity was 88 % occurred in November. The probability distributions of Q_1, Q_2, and Q_3 and the descriptive statistics of relative humidity in percentage (%) were presented in Figure 6 and Table 6, respectively.

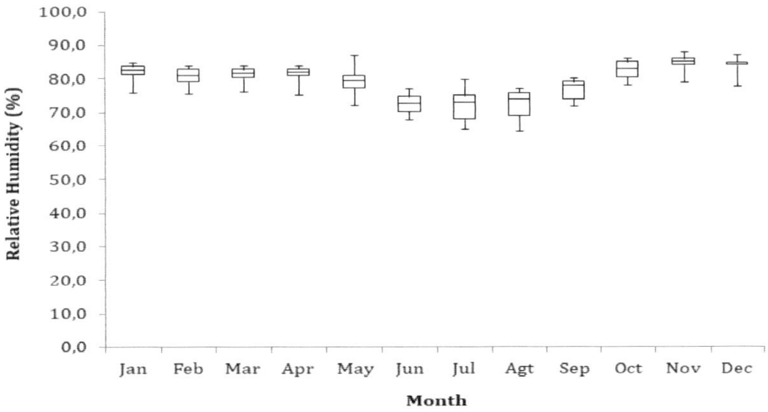

Figure 6. Probability Distributions of Relative Humidity in Aceh Besar District from 2011 to 2020.

Table 6. Descriptive Statistics of Relative Humidity in Aceh Besar District from 2011 to 2020.

	Jan	Feb	Mar	Apr	May	Jun
Max	85	84	84	84	87	77
Min	76	76	76	75	72	68
SD	3	3	2	3	4	3
Mean	82	81	81	82	79	72
CV	0.038	0.034	0.028	0.031	0.051	0.044

	Jul	Agt	Sept	Oct	Nov	Dec
Max	80	77	80	86	88	87
Min	65	64	72	78	79	78
SD	5	5	3	3	2	2
Mean	72	72	77	83	85	84
CV	0.065	0.067	0.043	0.036	0.028	0.028

The relationships between rainfall and temperature, relative humidity, and rice field productivity were analyzed using Pearson correlation coefficient (equation 2). Rainfall was assigned as the independent variable (x) and temperature, relative humidity, and rice field productivity were assigned as the dependent variables (y)s. The Pearson correlation coefficient for rainfall and mean temperature (r = -0.14) which indicated that the dependent variable y (mean temperature) was weakly correlated negatively with the independent variable x (rainfall). The similar indication also happened to the minimum temperature (r = -0.15). Meanwhile, the relationship between rainfall and maximum temperature resulted r = 0.27, indicating that the dependent variable (maximum temperature) was weakly correlated positively with the independent variable (rainfall). Some factors that influenced the temperature were solar radiation, altitude, wind speed, and latent heat. The relationships between solar radiation and temperature, wind speed and temperature, and pressure and temperature were also analyzed using Pearson correlation coefficient (Table 7). The climatology data and descriptive statistics of solar radiation, wind speed, and pressure were not shown.

Table 7. The Pearson correlation coefficients for the relationships between solar radiation, wind speed, pressure, and temperature.

	Temperature			
	Mean	Min	Max	
Solar Radiation	0.17			The dependent variable y was weakly correlated positively with the independent variable x
		-0.18		The dependent variable y was weakly correlated negatively with the independent variable x
			0.49	The dependent variable y was moderately correlated positively with the independent variable x
Wind Speed	-0.44			The dependent variable y was moderately correlated negatively with the independent variable x
		0.78		The dependent variable y was strongly correlated positively with the independent variable x
			-0.40	The dependent variable y was moderately correlated negatively with the independent variable x
Pressure	-0.48			The dependent variable y was moderately correlated negatively with the independent variable x
		0.69		The dependent variable y was moderately correlated positively with the independent variable x
			-0.36	The dependent variable y was moderately correlated negatively with the independent variable x

The Pearson correlation coefficient for rainfall and relative humidity relationship was 0.41, suggesting that the dependent variable (relative humidity) was moderately correlated positively with the independent variable (rainfall). Some factors that influenced the relative humidity were temperature, wind speed, pressure and air density, quality and quantity of solar radiation, water availability, vegetation density, and altitude. The relationships between relative humidity and temperature, wind speed, pressure, and solar radiation were presented in Table 8 and 9 respectively.

Table 8. The Pearson correlation coefficients for the relationships between relative humidity and temperature, wind speed, and pressure.

	Temperature			Wind Speed	Pressure	
	Mean	Min	Max			
Relative Humidity	-0.84					The dependent variable y was strongly correlated negatively with the independent variable x
		0.56				The dependent variable y was moderately correlated positively with the independent variable x
			-0.48			The dependent variable y was moderately correlated negatively with the independent variable x
				0.34		The dependent variable y was moderately correlated positively with the independent variable x
					0.20	The dependent variable y was weakly correlated positively with the independent variable x

Table 9. The Pearson correlation coefficients for the relationship between solar radiation and relative Humidity.

	Relative Humidity	
Solar Radiation	-0.07	The dependent variable y was weakly correlated negatively with the independent variable x

The Pearson correlation coefficient for rainfall and rice field productivity relationship was -0.26, reflecting that the dependent variable (rice field productivity) was weakly correlated negatively with the independent variable (rainfall). Although, the factors which influenced the rice field productivity in Aceh Besar District was not investigated yet, but Mahananto *et.al* (2009) suggested that the cultivated area, the quantity of effective labors, the quantity of fertilizers and pesticides, the distance between farmers' houses and the cultivated area, and the system of irrigation significantly increased the rice field productivity. The Pearson correlation coefficients for the above relationships was presented in Table 10.

Table 10. The Pearson correlation coefficients for the relationships between rainfall and temperature, relative humidity, and rice field productivity.

	Temperature			Relative	Rice Field	
	Mean	Min	Max	Humidity	Productivity	
	-0.14					The dependent variable y was weakly correlated negatively with the independent variable x
		-0.15				The dependent variabley was weakly correlated negatively with the independent variable x
Rainfall			0.27			The dependent variable y was weakly correlated positively with the independent variable x
				0.41		The dependent variable y was moderately correlated positively with the independent variable x
					-0.26	The dependent variable y was weakly correlated negatively with the independent variable x

4. Conclusion

Rainfall and rice field productivity were negatively correlated in Aceh Besar, indicating that rainfall did not impact the rice field productivity in Aceh Besar. This finding was not aligned with the reality in the field. Aceh Besar had experienced long drought, higher temperature, shifted seasons, and the decrease yield of rice field productivity due to climate change. The lack of climate data was assumed to be the cause of this contradictive results which required longer period of data preferably 30 to 50 years.

5. References

[1] Central Bureau of Statistics for Aceh Besar 2011 to 2020 *Aceh Besar in Number*

[2] Chowdury A 2014 Factors affecting productivity and efficiency of rice production in hoar area in Bangladesh: likely impact on food security *Thesis or Dissertation - Master of Science*

[3] Directorate General of Climate Change Control of Indonesia 2019 *Climate and Climate Change* available at ditjenppi.menlhk.go.id

[4] Government of Aceh Besar District 2016 *Map of Aceh Besar* available at acehbesarkab.go.id

[5] Hossain M K, Kamil A A, Masron T A and Baten M A 2013 Impact of Environmental Factors on Efficiency of Rice Production in Bangladesh *J. App. Sci.* **13** 564-571

[6] Kaletapwa F, Agwu A E and Egbule C 2012 Farmer's Adaptation Initiatives to the Impact of Climate Change on Agriculture in Northern Nigeria *J. Agricultural Education and Extension* **16** (1) 132-144

[7] Kasmin M O and Kartomo 2020 Factors Affecting The Production of Rice Farming in Polenga Village, Kecamatan Watubangga District, Kolaka Regency *Agribussiness J.* **3** (2) 2548-379X 2548-2211

[8] Mahananto S, Sutrisno, and Ananda C F 2009 Analysis of Influencing Factors in Increasing Rice Production (Case Study in the Nogosari Subdistric Boyolali Regency) Central Java Province *J. Wacana* **12** 179-191

[9] Mailena L, Shamsudin M N, Radam A and Latief S 2014 Rice Farms Efficiency and Factors Affecting the Efficiency in MADA Malaysia *J. App. Sci.* **14** 2177-82
[10] Profillidis V A and Botzoris G N 2019 *Modeling of Transport Demand* in *Statistical Method for Transport Demand Modeling* available at https://www.sciencedirect.com/topics/social-sciences/pearson-correlation-coefficient
[11] UNFCCC 2011 *Fact Sheet: Climate Change Science-The Status of Climate Change Science Today* United Nations Framework Convention on Climate Change New York

Acknowledgments
Author would like to acknowledge the Central Bureau of Statistics for Aceh Besar for providing data required in this research.

The pre-liminary study on surface modification of magnetite for microplastic removal from water

T Phengsaart*, P Phradub, W Tonabut, J Ekasin, P Katethol, P Sirisetpakdee, J Pimolrat, O Juntarasakul, N Soonthornwiphat and S Saisinchai

Department of Mining and Petroleum Engineering, Faculty of Engineering, Chulalongkorn University, Bangkok, 10330, Thailand
*Corresponding author's e-mail: theerayut.p@chula.ac.th

Abstract. Microplastics (MPs), the plastic particles with size fraction smaller than 5 mm, are recently becoming one important problem worldwide due to their ubiquitous distribution in the environment, especially in water. In this study, the separation techniques in mineral processing (i.e., magnetic separation and surface-based separation) were applied to remove MPs from contaminated water. Five kinds of model MPs samples (+38–2000 µm); polypropylene (PP), polyethylene (PE), polystyrene (PS), polyethylene terephthalate (PET), and polyvinyl chloride (PVC) were used and magnetite (Fe_3O_4) powders (−75 µm) were modified using hexadecyltrimethoxysilane (HDTMS) to change their surface from hydrophilic to hydrophobic for MPs recovery. The results showed that MPs could be attached on the surface of modified magnetite due to the hydrophobic interaction and could be recovered together with modified magnetite by magnetic separation.

1. Introduction

Recently, the United Nations (UN) including the microplastic contamination problem into "Sustainable Development Goals (SGDs)" as goal number 14 "Life below water" [1]. Microplastics (MPs, −5 mm plastic particles) are fast becoming a critical and serious environmental issue of our modern society due to their widespread distribution in various ecosystems (i.e., hydrosphere, lithosphere, atmosphere, and biosphere) [2-4].

The contamination of MPs can affect the environment, animal's health, and also human health by the food chain due to its surface structures and features. MPs' surface can adsorb heavy metals (e.g., Pb, Cd, Cr, Ba, Cu, Co, and As), persistent chemical pollutants (PCPs), and organic pollutants from the environment [3, 4]. Recently, there are some researches applied the separation techniques from mineral processing and resources recycling to separate and recover MPs from the environment, for example, gravity separation [1], magnetic separation [5], and surface-based separation [5].

In a previous study by Grbic et al., 2019 [5], the surface-based technique and magnetic separation were combined to recover MPs from contaminated water. The surface of Fe nanoparticles was modified using hexadecyltrimethoxysilane (HDTMS) to change their surface from hydrophilic to hydrophobic. Microplastic particles in water could be attached to the modified Fe nanoparticles due to their hydrophobic interaction. After that, Fe nanoparticles with attached microplastics could be recovered using magnetic separation.

In this study, the magnetite (Fe_3O_4) ores were used instead of Fe nanoparticles since magnetite is one of the important primary resources for steel production, the use of magnetite ores are more sustainable due to their lower cost and less environmental impacts. The surface modification and recovery process were investigated to understand the probability of this method for the development in the real environmental situation.

2. Material and methods

2.1. Microplastic samples

The model microplastic samples were used in this study. Five kinds of virgin plastic boards with the 2-mm thickness (Table 1) were crushed by a cutting mill (Orient mill, VH16, Seishin Enterprise Co., Ltd. Japan, ground at low temperature with liquid nitrogen, and sieved to obtain +212–2000 μm (course) and +38–212 μm (fine) size fractions.

Table 1. List of plastic samples used in this study.

Plastic samples	Abbreviation	Specific gravity (SG)
Polypropylene	PP	0.92
Polyethylene	PE	0.95
Polystyrene	PS	1.06
Polyethylene Terephthalate	PET	1.31
Polyvinyl Chloride	PVC	1.38

2.2. Magnetite samples

In this study, magnetite ores were used to remove the microplastics from water instead of Fe nanoparticles. The magnetite ores (laboratory samples) were firstly separated the magnetic and non-magnetic fraction using a permanent magnet. The magnetic fraction was fed to shaking table separation to increase the grade (purity) of magnetite (Fe_3O_4) then, was ground using a disc mill (Kawasaki-Siebtechnik) and sieve to obtained –75 μm size fraction. The elemental and mineral contents of feed and each product of shaking table separation were analyzed by X-ray fluorescence (XRF) spectrometer and X-ray diffractometer (XRD). The most suitable product (highest magnetite content) will be used for surface modification experiments.

2.3. Surface modification experiments

The surface modification experiments were carried out under the closed system in a beaker with a mechanical stirrer and paraffin as shown in Figure 1. The experiments were carried out under the following conditions: methanol (CH_3OH) of 100 ml, Hexadecytrimethoxysilane (HDTMS, $H_3C(CH_2)_{15}Si(OCH_3)_3$, ≥85% (GC), Merck KGaA, Darmstadt, Germany) of 1 ml, 10 g of –75 μm magnetite powders, conditioning time of 2 hr at room temperature. After the modification was done, the modified magnetite was separated from the solution by a permanent magnet, then keep in distilled water.

2.4. Microplastic removal experiments

1 g of microplastic sample (PP, PE, PS, PET, and PVC; coarse and fine fractions) was fed into a beaker with 200 ml of distilled water and 1 g of magnetite (w/ or w/o modification) and then manually mix using a glass stirring rod for a while. After that, magnetite with attached microplastics was recovered using a permanent magnet. The remained microplastics in water were dried and weight to determine the recovery rate.

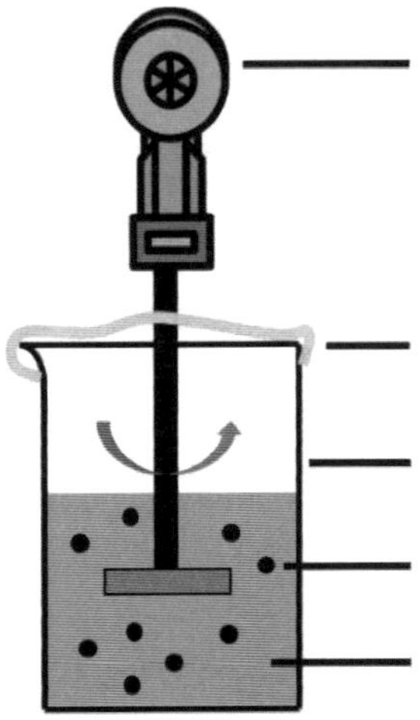

Figure 1. The schematic diagram of surface modification setup used in this study.

3. Results and discussion

3.1. Sample preparation and characterization

To improve the purity of magnetite ores for surface modification and MPs removal experiments, the sample preparation using magnetic separation and gravity separation was carried out. The XRF and XRD results of magnetic fraction from magnetic separation using a permanent magnet contains 84.05 and 61.2 % of Fe_2O_3 and magnetite, respectively. And after treated by a wet-type shaking table (optimum conditions), the Fe_2O_3 and magnetite contents in middling (suitable products) increased to 90.87 and 62.1 %, respectively and the impurities are mainly hematite (Fe_2O_3, 20.7%) and quartz (SiO_2, 17.1%).

3.2. Microplastic removal experiments

In preliminary experiments, various tests of surface modification experiments. The results showed that the surface modification processes need to be in the closed system since methanol can quickly evaporate at room temperature that could change the concentration of HDTMS solution then, magnetite powders could not be modified properly. After the modification, various surface wettability tests (i.e., visual observation, direct observation of bubble attachment using a dropper, floatability analysis using column flotation, and floatability analysis using mechanical flotation) were carried out. However, clear results could not be obtained then, the modified magnetite was tested in remove microplastic removal experiments directly by comparison with non-modified magnetite.

The results showed that the microplastics could attract with the modified magnetite better than the non-modified magnetite in all kinds of plastic indicating that surface modification succeeded and hydrophobic interaction occurred (Table 2). The high recovery rate was obtained for light plastics (i.e., PP, PE, and PS) while the recovery rate of heavy plastic (i.e., PET and PVC) indicating that SG of plastics could affect this attraction. Besides, the effects of particle sizes and surface wettability still unclear. Polypropylene (PP) with modified magnetite showed the highest recovery rate; 80.38% for coarse microplastics (+212–2000 μm) and 72.84% for fine microplastics (+36–212 μm) indicating this method might have a potential for development in the real environmental situation since PP is the plastic that constituted the highest consumption [6].

Table 2. Microplastic removal results

		Recovery rate of MPs (%)			
Size fraction of MPs (μm)		+212–2000		+36–212	
Surface modification of magnetite		w/o	w/	w/o	w/
Type of MPs	PP	10.35	80.38	–	72.84
	PE	0.95	32.09	–	39.30
	PS	0.00	51.25	–	32.69
	PET	0.47	1.13	–	4.58
	PVC	0.00	11.41	–	23.35

Note "–" means no data (The experiments of fine MPs using non-modified magnetite were not carried out since the good results could not be obtained from coarse MPs).

4. Conclusions

In this study, hydrophobic magnetite powders were used to recover MPs from the contaminated water. The results showed that MPs could attract with the modified magnetite better than the non-modified magnetite in all kinds of plastic and SG of plastics could affect the recovery rate. The highest recovery rate could be obtained from the recovery of PP with modified magnetite (80.38% for coarse microplastics (+212–2000 μm) and 72.84% for fine microplastics (+36–212 μm)).

References

[1] Coppock R L, Cole M, Lindeque P K, Queiros A M and Galloway T S 2017 A small-scale, portable method for extracting microplastics from marine sediments *Environ. Pollut.* **230** 827−837

[2] Bissen R and Chawchai S 2020 Microplastics on beaches along the eastern Gulf of Thailand – A preliminary study *Mar. Pollut. Bull.* **157** 111345

[3] Elkhatib D and Oyanedel-Craver V 2020 A critical review of extraction and identification methods of microplastics in wastewater and drinking water *Environ. Sci. Tech.* **54** 7037−7049

[4] Chai B, Wei Q, She Y, Lua G, Dang Z and Yin H 2020 Soil microplastic pollution in an e-waste dismantling zone of China *Waste Manage.* **118** 291−301

[5] Grbic J, Nguyen B, Guo E, You J B, Sinton D and Rochman C M 2019 Magnetic extraction of microplastics from environmental samples *Environ. Sci. Tech. Let.* **6** 68−72

[6] Ito M, Saito A, Takeuchi M, Murase N, Phengsaart T, Tabelin C B and Hiroyoshi N 2021 Development of the reverse hybrid jig: Separation of polyethylene and cross-linked polyethylene from eco-cable wire *Miner. Eng.* **174** 107241

Acknowledgments

The authors are grateful to the Laboratory of Mineral Processing and Resources Recycling, Hokkaido University for the sample preparation, and other members of the Department of Mining and Petroleum Engineering, Chulalongkorn University for their assistance in this research.

Spatial distribution and pattern of heavy metals in urban soils of Yogyakarta, Indonesia

J L Magno[1], **W Budianta**[1*]

[1]Department of Geological Engineering, Universitas Gadjah Mada, Jl. Grafika 2 Kampus UGM, Yogyakarta, Indonesia
*Corresponding author's e-mail: wbudianta@ugm.ac.id

Abstract. Heavy metals contamination is becoming a global problem in urban areas. With that, understanding spatial distribution and pattern of heavy metals in urban soils is a crucial step toward sustainable urban development. This study intends to assess the spatial distribution and pattern of four heavy metals (Pb, Cu, Zn, and Cd) in Yogyakarta City outward in the boundaries of Sleman and Bantul Regencies. In relation, we utilized geostatistical method Empirical Bayesian Kriging (EBK) then correlated to Land Cover/Use data for the spatial analysis of heavy metals concentration. The degree of contamination was quantified using indices - *PI* (Pollution Index), I_{geo} (Geo-accumulation Index), and *PLI* (Pollution Load Index). We show that by analysing the Pb, Cu, Zn, and Cd elements of 168 urban soils samples collected, Yogyakarta City, a densely populated area, serves as a place of heavy metals contamination hotspots. Pb and Cu is posed as moderately contaminated, whereas Zn and Cd is considered as uncontaminated (as majority of Zn and Cd values does not exceed the background values). But in overlapping consideration of four heavy metals contamination, they are classified as moderately contaminated *(PLI=1.10)*. In summary, heavy metals contamination in soils varies as a function of urbanization.

1. Introduction

Today, the world is facing environmental challenges due to anthropogenic activities. Anthropogenic activities, such as industry, aquaculture, agriculture, and mainly urbanization, remain the greatest threat to the sustainability of soils. Urbanization involves unprecedented spatial expansion of urban related area and conversion of natural physical landscape to more economic land uses, all of which involve human utilization of soil (i.e., Urban Soil). Most of the world biggest cities face challenges that accompany urbanization, soil degradation due to contamination (e.g., heavy metals) has not been averted by cities of any size [1]. Certainly, the impact of urban activities of cities on local, regional, and global-scale soil degradation trends is increasing [2]. Also, it is known that soil degradation due to contamination is becoming one of the global problems that endangers sustainable development [3].

Generally, the concentration and distribution of heavy metals as a contaminant in soils are influenced by the nature of parent material-mineralogy, climate, mobility, geological factors, and physio-chemical variable of the environmental matrices [4]. Yogyakarta Special Region is a complex urban setting expose to various environmental matrices: it has a nearby volcano (Mt. Merapi), prone to earthquake, and lastly, presence of contaminants [5]. In addition, Yogyakarta City is considered as one of the fast-growing urban areas in Indonesia [6]. As urbanization of Yogyakarta City expands, there is limited understanding of heavy metals spatial distribution in its soil. Through time, heavy metals may accumulate to contaminate the soils and influence the nearby municipalities/cities of Yogyakarta City.

This study aims to evaluate the relative importance of heavy metals (Pb, Cu, Zn, Cd) concentration in spatially expanding urban area (viz. Yogyakarta City), and the extent & pattern of contamination/pollution hotspots. Results contribute to the improvement of spatial expansion planning and to strengthen decision-making of authorities in selection/zoning of urban development. Importantly, spatial determination of contaminants may serve as a basis for more effective soil pollution mitigation strategies – such as identifying the contaminant specific-sources, and remediation implementations.

2. Materials and Methods

In this study, quantification of heavy metals in the soil samples collected from Yogyakarta Special Region was carried out using ICP-AES (Inductively Coupled Plasma – Atomic Emission Spectrometry). ArcGIS Pro 2.9.0 has been utilized for geostatistical mapping & analysis.

2.1. Study Area

Yogyakarta Special Region has five Regency (i.e., City), namely. Kulon Progo, Gunung Kidul, Bantul, Sleman, and Yogyakarta City. The study area has an emphasis in Sleman, Bantul, and Yogyakarta City as where all of the sample collection was conducted. The samples were collected from 168 unique location (Figure 1). Hence, this study will centralize in the idea of Yogyakarta City expanding urbanization, spreading outwards Sleman and Bantul Regency.

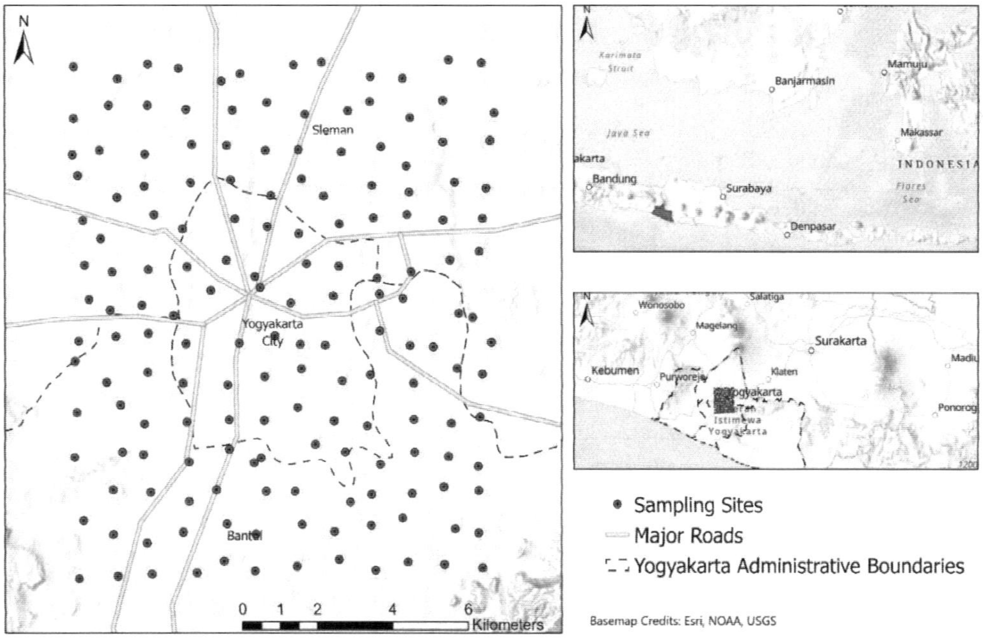

Figure 1. Location of Samples in Yogyakarta Special Region, Indonesia.

2.2. Sampling and Heavy Metal Analysis

The sampling sites was established using the gridding method, a grid with each other – 1.5 km x 1.5 km. The sampling was done by hand drill (i.e., hand auger) at a depth of 10 cm, with the objective to obtain topsoil layer representation. At each sampling site, 500g (0.5 kilogram) of soil samples were collected and placed in a sealed plastic bag for laboratory processing. The samples were then dried for two weeks at room temperature before being analysed in the lab. 1.5 g of soil sample was crushed in an agate mortar and pestle for laboratory preparation, then sieved to obtain fine size (2 millimeters).

Then, the sample digestion was done using triacid method in procedural mirror to EPA – Method 3050B [7]. HF-HCl and HNO_3 were mixed then was added to weighted sample in conical flask followed by microwave digestion [8]. After digestion, the samples were in the form of solution where it was analyzed using ICP-AES to determine heavy metal concentrations, namely, Lead (Pb), Copper (Cu), Zinc (Zn), and Cadmium (Cd).

2.3. Assessment of Heavy Metal in contaminated Urban Soils
This study used three parameters for the assessment of heavy metal contamination in the urban environment. They are geo-accumulation index (I_{geo}), pollution index (*PI*), and pollution load index (*PLI*). These three parameters were the indicators of degree of contamination in the urban environment.

2.3.1. Geo-accumulation Index (I_{geo}). Muller (1979) introduce the concept of Geo-accumulation Index (I_{geo}). The concentration of heavy metal accumulation in soils correlated to background concentration were demonstrated using this index. This index was calculated using the equation below [9]:

$$I_{geo} = log_2(C_n/1.5 * B_n) \quad (1)$$

where C_n is the concentration of heavy metal in the soil, B_n is the background value of heavy metal (i.e., natural concentration). To remunerate the lithogenic effects, a constant factor of 1.5 is consider as matrix correction. I_{geo} values were classified as follows: ≤0 – uncontaminated, 0-2 – moderately contaminated, 2-4 – Strongly contaminated, and ≥4 – very strongly contaminated [9].

In this study, the background values were based on the study of natural non-contaminated soil within the study area by Wulaningsish 2009 (Table 1) [10].

Table 1. Background value (B_n) for soil heavy metals from Wulaningsish, 2009.

Heavy Metals	B_n values (mg/kg)
Pb	25.5
Cu	23.6
Zn	86.1
Cd	0.1

2.3.2. Pollution Index (PI). The enrichment of heavy metals in soil samples was determined by calculating the Pollution Index (PI) and comparing it to soil background values (Table 1). This index was calculated using the equation below:

$$PI = C_s/B_n \quad (2)$$

where C_s is the heavy metal concentration in contaminated soils, and B_n is the background value of heavy metal in uncontaminated setting. In reference to the calculated values, PI values were categorized as follows: PI <1 – low contamination, PI 1-3 – moderate contamination, PI 3-6 – high contamination, and PI>6 – very high contamination [11].

2.3.3. Pollution load index (PLI). Pollution load index (PLI) is used for the determination of heavy metal pollution level in a specific area. It is also correlated as the overall soil assessment toxicity. This index was calculated using the equation below

$$PLI = (PI_{Pb} * PI_{Cu} * PI_{Zn} * PI_{Cd})^{1/4} \quad (3)$$

where PI is the calculated Pollution Index on each heavy metals, and ¼ exponent act as the metal assessing. The classification of pollution is based on four class: PLI < 1 – No pollution; 1 ≤ PLI < 2 – moderate pollution value; 2 ≤ PLI < 3 – high pollution value; and PLI ≥ 3 very high pollution value [12].

2.4. Geostatistical Mapping and Analysis

Empirical Bayesian Kriging (EBK) was used as a geostatistical method to demonstrate the spatial distribution of Pb, Cu, Zn, and Cd concentrations. EBK accounted the error introduced by estimating the underlying semivariogram involved – for an effective interpolation and simulations compared to other kriging methods [13]. The analysis was done with the help of ArcGISPro 2.9.0. by inputting Pb, Cu, Zn, and Cd concentration as z-field values.

The Land Use & Land Cover (LULC) Map created and processed using ArcGIS Pro 2.9.0 act as supplemental data for discussion analysis. Imagery in LULC Map was source out by utilizing ESA-Sentinel 2. There were seven (7) class definitions established, inputting major road for urban area network reference visualization. The cell's surface area to volume ratio (cell size) was set to 10 based on Esri published 2020 Global LULC dataset [14].

3. Results

3.1. Spatial Distribution of Heavy Metals

The spatial distribution of heavy metals concentration (*mg/kg*), namely, Pb, Cu, Zn, and Cd in vicinity of sampling sites covering City of Yogyakarta, Sleman Regency, and Bantul Regency is shown in Figure 2. The heavy metals concentration distribution is map out to determine the pattern of contamination. A higher (i.e., *hotspots*) heavy metals concentration is present in Yogyakarta City compared to rural areas of Sleman and Bantul (i.e., cold spots). Zn concentration compared to other heavy metals has less presence in Yogyakarta City, especially the southern part. Although Cd is concentrated in Yogyakarta City, it does not exceed the background concentration values, indicating that Cd as contaminant is not yet evident. In pattern, all heavy metals concentration hotspots exist and surrounds Yogyakarta City. These results confirm that urbanization is a major key player in soil contamination.

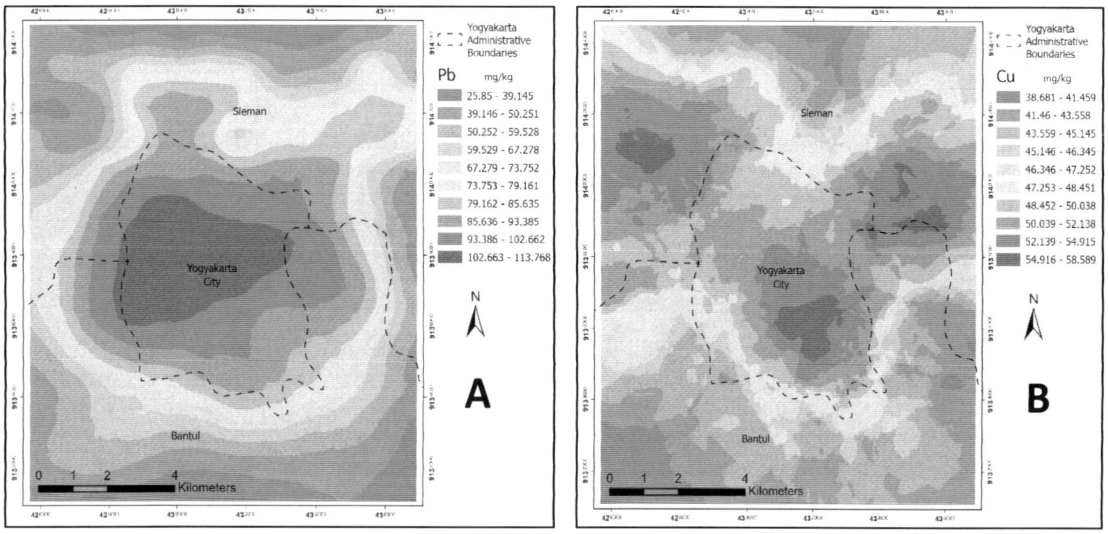

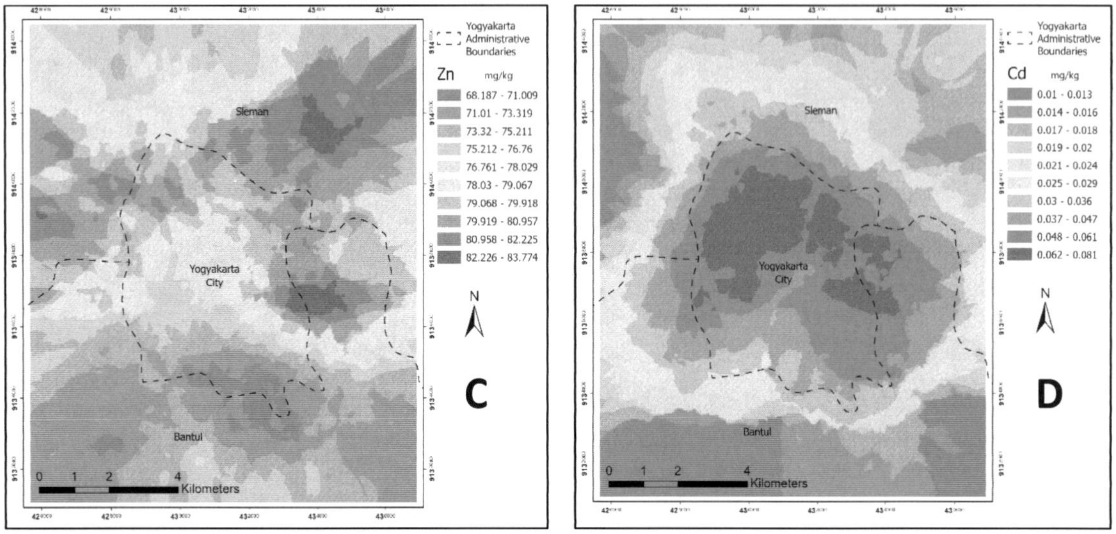

Figure 2. Spatial Distribution of A. Lead (**Pb**), B. Copper (**Cu**), C. Zinc (**Zn**), and D. Cadmium (**Cd**). For visualization, areas with high concentration values are represented as hotspots (red), and low concentration values are represented as coldspots (blue).

To further understand the urban soil contamination distribution (*in data perspective*), statistical summary of heavy metals concentration is given in Table 2. Concentration of Pb, Cu, and Zn by mean exceeds the background concentration. In comparison, Cd mean value is lower than the background concentration. Heavy metals concentration for Pb has negative skewness and kurtosis, indicating extremity of values that exceeds background concentration. Cu and Zn have skewness and kurtosis values closer to zero, illustrating a normal distribution trend of concentration values. Cd has positive skewness and kurtosis; it indicates that low concentration values is dominant. SD (Standard Deviation) demonstrates that there is a spread in wider range of Pb, Cu, and Zn values, except Cd. These statistical values bring up the importance of looking at how concentration values of heavy metals are being influence.

Table 2. Summary statistics for heavy metals concentration (mg/kg) in soils.

	Pb	Cu	Zn	Cd
Range	18-129	30-78	49-98	0.005-0.19
Mean	70.04	47.35	77.14	0.03
SD	29.26	10.82	8.35	0.04
Skewness	-0.21	0.96	-0.29	2.54
Kurtosis	-1.10	0.32	0.93	6.82

3.2. Heavy Metals Calculated Indices

Indices values are criterions to determine the degree of contamination of Urban Soils in all sampling sites (Table 3). *PI* and I_{geo} values of Pb is higher compared to Cu, Zn, and Cd. The *PI* values of Pb and Cu are considered as moderately contaminated, whereas Zn and Cd are low contamination level. In conjunction, Pb and Cu I_{geo} values are positive indicating moderate contamination, while Zn and Cd have negative I_{geo} values indicating an overall content that is below the detection limit or non-contaminated. Considering the 4 heavy metals, its *PLI* indicates that there is a moderate pollution in urban soils. These results suggest that indices may provide information to understand the trend of contamination.

Table 3. Summary for heavy metals (mg/kg) values of Pollution Index (*PI*), Geo-accumulation Index (I_{geo}), and Pollution Load Index (*PLI*).

Heavy Metals	Pollution Index (*PI*) Mean±SD	Geo-accumulation Index (I_{geo}) Mean±SD
Pb	2.75±1.14	0.71±0.75
Cu	2.01±0.46	0.38±0.31
Zn	0.90±0.01	-0.75±0.16
Cd	0.29±0.36	-3.05±1.24
	*Pollution Load Index (*PLI*) = 1.10±0.21	

4. Discussion

4.1. Spatial Distribution and Pattern Interpretation of Heavy Metals Concentration

Urbanization provides people with better standard of living. Urban growth generates revenues that fund projects related to infrastructures, transportation & communication access, and improving public health [15]. Hence, proper and scientific based spatial planning is important to pave way in establishing a sustainable urban development. Also, urbanization comes with costs, it causes soil degradation and environmental-health hazard [1]. In this study, we find that most of soil contamination hotspots are in highly urbanized areas – Yogyakarta City. It matches emerging findings on having soil contamination occurrences in Yogyakarta City, such as Pb [5][16][17].

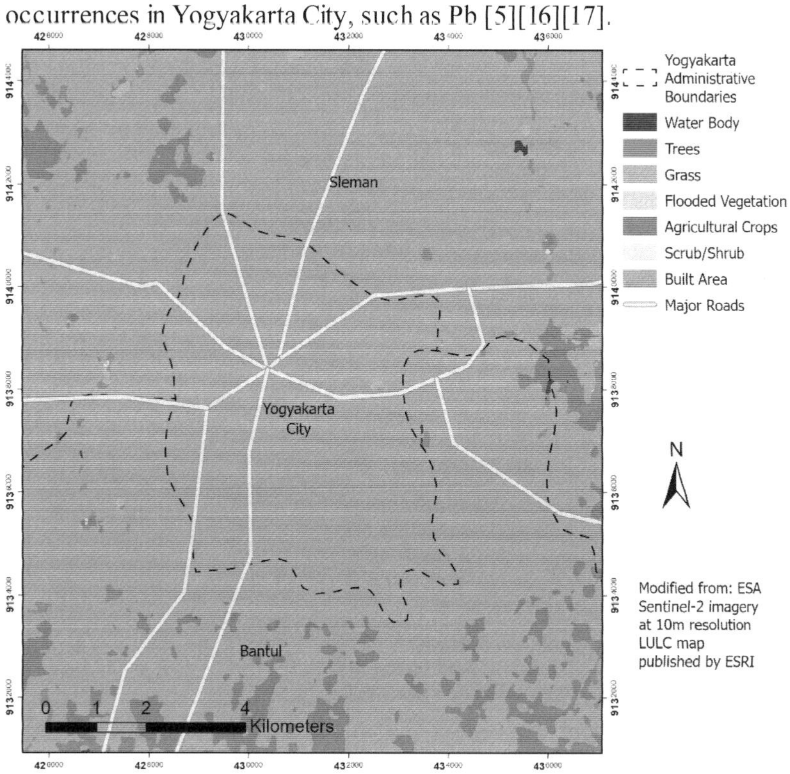

Figure 3. Generalized Land Use/Land Cover (LULC) for Yogyakarta City, Sleman, and Bantul (modified from: Esri 2020 Global LULC [14]).

In reference to rapid urbanization of Yogyakarta City, urban expansion moves outward Sleman and Bantul Regencies. The high heavy metals concentration pattern aligns with major roads existence where most of human activities happens (Figure 2 & 3). Built area, a densely populated area, most of which is developed as industrial, commercial, or residential area coincides with having high heavy metal concentration values in its soils. In southern part, it is not wholly occupied by built area, there are a lot of agricultural activities still exist. Due to less built area in southern part, it agrees with the observations that having lower heavy metal concentrations in its soils. (in ref. with zoning studies [16]). Interestingly, Zn does not follow the usual spatial pattern of Pb, Cu, and Cd in the core of Yogyakarta City, Zn is more concentrated in upper portion. Also, one noticeable spatial pattern is the heavy metal concentration accumulation of Cd, it is enriching and sourcing within Yogyakarta City (i.e., central areas) which on long-term (i.e., continuous urban activities) add up as a contaminant in par with Pb and Cu. Thus, heavy metals concentration trend increases as urbanization expands further.

4.2. Environmental Implication of contaminated Urban Soils

Heavy metals contamination in urban soils imposed a soil degradation problem globally [1]. Aside from health-hazard caused by heavy metals contamination in soils, presence of heavy metals ions might also trigger changing of physical and mechanical properties of soils [18]. In pro-environmental perspective, a good quality soil (i.e., less contaminated) limits erosion, and help improve air and water quality [19]. In relation, it is essential to monitor and understand soil condition in what degree it is contaminated using indices.

The results of PI, I_{geo}, and PLI indices computed values demonstrates the condition of urban soils contaminated by heavy metals as a direct or indirect results of urbanization. The I_{geo} of Pb and Cu are in moderate contamination class agreeing to the findings that Yogyakarta City urbanization causes contamination in soils (e.g., traffic density) [5][17]. In contrast, Zn and Cd I_{geo} was less than zero indicating that it is practically uncontaminated. In addition, Zn and Cd have low PI values compared to Pb and Cu where it is considered as moderately contaminated. The unusual heavy metal concentration values of Zn and Cd should be further assessed in correlation with geological factors because natural causes can also increase heavy metal contamination such as volcanic activity [4].

Based on PI and I_{geo} values, Zn and Cd does not pose alarming contamination but as these values correlated to spatial distribution of Zn and Cd, its hotspots are concentrated in Yogyakarta City (especially Cd). It implies that urbanization intensity in relation to time development progress is also crucial to understand (Time as a factor in contamination enrichment). Still, the PLI value of all heavy metals studied indicates an overall moderate pollution, these findings are significant to the fact that Yogyakarta City is still on its urban growth stage [6]. Hence, heavy metals contamination in urban soils is important to spatially correlate with Land-Use. But it is also noticeable to consider and study further the time since development of an urban area to understand soil contamination in the past, present, and future for detailed decision-making.

5. Conclusion

Acquiring insights into the spatial distribution and pattern of heavy metals in urban soils is an important step toward sustainable urban development. By analysing the heavy metal concentration in urban soils, this research shows that heavy metal contamination in soil vary as a function of urbanization. Yogyakarta City serve as a place of heavy metal contamination hotspots compared to nearby Sleman and Bantul Regencies. Pb and Cu is posed as moderately contaminated, whereas Zn and Cd is considered as uncontaminated. The mean heavy metal concentration of Zn and Cd does not exceed the influence of natural environment (background values). In addition, Zn and Cd spatial pattern should be given an attention. Zn is concentrating in upper urban parts of Yogyakarta City bounded by Sleman Regency, and Cd (though uncontaminated) implies accumulation in inner-upper Yogyakarta City. It is evident that urbanization play a major role in heavy metal contamination. However, more research is also needed to understand geological factors effects on urban soil contamination and determining the specific type of built area that contribute as a source to heavy metals contamination particularly Zn and Cd - which leads to unusual spatial pattern of heavy metals. In these contexts, strengthening regulations & policies related to heavy metals contamination (align with rapid urban growth/expansion of Yogyakarta City) might be necessary to make interventions in improving urban soil quality and remediation strategies.

References

[1] Hardoy, J. E., Mitlin, D., and Satterthwaite, D. 2001. *Environmental problems in an urbanization world.* London: Earthscan.
[2] McGranahan, G., Jacobi, P., Songsore, J., Surjadi, C., & Kjellen, M. 2001. *The citizens at risk: From urban sanitation to sustainable cities*. London: Earthscan.
[3] Jie, C., Jing-zhang, C., Man-zhi, T., and Zi-tong, G. 2002. Soil degradation: a global problem endangering sustainable development. *J. of Geographical Sci.* **12**, 243-252.
[4] Palumbo B, Angelone M, Bellanca A, Dazzi C, Hauser S, Neri R, and Wilson J, 2000. Influence of Inheritance and Pedogenesis on the Metal Distribution in Soils of Sicily, Italy. *Geoderma.* **95**. 247–266.
[5] Budianta W, 2012. Lead Contamination in Soil of Yogyakarta city, Indonesia. *J. SE Asian Appl. Geol.*, **4(2)**, pp 90-98.
[6] Pradoto W 2011. Dynamics of Peri-urbanization and socioeconomic transformation: Case of Metropolitan Yogyakarta, Indonesia. *Intern. J. Arts. Sci.***4(27)** 19.
[7] EPA 1996. *Method 3050B – Acid Digestion of Sediments, Sludges, and Soils.*
[8] Bettinelli M, Beone G M, Spezia S, and Baffi C 2000 Determination of heavy metals in soils and sediments by microwave-assisted digestion and inductively coupled plasma optical emission spectrometry analysis. *Anal. Chem. Acta.***424(2)**, 289-296.
[9] Müller G 1979 Schwermetalle in den Sedimenten des Rheins-Veränderungen seit 1971.
[10] Wulaningsih T 2009 Karakteristik Abu Vulkanik Gunung Merapi dan Pengaruhnya terhadap Kesehatan Masyarakat di Desa Kaliadem dan Sekitarnya, Kecamatan Cangkringan, Kabupaten Sleman Propinsi Daerah Istimewa Yogyakarta, Undergraduate Thesis, Department of Geological Engineering, Universitas Gadjah Mada (Unpublished – Indonesian version only).
[11] Loska K, Cebula J, Pelczar J, Wiechula D and Kwapulinski J 1997 Use of enrichment, and contamination factors together with geoaccumulation indexes to evaluate the content of Cd, Cu, and Ni in the Rybnik water reservoir in Poland. *Water Air Soil Pollution.* **93**(1) 347–365.
[12] Wang, X. Q., He, M. C., Xie, J., Xi, J. H., and Lu, X. F. 2010. Heavy metal pollution of the world largest antimony mine-affected agricultural soils in Hunan province (China). *J. Soil. Sediment.* **10**(5) 827–837.
[13] Gribov A and Krivoruchko 2020. Empirical Bayesian Kringing implementation and usage. *Sci. of the Total Env.* **722**.
[14] Karra, Kontgis, et al. 2021. Global land use/land cover with Sentinel-2 and deep learning. IGARSS 2021-2021 IEEE International Geoscience and Remote Sensing Symposium. *IEEE*
[15] Wan G 2013. *Why Urbanization may benefit the environment*. Asian Development Bank (ADB). Perspective, Op-eds.
[16] A T Sekarningsih *et al* 2021 *IOP Conf. Ser.: Earth Environ. Sci.* **851** 012026.
[17] Sekarningsih A T *et al* 2021. Assessing the spatial distribution of urban soil lead contamination in Yogyakarta City, Indonesia. *E3S Web Conf.* ***325*** (03001).
[18] Chen R.G., Ma F.R., Pang Y.B. 2014. Experimental study on strength characteristics of heavy metal contaminated soil. *J. Civil Engineering and Environmental Engineering*., **36**: 94-98.
[19] Brevik, E.C. & Burgess, L.C. eds 2013. *Soils and Human Health*. Boca Raton: CRC Press.

Acknowledgments
This research is financially supported by CUBE-ASEAN Program of JICA Project for AUN/SEED-Net together with Universitas Gadjah Mada. Researchers extend its gratitude to Tokyo Institute of Technology, Japan for the facility accommodation to conduct analysis of all 168 soil samples.

Soil expansion acting within geosynthetic-reinforced embankment over sinkhole

M T Pham* and D T Nguyen

Department of Geotechnics, Faculty of Geology and Petroleum Engineering, Ho Chi Minh University of Technology, 268 Ly Thuong Kiet street, District 10, Ho Chi Minh City, Vietnam
*Corresponding author's e-mail: pmtuan@hcmut.edu.vn

Abstract. The appearance of sinkholes which may not be accurately observed is one of the main problems that caused the stability of construction. Among geotechnical solutions, the geosynthetic-reinforced embankment has been commonly applied due to a variety of benefits. Numerous studies have recently improved the knowledge of the complicated mechanisms acting inside the reinforcement system once a cavity appears. Many present design methods have been established to propose the necessary parameters to apply the technic to potential areas that are prone to subsidence. However, there are many calculated hypotheses based on simple assumptions that may cause the actual effect of the design including the soil expansion inside the embankment. A numerical model has been developed in the present study by using the finite element method and the soil expansion is the main objective to investigate. Considering the deformation, axial forces, and stress applied to the geosynthetics, the expansion factor is determined by two methods and it is confirmed not to be a global and uniform phenomenon.

1. Introduction

Among numerous methods to support constructions over areas prone to subsidence, geosynthetic-reinforced soil is a commonly applied technic due to several benefits such as simple and speedy construction, a cost-effective application, or a minor environmental effect. Nevertheless, the knowledge of some particular mechanisms for this solution is needed to increase. One of the most important gaps is the argument of the complicated mechanisms acting within the embankment above cavities: the soil expansion. As one of the most complicated mechanisms acting within the reinforcement solution, the understanding of soil expansion within granular embankments is still inconsistent, especially the expansion coefficient which is important to the design method.

When a granular material is deformed, its particles are reorganized, increasing its apparent volume. The expansion of soil is underestimated even in such design guidelines as [1] and [2]. The increase in soil volume during the deformation process is related to the material properties, granular delivery, initial density, and stresses that affected the material. The expansion coefficient (Ce) can be defined by the volume change of the filling material within the soil mass, $Ce = V_f/V_i$, which occurs during the appearance of the sinkhole (Fig. 1). In which, V_i is the initial volume of soil and V_f is the soil volume after the disturbance; whereas, Vs and Vg are the volume of deformed soil on the surface and bottom of the embankment, respectively. The evolution of the soil expansion process causes the surface deformation which is affected by the rigidity of the reinforcement applied and the detected cavity development mechanism (gradual or rapid collapse). Beneath the mass of the expanded soil, the geosynthetic sheet acts as a membrane. So far, there has been no testing method that allows the precise measurement of an exact evaluation of the value of the coefficient of soil expansion.

In this field, the numerical modeling of geosynthetic reinforcement is a complicated mission that needs need to precisely define the large deformation of the granular soil embankment, the mechanical behavior of the geosynthetic, and the interaction between the geosynthetic and the adjacent soil. Many recent studies have used numerical solutions, especially the finite element method to well reproduce the use of geosynthetics to reinforce soil such as [4], [5], and [7].

This study is developed based on typical research of [5] in which physical model tests and numerical investigation were combined. The behavior of soil acting within the granular embankment during the process of cavity opening has been studied especially the change in void ratio. In addition, the responses acting along geosynthetic have been illustrated considering the factors affecting the activity of reinforcement material during the phenomenon.

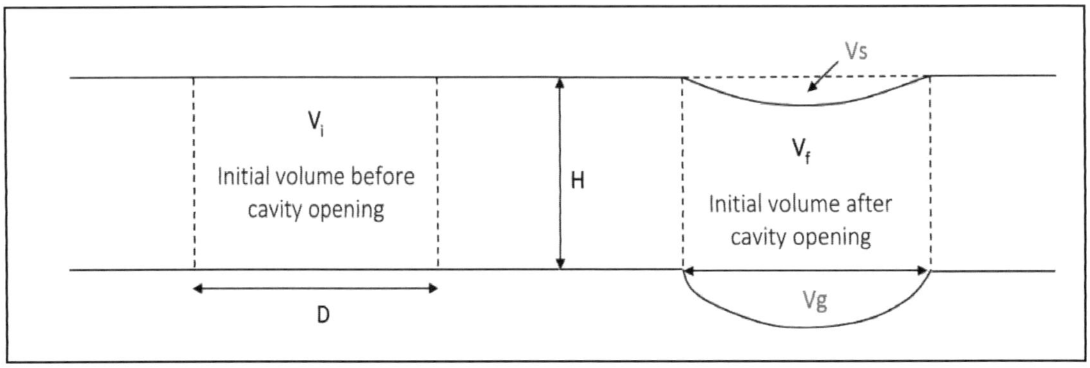

Figure 1. Illustration of the increase in soil [5]

2. Background

In some current design methods, several mechanisms acting during the appearance duration of a cavity beneath a geosynthetic-reinforced embankment have been taken into account. These two first phenomena are the membrane effect of the geosynthetic sheet and the friction between the soil and the reinforcement material in the anchorage areas outside the cavity. The other complex mechanisms are the load distribution within the filing embankment and the expansion of the granular material above the cavity.

Regarding the soil expansion, various shortcomings are still existing due to the basic assumptions that had been implemented [10] and have been continuously studied by [3] and [9]. It can be noted as the movement of granular material causes an increase in the soil mass during the cavity opening. Regarding the cause of surface settlement of the embankment above, the soil expansion appears in a truncated or cylindrical shape that leads to a considerable decrease in the soil surface. Currently, the soil expansion coefficient has been considered a global expansion factor, and this is a significant reason to overestimate the expansion mechanisms. In order to define the coefficient, the calculation of the volume of deformed soil mass is required as it is affected by the deformation of geosynthetic and surface settlement. Many recent studies ([5], [6], and [11]) assumed a parabolic function to fit the shape of deformed curves, and therefore it is potential to estimate Ce by using Eq. 1, in which ds is the surface settlement, dg is the maximum deflection of geosynthetics, and H is the embankment height.

$$Ce = 1 + \frac{(dg - ds)}{2H} \tag{1}$$

However, this assumption is not appropriate, because [6] and [9] have confirmed that paraboloid is not an accurate curve to match the shape of both surface soil and geosynthetic are not exactly. Moreover, the relevance between influences has not yet been mentioned in current existing design methods, especially the geometrical parameters or physical properties of filling soil, such as cohesion, internal friction angle, and dilation angle.

Considering the influence of the void ratio of soil on the deformed volume of the embankment, [12] suggested a formulation to estimate the expansion coefficient Ce, which is defined by a relation between the maximum void ratio (e_{max}) and the initial void ratio (e_0) ratio (Eq. 2).

$$Ce = \frac{1+e_{max}}{1+e_0} \tag{2}$$

3. Numerical analysis

3.1. Model description

As presented above, the study has been developed based on typical research conducted by a combination of the model experimental tests and the finite element method [5]. The testing of the cavity was performed by a laboratory experiment, in which the opening process was controlled. In fact, two types of opening are considered in the experiment: a gradual or a progressive process corresponding to several embankment heights. The complicated mechanisms were studied by determining the vertical deformation of the geosynthetic, the settlement at the embankment surface, and the load applied along the geosynthetic. Then, a series of numerical models were developed based on the finite element method to clarify and to expand the experimental results. The overlying soil in the study is used as fine sand and the significant parameters can be noted as the unit weight is 18 kN/m3, the cohesion is 0 kPa, the friction angle is 37° and the void ratio is 0.8.

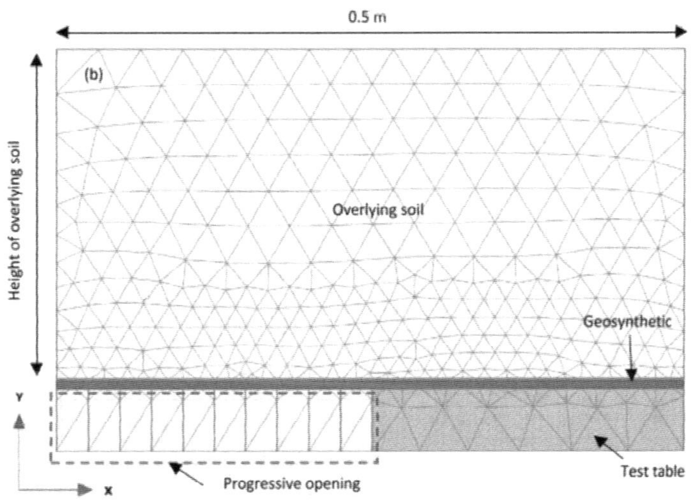

Figure 2. Geometrical configuration for the numerical calculation [5]

In the present study, the cavity is considered a circular shape which corresponds to the symmetric assumption. Therefore, the numerical simulation was conducted using PLAXIS 2-D program [8] in axisymmetric and drained conditions. The 0.25-m-radius cavity was reproduced in a fine-mesh model and the relative element dimension factor is 0.67 as presented in Fig. 2. In this study only one opening mode, a progressive procedure is considered due to the higher effect on the deformation as confirmed by [6] and [9]. The cavity opening is simulated by using an increased diameter opening. The geosynthetics are positioned under the embankment and above the rigid bodies. The length of the model is 0.5 m and the number of elements depends on the embankment height and the cavity width.

Considering the deformation from the previous incremental movement during the calculation, the mesh is updated between each phase, and the mesh distribution is refined along the geosynthetic and the cavity area. The material behavior of the reinforcement material is elastic stiffness (kN/m) and the geosynthetic behavior is modeled following the function $T = J \times \varepsilon$ in which, J is the geosynthetic stiffness (kN/m) and ε is a material strain (%). Additionally, the interfaces are defined in the model to demonstrate the interactions between embankment fill, rigid bodies, and geosynthetics. Defining based on the relevant material parameters, interfaces were positioned along the geosynthetics in order to allow the sheet to move separately from the surrounding soils.

3.2. Numerical results

The deformed volume of the granular embankment is presented in Fig. 3. It can be noted as the surface settlement (ds = 0.03 mm) is lower than the geosynthetic deflection (dg = 0.04 mm). The difference between ds and dg is not significant due to the fact that the embankment height is not large. Since the embankment height is higher, the higher variation can be noticed more clearly [6]. Nevertheless, the important difference between ds and dg confirms the soil expansion during the opening process of the cavity. Furthermore, the form of the deformed volume of granular soil is funnel-shaped, although the opening angle is very small.

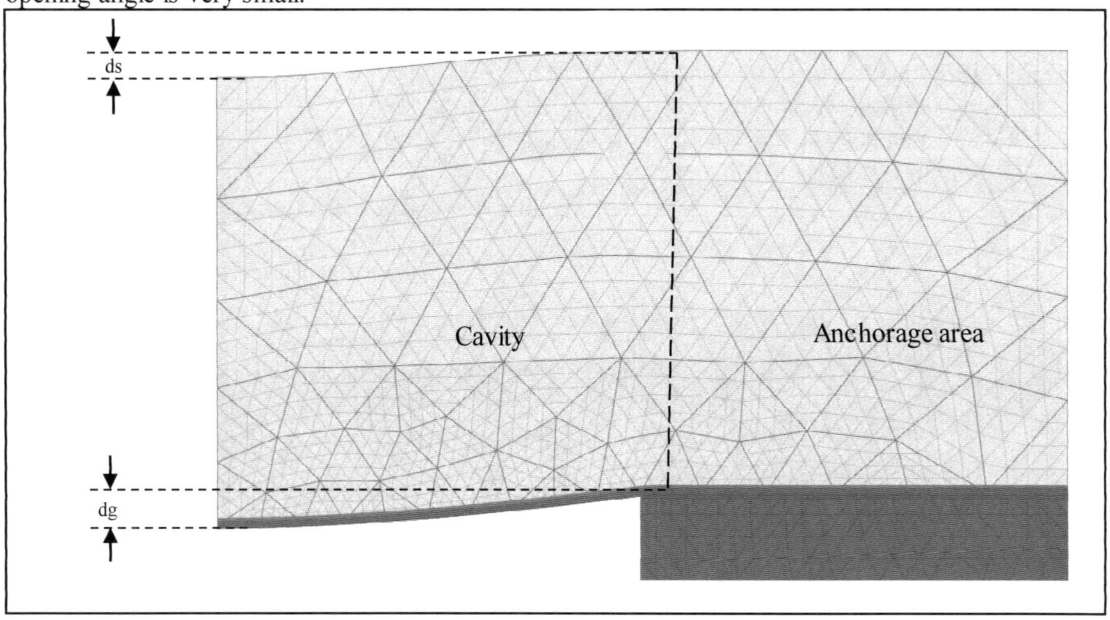

Figure 3. Geosynthetic deflection and surface settlement

Fig. 4 presents the axial forces applied along the geosynthetic after the cavity is totally formed. As can be seen, the significant values of forces are obtained at the cavity and the maximum force is 1.46 kN/m. Meanwhile, inconsistency of the forces can be noticed in the anchorage area. Indeed, the forces at the edge of the cavity are also high, like inside the cavity, but the force reduces in the areas farther away from the sinkhole. This is an interesting point due to the fact that the membrane mechanisms of the geosynthetic reinforcement are not uniform along with the material, and the closer to the center of the sinkhole, the force is stronger.

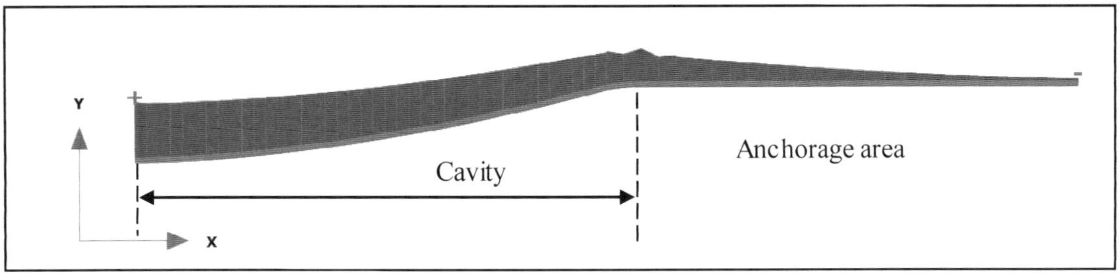

Figure 4. Axial forces of geosynthetic after the cavity opening

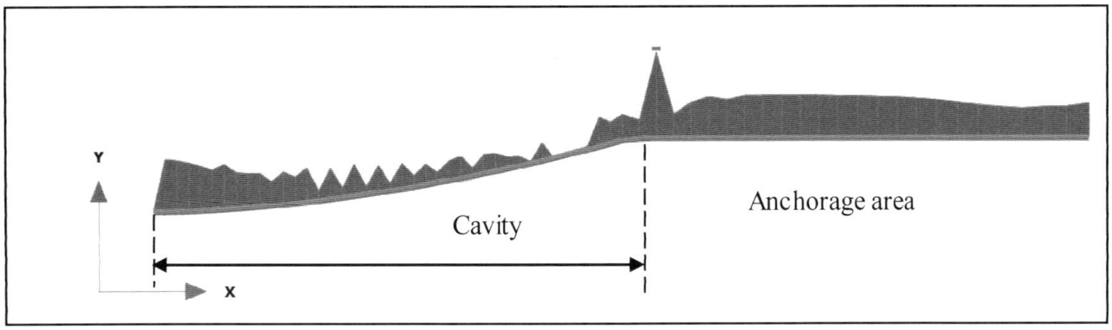

Figure 5. Stress applied along geosynthetic after the cavity opening

Regarding the load transfer mechanisms, Fig 5 illustrates the vertical load applied along with the geosynthetic sheet. There is a significant variation in the distributed load between cavity and anchorage areas. In fact, the vertical load increases in the anchorage area much more than in the cavity area. A maximum load of 9 kN/m^2 is obtained at the cavity edge. This confirmed the phenomenon of load distribution as the embankment stress applied on the geosynthetic is not uniform and it is transferred from the cavity area to the adjacent soil. The reason for the load transfer can be explained by the presence of soil arching inside the granular embankment.

In order to clarify the expansion mechanisms within granular embankment during the cavity opening, the change of void ratio is provided in Fig. 6, in which red colors represent higher void ratios; meanwhile, the blue colors correspond to lower rates. Note that the initial void ratio of overlying soil is 0.80. After the cavity is fully opened, the maximum void ratio reaches 0.93. Indeed, the change of void ratio is not uniform within the embankment. At the anchorage area, the void ratio seems unchanged. Meanwhile, the increment of void ratio above the cavity area is significant, especially in the area near the cavity edge to the geosynthetic sheet.

Then the efficient coefficient has been investigated considering two formulations mentioned in the background. Based on the common function to determine the value of Ce (Eq. 1), the value is 1.02 as the ds and dg are obtained as 0.03 m and 0.04 m, respectively (see Fig. 3). In addition, the expansion coefficient is computed from Eq. 2 considering the numerical results obtained by stress points from the finite element method. Fig. 7 presents the values of Ce obtained by two methods, which are determined by the stress points in the finite element model. It can be noted that Ce obtained by Eq. 2 fluctuates from 1.0 to 1.06 that confirming the inhomogeneous of the soil expansion within the embankment during the opening process. However, the mean value is 1.028 and it is quite similar to the value of 1.02 computed by Eq. 1.

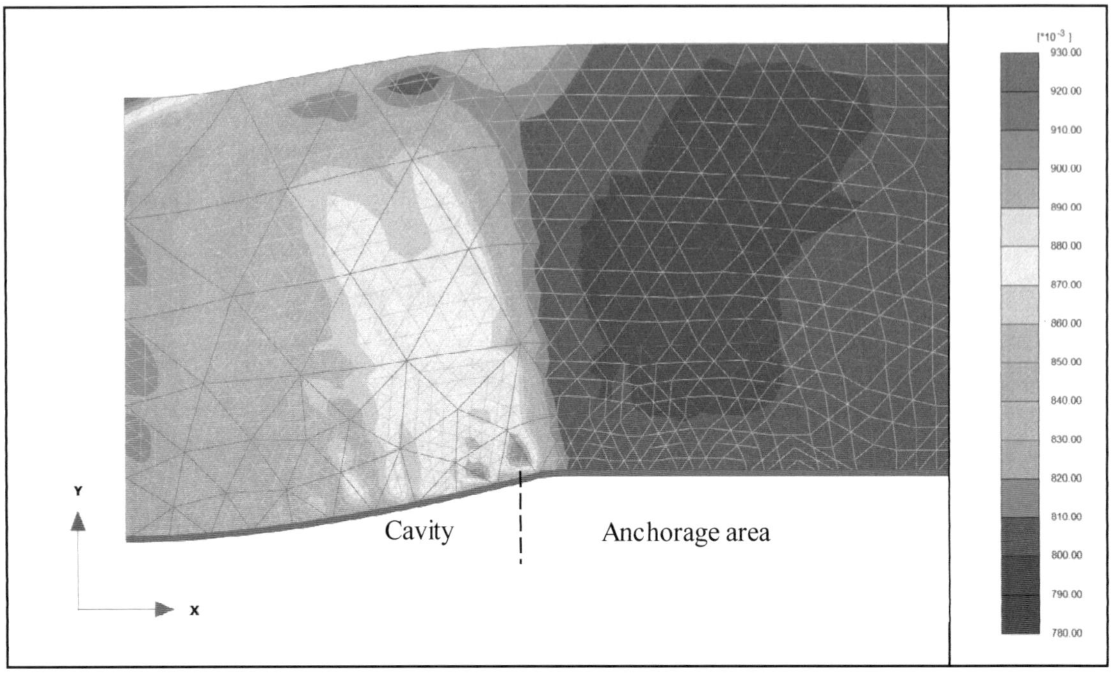

Figure 6. Void ratio changes after cavity opening

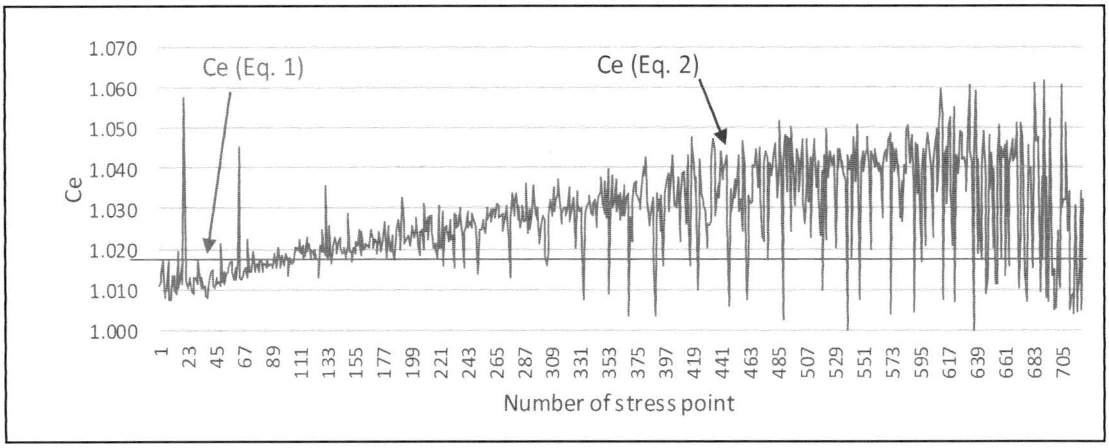

Figure 7. Expansion coefficient obtained by the finite element model

4. Conclusion

Based on the typical research performed by [5], a numerical analysis has been developed to investigate the expansion mechanisms of soil acting within a granular embankment reinforced by a geosynthetic above cavity. The numerical calculation has shown the relevance between surface settlement, geosynthetic deflection, and the expansion coefficient. The change in void ratio is taken into account to investigate the behavior of granular material forming embankment during the process of the cavity opening. Indeed, the expansion mechanisms have been demonstrated by the variation of void ratio, especially the difference between the cavity and adjacent soils are represented. Furthermore, the expansion coefficient has been calculated considering two different methods, the similarity can be noticed in the calculated mean values, and the inhomogeneous ratio of the soil expansion inside the embankment is also confirmed.

Moreover, the behavior of the reinforced platform on the geosynthetic is also illustrated by the axial forces and vertical stress applied to the sheet. In fact, differences between the two areas of cavity and anchorage were well noticed.

References

[1] British Standard Institute BS8006 2010 Code of practice for strengthened/reinforced soils and other fills.
[2] EBGEO 2011 Recommendations for design and analysis of earth structures using geosynthetic reinforcements, ISBN 978-3-433-60093-1.
[3] Huckert A, Briançon L, Villard P and Garcin P 2016 Load transfer mechanisms in geotextile-reinforced embankments overlying voids: Experimental and analytical approaches. Geotext. Geomembranes 44, 442–456.
[4] Girout R, Blanc M, Dias D and Thorel L 2014 Geotextiles and Geomembranes Numerical analysis of a geosynthetic-reinforced piled load transfer platform e Validation on centrifuge test. Geotext. Geomembranes 42, 2014, 525–539. https://doi.org/10.1016/j.geotexmem.2014.07.012
[5] Pham M-T 2019 Granular platform reinforced by geosynthetics above cavities: Laboratory experiments and numerical modeling of load transfer mechanisms PhD thesis, Grenoble Alpes University, France.
[6] Pham M-T, Briançon L and Dias D 2018 Experimental study of the strain mechanisms developed inside geosynthetic reinforced granular platforms over cavities 536–537. https://doi.org/10.1007/978-3-319-73568-9_174
[7] Pham M-T, Vo D-N, Nguyen D-T and Do N-A 2022 Numerical Analysis of Load Transfer Mechanisms Within Embankment Reinforced by Geosynthetic Above Cavity. Transp. Infrastruct. Geotech.
[8] PLAXIS 2022 PLAXIS 2-D Manuals. ISBN-13: 978-90-76016-20-7.
[9] Villard P, Huckert A and Briançon L 2016 Load transfer mechanisms in geotextile-reinforced embankments overlying voids: Numerical approach and design. Geotext. Geomembranes 44 381–395. https://doi.org/10.1016/j.geotexmem.2016.01.007
[10] Villard P, Chevalier B, Le Hello B and Combe, G 2009 Coupling between finite and discrete element methods for the modelling of earth structures reinforced by geosynthetic. Comput. Geotech. 36, 709–717.
[11] Villard P, Gourc J P and Giraud H 2000 A geosynthetic reinforcement solution to prevent the formation of localized sinkholes. Can. Geotech. J. 37, 987–999.
[12] Feng S-J, Ai S-G, Chen H-X, Xie H-J 2017 An analytical method for predicting load acting on geosynthetic overlying voids Geotextiles and Geomembranes 45 570-579.

Assessment of Agricultural Irrigation water from Groundwater in Lobalain District, Rote Ndao Regency, East Nusa Tenggara Province, Indonesia

Rony Rahmad Riady[1], **Wahyu Wilopo**[2,3*]

[1]Master Program in Geological Engineering, Gadjah Mada University, Grafika St. No. 2, Bulaksumur, Yogyakarta, 55281 Indonesia
[2]Department of Geological Engineering, Gadjah Mada University, Grafika St. No. 2, Bulaksumur, Yogyakarta, 55281 Indonesia
[3]Center for Disaster Mitigation and Technological Innovation (GAMA-InaTEK), Gadjah Mada University, Grafika St. No. 2, Bulaksumur, Yogyakarta, 55281 Indonesia
*Corresponding author: wilopo_w@ugm.ac.id

Abstract. Determination of water availability for irrigation consists of quantity and quality. However, water quality is generally often neglected. This study aims to determine the groundwater quality in order to meet the standard of agricultural irrigation quality standard in the Lobalain District, East Nusa Tenggara Province, Indonesia. The results showed that the groundwater quality with a predominantly pH value did not meet the standard. The none restriction on use class dominated the total dissolved solids (TDS) value. The average electrical conductivity (EC) value is less than 750 μs/cm. The chloride is dominated by less than the tolerance threshold (Cl^-<4.0 meq/l), for sulfate (SO_4), which is within the threshold (<20 meq/l), and the nitrate (NO_3N) dominated as slight to moderate. The boron (B) showed positive results, which in all areas was very satisfactory because it is below the tolerance threshold (B<0.7 mg/l). The sodium percentage (Na%) analysis was dominated by excellent and good class, and the excellent category dominated the sodium adsorption ratio (SAR). Eight locations are highly recommended as agricultural irrigation sources due to their groundwater quality. Only one location is not recommended in Kuli Village because only three parameters (Na%, EC, and Cl^-) meet the irrigation water quality standard.

Keywords: Groundwater, agricultural irrigation, groundwater quality

1. Introduction

Rapid population growth and industrialization have a negative impact on natural resources, especially in the agricultural sector. Groundwater is one of the most important water resources for the life of creatures on this earth. The quantity and quality of groundwater are equally important because they are essential in determining these resources for household, industrial and agricultural purposes [1]. Therefore, groundwater sources need to be protected and appropriately conserved. Currently, the main problem in meeting water needs in the agricultural sector is the availability of irrigation water. The main problems with groundwater are health risks, industry, land subsidence due to excessive groundwater extraction, and ecological and agricultural effects [2].

One of the production centers for food supply in Rote Ndao Regency, East Nusa Tenggara Province, Indonesia, is Lobalain District. Therefore, to maintain or increase agricultural production in this area, it is necessary to support irrigation water sources both in quantity and quality. Thus, this study aims to determine groundwater quality in Lobalain District.

2. Methodology

The research location is in Lobalain District, Rote Ndao Regency, East Nusa Tenggara, Indonesia, as presented in Figure 1. The field survey includes direct measurement of the groundwater properties, measuring the depth of the groundwater table, and taking groundwater samples to test chemical composition in the laboratory. Field measurements include electrical conductivity (EC), pH, and total dissolved solids (TDS) using the Hanameter brand portable pH meter. The pH meter before use is calibrated according to the tool manual. Using purposive sampling, groundwater sampling was carried out with nine samples (Table 1). Laboratory analysis for groundwater samples includes K^+, Na^+, Mg^{2+}, Ca^{2+}, NO_3N, SO_4, Cl^-, and B, referring to the Indonesian National Standard (SNI) for water quality for irrigation [4]. The chemical test results (K^+, Na^+, Mg^{2+}, Ca^{2+}) were then used to analyze the sodium adsorption ratio (SAR) and the percent sodium (Na%).

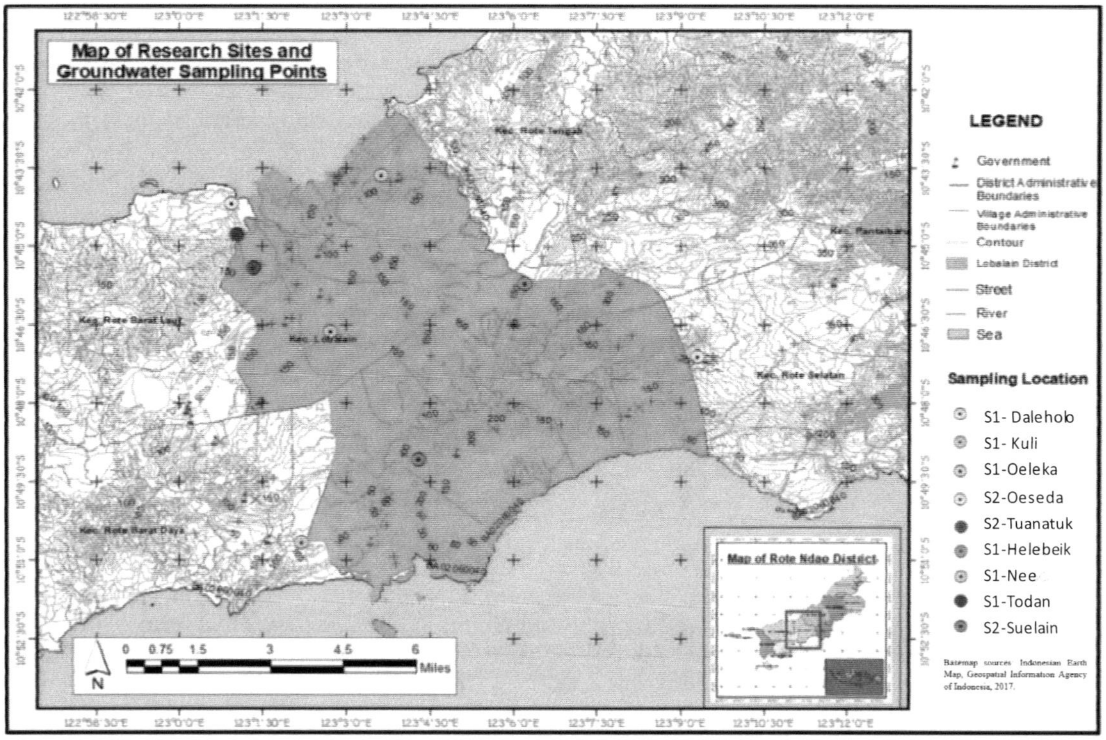

Figure 1. Research Locations and measurement-sampling points.

Table 1. Groundwater measurement and sampling location

No	Village	Code of Sample	Coordinates		Groundwater table
			Latitude	Longitude	(masl)
1	Daleholo	S 1-Daleholo	-10.78518466	123.1553623	259.50
2	Oeleka	S 1-Oeleka	-10.72735683	123.0608700	101.90
3	Tuanatuk	S 2-Tuanatuk	-10.75684305	123.0228453	103.70
4	Oeseda	S 2-Oeseda	-10.73549511	123.0168176	24.90
5	Helebeik	S 1-Helebeik	-10.76188175	123.1037479	187.70
6	Todan	S 1-Todan	-10.7460833	123.0179399	59.30
7	Suelain	S 2-Suelain	-10.81790918	123.0719387	110.60
8	Kuli	S 1-Kuli	-10.84423521	123.0372714	94.50
9	Nee	S 1-Nee	-10.77708238	123.0457799	165.10

pH is an indicator of the basicity or acidity of liquid, solid and gaseous materials. The normal pH for irrigation water ranges from 6.5 to 8.4 [4]. pH that is too high or too low causes an imbalance, and contamination with toxic ions is possible. Very fine inorganic and organic matter determine the concentration of TDS in groundwater. TDS values were divided into several groups based on use, namely severe (> 2,000 mg/L), slight to moderate (450-2,000 mg/L), and none (< 450 mg/L) [4]. Water salinity greatly affects plant productivity which is indicated by the EC value. High EC causes plant productivity to fall due to the inability of vegetation to compete with ions in the soil solution for air [5]. EC classification of irrigation water based on the level of use, including unsuitable (>3,000 μs/cm), doubtful (2,000-3,000 μs/cm), permissible (750-2,000 μs/cm), good (250-750 μs/cm), and excellent (< 250 μs/cm) [6].

Chloride is an ion that is quite abundant in irrigation water. Chloride is easily soluble in groundwater, absorbed by plants, absorbed into transpiration streams, and accumulated in leaves. If the chloride content in the leaves is more than standard, injury symptoms will occur, such as leaf burn [4]. Chloride concentrations based on level of use consisted of severe (> 10 meq/L), slight to moderate (4-10 meq/L), and none (< 4 meq/L) [4]. Nitrates and sulfates are important plant nutrients that require an appropriate amount [4]. The concentration of NO3N is divided into severe (> 30.0 mg/l), slight to moderate (5.0-30.0 mg/l) and none (< 5.0 mg/l), while SO_4 in irrigation water generally ranges from 0 to 20 meq/l [4]. Boron is required in a relatively small concentration, but it will be poisonous if the concentration is high [4]. Classification of boron according to the level of use for irrigation includes severe (> 3 mg/l), light to moderate (0.7-3 mg/l) and none (< 0.7 mg/l) [4] .

Sodium Adsorption Ratio (SAR) governs the relation Na^+ to (Ca^{2+} + Mg^{2+}) in the materials. The connection is vital because sodium splits calcium from soil elements to prevent soil formation and flow in the soil [7]. The SAR value classification of agricultural irrigation is divided into poor (>26), fair (18-26), good (10-18), and excellent (<10) [8]. The SAR value can be calculated using equation 1 with the unit milliequivalent ion concentration per liter.

$$SAR = \frac{Na}{\sqrt{\frac{1}{2}(Ca + Mg)}} \qquad (1)$$

A high sodium (Na) concentration in irrigation water becomes dangerous to the soil. Excess sodium concentration in water reduces soil permeability and changes soil properties [9]. Groundwater quality based on the Na% value is divided into: excellent (< 20%), good (20-40%), permissible (40-60%), doubtful (60-80%), and unsuitable (>80%) [10]. Na% can be calculated using equation 2 with the unit in milliequivalent ion concentration per liter.

$$Na\% = \frac{(Na + K)100}{(Ca + Mg + Na - K)} \qquad (2)$$

3. Results and Discussion

The field measurements of pH presented that the pH value primarily did not meet the standard for agricultural irrigation water. Only two samples from nine met the quality standard of irrigation water (pH 6.5-8.5), as shown in Table 2 and Figure 2(a). The EC value of groundwater showed dominated by class excellent to good with EC values less than 750 μs/cm. Only one location with an EC value is relatively high, reaching 980 μs/cm, and classified as permissible, as shown in Table 2 and Figure 2(b). The groundwater quality based on the EC values can still be used for irrigation water due to below the threshold. The TDS values can be divided into two classes of light to moderate (450-2,000 mg/l) for location SGW 01-Nee, and it does not meet the requirement for irrigation water. The high TDS value was influenced by soil run-off and rock weathering [11]. The remaining samples had a TDS value of less than 450 mg/l, as shown in Table 2 and Figure 2(c). Therefore, most groundwater in the research area is suitable for irrigation water based on the TDS values.

A high concentration of Chloride (Cl^-), more than the standard of 4.0 meq/l, dominates the groundwater samples, as shown in Table 1 and Figure 3(a). It shows that the groundwater has been polluted or from the aquatic seepage and marine environment. A high concentration of Cl^- may source toxicity to some plants [5]. The sulfate (SO_4) concentration from all groundwater samples has a value below the standard (< 20 meq/l), as shown in Table 2 and Figure 3(b).

The nitrate (NO_3N) concentration of three samples was classified as none, including S 1-Kuli, S 1-Helebeik, and S1-Nee. The remaining samples were categorized as slight to moderate to severe, as shown in Table 2 and Figure 4(a). High nitrate concentration is probably from fertilizers, livestock manure, organic soil mineralization, soil, plants, and bacterial biomass [12]. The concentration of boron (B) in groundwater shows good results because all locations are below the standard (< 0.7 mg/l), as shown in Table 2 and Figure 4(b). High boron content in water makes specific plants die, but some plants tolerate the boron [10].

Table 2. Analysis result of groundwater quality in the research area.

No	Sample Location	pH		EC (μS/cm)		TDS (mg/l)		Chloride (Cl^-) (meq/l)	
		Value	Classification	Value	Classification	Value	Classification	Value	Classification
1	S1-Daleholo	5.6	***Abnormal***	090	*Excellent*	20	*None*	1.85	*None*
2	S1-Oeleka	5.4	***Abnormal***	100	*Excellent*	40	*None*	4.24	***Slight to Moderate***
3	S2-Tuanatuk	5.9	***Abnormal***	270	*Good*	30	*None*	4.32	***Slight to Moderate***
4	S2-Oeseda	5.9	***Abnormal***	150	*Excellent*	0	*None*	2.51	*None*
5	S1-Helebeik	7.8	*Normal*	510	*Good*	240	*None*	1.73	*None*
6	S1-Todan	6.1	***Abnormal***	070	*Excellent*	40	*None*	3.85	*None*
7	S2-Suelain	6.5	*Normal*	130	*Excellent*	30	*None*	1.97	*None*
8	S1-Kuli	6.1	***Abnormal***	820	***Permissible***	415	*None*	24.27	*Severe*
9	S1-Nee	6.1	***Abnormal***	980	***Permissible***	480	***Slight to Moderate***	7.12	***Slight to Moderate***

No	Sample Location	Sulfate (SO_4) (meq/l)		Nitrate (NO_3N) (mg/l)		Boron (B) (mg/l)		*Sodium Percentage (Na%)*		*Adsorption Ratio (SAR)*	
		Value	Classification	Value	Classification	Value	Classification	Value	Classification	Value	Classification
1	S1-Daleholo	0.49	*Usually*	24.90	***Slight to Moderate***	<0,006	*None*	10.81	*Excellent*	0.41	*Excellent*
2	S1-Oeleka	1.23	*Usually*	9.76	***Slight to Moderate***	0.06	*None*	20.58	*Good*	0.83	*Excellent*
3	S2-Tuanatuk	1.73	*Usually*	10.89	***Slight to Moderate***	0.06	*None*	13.42	*Excellent*	0.54	*Excellent*
4	S2-Oeseda	1.15	*Usually*	7.79	***Slight to Moderate***	0.02	*None*	17.12	*Excellent*	0.64	*Excellent*
5	S1-Helebeik	0.91	*Usually*	1.85	*None*	0.16	*None*	13.90	*Excellent*	0.48	*Excellent*
6	S1-Todan	2.22	*Usually*	8.54	***Slight to Moderate***	0.03	*None*	22.74	*Good*	0.92	*Excellent*
7	S2-Suelain	1.98	*Usually*	6.48	***Slight to Moderate***	0.04	*None*	17.72	*Excellent*	0.66	*Excellent*
8	S1-Kuli	18.35	*Usually*	0.58	*None*	0.26	*None*	64.45	***Doubtful***	6.03	*Excellent*
9	S1-Nee	9.13	*Usually*	0.50	*None*	0.12	*None*	44.09	*Permissible*	2.45	*Excellent*

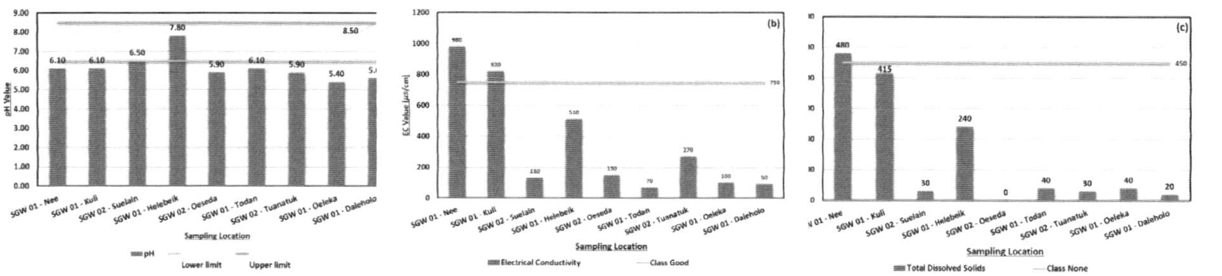

Figure 2. The result of field measurements (a) pH, (b) EC, and (c) TDS.

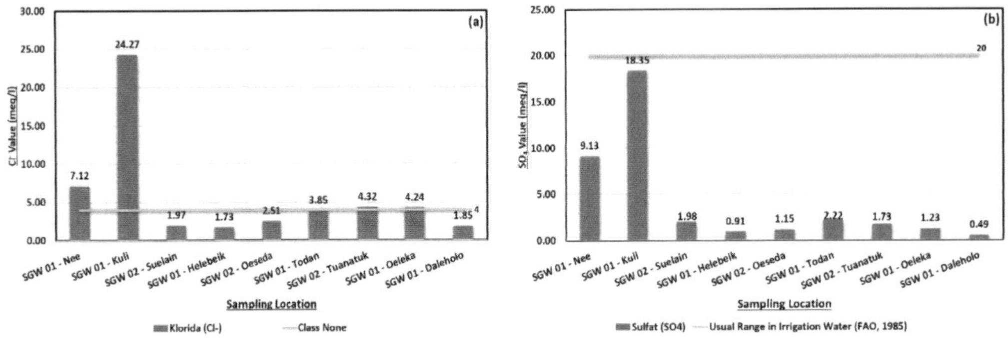

Figure 3. The result of chemical analysis is (a) Chloride (Cl^-) and (b) Sulfate (SO_4).

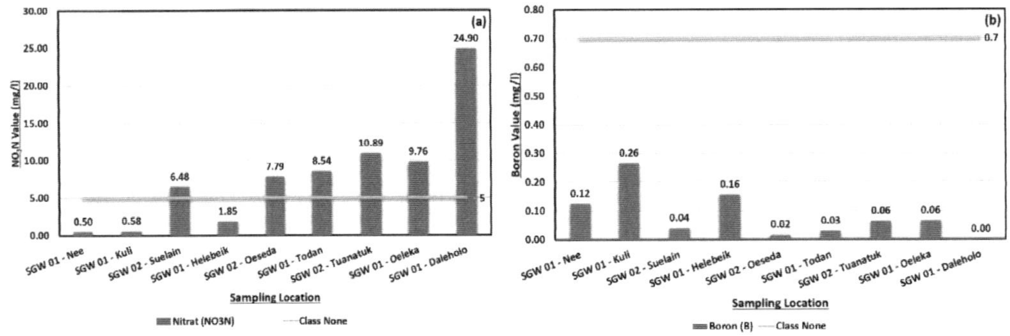

Figure 4. The result of chemical analysis is (a) Nitrate (NO_3N) and (b) Boron (B).

SAR calculations using Equation 1 show that the SAR value ranges from 0.41 to 6.03, as shown in Table 2 and Figure 5(b). A high SAR value causes decreasing soil infiltration capacity due to clay clogging the pores and crusting, which disturbs water supply to the plant or sodicity conditions. Therefore, all groundwater is classified as excellent (< 10). All the groundwater is safe and meets the specifications for agricultural irrigation.

According to Equation 2, the value of the Na% ranges from 10.81% to 64.45%, as shown in Table 2 and Figure 5(a), which is categorized as good and excellent. Only two locations are in the allowed category located in Kuli village. Due to the low Na% value, most groundwater meets the agricultural irrigation standard. The low content of Na% in the water does not hinder plant growth.

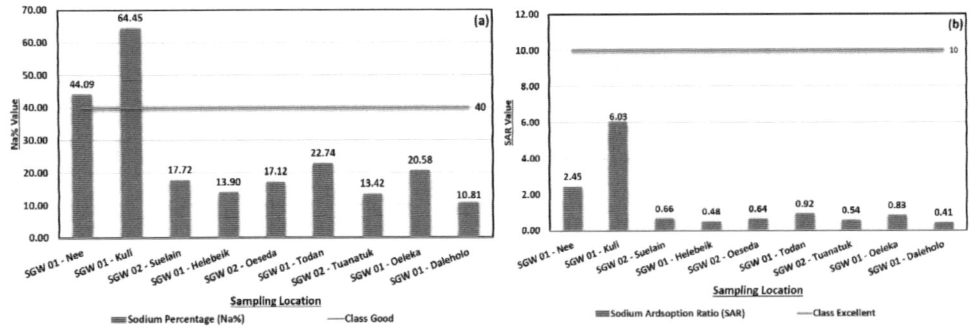

Figure 5. Data of (a) sodium percentage (Na%) and (b) sodium carbonate ratio (SAR) in each location.

4. Conclusion

The groundwater measurements and analysis were performed and obtained values for several parameters from Lobalain District, Rote Ndao Regency, East Nusa Tenggara, Indonesia. pH values show that seven of the nine locations do not meet the standard of water irrigation (pH 6.5-8.5). The TDS values show that the none category (< 450 mg/l) dominates the groundwater. EC values of groundwater have excellent and good categories with values less than 750 μs/cm. Six locations were slight to moderate (5-30 mg/l) for nitrate (NO_3N). The boron (B) concentration in all groundwater below the meet the irrigation water standard (< 0, 7 mg/l). Almost all groundwaters have chloride content more than the standard (> 4.0 meq/l), in the opposite condition for the sulfate (SO_4) that all were below the tolerance threshold (< 20 meq/l). The percent sodium (Na%) of groundwater was classified as excellent (5 locations) and good (2 locations), and permissible (1 location). The SAR value indicates that almost all groundwater is categorized as excellent (8 locations), and only one sample is categorized as poor. Due to their groundwater quality, 8 locations are highly recommended as a source of agricultural irrigation. Those area are Nee, Helebeik, Suelain, Todan, Oeseda, Oeleka, Daleholo, and Tuanatuk Villages. Only Kuli Village is not recommended because only 3 components (Na%, EC, and Cl^-) meet the irrigation water quality standard.

5. References

[1] Kaur T, Bhardwaj R, & Arora S. 2017. *Assessment of groundwater quality for drinking and irrigation purposes using hydrochemical studies in Malwa region, southwestern part of Punjab,* India. Appl Water Sci **7**: 3301-3316.

[2] Harter T, *Groundwater Tour: Groundwater Quality & Contaminants. Water Education Foundation. University of California*, Davis. http://groundwater.ucdavis.edu.

[3] *Laporan Akhir: Review Masterplan dan Perencanaan PSKPT Rote Ndao. SKPT Rote Ndao*, Kementerian Kelautan dan Perikanan. Page 4-36. http://kkp.go.id.

[4] Ayers R. S, & Westcot D. W, 1985. *Water Quality for Agriculture* (FAO Irrigation and Drainage Paper No. **29 rev.1**). Rome, Italy: Food and Agriculture Organization of United Nations, 163p.

[5] Bauder T. A, Waskom R. M, Sutherland P. L, & Davis J. G, 2011. *Irrigation Water Quality Criteria. Crop Series Irrigation*, Fact Sheet No. 0506. Colorado State University.

[6] Chandrasekar N, Selvakumar S, Srinivas Y, Wilson J.S.J, Peter T.S., Magesh N.S., 2013. *Hydrogeochemical assessment of groundwater quality along the coastal aquifers of southern Tamil Nadu*, India. J Environ Earth Sci **71**(11):4739–4750.

[7] Moss J. Q & Kress M, *Turf Irrigation Water Quality: A Concise Guide*. Oklahoma Cooperative Extension Service. HLA-6612. Division of Agricultural Sciences and Natural Resources, Oklahoma State University, 4p.

[8] Todd, D. K, 1980. *Groundwater Hydrology, Second Edition*. John Wiley & Sons Inc. New York, 535 p.

[9] Nishanthiny S.C, Thushyanthy M., Brathithasan, T., Saravanan S. 2010. *Irrigation Water Quality Based on Hydro Chemical Analysis, Jaffna*, Sri Lanka. American-Eurasian J. Agric. & Environ. Sci., **7** (1): 100-102.

[10] Wilcox L.V, 1948. *The quality of water for agricultural use*. Edit. U.S Department of Agriculture, Technical Bulletin, vol. **962**, Washington (USA), 40p.

[11] Effendi H, 2003. *Telaah Kualitas Air: Bagi Pengelolaan Sumber Daya dan Lingkungan Perairan*. Penerbit: Kanisius. Yogyakarta. 257 p.

[12] Bhat M. A, Wani A. S, Vijay K, Jyotirmaya S, Dinesh T, Sanswal R., 2018. *An Overview of the Assessment of Groundwater Quality for Irrigati*on. J Agri Sci Food Res **9**: 209. 9p.

IOP Conf. Series: Earth and Environmental Science 1071 (2022) 012035 doi:10.1088/1755-1315/1071/1/012035

Influence of coastal alluvial aquifer on groundwater chemistry in Wates, Yogyakarta, Indonesia

P P Thin[1], H Hendrayana[2]* and S Kawasaki[3]

[1]Department of Engineering Geology, Yangon Technological University, Gyogone, Insein 11011, Yangon, Myanmar.
[2]Department of Geological Engineering, Faculty of Engineering, Universitas Gadjah Mada, Yogyakarta 55281, Indonesia.
[3]Division of Sustainable Resources Engineering, Faculty of Engineering, Hokkaido University, Sapporo 60-8628, Japan.
*Corresponding author's e-mail: heruha@ugm.ac.id

Abstract. Saline groundwater is found far from the coastline, whereas closer groundwater has low salinity in the Wates area, Indonesia. This situation which has led to problems linked to the coastal aquifer, shows different groundwater chemistries. This connection needs to highlight the geochemical evolution of aquifers. This research aims to evaluate the hydrochemical characteristics in Wates Sub-District, Kulon Progo District, Indonesia. The study area is located in the southern part of the Kulon Progo with an aquifer of marine and fluvial deposits. The boundary of the south is the Indian Ocean. The approach is based on analyzing the major ions of 27 groundwater samples, and the XRD results include the Stiff diagram and the Stuyfzand classification. The Stiff diagram showed in agreement with that of the Stuyfzand classification. Results showed that the F2-(CaHCO3)2Ø and F1-(NaMix)Ø types reflect the recharge process and dominate the dissolution of silicate minerals. Halite dissolution originates in topographically flat areas of the Bs1-NaCl (-) water. Total hardness is softer than others as a result of cation exchange. The groundwater chemistry's geochemical processes are the dissolution of silicate and halite minerals, ion exchange, and fresh and seawater mixing. This research indicates to focus on pumping problems in the aquifer.

1. Introduction

The problem of groundwater in the coastal area is mostly related to quality. Natural processes and anthropogenic activities influence groundwater quality [1]. In coastal aquifers, groundwater quality is primarily controlled by the interaction between the sea and the bordering aquifer systems. This mutual interaction can result in freshening or salinization of water bodies. While the mechanism for freshening or salinization is hydrodynamically driven (groundwater flows), chemical reactions with the aquifer matrix will alter groundwater chemistry [2]. Although salinization in groundwater can result from marine water intrusion induced by groundwater pumping [3], the rest sources are related to the interaction of groundwater with formations deposited in the marine transgression/regression periods [4, 5], upcoming of connate saline water [4, 6], and geothermal origin [4,7]. Also, the hydrochemical processes like the dissolution of naturally occurring soluble salts, fresh and seawater mixing, cation exchange, and evaporation play an important role in the variation of groundwater quality. When the groundwater is mixed with seawater, the salinization and cation exchange process will occur [8, 9]. Especially in complex cases, cation exchange is a critical issue to consider. It is a spectrum of groundwater facies reflecting the processes controlling for aquifer's geochemical evolution concerning local lithologies [2]. Coastal aquifers are thus complex due to all the aforementioned processes [10]. Also, the present fresh/salt distribution of a coastal aquifer is very often determined by the morphological history of the area [2].

The study area is located in the southern part of the Kulon Progo District with the aquifer of Quaternary fluvial sediment, underlain by marine deposits [11, 12, 13]. The boundary of the south is the Indian Ocean. Saline groundwater is reported far from the coastline, whereas closer groundwater to the coastline has low salinity. The condition has controlled alarming problems related to the aquifer, showing different water chemistries and disturbance of seawater intrusion. This connection needs special attention to highlight the geochemical evolution of aquifers, which is sadly lacking. Therefore, this research aims to evaluate the hydrochemical evolution of the Wates Sub-District, Kulon Progo District, Indonesia. Based on the previous research [12, 13], low resistivity values ($< 10\ \Omega m$) are found in the upper layers of the aquifer far away from the coastline. As a result, the groundwater is regarded as brackish or saline [14]. Closer to the coastline, high resistivity values ($> 10\ \Omega m$) indicate freshwater. However, this will increase the interpretation's complexity and uncertainty when the seawater within a clay-based formation is delineated, which may also consist of freshwater. Because seawater, just like marine clays, also has low resistivity ($< 10\ \Omega m$) [15]. Considering anthropogenic origins, the integration of X-ray diffraction (XRD) of aquifer sediments combined with chemical analyses of groundwater samples was used to identify the hydrochemical characteristics and origin. Although few hydrogeochemical studies have been previously carried out, the combination of graphical methods, including the Stiff diagram and Stuyfzand classification, has not been reported. The inference drawn from the present study is intended to fill this gap and mostly resolves the resistivity assessment's complexity and uncertainty.

2. Study area

Wates Sub-District is characterized as a coastal aquifer with approximately 32 km^2. It is situated in the Kulon Progo District and is limited by the Indian Ocean to the south, as in figure 1. Geographical coordinates of the study area are latitude 7° 50' 00" S to 7° 56' 00" S and longitude 110° 05' 00" E to 110° 07' 00" E. The study area falls under the survey of Indonesia topographic sheet number- 49 S. The boundary of the study area from the map was assigned the blue color. Other sub-districts and Wates Groundwater Basin are associated with yellow and red colors. Thus the study area among boundaries could easily be identified. Physiographically, the site is flat with gentle slopes and a maximum elevation of 50 m above sea level. The Serang River is crossing the research area.

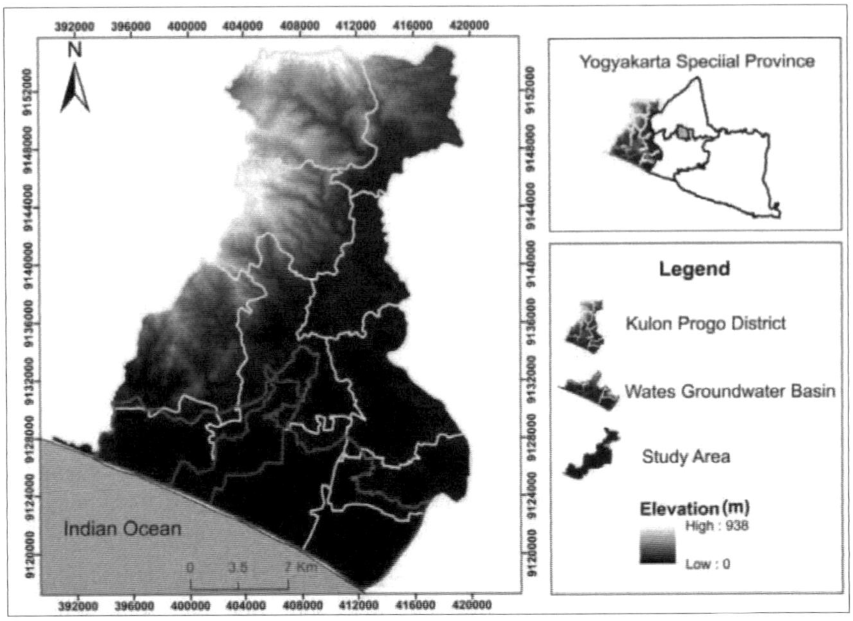

Figure 1. Map of the study area (Basemap Source: Geospatial Information Agency, 2004) [16].

3. Geomorphic features

According to Santosa (2010), the study area belongs to two geomorphic units, named the Fluviomarine Plain (F_m) and Sand Dunes, as in Figure 2. The borehole drilling [11], together with geophysical data [12, 13], is mainly composed of marine and fluvial-deposited sediments dating from the late Tertiary to Quaternary periods. The Fluviomarine Plain (F_m) has a flat relief with a slope of 0-3%, formed due to the past marine activity and fluvial (river) deposition. In the Pliocene-Pleistocene periods, the study area underwent a succession of transgressions and regressions. During transgressions, the coastline moves toward a landward (higher ground), which results in flooding, and the sediments are deposited in a marine environment. That was later buried by fluvial (river)-deposited sediments. The materials are mostly marine clay deposits containing mollusks' fossils, which alternate with silts and fine marine sands. These clues provide strong evidence that the study area is predicted as a previous lagoon and shallow marine zone. At the end of the Pleistocene age, the ocean was located in the southern part of the area and exposed the former sea bottom (submerged seafloor) above the sea level. The fluvial activity caused significant erosion in the Holocene period. The fluvial sediments are the products of the sedimentation process of the Serang River, which naturally occurs through erosion from the northern hills. The deposits of fluvial (river) origin are noticed as a flat terrain that extends around the river flow. Topographically it can be referred to as level land occurring at a slightly higher elevation than the rest of the plain. The highest areas were shaped by fluvial erosion in the surroundings. Consequently, in Holocene time, sediments that accumulated on the previous shallow marine and lagoon zone were established as the Fluviomarine Plain (F_m).

The southernmost is the Sand Dunes Unit, which borders the Indian Ocean. The characteristics of this unit include distinct-morphology relief with a slope of 3-8%. It is further subdivided into three sub-units covered by Holocene sediments. Firstly, the Old Beach Ridges (M_2) are formed by the sedimentation process of marine sands with the activity of waves along the coast. Secondly, the Sand Dunes (E) comprise the accumulation of marine sands with the activity of winds on the previous beach shelf. Thirdly, the Young Beach (M_1) is composed of marine sands and is also influenced by tidal activity. Based on the interpretation of geoelectric data [12, 13], the lower layers of Old Beach Ridges (M_2) are composed of marine clays mixed with silts and fine marine sands, as well as the units of Sand Dunes (E) and Young Beach (M_1). It can be explained that all sub-units were influenced by ancient marine activities (transgression and regression), which was possible on a former shallow marine (littoral) zone and lagoon with the intensive marine sedimentation process to form a new beach.

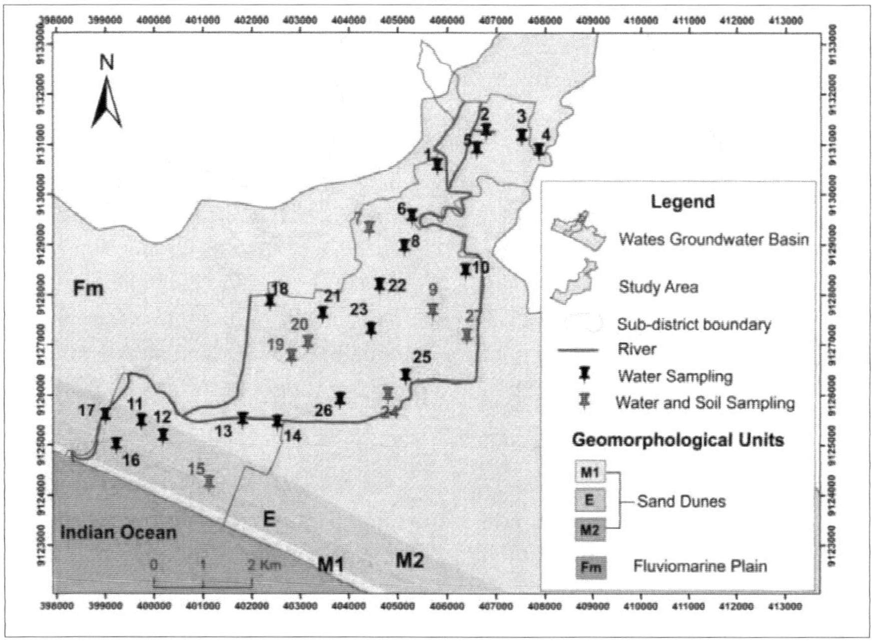

Figure 2. Geomorphological map of the study area with the sample points (Basemap Source: Geospatial Information Agency, 2004)[16].

4. Research methodology

4.1. Groundwater sampling

Initially, a water sampling site inventory survey was carried out to cover the study area concerning groundwater flows, lithologies, and morphology. Sampling locations were taken in the field through a GARMAN Global Positioning System. Groundwater levels were measured using the Roller meter. The Kriging interpolation technique constructed the water level contour map with the help of ArcMap 10.3. Kriging is better than other methods because of its accuracy [17]. The temperature (T), pH, electrical conductivity (EC), and total dissolved solids (TDS) were measured on-site with the use of a water test kit (GTL-HG-002) that was calibrated beforehand. Twenty-seven water samples were collected from well tapping the alluvial aquifer as in figure 2 in clean plastic bottles. Before sampling, the bottles were rinsed with the water to be taken for analysis to minimize the chance of any contamination. The bottles were tightly capped to protect samples from atmospheric carbon dioxide (CO_2), properly labeled, and sent to the laboratory for major cations and anions concentrations. The following major ions were checked in the study: sodium (Na^+), calcium (Ca^{2+}), potassium (K^+), magnesium (Mg^{2+}), bicarbonate (HCO_3^-), chloride (Cl^-), carbonate (CO_3^{2-}), sulfate (SO_4^{2-}), nitrate (NO_3^-) and phosphate (PO_4^{3-}) content. Ionic balance error was within ± 5%.

4.2. Soil sampling

A soil sampling survey was scheduled to coincide with the concentrations of EC and TDS in groundwater, which had experienced the salinity. Importantly, this was designed to ensure that the geochemical characterizations of groundwater were being captured. Samples from depths as much as 2 m were manually dug and packed in three layers of plastic to prevent drying and then appropriately labeled and sent to the laboratory for identifying mineral phases by using X-ray diffraction (XRD). Sampling location numbers to eight-point as in Figure 2.

5. Data processing

5.1. Mineralogical interpretation
The diffractograms were checked manually using a table of key lines for X-ray diffraction patterns [18]. Firstly, a line (peak) of the strongest (highest) intensity was selected from the XRD pattern. The first column is the value of the key line used for identifying a mineral. The corresponding values of the unidentified peak were checked at a 2θ angle and d-spacing value on the XRD pattern. The name of mineral species corresponding to that value was used from the fourth column. More than five diffraction lines for a mineral were used because of the d-spacing values for the same mineral in the next entry. The table did not provide a matching value for unidentified peaks, so the procedure was repeated to identify other components (mineral phases).

5.2. Stiff diagram
The Stiff diagram is a polygon with three parallel horizontal axes spreading on either side of vertical zero axes. Major cations were plotted on the left side of the polygon, and major anions were plotted on the right side in units of milliequivalents per liter (meq/L). It helps a visual evaluation of waters with different characteristics. Its pattern was used to trace the groundwater flow on the map based on the variations in its major ions responsible for the EC and TDS in the groundwater [19]. Microsoft Excel files were uploaded into the Grapher 12 software, automatically producing the Stiff diagram.

5.3. Stuyfzand classification
The Stuyfzand's classification [9] used here results in easier identification of water types and verification if the cation exchange took place. Indeed, cation exchange is the best criterion to prove the fresh- or salt-water intrusion [20]. This classification involves subdividing chemical water features at four (4) levels: the main type, type, subtype, and class, as in table 1. The main type is determined based on the chloride content. The type (code) is determined based on the total hardness. The subtypes sorting is determined according to the dominant cations and anions. The class of water is calculated on the sum of $Na^{+} + K^{+} + Mg^{2+}$ (meq/L), corrected for seawater contribution, determined from the Cl^{-} content as in equation (1).

$$(Na^{+} + K^{+} + Mg^{2+})\ \text{corrected} = (Na^{+} + K^{+} + Mg^{2+})\ \text{measured} - 1.061\ Cl^{-} \tag{1}$$

The cation exchange is defined according to $(Na^{+} + K^{+} + Mg^{2+})$ corrected by considering a margin equal to $\sqrt{0.5\ Cl^{-}}$ as in Table 1. The cation exchange code will be "-" due to a (former) saltwater intrusion in freshwater bodies, whereas the code will be "+" due to a (former) freshwater intrusion in a marine environment. The Ø shows no evidence of cation exchange [9, 20].

Table 1. Water classification according to the chloride, hardness, and cation exchange code [9].

Classification	Critical parameter	Description criterion	Code
Main type	Cl^- (mg/L)	≤ 150	Fresh (F)
		150 - 300	Fresh-brackish (Fb)
		300 - 1,000	Brackish (B)
		1,000 - 10,000	Brackish-salt (Bs)
		10,000 - 20,000	Salt (S)
		> 20,000	Hypersaline (H)
Type	Total hardness (meq/L) ($Ca^{2+} + Mg^{2+}$)	< 5	Very soft (*)
		5 -10	Soft (0)
		10 - 20	Moderately hard (1)
		20 - 40	Hard (2)
		40 - 80	Very hard (3)
		80 - 160	Extremely hard (4)
		160 - 320	Extremely hard (5)
		320 - 640	Extremely hard (6)
		640 - 1280	Extremely hard (7)
		1280 - 1560	Extremely hard (8)
		> 1560	Extremely hard (9)
Class	($Na^+ + K^+ + Mg^{2+}$) deficit	($Na^+ + K^+ + Mg^{2+}$) corrected < $-\sqrt{0.5}\ Cl^-$	Salinized (-)
	($Na^+ + K^+ + Mg^{2+}$) equilibrium	$-\sqrt{0.5}Cl^- \leq$ ($Na^+ + K^+ + Mg^{2+}$) corrected $\leq +\sqrt{0.5}Cl^-$	No base exchange (Ø)
	($Na^+ + K^+ + Mg^{2+}$) surplus	($Na^+ + K^+ + Mg^{2+}$) corrected > $\sqrt{0.5}Cl^-$	Freshened (+)

6. Results

6.1. Mineralogy

In contact with the Indian Ocean, the study area oriented north-south shows up in two different geomorphic features. In the Fluviomarine Plain (F_m), aquifer media is mainly composed of various lithologies in association with marine and fluvial-deposited sediments, which are explored using drilling [11] and geophysics [12, 13]. Marine (littoral) clays dominate this unit with silts and fine marine sands at the bottom and fluvial (river) clays, silts, and fine sands lying on top. Groundwater chemistry is tightly linked to the aquifer mineral composition in which it flows [2, 21]. Therefore, the identification of minerals was undertaken through XRD. In the Fluviomarine Plain (F_m), the dominant lithologies are likely to associate the quartz, plagioclase feldspars (anorthite and albite), augite hornblende, clay (montmorillonite), and evaporite (halite). They are related to different sedimentation processes: one coming from the deposition of the Serang River and the another coming from the marine transgression. Based on geophysical data [12, 13], the upper layers of Sand Dunes Units are composed of marine sands, overlying the eolian (windblown) sands on the surface as the aeration zone. Here, XRD analysis results are nearly similar to the Fluviomarine Plain (F_m) in mineralogical composition, consisting mainly of quartz, anorthite, hornblende, and hematite. The results of XRD were confirmed by the weathering degrees and their products of locations 1, 7, 9, 15, 19, 20, 24, and 27 in both the A-CN-K and A-CNK-FM ternary diagram. These minerals have been weathered to different degrees [22, 23]. Accordingly, the mineral compositions of sediments are also in line with the chemical weathering products in the Wates coastal aquifer [24].

6.2. Groundwater flow

Figure 3 shows the groundwater flow patterns in two morphological units. The groundwater table elevations from 2.2 to 10.6 m above mean sea level in the Fluviomarine Plain (F_m) and 1.3 to 6.3 m above mean sea level in the Sand Dunes Units. The groundwater level map shows local groundwater recharges in various directions at locations of both units. The groundwater has closely distance equipotential lines representing a steep hydraulic gradient in the northern and eastern parts of the Fluviomarine Plain (F_m). This would be high the permeable nature of the prevailing lithology [24]. The presence of tight equipotential lines shows the marked local recharge area. The groundwater in the central and south-western parts of the plain is situated at a lower elevation than the recharge area. It has widely spaced the equipotential lines indicating a flat hydraulic gradient and less permeability of lithology [25]. Due to the low topographic position, these depressions receive the groundwater from the surrounding recharge areas through the less permeable layers. Consequently, the groundwater is expressed by salty pore solutions on the one side and indicated by high EC and TDS on the other hand. The aquifer is thus unconfined to semi-confined. Local recharge to the unconfined aquifer infiltrates rainfall and river water. The aquifer and the Serang River constitute a hydraulic continuity, where depending on the location either the river water infiltrates into the aquifer or groundwater flows towards the river.

The Sand Dunes Unit is the relict features formed by wave and wind acting on the marine floor sediments exposed to the regression [12]. This unit is an old beach with the movement of the coastline to the ocean. The development of Sand Dunes induced a new flow regime. It has a short circulating period with a near exit through a locally-fed groundwater system. Therefore, groundwater flows move in all directions, and the water table is shallow. It can be proved that the sudden change in river flow direction follows the former shoreline. On the other hand, the concentrations of EC and TDS in groundwater are lower than that of the Fluviomarine Plain (F_m). Therefore, a close relationship between groundwater flow and prevailing lithology can be expected.

6.3. Groundwater types

The groundwater types were determined through the Stiff diagram [19]. Groundwater is available in unconfined and semi-confined aquifer conditions [11, 12, 13]. Major ionic concentrations of groundwater are not under uniform coverage. Based on the average values of chemical parameters, the cations are in order of abundance as $Na^+ > Ca^{2+} > Mg^{2+} > K^+$ concentrations. The general order of dominant anions is as $HCO_3^- > Cl^- > SO_4^{2-} > NO_3^- > CO_3^{2-} > PO_4^{3-}$. Accordingly, Na^+, Ca^{2+}, HCO_3^- and Cl^- are the most striking major ions present in the groundwater of the study area. Consequently, three different groundwater types are popular here, as in Figure 3. The shape and size of the Stiff diagrams are proportional to their EC and TDS.

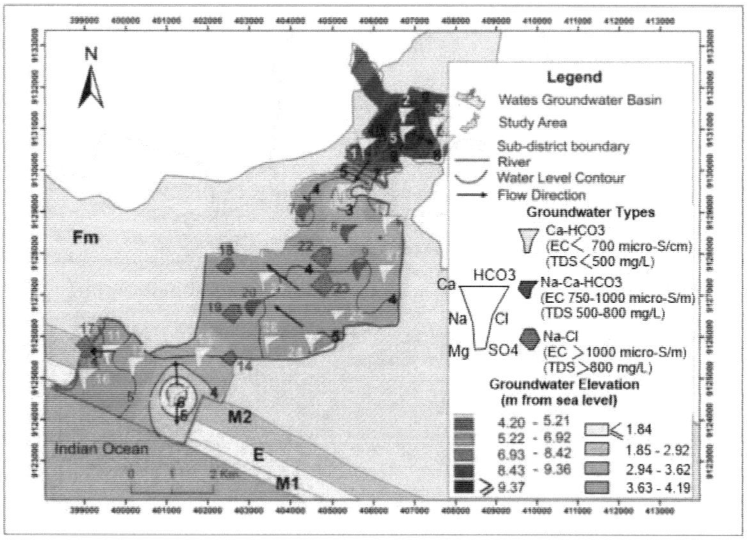

Figure 3. The piezometric level map and spatial distributions of water types according to the mean concentrations of major ions in the groundwater of the study area (Basemap Source: Geospatial Information Agency, 2004) [16].

6.3.1. Ca-HCO$_3$ Water. In locations 1, 2, 3, 4, 5, 6, 10, 11, 12, 13, 15, 16, 21, 24, 25, 26 and 27, the calcium (Ca^{2+}) is the highest among the cations whilst the bicarbonate (HCO_3^-) is the predominant anion. The main water is the Ca-HCO$_3$ type, which has relatively low EC (< 750 µS/m) and TDS (< 500 mg/L). This water includes about 63% of the water types. This can be regarded as recently recharged water. The principal source of water is the percolation of rainfall on highlands. Groundwater flows through high permeable fluvial sands, which serve as the unconfined aquifer. Based on the geochemical calculation with PHREEQC Software, calcium-bearing silicate minerals (anorthite, augite) were dissolved in fluvial sandy recharge areas [22, 23]. Evidence of these minerals occurs through XRD patterns. The low EC and TDS values could be due to the fast circulation.

6.3.2. Na-Ca-HCO$_3$ Water. In locations 7, 8, 9, and 20, calcium (Ca^{2+}) and bicarbonate (HCO_3^-) are the most concentrated ions. Here, sodium (Na^+) is superimposed as the predominant cation. Main water is the Na-Ca-HCO$_3$ type, with a moderate EC of 750-1000 µS/m and TDS of 500-800 mg/L. This water constitutes about 15 % of the water types. The aquifer obtains water from rainfall. Groundwater flows through permeable fluvial silts. These materials built an unconfined aquifer. According to geochemical calculation, the sodium (Na^+) ion comes from silicate minerals dissolution (albite, hornblende) [22, 23]. Here, the XRD results show up to the confirmation of these minerals. The Na-Ca-HCO$_3$ water is very close to the Ca-HCO3 type as both drain similar lithologies. The moderate EC and TDS values show the fast flow systems of the aquifer.

6.3.3. Na-Cl Water. In locations 14, 17,18, 19, 22, and 23, sodium (Na^+) is the predominant cation, but chloride (Cl^-) is the highest among the anions. Main water is the Na-Cl type, with the most mineralization (EC > 1000 µS/m; TDS > 800 mg/L), leading to slow groundwater velocity and long residence time. This water constitutes about 22% of the water types. The aquifer comprises marine clays with silts and fine sands developed by former maritime activity [12]. It becomes semi-confined. The aquifer is being recharged by surrounding high areas and the river water. According to geochemical calculation, ion exchange and halite dissolution are key factors for the increase of sodium (Na^+) and chloride (Cl^-) [22, 23]. The XRD results confirm the presence of halite. The Stuyfzand classification recognizes if a cation exchange has occurred in an aquifer [9]. Therefore, to confirm the Stiff diagram results, this is necessary to continue the Stuyfzand water classification.

7. Discussion

7.1. Stuyfzand water classification

According to the Stuyfzand classification [9], the water samples can be ranked in the following three main categories: freshwater, mixed sodium water, and salinized water, as in Figure 4.

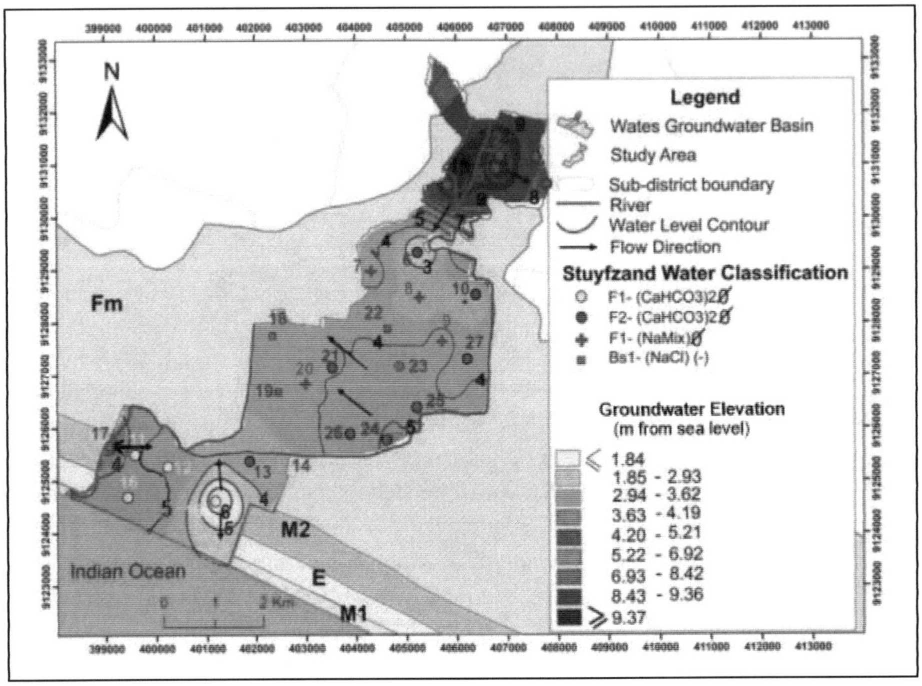

Figure 4. Determination of the water types in the study area according to the Stuyfzand method (Basemap Source: Geospatial Information Agency, 2004) [16].

7.1.1 Freshwater on the Fluviomarine Plain. The variations of freshwater types are in two geomorphic units of these forms: F1-$(CaHCO_3)_2$Ø and F2-$(CaHCO_3)_2$Ø with EC < 750 µS/m and TDS < 500 mg/L. In the Fluviomarine Plain (F_m), the F2-$(CaHCO_3)_2$Ø shows fresh groundwater with a hard condition, characterized by very low Cl^- content (< 150 mg/L) and Ca^{2+}+ Mg^{2+} between 21.9 and 28.2 meq/L. This water type shows slightly higher hardness than F1 in the Sand Dunes. The Ø sign indicates no evidence of cation exchange has occurred. This water type is derived from the dissolution of silicate minerals (anorthite, augite). This water is typically found in the recharge area, where water renewal is frequent and fast. Bicarbonate (HCO_3^-) is formed by the reaction between CO_2 gas from air and H_2O from rainwater. This type covers almost throughout the Fluviomarine Plain.

7.1.2 Mixed sodium water on the Fluviomarine Plain. Moreover, the groundwater in the Fluviomarine Plain (F_m) is dominated by the composition of sodium mixed water type (Na Mix), with a fresh (F) range. This is indicated by moderate values of EC (750-1000 μS/m) and TDS (500-800 mg/L) and very low Cl^- content (< 150 mg/L); the concentration of Ca^{2+} + Mg^{2+} ions is moderately hard from 10.1 to 14.6 meq/L; next incorporated dominant ion is the sodium (Na^+). The Ø sign mentions that the water gives no evidence of cation exchange. The hydrochemical variation is F1-(NaMix)Ø. This water composition is controlled by the dissolution of silicate minerals (albite, hornblende). This water type corresponds to an intermediary stage between freshwater with calcium bicarbonate facies, typically found in the recharge area, and brackish to saltwater containing marine ions. This water is primarily localized in the lower topographic areas than recharge. These areas have also favored the infiltration and circulation of water.

7.1.3 Salinized water on the Fluviomarine Plain. Again, the Fluviomarine Plain (F_m) also has salinized water with the hydrochemical variation. This type is from brackish (B) to salt (s), with EC (> 1000 μS/m) and TDS (> 800 mg/L). The chloride (Cl^-) ions are between 1,004.1 and 11,48.6 mg/L. The total hardness is 8.9 to 13.6 meq/L with moderately hard conditions. The corrected salt index gives the values (-), which means that a deficit (-) indicates cation exchange phenomena due to the salinization of freshwater bodies. The fresh and seawater will mix in the aquifer when the halite dissolves. The cation exchange process will happen on the clay surface of the substratum. The sodium (Na^+) and chloride (Cl^-) levels will increase significantly. This is typical for the Bs1-NaCl (-) waters. These water types occur at deeper depths than the former two. The topographically flat area does not favor the fast water circulation and short residence time; this permits the slow groundwater velocity and long residence time.

7.1.4 Freshwater on the Sand Dunes. The F1-$(CaHCO_3)_2$Ø water type is fresh and moderately hard in the Sand Dunes area. It is illustrated by a very low Cl^- content (< 150 mg/L), and Ca^{2+}+Mg^{2+} between 14.1 and 18.8 meq/L. The Ø sign indicates that cation exchange has not been active in this water. Referring to the concept of geology [11] and hydrostratigraphy [12, 13] of the study area, the units of Sand Dunes are a local aquifer system separated from the Fluviomarine Plain (F_m). It does not show the existence of groundwater evolution. This is indicated by the uniformity of hydrochemical conditions, grouped as freshwater type, in F1-$(CaHCO_3)_2$Ø. This condition is probably caused by the saturated marine sands formed by wave and wind activity on the exposed marine sediments. This is attributable to freshwater's nature, which has resulted in low EC and TDS values. The summary of the hydrochemical interpretation results is shown in Table 2. In this table, the Stuyfzand classification agrees with the Stiff diagram.

7.2. Salinization processes

In this well-document aquifer, the geochemical characterization of the groundwater will differ. The groundwater of successively different qualities results from silicate and halite minerals dissolution, cation exchange, and fresh and seawater mixing. The latter three are principle processes controlling the groundwater salinity in the study area. The major ionic ratio (Na/Cl versus Cl) supports such conclusions. These are well documented in previous research [22, 23, 25, 26].

Table 2. Summary of the hydrochemical interpretation results according to the Stiff diagram and Stuyfzand classification.

		Water type classification	
Geomorphological units	Lithology	Stiff diagram	Stuyfzand classification
Fluviomarine Plain (F_m)	Fine sand	$Ca-HCO_3$	F2 - $(CaHCO_3)_2$Ø
Fluviomarine Plain (F_m)	Silt	$Na-Ca-HCO_3$	F1 - (NaMix)Ø
Fluviomarine Plain (F_m)	Clay	Na-Cl	Bs1 - NaCl (-)
Sand Dunes Units	Fine to coarse sands with iron oxide	$Ca-HCO_3$	F1 - $(CaHCO_3)_2$Ø

The fluvial (river)-deposited sediments (sands and silts) have been dissolved to a considerable amount in the steep hydraulic gradient (recharge area), where the infiltration through them is intense. The water types concerned here are the F2-$(CaHCO_3)_2$Ø and F1-(NaMix)Ø types. The former type occurs at a shallower depth than the latter. These waters dominate the mineral dissolution, whereas the cation exchange is practically absent. Last follows the Bs1-NaCl (-) water type, displayed in topographically flat areas. This water reflects cation exchange. The order of observed water types reflects no sea impact as more freshwater is supplementary in the recharge area. The preceding is explained because the Bs1-NaCl (-) types are located far from the recharge areas, and the water in the semi-confined aquifer is more profound than in the unconfined aquifer, showing signs of salinization.

The newly exposed deposits balanced the previous marine conditions like the sea during the Pleistocene. A saline pore liquid is expressed on one side, and cations sorption on the clay minerals. Percolating freshwater infiltrated the sediments in the chemical equilibrium, and the sediment pore solutions were gradually washed out at first. First, the saline pore solution has been diluted. The early diluted water flowed to the deep layer as more freshwater infiltrated the marine deposits (mainly composed of clays). When the dilution process was finished, cation exchange started to play a role between the freshwater and the clay particles. Clay minerals primarily adsorbed the sodium (Na^+) cations at the deposition time. Clay forms the exchanger because of its ability [9]. The calcium (Ca^{2+}) expels the previously adsorbed Na+ ion from the exchanger with salinization, leading to a significant (Na^+) deficit. Here, the influenced water is of the Bs1-NaCl (-) type, evidence of cation exchange. Another evidence of cation exchange is the total hardness ($Ca^{2+} + Mg^{2+}$) in meq/L. There is a difference in striking total hardness between the groundwater variations in the prevailing lithology. The cation exchange is far in the deposits of fluvial origin. The typical water is the F2-$(CaHCO_3)_2$Ø and (NaMix)Ø. The calcium (Ca^{2+}) was exchanged for the sodium (Na^+) on the clay mineral surfaces, decreasing the water's hardness. This explains why the F2-$(CaHCO_3)_2$Ø water type is hard (21.9 - 28.2); the F1- (NaMix)Ø water type is moderately hard (10.1-14.6); the Bs1-NaCl (-) water type is moderately hard (8.9 - 13.6). Total hardness is lowered later as a result of cation exchange.

8. Conclusions

The present study uses the major ion chemistry of groundwater together with the mineralogical composition of sediments to investigate the hydrogeochemical evolution and their origin. Results were interpreted using the Stiff diagram and the Stuyfzand classification. The Stiff diagrams showed the $Ca-HCO_3$, $Na-Ca-HCO_3$, and Na-Cl types. The Stuyfzand classification validates the findings of the Stiff diagram. In the recharge area, the F2-$(CaHCO_3)_2$Ø type derives from the dissolution of anorthite and augite. As an intermediary stage, the F1- (NaMix)Ø type comes from the dissolution of albite and hornblende. The Bs1-NaCl (-) water is typical in the topographically flat area. The deficit (-) indicates cation exchange phenomena due to salinization. Halite will dissolve in the aquifer. The infiltrated freshwater slowly dilutes the marine pore liquid. The Ca^{2+} expels the previously adsorbed Na^+ ion from the exchanger. This explains why the F2-$(CaHCO_3)_2$Ø water is hard (21.9 - 28.2 meq/L); the F1-(NaMix)Ø water is moderately hard (10.1-14.6 meq/L); the Bs1-NaCl (-) water is moderately hard (8.9 - 13.6 meq/L). Total hardness in the Bs1-NaCl (-) water is softer due to cation exchange. The distribution of groundwater types results from minerals dissolution, ion exchange, and fresh and seawater mixing. Therefore, this research points out to shift focus to indiscriminate pumping problems in the coastal aquifer.

References

[1] Selvam S 2015 *Environ. Monit. Assess.* 187 p 106
[2] Walraevens K and Van Camp M 2004 *Proc. on the 18th Salt Water Intrusion Meeting* (Cartagena: Instituto Geologico Y Minero De Espana) p 449
[3] Anders R, Mendez G O, Futa K and Danskin W R 2013 *Groundwater* 52 p 756
[4] Li C, Gao X, Li S and Bundschun J 2020 *Environ. Sci. Pollut. Res.* 27 pp 41-57
[5] Kooi H J, Groen J and Leijnse A 2000 *Water Resour. Res.* 36 p 3581
[6] Gondwe E 1990 *Geoexplor.* 27 p 25
[7] Baba A and Tayfur G 2011 *Environ. Monit. Assess.* 183 p 77
[8] Appelo C A J and Postma D 1999 *Geochemistry, Groundwater and Pollution* (Netherland: Taylor & Francis) p 241
[9] Stuyfzand P J 1986 *Proc. on the 9th Salt Water Intrusion Meeting* (Delft: Water Management Group, Department of Civil Engineering, Delft University of Technology) p 641
[10] El Yaouti F, El Mandour A, Khattach D, Benavente J, and Kaufmann O 2009 *Appl. Geochemistry* 24 p 16
[11] MacDonald M and Partners 1984 *Greater Yogyakarta Groundwater Resources Study: Groundwater* vol 3 (London: Overseas Development Administration)
[12] Santosa L W 2012 *Proc. on the Forum Geografi* (Yogyakarta: Faculty of Geography, Universitas Gadjah Mada)
[13] Hendrayana H and Ramadhika R 2016 *Proc. on the 9th National Conference on Geoweek* (Yogyakarta: Department of Geological Engineering, Universitas Gadjah Mada)
[14] Kura N U, Ramil M F, Ibrahim S, Sulaiman W N A and Aris A Z 2014 *Environ. Sci. Pollut. Res.* 21 p 11
[15] Rao V G, Surinaidu L, Rajesh R, and Mahesh J 2011 *Wat. Air and Soil Poll.* 217 p 503
[16] Geospatial Information Agency (BIG) 2004 Map of Kulon Progo Regency, Yogyakarta Province
[17] Gorai A K, Jain K G, Shaw N, Tuluri F, and B.Tchounwou P 2015 *Asian J. Atmos. Environ.* 9 pp 47-258
[18] Chen P Y 1977 *Table of Key Lines in X-ray Powder Diffraction Patterns of Minerals in Clays and Associated Rocks* (Indiana: Geological Survey, Bloomington) pp 1-67
[19] Stiff H A 1951 *J. Pet. Technol.* 3 pp 376-379
[20] Walraevens K and Lebbe L 1989 *Proc. on the 10th Salt Water Intrusion Meeting* (Ghent: Natuurwetenschappelijk Tijdschrift) 70 pp 30-44
[21] Tran D A, Tsujimura M, Vo L P, Nguyen V T, Kambuku D and Dang T D 2019 *Environ. Geochem. Health* 42 pp 661-680
[22] Hendrayana H, Thin P P, Wilopo W, and Kawasaki S 2020 *Symp. on the 1st Geosciences and Environmental Sciences Symposium (ICST)* (Yogyakarta: Department of Geological Engineering, Universitas Gadjah Mada) 200

[23] Thin P P 2019 *Assessment of Groundwater Chemistry in Wates Coastal Aquifer, Kulon Progo, Yogyakarta Special Province, Indonesia, Thesis,* Department of Geological Engineering, Universitas Gadjah Mada, Indonesia (unpublished)
[24] Thin P P, Hendrayana H, Wilopo W, and Kawasaki S 2018 *Proc. on the 12th South East Asian Technical Univesity Consortium (SEATUC)* (Yogyakarta: Department of Geological Engineering, Universitas Gadjah Mada) p 70
[25] Batabyal A K and Chakraborty S 2015 *Water Environ. Res.* 87 pp 607-617
[26] Thin P P, Hendrayana H, Wilopo W and Kawasaki S 2018 *J. Degrad. Min. Lands Manag.* 5 pp 1389-1401

Acknowledgments
The authors thank AUN-Seed/Net Program for supporting research and the Department of Geological Engineering, Gadjah Mada University, providing the required equipment.

Characteristics of Engineering Geology in Talegong Road, Talegong, Garut, West Java, Indonesia

M H Naufal[1], **and H Setiawan**[1*]

[1]Departement of Geological Engineering, Faculty of Engineering, Universitas Gadjah Mada. Jalan Grafika no 2, Yogyakarta 55281, Indonesia
*Corresponding author's e-mail: hendy.setiawan@ugm.ac.id

Abstract. Talegong road, Garut, West Java, is provincial road with a length of 7 kilometers that connects Garut and Bandung. The road segment KM 2.3 – KM 7 is located between the steep slope of high hills and has potential for landslide. To find out the landslide potential, determining the characteristics of geology and engineering geology at this road segment are needed. The methods used in this research are engineering geology mapping (geomorphology, lithology, structural geology and rock mass quality) and laboratory analyses of soil and rock properties index, shear strength test, petrographic analysis and uniaxial compressive strength test. The results of geological mapping show the aspect geomorphological in the area consists of high hills with rather steep to very steep slopes. Lithological, the area consists of aphanitic andesite and porphyritic andesite. The area is dominated by Northeast – Southeast stress regime. In term of rock mass quality, the study area consists of three class with value in Rock Mass Rating classification, which are good rock, fair rock, poor rock and no value rock (soil). The study area is divided into three units in the engineering geological map: (1) unit A has fair – good rock (II – III) with rather steep - very steep slope with low landslide potential, (2) unit B has fair rock (III) with steep – very steep slope with medium landslide potential, and (3) unit C has poor – no value rock or soil (IV – 0) with a very steep slope with high landslide potential.

1 Introduction

The southern part West Java is prone to landslides, especially Garut Regency [1]. In December 2020, a landslide occurred at southern route Garut Regency, precisely at Talegong Street, Sukalaksana, Talegong which cut-off access between Garut and Bandung [2]. This landslide might have been caused by several factors: (1) steep slope and (2) highly weathered rock due to high rainfall [1]. This road segment is classified as medium (yellow) and high (purple) in Vulnerability Ground Movement Zone on Land Movement Vulnerability Zone Map (Figure 1) [3].

Land Movement Vulnerability Zone map [3] cannot be used as a sole reference in investigations of potential landslides, therefore further investigation of engineering geology is necessary [1]. The investigation covers both geological and engineering geology aspect as the basis for determining potential landslide. Some of the parameters that are considered in this study are including geomorphology (slope morphometry), lithology (rock types), structural geology (such as faults, joints, and folds), rock mass quality and index properties of soil and rock. [4]

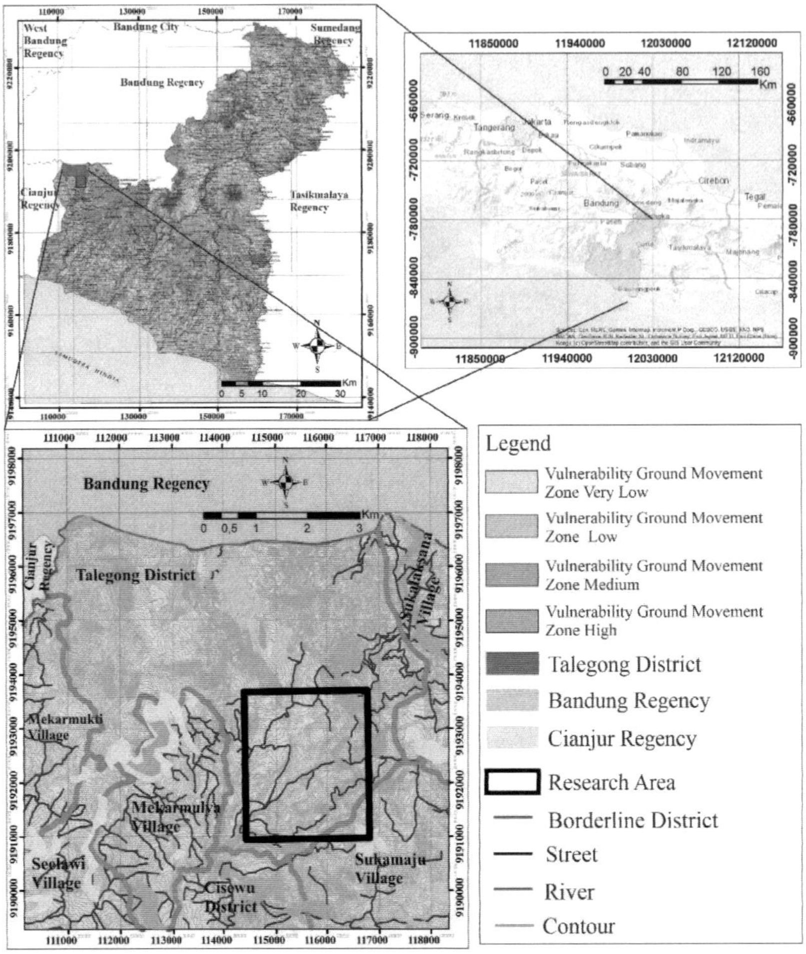

Figure 1. Research area location and vulnerability ground movement [3]. Basemap and digital elevation model acquired from Indonesian Geospatial Information Agency (https://tanahair.indonesia.go.id/).

2 Geology of Talegong

The research area is located at the Southern Mountain Zone of West Java [5]. The stratigraphy in research area is consists of two formations: (1) the Andesit Waringin Bedil Malabar Tua and (2) Endapan Rempah Lepas Gunungapi Tak-Teruraikan [4]. The lithology of Andesit Waringin Bedil Malabar Tua consists of breccia and tuff, interlayering with andesite lava rich in pyroxene and hornblende. The lithology of Endapan Rempah Lepas Gunungapi Tak-Teruaraikan consists of crystalline tuff with fine-coarse grain size consisting of dacite, tuffaceous breccia containing pumice, and old lava andesite-basalt deposits. Both formation area thought to have been deposited in Early Pleistocene [4].

3 Methods

The research method was determined into three phases: (1) Lithological, slope, geomorphological domains, hydrogeological and structural mapping. (2) Geomechanics mapping refers to Rock Mass Rating classification [9] with field tests and laboratory analysis. (3) Superposition of both phases for geothecnical mapping. Rock samples are used for petrographic analysis to know the types of rocks and geological processes that occurred in the past and index properties. The soil sample is used to obtain a soil shear test analysis, grain size distribution, atterberg limit, and soil index properties.

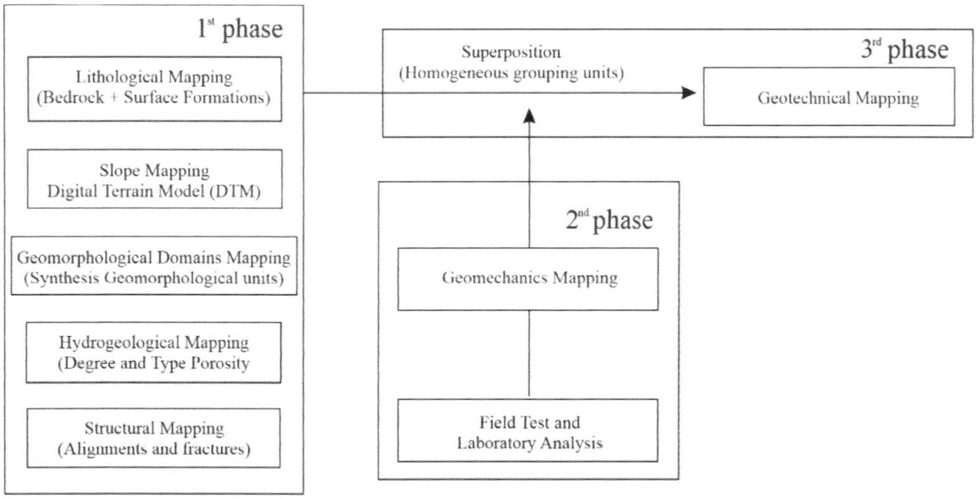

Figure 2. Methodological scheme for the production of the engineering geology map [6].

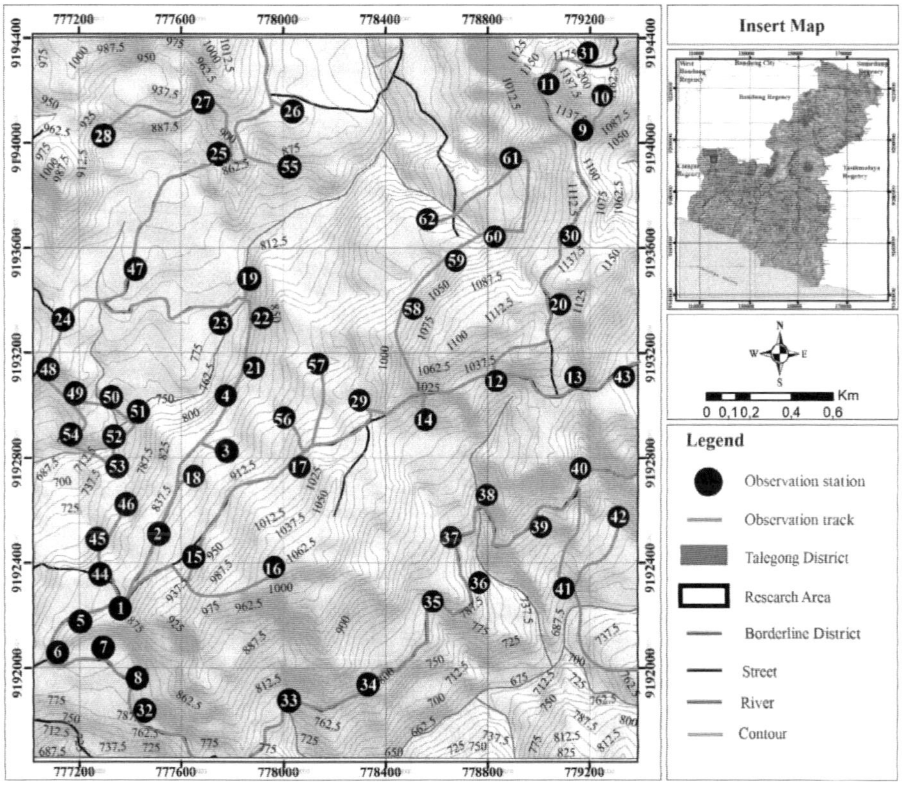

Figure 3. Traverse map of a research area. Basemap and digital elevation model acquired from the Indonesian Geospatial Information Agency (https://tanahair.indonesia.go.id/).

4 Result and discussion

4.1. Geomorphology and slope aspect

Morphometric and morphogenetic analyses are using the classification from [7]. The classification suggests, that the research area can be divided into three units based on the slope gradients. (Figure 4 and 5). The three units: (1) the high hills of volcanic rock with rather steep slopes (8°-16°) and altitude of 737.5 m – 900 m, (2) high hills of volcanic rock with steep slopes (16°-35°) and altitude of 662.5 m – 950 m and (3) the high hill of volcanic rock with very steep slopes (35°-55°) and altitude of 675 m – 1200 m.

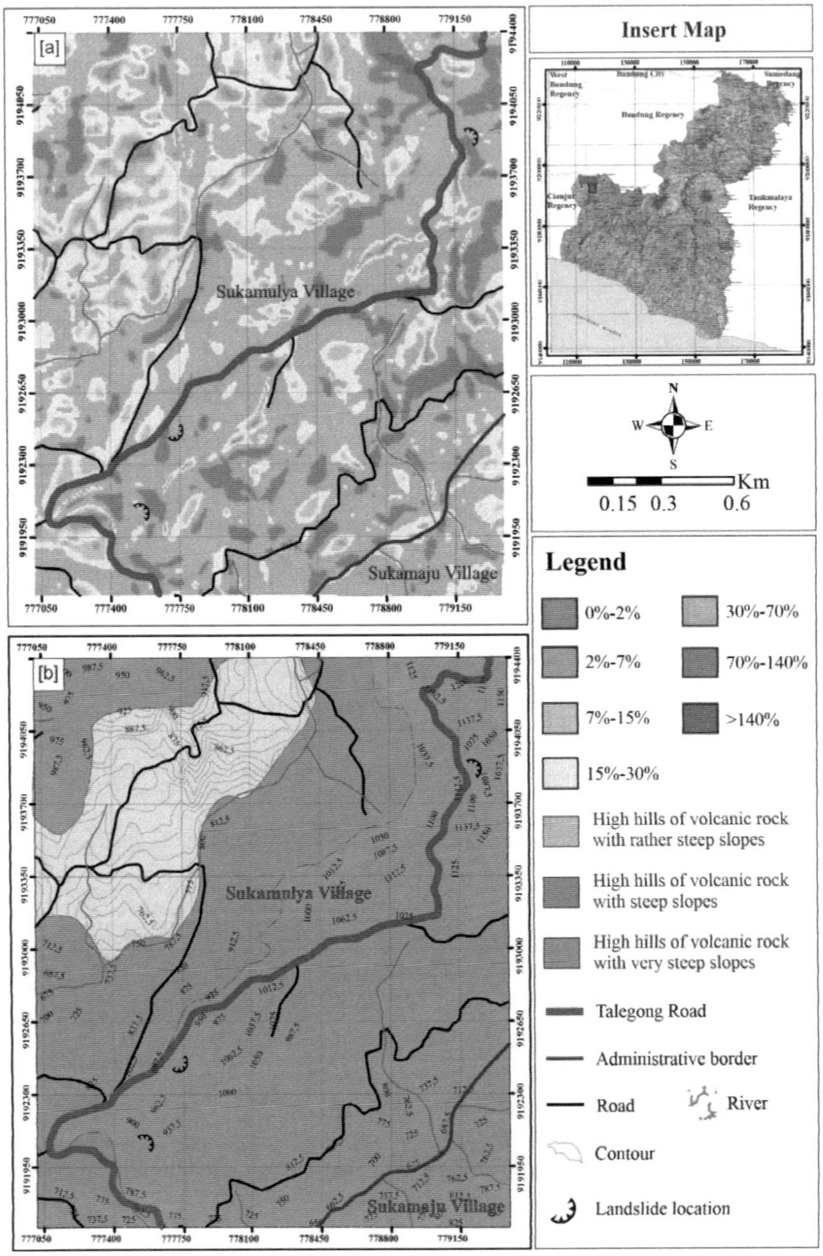

Figure 4. (a) Slope map of the research area and (b) Geomorphology map of the research area.

4.2. Lithology aspect

Figure 5, 6, and 7 show the representative outcrops. The research area dominated by two lithological units: (1) The aphanitic andesite rock unit which largely composes the N-NW of mapping area. This unit has black color in fresh appearance and color of blackish-brown in weathered appearance. This unit shows sheeting joints structure in the field, aphanitic texture, and holocrystalline degree of crystallization. Based on petrographic analysis, the mineral composition by in this unit is divided into two by grain size, first is phenocryst with mineral composition plagioclase (25%), hornblende (20%) and opaque (5%). The second is groundmass with mineral composition microlite plagioclase (40%), hematite (5%) and chlorite (5%). (2) The porphyritic andesite rock unit which largely composes the remaining of mapping area. This unit has a gray to black color in fresh appearance and a blackish-brown color in weathered appearance. This units shows shear joints structure in the field, porphyritic texture, holocrystalline degree of crystallization. Based on petrographic analysis, the mineral composition by in this unit is divided into two by grain size, first is phenocryst with mineral composition plagioclase (25%), hornblende (20%) and opaque (5%). The second is groundmass with mineral composition microlite plagioclase (50%).

4.3. Structural geology aspect

Figure 5 shows the lineament pattern in the research area. The lineament analysis suggests that the dominant force in the research area is north-northwest (NNW) to south-southeast (SSE), as shown by rose diagram in Figure 5.

Figure 5. The presence of joint in the research area (a) Sheeting joint at STA 1, (b) Shear joint at STA 7, and (c) Extension joint at STA 1.

Figure 6. (a) Photo of outcrop STA 1 with lithology andesite rock [8] has appearance sheeting joint with a geological hammer as comparison scale. In the thin section, minerals composition in this unit is plagioclase (25%), hornblende (20%), opaque (5%), micro-plagioclase (50%). (b) Photo of outcrop STA 28 with lithology andesite rock [8] with a geological hammer as comparison scale. In the thin section, minerals composition in the rock is plagioclase (25%), hornblende (20%), opaque (5%), micro-plagioclase (40%), hematite (5%) and chlorite (5%).

4.4. Geomechanics aspect

The research area is divided into four classes following the RMR classification [9] with parameters intact rock strength (IRS) [10], Rock Quality Designation (RQD) [11], Joints Spacing (JS), Joints Condition (JC) and Groundwater Condition (GWC). (Figures 7): (1) Good rock class which largely composes the N – NW of the mapping area. This class has intact rock strength of R5 – R6 with Uniaxial Comprehensive Strength (UCS) value of 34.91 MPa; Poisson ratio of 0.1; modulus elasticity of 9426.75 MPa; Rock quality designation of 50% – 75%; Joint spacing of <60 mm – 200 mm; Joint condition of slightly rough surfaces with separation less than 1 mm and highly weathered; Groundwater condition completely dry. (2) Fair rock class which largely composes the NE – SE of the mapping area. This class has intact rock strength of R3 – R5 with Uniaxial Comprehensive Strength (UCS) value of 28.20 MPa; Poisson ratio of 0.14; modulus elasticity of 17861.21 MPa; Rock quality designation of 25% – 50%; Joints spacing of <60 mm – 200 mm; Joints condition of slightly rough – slickenside surfaces with separation 1 mm – 5 mm and highly weathered; Condition groundwater wet – damp. (3) Poor rock class largely composes the SW of the mapping area. This class has intact rock strength of R2 – R4 with Uniaxial Comprehensive Strength (UCS) value of 7.14 MPa – 27.70 MPa; Poisson ratio 0.27 – 0.43; modulus elasticity 1855.7 MPa – 16032.90 MPa; Rock quality designation of 25% – 50%; Joints spacing of <60 mm – 200 mm; Joints Condition of slickenside with separation 1 mm – <5 mm and highly weathered; Condition groundwater wet. (4) No value rock largely composes the NE of the mapping area. This class has no value because it is soil.

4.5. Engineering Geology aspect

Our results show that the engineering geology unit in the research area is divided into three classes (Figure 8): (1) Engineering geology unit A has a rather steep slope – very steep slopes [7]. The structural aspect of this unit has a rare presence of joints. The lithology is porphyritic andesite rock with massive structures and boulders. For the geomechanics aspect [9], the unit has rock mass quality class poor rock on rather steep and good rock on very steep rock. (2) Engineering geology unit B has a steep slope – very steep slopes [7]. The structural aspect of this unit has an abundant presence of joints. The lithology is porphyry-aphanitic andesite and aphanitic andesite rock with massive structure and sheeting joints. For the geomechanics aspect [9], the unit has rock mass quality class fair rock. (3) Engineering geology unit C has a steep slope – very steep slopes [7]. The structural aspect of this unit has an abundance presence of joints. The lithology is porphyry-aphanitic andesite and aphanitic andesite rocks with massive structures and shearing joints. For the geomechanics aspect [9], this unit is in the rock mass quality class very poor – poor rock.

Figure 7. Appearance outcrop of rock mass rating (a) Appearance good rock quality at STA 28, (b) Appearance fair rock quality at STA 40, (c) Appearance poor rock quality at STA 6, and (d) Appearance no value rock quality at STA 32. [SR: Strength of intact rock; RQD: Rock quality designation; JS: Joints spacing; JC: Joints condition and GWC: Groundwater condition.

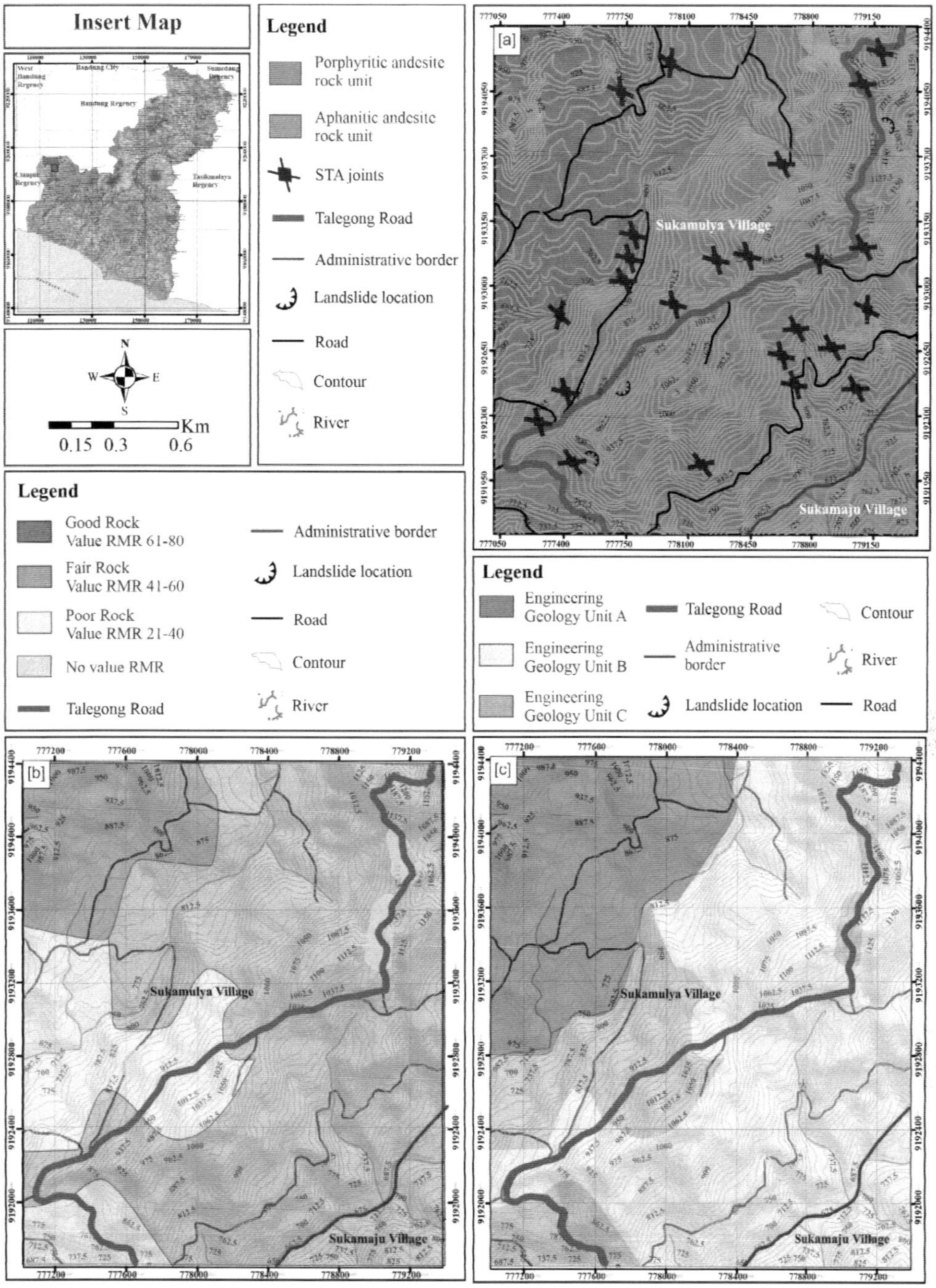

Figure 8. (a) Geology map in the research area, (b) Geomechanics map in the research area, and (c) Engineering geology map in the research area.

5 Conclusion

Our study shows that the research area can be divided into three units based on morphometric and morphogenetic. They are the high hills of volcanic rock with rather steep slopes (8°-16°) and altitude 737.5 m – 900 m, high hills of volcanic rock with steep slopes (16°-35°) and altitude 662.5 m – 950 m, and the high hill of volcanic rock with very steep slopes (35°-55°) and altitude 675 m – 1200 m. The research area is composed of two main lithologies: (1) aphanitic andesite rock unit and (2) porphyritic andesite rock unit. Our finding also suggested that dominant force in the area is north-northwest (NNW) to south-southeast (SSE). The rock mass quality following RMR classification in the research area is divided into four classes. They are no value rock, poor rock, fair rock, and good rock. The combination between geomorphology, lithologies, structural geology and geomechanics suggest that the engineering geological aspect is divided into three units, they are engineering geology unit A with rock mass quality class poor rock on rather steep, good rock on very steep rock and has a rare presence of joints. engineering geology unit B has a steep slope and very steep slopes with an abundant presence of joints. This unit has massive structure and sheeting joints as well as rock mass quality class fair rock. Engineering geology unit C has a slope steep to very steep slope with an abundance presence of joints. This unit has a massive structure with shearing joints while their rock mass quality is classified into very poor – poor rock classes.

Land use in the research area can be determined with refers to engineering geology map. Engineering geology unit A can be used for settlement and public facilities, engineering geology unit B can be used for settlement and plantation, and engineering geology unit C can be used for plantation. The result of the study can be put into a reference for further susceptibility mapping/analysis of landslide disasters in this research area.

Reference

[1] PVMBG 2020 *Tanggapan Bencana Gerakan Tanah Di Kecamatan Talegong, Kab. Garut Jawa Barat* accessed 2 Mei 2021 <https://vsi.esdm.go.id/index.php/kegiatan-pvmbg/berita-harian-kebencanaan-geologi/3353-tanggapan-bencana-gerakan-tanah-di-kecamatan-talegong-kab-garut-jawa-barat>

[2] Ghani H 2020 *Jalur Garut-Bandung Terputus Akibat Longsor di Kawasan Talegong* accessed 11 January 2021. <https://news.detik.com/berita-jawa-barat/d-5308503/jalur-garut-bandung-terputus-akibat-longsor-di-kawasan-talegong>

[3] PVMBG 2020 *Peta Prakiraan Wilayah Terjadinya Gerakan Tanah Pada Bulan Desember 2020, Provinsi Jawa Barat* accessed 11 January 2021. <https://vsi.esdm.go.id/gallery/index.php?/category/15>

[4] Alzwar M., Akbar N and Bachri S. 1992 *Geologi lembar Garut dan Pamengpeuk, Jawa, Lembar 1208-6 dan 1208-3, Skala 1: 100.000* (Bandung: Pusat Penelitian dan Pengembangan Geologi. Departemen Pertambangan dan Energi)

[5] van Bemmelen R W 1949 *The Geology of Indonesia* General Geology of Indonesia and Adjacent Archipelagoes **1A** Hal: 732.

[6] Geological Society Engineering Group Working Party 1972 *The preparation of maps and plans in terms of engineering geology* Q. J. Eng. Geol. **5** pp: 293–381.

[7] van Zuidam R A 1985 *Aerial Photo – Interpretation in Terrain Analysis and Geomorphologic Mapping* (The Hague, ITC: Smith Publisher)

[8] Streckeisen A L 1976 *To Each Plutonic Rock Its Proper Name* Earth-Sci Rev **12** pp 1-33.

[9] Bieniawski Z T 1989 *Engineering Rock Mass Classifications* (New York: John Wiley & Sons) pp 51-55.

[10] ISRM 1978 *International Society for Rock Mechanics Commission on Standardization of Laboratory and Field Tests* Int. J. Rock Mech. Min. Sci. & Geomech. Abstr **15** pp 319-368.

[11] Palmström A 2005 *Measurements of and correlations between block size and rock quality designation (RQD)* Tunnelling and Underground Space Technology **20** pp 362–377.

Acknowledgment

This work is fully funded by the Department of Geology through the *Dana Hibah Kampus* funding program for year 2021 and also thanks to Get in Cicero Laboratory and *Geologi Tata Lingungan* Laboratory for their support.

Integrated SWAT-MODFLOW model to study saltwater intrusion in Da Nang coastal city

Bach Thao Nguyen[1]* and Thi Van Le Khoa[2]

[1]Hanoi University of Mining and Geology
[2]Hanoi University of Natural Resources and Environment
*Corresponding author's email: nguyenbachthao@humg.edu.vn

Abstract. This study applied an integrated surface water-groundwater (SW-GW) model, SWAT-MODFLOW to study interaction of surface and groundwater than simulate saltwater intrusion at a regional scale in Da Nang coastal city, center of Vietnam. Model components were calibrated and validated using monthly river flow data and hydraulic head data for the 2005-2020 period and showed good agreement with observed data. The results demonstrated that GW-SW exchange in the upstream areas had the most pronounced fluctuation between the wet and dry months under historical conditions. The combined potential impact is that intensive groundwater use may have more immediate effects on river flow than those of climate change, which has important implications for water resources management and supply in the future. The results showed that the total average recharge to groundwater from the rivers was varied from 12046 to 23147 m^3/d relative to 50 74% of water resources in rain and dry season. Saltwater intrusion intrusion in 2020 is about 100 km^2. In the area of Son Tra and Ngu Hanh Son districts, groundwater pumping must be reduced to zero, in some other areas, pumping rate do not exceed 250 to 1000 m^3/day with distance to coastline at least 500 to 1400m depends on each area.

1. Introduction

These days, water scarcity and saltwater intrusion is a widespread problem around the world. Water availability becomes a matter of interest in everywhere, especially in arid or semiarid areas. Traditionally, management of water resources has concentrated on surface water or ground water as if they were separate entities [1]. However, surface water and ground water are not separate components in the hydrological cycle [2]. In Vietnam, water resources management has mainly focus on the surface water [3, 4], while problems related to groundwater have not been managed in a rigorous manner. In most of the studies have been done, modelling is the most suitable method for simulating surface and groundwater availability.

The Soil and Water Assessment Tool (SWAT) and MODFLOW are 2 well-known and widely used surface and groundwater models, respectively. These two models represent two different environments, and each is limited in its simulation domain with their corresponding strong points and drawbacks. In one side, SWAT is a basin scale, semi-distributed model and is often used to simulate hydrological processes in surface and in shallow aquifer. Its calculation is based on hydrological response units (HRUs), which are conceptual units of homogeneous land use, management, slope, and soil characteristics that extend below the surface to a soil profile depth [5]. SWAT model can only simulate shallow ground water flow in a restricted layer, around 6 m below ground surface, in which the seepage below it is assumed to be lost and out of the system [6]. In the other side, MODFLOW presents as a three-dimensional, distributed finite-difference ground water model and it can simulate ground water flow for variably saturated subsurface systems including shallow and deep aquifers. However, the model is limited to investigating ground water-surface interaction, as it cannot simulate surface process. On the other words, the ground water model was not adequately linked to surface water model [7]. In those study, ground water recharge, an important input for ground water model, could not be calculated from hydrological components, which are precipitation, evapotranspiration, and surface runoff, however it was determined through trial-and-error method during calibration process.

In recent decades, there were some conjunctive simulations of surface water and ground water using SWAT and MODFLOW [2, 8] . In those studies, there were several methods to integrate SWAT and MODFLOW, however the integration through recharge rate between HRUs in SWAT and cells in MODFLOW is the most feasible method. Those studies were successful in evaluation of water availability in various regions of the world and became a useful data to support the water management policy.

Da Nang is a large coastal city in central Vietnam with 74km of coastline, plays an important role in tourism, industrial production, and agriculture with a population density of 828 people/km^2. Therefore, Da Nang city always faces problems such as insufficient water to meet demand due to overexploitation causing depletion, water pollution and especially saltwater intrusion. Da Nang city has a long coastline, so the aquifers in this area are greatly affected by the saltwater intrusion process, however, there is no work to evaluate saline intrusion using numerical models. Based on the results of survey to establish hydrogeological map of Da Nang city , about 30% of aquifers area is affected by salinity out of total area [9]. Therefore, it is very urgent to assess the current status of saline intrusion into the ground and propose a reasonable exploitation solution. The article uses the method of combining surface water model and underground water model to assess the current status of groundwater salinity intrusion in Da Nang area, the model correction is based on the results of water level measurement, total mineralization at the boreholes, thereby providing a reasonable mining solution.

2. SWAT, MODFLOW and their integrated structure

2.1. SWAT model

Soil & Water Assessment Tool (SWAT) is a continuous-time, semi-distributed, process-based rainfall-runoff model [10]. As its name implies, SWAT was developed to simulate the water resource system and assess the impacts of single or multiple changes triggered by natural or human-induced factor to both water quantity and quality.

SWAT is the result of various efforts. In 1980, Knisel developed a model called CREAMS (Chemicals, Runoff, and Erosion from Agricultural Management Systems) to simulate the non-point sources [11]. Later, several new models were developed based on CREAMS such as EPIC (Erosion-Productivity Impact Calculator) [12] or GLEAMS (Groundwater Loading Effects on Agricultural Management Systems) [13]. While EPIC was launched to evaluate the efficiency of agricultural practices and non-point source pollutions, GLEAMS demonstrated the ability to assess pesticides and nutrients to groundwater [10].

Also, in 1987, Arnold & Williams introduced the model SWRRB (Simulator for Water Resources in Rural Basins). Their idea of dividing a large watershed into sub-basins became a highlight of SWAT later on. The SWRRB was upgraded from CREAMS, EPIC and GLEAMS aiming to understand the impacts of management practices on water and sediment transport in poor data river basins.

In the early 1990s, the SWAT model officially came into being after undergoing a merger between ROTO (Routing Outputs to Outlet) [14] and a GRASS (Geographic Resources Analysis Support System) GIS interface [15] and SWRRB. Subsequently, then there is a series of reviews and upgrades. There were even a number of other versions of SWAT models, tailored for specific goals and different data formats.

SWAT calculates the water balance for each Hydrologic Response Units (HRUs) by basing on the following equation:

$$SW_t = SW_0 + \sum_{i=1}^{t}\left(R_{day} - Q_{surf} - E_a - W_{seep} - Q_{qw}\right)_i \quad (1)$$

Where SW_t is the final soil water content at time t (mm), SW_o is the initial soil water content (mm), Rday is precipitation in day i (mm), Q_{surf} is the amount of surface runoff in day i (mm), Ea is the amount of return flow in day i (mm), Q_{seep} is the amount of water entering the vadose zone from soil profile in day i (mm) and Q_{qw} is the amount of return flow in day i (mm)

The inputs required by the SWAT model are topography, land use map, soil map and weather data. The SRTM 1 Arc-Second (30 meter) Global elevation data offered by the United States Geological Survey (USGS), 0.5 km MODIS-based Global Land Cover Climatology, the FAO-UNESCO Soil Map of the World, and the ground-truth meteorological data of the stations located in the study area were used.

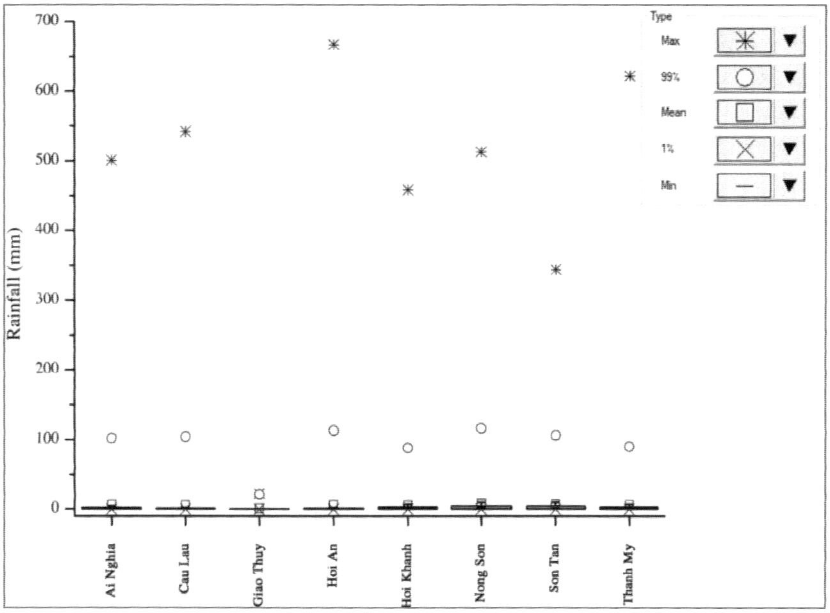

Figure 1. Rainfall variation in the study area (daily data is collected from the records of hydrological and meteorological monitoring stations).

It can be seen from the Figure 1 that the daily rainfall distribution in this area was quite uniform at the monitoring stations.

2.2. MODFLOW model

MODFLOW is a three-dimensional finite-difference ground water model developed by Waterloo Hydrogeologic. It can simulate steady and non-steady flows in a saturated system, in which aquifer layers can be confined, unconfined, or a combination of confined and unconfined [2]. The model can consider all common boundary conditions including fixed pressure head, groundwater recharge, variable or constant fluxes etc. In MODFLOW, the aquifer system is meshed by a discretized domain consisting of an array of node and associated finite difference cells [16]. Its governing equation is based on Darcy's law which is described by the following partial differential equation:

$$\frac{\partial}{\partial x}\left(K_{xx}\frac{\partial h}{\partial x}\right) + \frac{\partial}{\partial y}\left(K_{yy}\frac{\partial h}{\partial y}\right) + \frac{\partial}{\partial z}\left(K_{zz}\frac{\partial h}{\partial z}\right) - W = S_S\frac{\partial h}{\partial t} \quad (2)$$

Where K_{xx}, K_{yy} and K_{zz} are the hydraulic conductivities along the x, y and z axes parallel to the major axes of hydraulic conductivities, h is the piezometric head, W is a volumetric flux per unit volume representing sources/sink of water, Ss is the specific storage of the porous medium, and t is time.

2.3. Integrated SWAT and MODFLOW modelling

SWAT-MODFLOW is capable of simulating spatial-temporal groundwater recharge and surface water-groundwater (river-aquifer) interaction [8, 17]. Figure 2a shows the schematic diagram of combined surface water model (SWAT) and groundwater model (MODFLOW). The upper layers including root zone, vadose zone and shallow aquifer are belong to SWAT model, and the lower layer – deep aquifer is belonging to MODFLOW model.

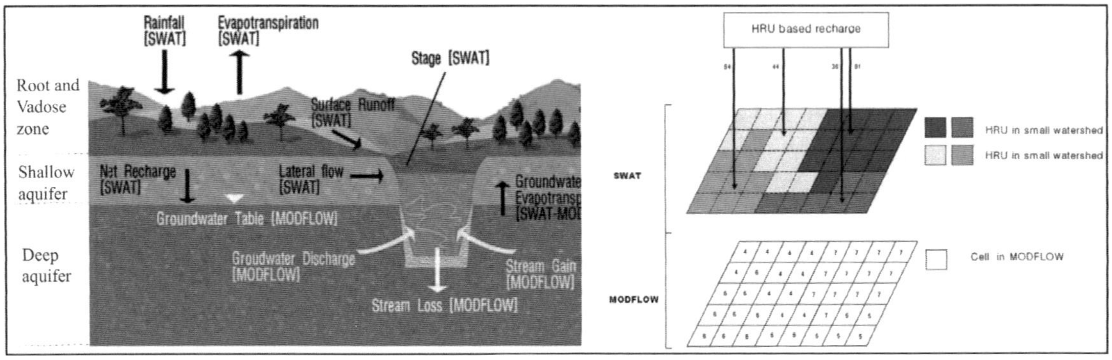

Figure 2. Schematic diagram of a) combining SWAT and MODFLOW b) exchange recharge rate from SWAT to MODFLOW [8].

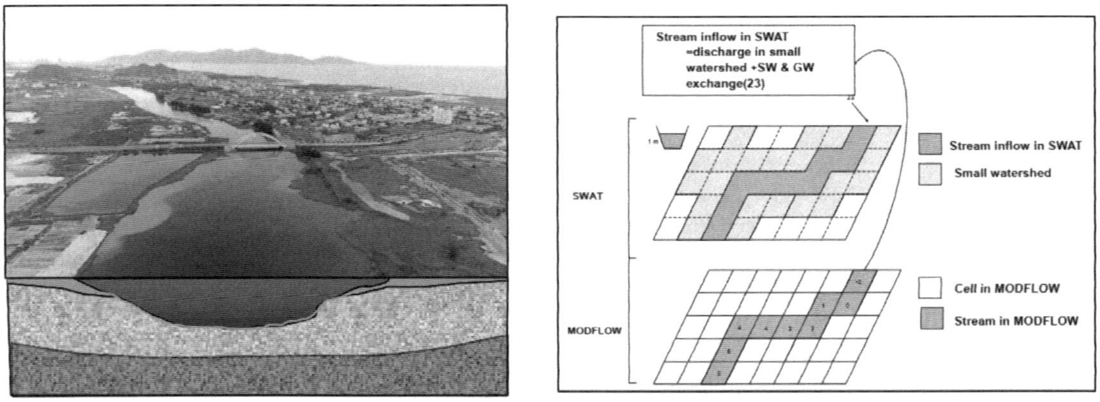

Figure 3. Intergrated surface-groundwater analysis for Da Nang coastal area (a) and linking river in Modflow with SWAT (b) [8].

In this study, SWAT and MODFLOW were setup to run individually and integrated through the recharge rates. These recharge rates were firstly estimated by SWAT model and presented as groundwater recharge values in HRUs level. In the integration process, the recharge rate of the HRU should be exchanged with cells and used as input data for MODFLOW (Figure 3b). Due to the semi-distributed features of SWAT, spatial location of each HRU in sub basins cannot be determined. Thus, to reflect HRU locations, one HRU is created for each sub basin by dominant land use, soil and slope option [2].

3. Geological and hydrogeological conditions of Da Nang coastal area

The study area includes three types of terrain including mountains, plains, and coastal sand dunes. The North is Bach Ma mountain range with an average height of over 700m, which is the natural border between Da Nang city and Thua Thien Hue; In the northwest, there is Mang mountain 1712m high, adjacent to the three provinces of Thua Thien Hue, Quang Nam and Da Nang; In the southwest, there is Ba mountain 1487m high; To the east is the Son Tra mountain range. The coastal plain and the plain south of the city are divided by the Han River. Terrain due to wind reaccumulation of sea sand is found in sand dunes in Ngu Hanh Son and Nam O areas.

There are five aquifers in the study area, which are the undivided Quaternary pore aquifer (q), Holocene fissile aquifer (qh), Pleistocene fissile aquifer (qp) is divided into upper layer (qp2) and lower layer (qp1), Neogene rift aquifer (n), karst fracture aquifer, terrigenous sedimentary fault and metamorphic sediment (Є-C).

The aquifer q is scattered on the surface with a small outcrop of about 3-5 km^2. The lithological composition is mixed, including clay, mixed clay, mixed sand, gravel and original rock. According to the results of the experimental water suction pump at the borehole LK710, the aquifer is classified as poor water with flow Q=0.84 l/s, flow rate q = 0.05 l/sm. The move changes strongly according to the season, in the dry season many dug wells are exhausted.

The qh aquifer consists of mixed sediments of rivers, seas, swamps, and winds. They are unevenly distributed with a total area of about 160 km2, hydrogeological characteristics vary by area. The petrographic composition is mainly fine to medium-grained quartz sand, mixed sand, mixed clay, discrete structure with good water capacity. Groundwater dynamics are strongly seasonal with the amplitude varying from 0.45-3.0m in Lien Chieu Hai Chau to 1.8 -11.2m in Son Tra - Ngu Hanh Son area. According to the test results of the water pump in 10 projects, the water storage level of the strata is from medium to rich, the flow rate varies from 0.04 to 3.65l/sm. Although the level of water storage is large, it is a surface aquifer, so it is significantly affected by the process of salinization due to tidal activity; especially in the area of Cau Do and Cam Le rivers, more than 50% of the aquifer area is saline. The chemical type of groundwater in this aquifer is mainly chloride-sodium.

The qp aquifer is distributed over an area of about 360 km2, the lithological composition includes quartz sand from fine to coarse, sand containing gravel, pebbles. The qp2 water storage layer has a flow rate q=0.32 l/sm, the average water level. Whereas the qp1 aquifer has a more abundant water content from moderate to rich, the discharge varies from 0.03 to 3.32 l/s.m.

The fissure aquifers in the study area have a distribution area of about 40 km2, most of which are covered by younger sediments. The petrographic composition is mainly composed of calcareous limestone, cericite schist, shale and quartz schist. Based on the research results in this area , the aquifer is ranked from poor to moderate water holding level.

With the hydrogeological characteristics mentioned above, the aquifer that is significant in exploitation is the Holocene and Pleistocene porous aquifer. However, both aquifers have saline intrusion, so a reasonable exploitation plan is required.

4. Results of simulation and discussion

4.1. Surface water availability

Discharge data at the two hydrological stations of Nong Son and Thanh My were used for the calibration and validation process. The calibration period was from 2005 to 2020, and the validation was from 2013 to 2020. The simulation results are evaluated through two coefficients namely Nash – Sutcliffe coefficient (E) and determination coefficient (r). The values of E and r are calculated as following:

$$E = 1 - \frac{\sum_{i=1}^{n}(X_{obs,i}-X_{sim,i})^2}{\sum_{i=1}^{n}(X_{obs,i}-\overline{X_{obs}})^2} \quad (3)$$

$$r = \frac{\sum_{i=1}^{n}(X_{obs,i}-\overline{X_{obs}})\cdot(X_{sim,i}-\overline{X_{sim}})}{\sqrt{\sum_{i=1}^{n}(X_{obs,i}-\overline{X_{obs}})^2\cdot\sum_{i=1}^{n}(X_{sim,i}-\overline{X_{sim}})^2}} \quad (4)$$

Where: $X_{obs,i}$ is observed discharge, $\overline{X_{obs}}$ is average observed discharge, $X_{sim,i}$ is simulated discharge, $\overline{X_{sim}}$ is average simulated discharge, n is number of samples.
According to Moriasi [18], with the value of r is larger 0.5 and E is greater than 0.75 the simulation results can be judged very well. Results of calibration and validation from SWAT model show in Table 1.

Table 1. Results of calibration and validation.

Station	r		E	
	Calibration	Validation	Calibration	Validation
Nong Son	0.73	0.75	0.77	0.76
Thanh My	0.74	0.71	0.78	0.80

Table 2. Calibrated SWAT parameters, their description and best range value.

No	Parameters	Definition	Range
1	ALPHA_BF	Base flow alpha factor (*days*)	0.1-0.2
2	GW_DELAY	Ground water delay time (*days*)	31-51
3	CN2	SCS runoff curve number of moisture condition II	60-70
4	ESCO	Soil evaporation compensation factor	0.5-0.9
5	REVAPMIN (*mm*)	Threshold water depth in the shallow aquifer for revap to the deep aquifer	300-500
6	GW_REVAP	Ground water revap coefficient	0.02-0.2
7	QWQMIN (*mm*)	Threshold water depth in shallow aquifer required for return flow to occur	600-800
8	SOL_AWC	Soil available water storage capacity (*mm H_2O/mm soil*)	0.2-0.4
9	R_RCHRG	Ground water recharge coefficient for deep aquifer	0.05-0.4
10	SOL_K	Soil conductivity (*mm/hr*)	15-50

4.2. Ground water availability in Da Nang coastal aquifers
The ground surface of basin has been created by using the 30 m resolution Digital Elevation Map (DEM) (Figure 4a). The main geometric-structure and hydrogeological characteristics of the study area were based on the geological and lithological descriptions of 41 boreholes located in Da Nang area. The grid size of the model is 1 km x 1 km and the boundary conditions are river network, recharge rate and pumping wells (Figure 4b).

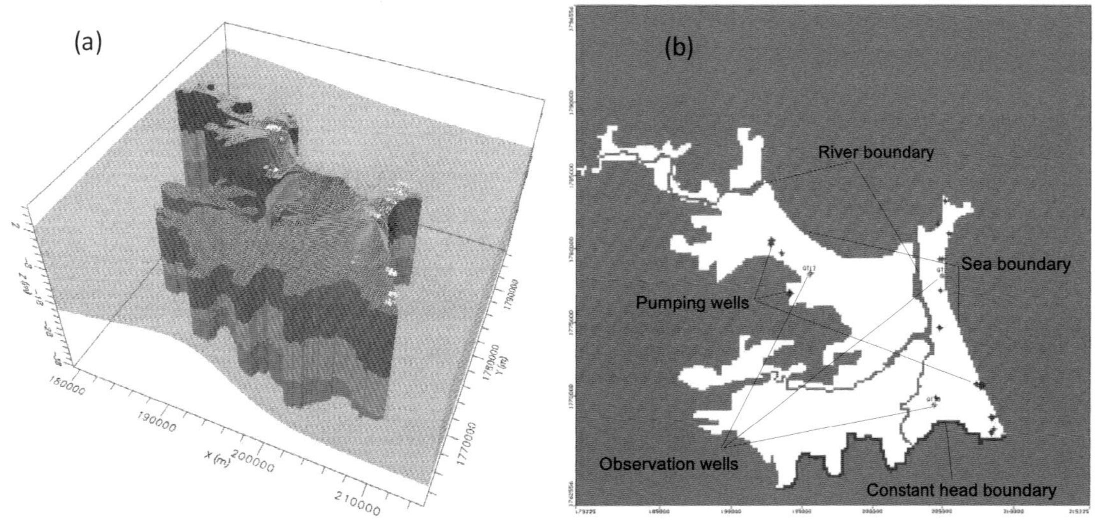

Figure 4. Three-dimensional visualization of model and boundary conditions

Their characteristics are very complex, however they can be categorized in to three main geological layers (Table 3).

Table 3. Geometric-structure and hydrogeological characteristics of basin

Layer in model/ Geological type	Lithological description	Average Thickness (m)	Hydraulic Conductivity (K)		Storage (S)			
			Range of *K* (m/day)	Average *K* (m/day)	Specific Yield: S_y (-)	Specific Storage Coefficient: *Ss* (1/m)	Effective porosity (-)	Total porosity (-)
Layer1: Holocene aquifer (qh)	Fine to coarse sand with gravel	16.8-20.4	0.14-7.02	4.20	6E-02	1.00E-05	7.5E-02	9.4E-02
Layer 2: Non-continuously Aquitard (maQ_2^1)	Clay with orgarnic	0-13.5	8.6E-03	8.6E-03	8.80E-02	1.00E-05	7.10E-02	8.90E-02
Layer 3: Pleistocene aquifer (qp)	Fine to coarse sand with gravel	15.0-29.0	1.41-20.2	9.27	18E-02	1.00E-05	7.00E-02	8.80E-02

The ground water model was setup to run in turn in 2 conditions of flow a) steady state to get the initial water head for transient state and b) transient state to get ground water availability. The model was first calibrated to fit the observed ground water levels until it reached to an acceptance normalized root mean square (RMS). Figure 5 shows the scatter diagram of calculated and observed head. The RMS was 2,421%, indicated a good simulation result. Figure 6a and Figure 6b shows the comparison between simulated and observed ground water level from 2013 to 2020 in borehole QT3A and QT3B, respectively. The graphs showed a good match between observed and simulation result in term of the

the graph's shape and their corresponding peaks. Figure 7a and Figure 7b shows the simulated groundwater level in Holocene and Pleistocene aquifer.

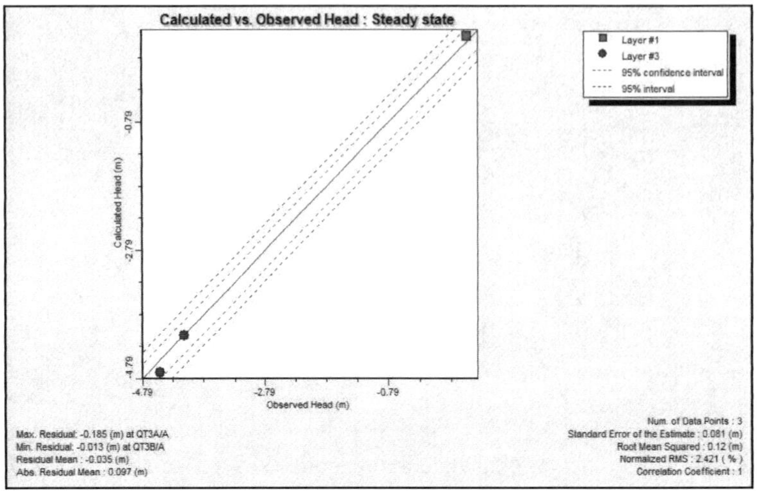

Figure 5. Calibration graph for MODFLOW performance, comparing simulated hydraulic head values to the observed hydraulic head values at 3 observation wells. Similar graphs can be displayed at any time step throughout the simulation period.

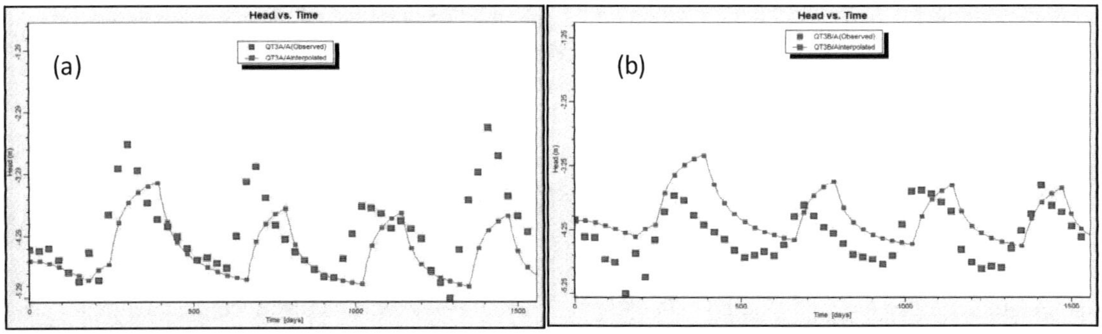

Figure 6. Comparison of the daily simulated vs. observed groundwater head at observation well QT3A and QT3B

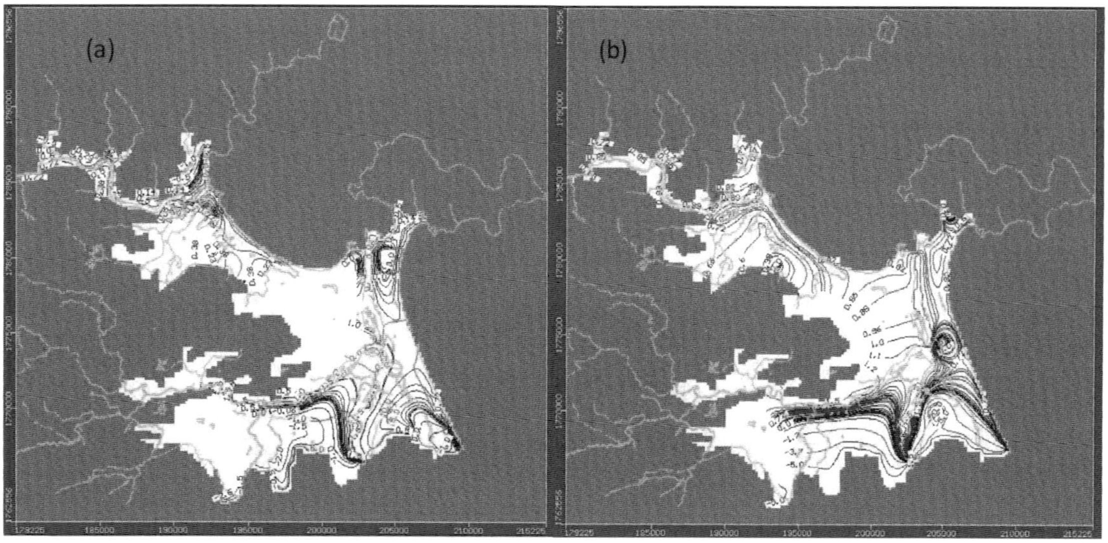

Figure 7. Simulated spartial map of groundwater level (a) in Holocene aquifer and (b) in Pleistocen aquifer in Da Nang coastal area (April 2020)

4.3. Saltwater intrusion in Da Nang coastal area

The results from saltwater intrusion numerical modeling in qh and qp aquifer at 04/2020 show that the initial salinity area of the qh aquifer decreases faster than the qp aquifer because of the precipitation recharge. During the extraction process, flowing to other areas, the surface aquifer is also affected by the replenishment process due to rain, which causes the aquifer to pale. However, the surface aquifer is heavily influenced by rivers and sea, so in the areas along the Han River, Cu De River, Do Toan River and coastal areas, the salinity area increases according to the calculated time series. of the model. Figure 8 shows areas with TDS greater than 1g/l of two aquifers Holocene and Pleistocene at the present time. Accordingly, the total saline aquifer area qh is 41.8km^2 and qp aquifer is 53.4km^2.

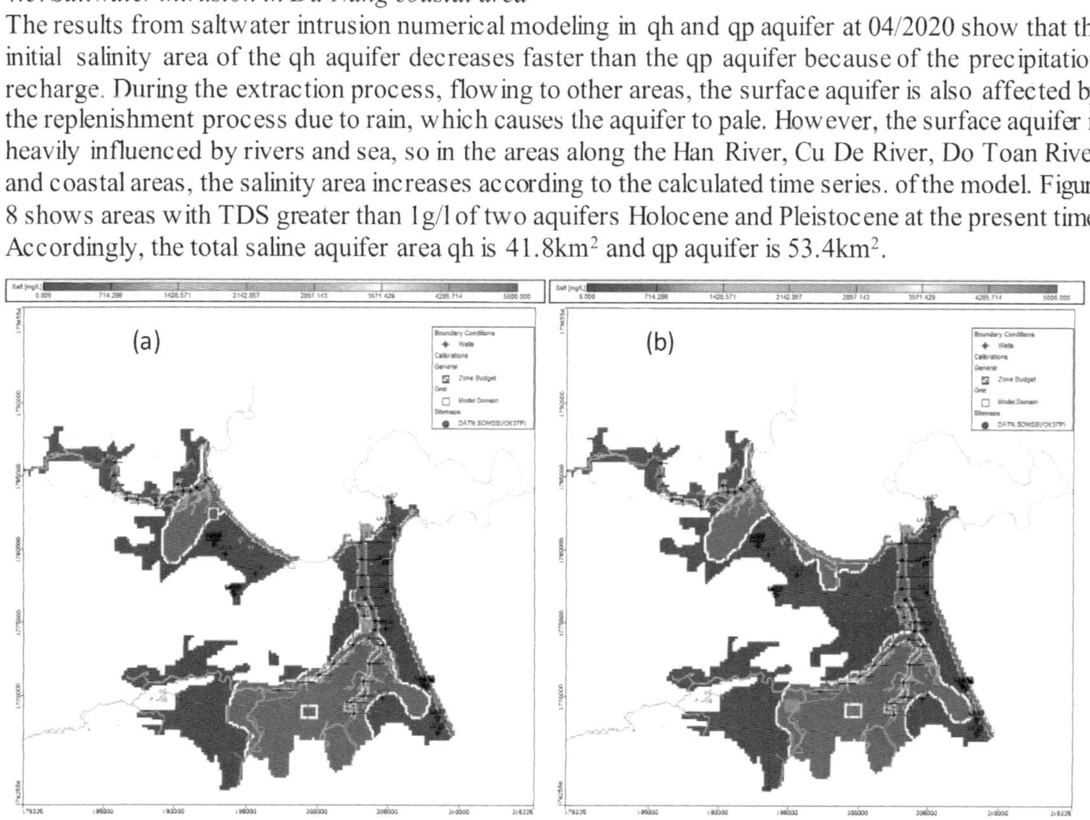

Figure 8. Spatial distribution of saltwater intrusion in groundwater simulated by Visual MODFLOW in April 2020. (a) in Holocene aquifer (b) in Pleistocene aquifer

5. Conclusion

In this study, an intergrate SWAT and MODFLOW models were conceptualized and developed for simulation of surface and ground water then simulate saltwater intrusion in the Da Nang coastal area. The SWAT model was calibrated and validated at two stream gauging stations Nong Son and Thanh My with Nash – Sutcliffe coefficient (E) values of 0.76 to 0.80. The simulated deep aquifer recharge from SWAT were used as recharge rates for MODFLOW model. The recharge values extracted from the HRU (hydrologic response units) cell conversion interface of SWAT model were used in the cells of MODFLOW. The simulation results including the stream flow and ground water level of two corresponding models were then compared and showed good agreements with observed data. It is demonstrated that the SWAT-MODFLOW is useful tool for simulate of surface- groundwater interaction and saltwater intrusion. This study demonstrated that the recharge from surface to groundwater in Da Nang coastal area is a resource to decrease saltwater area.

References

[1] Winter, T.C., et al., Ground water and surface water: A single resource, in *Circular*. 1998.
[2] Dowlatabadi, S. and S. Zomorodian, Conjunctive simulation of surface water and groundwater using SWAT and MODFLOW in Firoozabad watershed. *KSCE Journal of Civil Engineering*, 2015. **20**.
[3] Chau, T.K.a.K., D. X. , Study on water balance in Sesan river basin in drought year 2015/2016. *Journal of Meteorological*, 2017. **678**: p. 44-53.
[4] Au, N.T.T., Liem, N. D., Loi, N. K. , Applying GIS technique and SWAT model to assessing water discharge in Dakbla watershed. *Journal of National University*, 2013. **29**(3): p. 1-13.
[5] Arnold, J.G., et al., LARGE AREA HYDROLOGIC MODELING AND ASSESSMENT PART I: MODEL DEVELOPMENT1. *JAWRA Journal of the American Water Resources Association*, 1998. **34**(1): p. 73-89.
[6] Neitsch, S., et al., Soil and water assessment tool theoretical documentation. Version, 2011.
[7] Hiep, H.V., Ty, T. V., Đánh giá tài nguyên nước dưới đất tỉnh Trà Vinh sử dụng mô hình MODFLOW. *Tạp chí khoa học DH Cần Thơ*, 2012. **23**: p. 42-51.
[8] Kim, N.W., et al., Development and application of the integrated SWAT–MODFLOW model. *Journal of Hydrology*, 2008. **356**(1): p. 1-16.
[9] Nguyen, B.T., et al., Saltwater intrusion in aquifers of Danang area and solution for sustainable groundwater development. *Journal of Mining and Geology Science and Technology*, 2018: p. 10-20.
[10] Krysanova, V. and J.G. Arnold, Advances in ecohydrological modelling with SWAT—a review. *Hydrological Sciences Journal*, 2008. **53**(5): p. 939-947.
[11] Knisel, W.G. CREAMS: a field scale model for Chemicals, Runoff, and Erosion from Agricultural Management Systems [USA]. 1980.
[12] Williams, J.R., K.G. Renard, and P.T. Dyke, EPIC: A new method for assessing erosion's effect on soil productivity. *Journal of Soil and Water Conservation*, 1983. **38**(5): p. 381-383.
[13] Leonard, R., W. Knisel, and D. Still, GLEAMS: Groundwater loading effects of agricultural management systems. *Transactions of the ASAE*, 1987. **30**(5): p. 1403-1418.
[14] Arnold, J.G., J.R. Williams, and D.R. Maidment, Continuous-Time Water and Sediment-Routing Model for Large Basins. *Journal of Hydraulic Engineering*, 1995. **121**(2): p. 171-183.
[15] Srinivasan, R. and J.G. Arnold, INTEGRATION OF A BASIN-SCALE WATER QUALITY MODEL WITH GIS1. *JAWRA Journal of the American Water Resources Association*, 1994. **30**(3): p. 453-462.
[16] Chiang, W.-H. and W. Kinzelbach. Processing Modflow A Simulation System for Modeling Groundwater Flow and Pollution. 1998.

[17] Nguyen, B.T., et al., Swat-Modflow intergration in assessing groundwter potential in the Ba River basin. *International Conference on Sustainable Groundwater Development,* 2017: p. 53-63.
[18] Moriasi, D., et al., Model Evaluation Guidelines for Systematic Quantification of Accuracy in Watershed Simulations. *Transactions of the ASAB*E, 2007. **50**.

Acknowledgments
The authors express thanks to the research project code B2019-MDA-562-16 from Ministry of Education and Training (Program 562).

AUTHOR INDEX